FLORIDA STATE
UNIVERSITY LIBRARIES

APR 10 1995

TALLAHASSEE, FLORIDA

GaAs HIGH-SPEED DEVICES

GaAs HIGH-SPEED DEVICES

PHYSICS, TECHNOLOGY, AND CIRCUIT APPLICATIONS

C. Y. CHANG
FRANCIS KAI

A Wiley-Interscience Publication
JOHN WILEY & SONS, INC.
New York • Chichester • Brisbane • Toronto • Singapore

This text is printed on acid-free paper.

Copyright © 1994 by John Wiley & Sons, Inc.

All rights reserved. Published simultaneously in Canada.

Reproduction or translation of any part of this work beyond that permitted by Section 107 or 108 of the 1976 United States Copyright Act without the permission of the copyright owner is unlawful. Requests for permission or further information should be addressed to the Permissions Department, John Wiley & Sons, Inc., 605 Third Avenue, New York, NY 10158-0012.

Library of Congress Cataloging in Publication Data:

Chang, C. Y., 1937–
    GaAs high-speed devices : physics, technology, and circuit applications / by C. Y. Chang, Francis Kai.
       p.   cm.
    ISBN 0-471-85641-X
    1. Gallium arsenide semiconductors. 2. Integrated circuits.
I. Kai, Francis, 1956–   . II. Title.
TK7871.15.G3C47   1994   92-20705
621.38152 -- dc20

Printed in the United States of America

10 9 8 7 6 5 4 3 2 1

CONTENTS

PREFACE xiii

1 The Development of Gallium Arsenide Devices and Integrated Circuits 1

   1.1 Gallium Arsenide Device Development, 1
   1.2 GaAs Foundry, 3
       1.2.1 Private/Commercial Requirements, 3
       1.2.2 Analog/Digital Designs, 3
       1.2.3 Discrete Device/IC Production, 4
       1.2.4 Process Capability, 4
       1.2.5 Commercial Business, 4
       1.2.6 Unique Characteristics, 5

2 Gallium Arsenide Crystal Structure and Growth 7

   2.1 Introduction, 7
   2.2 Crystal Structure, 7
   2.3 Electron Energy-Band Structure, 13
       2.3.1 Energy-Band Calculations and Experimental Results, 13
       2.3.2 Direct and Indirect Intrinsic Gaps, 16
       2.3.3 Characterization of Zone–center-Band Extrema, 18
       2.3.4 Electron Effective Masses: Conduction-band System Statistical Weight, 21
       2.3.5 Hole Effective Masses, 25
   2.4 Electron and Hole Transport, 27
       2.4.1 Hole Mobility, 27
       2.4.2 Electron Velocity and Mobility, 29
       2.4.3 Electrical Conductivity, 33

vi   CONTENTS

    2.5 Crystal Growth of Semi-Insulating GaAs from Melt, 36
        2.5.1 Horizontal Bridgman Growth, 36
        2.5.2 Horizontal Gradient Freeze, 37
        2.5.3 Czochralski Crystal Growth, 38
        2.5.4 Defect Densities in SI LEC GaAs, 42
    2.6 EL2 Centers in GaAs, 43
        2.6.1 Electrical Characteristics of EL2, 44
        2.6.2 Experimental Properties, 48
        2.6.3 Theoretical Models, 50

3 Epitaxial Growth Processes   55

    3.1 Introduction to Epitaxial Growth Processes, 55
    3.2 Molecular Beam Epitaxy, 56
        3.2.1 MBE Growth Systems and Deposition Sources, 57
        3.2.2 In situ Analysis, 60
        3.2.3 Growth of III–V Compounds, 65
    3.3 Metal–Organic Chemical Vapor Deposition, 75
        3.3.1 Metal–Organic Sources, 75
        3.3.2 Nonhydride Group-V Sources for MOCVD, 77
        3.3.3 Basic Reactions, 82
        3.3.4 Purity and Dopants, 84
    3.4 Chemical Beam Epitaxy, 87
        3.4.1 Chemical Beam Epitaxy System, 88
        3.4.2 The Advantages of CBE, 90
        3.4.3 Growth Kinetics of CBE, 90
        3.4.4 Epilayer Morphologies of GaAs-Grown CBE, 92
        3.4.5 Photo-CBE, 93
    3.5 Atomic Layer Epitaxy and Molecular Layer Epitaxy, 94
        3.5.1 Growth Process and Growth Apparatus, 94
        3.5.2 Experiment, 96
        3.5.3 Characteristics of Chloride ALE, 98
        3.5.4 Atomic-Plane Doping, 98
        3.5.5 Photo-Assisted Molecular Layer Epitaxy, 99
    3.6 Vapor Transport Epitaxy, 104
        3.6.1 Principle of Vapor Transport Epitaxy, 104
        3.6.2 Experiment and Results, 106

4 Process Techniques   115

    4.1 Introduction to Process Techniques, 115
    4.2 Cleaning and Cleanliness, 116
        4.2.1 Environment and Handling, 116
        4.2.2 Cleaning Techniques, 117
        4.2.3 Acids, Bases, and Pure Water Systems, 117
    4.3 Etching, 119
        4.3.1 Wet Etching, 119
        4.3.2 Dry Etching, 123

CONTENTS vii

4.4 Plasma-Enhanced Deposition, 140
    4.4.1 dc/ac Glow Discharges, 141
    4.4.2 Frequency Effects on RF Plasma Reactor Behavior, 143
    4.4.3 Equipment for PECVD, 143
    4.4.4 Electron-Cyclotron-Resonance (ECR) Plasma CVD, 145
    4.4.5 Thin Film Deposition, 147
4.5 Rapid Thermal Process, 153
    4.5.1 n-Type Channel Implants, 154
    4.5.2 n^+ Contact Implants, 154
    4.5.3 Slip Lines and Wafer Distortion, 155
    4.5.4 Rapid Thermal Processing Equipment, 157

5 Lithography   164

5.1 Introduction to Lithography, 164
5.2 Photolithography, 165
    5.2.1 Photolithographic Techniques, 165
    5.2.2 Photoresists, 167
    5.2.3 Advanced Image Reversal Techniques, 173
5.3 Nonoptical Lithography, 178
    5.3.1 Electron-Beam Lithography, 180
    5.3.2 X-Ray Lithography, 189
    5.3.3 Ion-Beam Lithography, 193

6 Device-Related Physics and Principles   202

6.1 Some Basic Electronic Concepts in Low-Dimensional Physics, 202
6.2 Band Structure, Impurities, and Excitons in Superlattices, 205
    6.2.1 The Tight-Binding Approximation, 206
    6.2.2 The Envelope-Function Approximation, 208
    6.2.3 Bond-Orbital Models, 215
    6.2.4 Applications of the Bond-Orbital Model to Superlattices, 217
    6.2.5 Impurity and Exciton States in Heterostructures, 219
    6.2.6 Miniband Structure of a Superlattice, 222
6.3 Scattering Theory, 228
    6.3.1 Basic Scattering Mechanisms, 228
    6.3.2 Scattering Theory from the Golden Rule, 229
    6.3.3 Scattering Rates and Mobility, 233
    6.3.4 Intravalley Scattering at High Fields, 234
    6.3.5 Intervalley Scattering, 237
6.4 Ballistic Transport in GaAs Scattering, 239
    6.4.1 Hot-Electron Concept, 240
    6.4.2 Transport of a Single Particle, 242
    6.4.3 The Boltzmann Equation, 245
    6.4.4 Density Matrix Formulation, 252
    6.4.5 Feynman Path-Integral Formulation, 256
    6.4.6 Wigner Distribution Function, 259
    6.4.7 Moment Equation Approach, 261

viii CONTENTS

- 6.5 The Monte Carlo Method, 263
  - 6.5.1 The Initial Conditions of Motion, 265
  - 6.5.2 Flight Duration and Self-scattering, 265
  - 6.5.3 The Choice of the Scattering Mechanism, 266
  - 6.5.4 Time Average for the Collection of Results under Steady State Conditions, 269
  - 6.5.5 Quantum Monte Carlo Approach, 269

7 Metal-to-GaAs Contacts 279

- 7.1 Electrical Properties of Metal–Semiconductor Contacts, 279
- 7.2 The Physics of Metal–GaAs Systems, 280
  - 7.2.1 Classical Models of the Interface, 280
  - 7.2.2 Schottky Contacts, 284
  - 7.2.3 Schottky Barrier Metals, 288
  - 7.2.4 Techniques for Barrier Height Measurement, 294
- 7.3 Unpinned Schottky Barrier Formation, 296
  - 7.3.1 Experiment, 297
  - 7.3.2 Results, 297
  - 7.3.3 Discussion, 300
  - 7.3.4 Implications for Schottky Barrier Models, 302
- 7.4 Ohmic Contact, 302
  - 7.4.1 Methods of Forming Ohmic Contacts, 303
  - 7.4.2 Alloyed Ohmic Contacts, 304
  - 7.4.3 Nonalloyed Ohmic Contacts, 310
  - 7.4.4 Ohmic Contact to p-GaAs, 312
  - 7.4.5 Measurement of the Specific Contact Resistance r_c, 315
- 7.5 Interconnect Metal Systems, 318
  - 7.5.1 First-Level Metal, 318
  - 7.5.2 Vias, 318
  - 7.5.3 Second-Level Metal, 319

8 GaAs Metal–Semiconductor Field-Effect TRANSISTOR 323

- 8.1 Introduction to MESFET, 323
- 8.2 Fabrication Technology, 324
  - 8.2.1 Self-aligned Ion Implantation, 324
  - 8.2.2 Recessed Channel Technology, 327
  - 8.2.3 Submicrometer Gate MESFET Fabrication, 329
- 8.3 Models, 331
  - 8.3.1 The Shockley Model, 331
  - 8.3.2 Analytic Models of GaAs MESFETs, 335
- 8.4 Parameter Extraction, 343
  - 8.4.1 Determination of FET Parameters, 343
  - 8.4.2 Optimization of MESFET Models, 345
- 8.5 Parasitic Effects, 351
  - 8.5.1 $1/f$ Noise Component, 351
  - 8.5.2 Backgating Effects, 351
  - 8.5.3 Low-frequency Variations in Drain Conductance, 351

CONTENTS

8.6 Noise Theory of GaAs MESFETs, 352
    8.6.1 Noise Equivalent Circuit of GaAs MESFETs, 352
    8.6.2 Minimum Noise Figure of the GaAs FET, 354
    8.6.3 Noise Modeling of MESFET, 359

9 High Electron-Mobility Transistor (HEMT)   365

9.1 Introduction to HEMT, 365
9.2 The Basic HEMT Structure, 366
    9.2.1 Principles of Modulation Doping, 366
    9.2.2 The Structure of a HEMT, 369
9.3 Heterojunction Interface Sheet Carrier Concentration, 372
9.4 Transport in HEMT Structures, 374
    9.4.1 Low Field Mobility in 2DEGs, 374
    9.4.2 2DEG Transport in Moderate Electric Fields, 374
9.5 Capacitance-Voltage and Current-Voltage Characteristics, 375
    9.5.1 Charge Control Model, 377
    9.5.2 Equivalent Circuit and Figure of Merits, 379
    9.5.3 Transmission-Line Model, 379
9.6 Persistent Photoconductivity and Drain I–V Collapse, 383
9.7 Inverted HEMT, 383
    9.7.1 Crystal Growth, 384
    9.7.2 Device Characteristics, 385
9.8 Pseudomorphic HEMT, 386
    9.8.1 Materials Consideration, 387
    9.8.2 Fabrication of a 0.1-μm Pseudomorphic HEMT, 389
9.9 Pulsed-Doped HEMT, 391
9.10 Subthreshold Current in HEMT, 392
9.11 VLSI GaAs HEMT ICs, 394

10 Heterojunction Bipolar Transistors   399

10.1 Introduction to Heterojunction Bipolar Transistors, 399
10.2 The Structures of Heterojunction Bipolar Transistors, 400
    10.2.1 Homojunction and Heterojunction Bipolar Transistors, 400
    10.2.2 The Collector-up Heterojunction Bipolar Structure, 401
    10.2.3 Double Heterojunction Bipolar Transistors, 402
    10.2.4 The GaInP/GaAs Heterojunction Bipolar, Transistor, 406
10.3 Device Technology, 407
    10.3.1 Material Qualification, 408
    10.3.2 Surface Defects, 408
    10.3.3 Recombination Centers, 410
    10.3.4 Etch Control, 410
    10.3.5 Nonplanarity, 410
    10.3.6 Abrupt and Graded Band-gap HBTs, 411
    10.3.7 Carbon-Doped Base HBTs, 413
10.4 Characteristics of Heterojunction Transistors, 415
    10.4.1 Basic Characteristics, 415
    10.4.2 Efficiency versus Collector Layer Thickness, 417

- 10.5 Figures of Merit for High-Frequency Transistors, 418
- 10.6 Power Density in the HBT, 420

11 Resonant-Tunneling Transistors — 423

- 11.1 Introduction to Resonant-Tunneling Transistors, 423
- 11.2 Wave Property of Electrons and Resonant Tunneling, 424
- 11.3 Structure of the Resonant-Tunneling Diode, 426
  - 11.3.1 Resonant Tunneling through Parabolic Quantum Wells, 427
  - 11.3.2 The Double-Barrier Resonant-Tunneling Structures, 429
- 11.4 The Realization of Resonant-Tunneling Transistors, 433
  - 11.4.1 Resonant-Tunneling Hot-Electron Transistor, 433
  - 11.4.2 Asymmetric Barrier Structure, 435
  - 11.4.3 Resonant-Tunneling Bipolar Transistor, 436
  - 11.4.4 RHET Using InGaAs-Based Materials, 438
  - 11.4.5 Millimeter-Band Oscillations in a Resonant-Tunneling Device, 441

12 Hot-Electron Transistors and Novel Devices — 445

- 12.1 Ballistic-Injection Devices, 446
  - 12.1.1 Metal-Base Transistors, 446
  - 12.1.2 Doped-Base Transistors, 447
  - 12.1.3 The Hot-Electron Camel Transistor, 448
  - 12.1.4 The Tunneling Hot-Electron Transfer Amplifier, 448
  - 12.1.5 Quantum Well Base Transistors, 451
- 12.2 Real-Space Transfer Devices, 454
  - 12.2.1 Charge-Injection Logic, 456
  - 12.2.2 Light-Emitting Charge-Injection Transistor, 456
- 12.3 Quantum Devices, 459
- 12.4 Quasi-One-Dimensional Channel Devices, 463
- 12.5 Quantum Interference Devices, 468
  - 12.5.1 Novel Properties and Geometrical Properties of LSSL, 469
  - 12.5.2 Mesoscopic Structures, 469
  - 12.5.3 Lateral-Surface-Superlattice Devices, 471
- 12.6 Quantum Point Contacts, 473
  - 12.6.1 Angular Distribution for a Single Quantum Contacts, 474
  - 12.6.2 Electron Wave Interference, 478

13 GaAs FET Amplifiers and Monolithic Microwave Integrated Circuits — 483

- 13.1 Introduction to Monolithic Microwave Integrated Circuits, 483
- 13.2 Comparison of the Hybrid and Monolithic Approaches, 484
- 13.3 General Design Considerations, 486
- 13.4 Low-Noise Amplifier, 488
  - 13.4.1 Low-Noise Design, 488
  - 13.4.2 Noise, 495
  - 13.4.3 Low-Noise Amplifier Using HEMT, 496
- 13.5 High-Power GaAs MMIC, 500
  - 13.5.1 Nonlinearity in Class A Operation, 501

                13.5.2 Dynamic Load Line and Thermal Effects, 502
                13.5.3 RF Characteristics of Power GaAs FETs, 504
                13.5.4 Design and Performance of GaAs Power FET Amplifiers, 505
                13.5.5 Ka-Band Monolithic GaAs FET Power Amplifier Modules, 507
                13.5.6 A HBT MMIC Power Amplifier, 511
        13.6   Large-Signal Circuit Model for Nonlinear Analysis, 512
                13.6.1 Harmonic-Balance Method, 513
                13.6.2 Harmonic-Balance Simulation and Sensitivity Analysis, 514
                13.6.3 Load-Pull Method, 518
                13.6.4 Parameter-Extraction Program, 518
        13.7   Applications of GaAs ICs, 520
                13.7.1 Electronic Warfare Applications, 521
                13.7.2 Commercial Applications, 522

14 **GaAs Digital Integrated Circuits**   **528**

        14.1   Introduction, 528
        14.2   High-Speed GaAs Devices and Integrated Circuits, 530
        14.3   GaAs Logic Families, 532
                14.3.1 Buffered FET Logic, 532
                14.3.2 Schottky Diode FET Logic, 533
                14.3.3 Direct Coupled Field-Effect Transistor Logic, 534
                14.3.4 Source Coupled FET Logic, 535
                14.3.5 Capacitive Coupled Logic, 538
                14.3.6 Low Pinch-Off Voltage FET Logic, 539
                14.3.7 Heterojunction Bipolar Logic, 539
                14.3.8 GaAs Gate Array, 540
        14.4   Gallium Arsenide Circuits, 542
                14.4.1 GaAs Static Random Access Memory, 542
                14.4.2 Data Conversion Circuits, 547
                14.4.3 Data Communication Chip Set, 549
        14.5   GaAs Microprocessor, 549
        14.6   Digital Packaging, 554
                14.6.1 Multilayer Ceramic Packaging, 554
                14.6.2 Multichip Packaging, 554
                14.6.3 Special Packaging, 558

15 **High-Speed Photonic Devices**   **562**

        15.1   Introduction to Photonic Devices, 562
        15.2   Light-Emitting Diodes, 563
                15.2.1 Device Structures, 563
                15.2.2 Modulation Characteristics, 565
        15.3   Semiconductor Lasers, 566
                15.3.1 Basic Semiconductor Laser Physics, 568
                15.3.2 Laser Structures, 573
                15.3.3 Strained-Layer Quantum Well Heterojunction Lasers, 576
                15.3.4 Surface-Emitting Laser, 577

- 15.4 Pin Photodetectors, 578
  - 15.4.1 Basic Principles, 578
  - 15.4.2 Stability and Output Power of Pin-Avalanche Diodes, 582
- 15.5 Avalanche Photodiodes, 587
- 15.6 Hybrid Integration, 589
  - 15.6.1 Grafted-film Process, 589
  - 15.6.2 Device Fabrication, 590
  - 15.6.3 Bonding by Atomic Rearrangement, 591
- 15.7 Optical Interconnects, 592
  - 15.7.1 Network Requirements, 593
  - 15.7.2 Optoelectronic Transducers, 593
  - 15.7.3 Optoelectronic Interfaces, 594
  - 15.7.4 Monolithic Integration of Functions, 594
  - 15.7.5 High-Density Packaging, 595
  - 15.7.6 Future Directions, 598
- 15.8 Quantum Well Optical Modulators, 599
  - 15.8.1 Principles of Quantum Well Optical Modulators, 599
  - 15.8.2 Symmetric Self-electrooptic Effect Devices, 601

Index   **605**

PREFACE

High-speed semiconductor devices are the essential components of digital computers, telecommunication systems, optoelectronics, and advanced electronic systems as they can handle analog and digital signals at high frequencies and high bit rates. The design and development of these devices are vital to the continued growth of the high-tech industries. This book looks at the process advancements in GaAs device fabrication and offers insights into the design of devices, their physical operating principles, and their use in integrated circuits as well as other applications.

The book is organized into five parts: The first part, Chapter 2, discusses gallium arsenide materials and their crystal properties, the electron energy-band structures, hole and electron transport, the crystal growth of GaAs from the melt and the defect density analysis.

The second part consider the fabrication process of gallium arsenide devices and integrated circuits. Chapter 3 covers the epitaxial growth processes, molecular beam epitaxy, and the metal–organic chemical vapor deposition techniques used to grow a single atomic layer. An important feature of the chapter is the research on low-substrate temperature growth epitaxy systems which have been developed for better device fabrication. Chapter 4 gives an introduction on wafer-cleaning techniques and environmental control, wet etching methods and chemicals, and dry etching systems consisting of reactive ion etching and reactive ion-beam etching methods. The rapid thermal process is covered briefly, since it has captured much attention in recent years.

Patterning techniques have become hot issues in silicon as well as in gallium arsenide integrated circuit fabrication. Chapter 5 gives an overview of photolithography and nonoptical lithography techniques that include electron-beam, X-ray, and ion-beam lithography systems.

The third part, Chapter 6, discusses device-related physics. Advancements in the fabrication techniques described in the earlier chapters call for more understanding of low-dimensional device physics. The epitaxial processes make gallium arsenide and its related group III–V compounds and solid solutions band structure complex.

Scattering theory and ballistic transport are also discussed, and recent studies using the Monte Carlo method are presented.

The fourth part forms the core of the book. Chapters 7 and 8 develop the ideas on innovative device design and operating principles sketched out in Chapter 1. Chapter 7 considers metal–semiconductor contact systems, the Schottky barrier, ohmic contact formation, and reliability studies. Chapter 8 looks at metal–semiconductor field-effect transistors, the fabrication technology, and the models and the parameters for device analyses. The parasitic effects and noise theory are covered briefly here and developed later in Chapters 13 and 14, since MESFETs are the most popular devices in integrated circuits and integrated circuits applications. Chapter 15 concludes the book with a discussion of high-speed photonic devices and optoelectronic integrated circuits.

The fifth part, Chapters 9 through 12, discusses the heterostructure field-effect (HEMT in Chapter 9), potential-effect (HBT in Chapter 10), and quantum-effect (Chapters 11 and 12) devices. These new devices will have a large impact on high-speed integrated circuits and optoelectronic integrated circuits (OEICs) applications.

In summary, the most effective way to use the book is as follows:

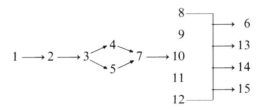

In all of the chapters we have tried to give the reader some idea of the history, even at this early stage of the development of the field. We have included some materials that are not commonly found in standard textbooks nor in collections of professional papers. In doing so, we wanted to give the reader as complete as possible a view of this fast-growing technology.

We are grateful to the many researchers who provided us with information and illustrations from their works. The comments from several reviewers were particularly helpful. Special thanks are due to Drs. Ta-Nien Yuan, Nan-Hong Kuo, Chi-Fu Deng, Simon S. Sze, Han-Ming Hsia, M. Feng, M. F. Chang, P. C. Chao, M. Pilkuhn, M. Razeghi, C. P. Lee, J. P. Duchemin, L. P. Chen, A. Y. Cho, K. Nakamura, J. Nishizawa, N. Yokoyama, A. Usui, H. Tanaka, P. S. D. Lin, T. C. L. G. Sollner, H. I. Smith, C. H. Liu, and S. S. Li for their encouragement, good will, and various assistances, and to the staff who worked with us at John Wiley and Sons, especially George Telecki, Cynthia Hess, and Rosalyn Farkas. Last but not least, we wish to thank our wives Shenn-May Lee and Lih-Nah Hwang.

Hsin-chu, Taiwan
Austin, Texas
June 1994

C. Y. CHANG
FRANCIS KAI

GaAs HIGH-SPEED DEVICES

1

THE DEVELOPMENT OF GALLIUM ARSENIDE DEVICES AND INTEGRATED CIRCUITS

1.1 GALLIUM ARSENIDE DEVICE DEVELOPMENT

In modern computer and telecommunication applications the most important semiconductors for high-speed devices are silicon and gallium arsenide and its related III–V compounds and solid solutions. Recent advancements in fabrication technology have produced the superlattice semiconductor, which is an artificial one-dimensional periodic structure made up of different semiconductor materials with a period of about 100 Å. Superlattice semiconductors include silicon-based materials (e.g., GeSi/Si) and gallium arsenide–based materials (e.g., AlGaAs/GaAs or InGaAs/GaAs) [Sze90].

Silicon's high density and high speed make it a popular material for very large-scale integrated (VLSI) circuit devices. But the III–V compounds have certain speed advantages over silicon in their higher carrier mobilities and effective carrier velocities. The semi-insulating substrates of the III–V compounds provide lower interconnection capacitances. Research on hybrid material systems based on a heteroepitaxial process has shown that an advanced architecture can be developed whereby high-speed GaAs/AlGaAs devices are located on heteroepitaxially grown islands on a silicon wafer (GaAs on Si), integrated with the silicon VLSI circuits by a suitable metallization process. There are promising developments as well in optical communications for silicon wafers, for example, in electronically triggered compound-semiconductor lasers or light-emitting diodes made in heteroepitaxial materials to avoid RCL delays due to on-chip interconnecting lines.

The most popular III–V compound high-speed devices are field-effect transistors. These are voltage-controlled devices. The control electrode is capacitively coupled to the active region of the device, and the charge carriers are separated by an insulator or a depletion layer [Sze90]. It is impossible to grow a good oxide layer on top of the GaAs surface to form the MOSFET as in silicon case. The junction field-effect transistor (JFET) and metal-semiconductor field-effect transistor the MESFET, which were proposed in 1952 and 1966, respectively, are both devices made of a homogeneous semiconductor material such as Si or GaAs.

The JFET is basically a voltage-controlled resistor that employs a *p–n* junction as a "gate" to control the resistance; thus the current flows between two ohmic contacts. JFETs have a lower switching speed than MESFETs, mainly because of the higher input edge capacitances in planar JFET processes. Complementary logic is possible because *p–n* and *n–p* structures can readily be fabricated on the same wafer. This enables us to design SRAM blocks on chip [Roc90]. A 32-bit RISC microprocessor [Ras86] was developed using JFET technology. Later McDonnell Douglas Astronautics Company [Gei87] developed an all-GaAs JFET vector signal processor. The architecture of this processor is optimized for the matrix–vector arithmetic operations for digital signal-processing applications.

The MESFET, however, uses a metal–semiconductor rectifying contact (the Schottky contact) instead of a *p–n* junction for the gate electrode. Another homogeneous FET is the permeable-base transistor (PBT), whose fine metal grids are covered with the semiconductor's epitaxial overgrowth, as will be discussed in Chapter 12. The PBT can be operated at high-current density with high transconductance, as is characteristic of high-speed power devices.

The development of advanced epitaxial growth techniques such as molecular beam epitaxy (MBE) and metal–organic chemical vapor deposition (MOCVD) techniques in the 1970s has enabled the fabrication of high-quality semiconductor heterostructures to be successful. Many other growth techniques, including low-temperature epitaxial growth, are currently being developed in order to facilitate the fabrication of high-quality semiconductor heterostructural devices. Heterojunction FETs include now a large number of family members with different applications. Donor layer devices have one or more *n*-type doped layers. The most extensively investigated donor layer device is the modulation-doped field-effect transistor (MODFET), also called *high-electron-mobility transistor* (HEMT). The pseudomorphic HEMT (discussed in Chapter 9) has a higher mobility than the conventional HEMT, which has a higher cutoff frequency.

The heterojunction bipolar transistor (HBT) was conceived of in 1957, but its implementation was delayed until the early 1980s due to technological difficulties of obtaining a perfect interface between dissimilar semiconductors. It offers substantial improvements in performance over the silicon bipolar transistor. In recent years there have been rapid advancements in HBTs for high-speed SSI and MSI circuits and power device applications. Higher g_m's in HBT are available. Short-channel effects essentially do not exist in HBTs. The theory and application of HBTs will be discussed in Chapter 10. The materials systems most studied for HBTs are semiconductors that have identical lattice constants, such as AlGaAs/GaAs and InGaAs/InP systems.

The III–V compound semiconductors can be used to fabricate quantum-effect transistors and photonic devices. A typical quantum-effect device is a resonant-tunneling transistor where the operation distance is comparable to a de Broglie wavelength, on the order of 200 Å at room temperature [Sze90]. These small dimensions give rise to a quantum size effect that alters the band structures and the densities of state and enhances the device's transport properties.

The basic building block of the quantum-effect device is the resonant-tunneling diode. This diode has a double-barrier structure with four heterojunctions and one quantum well. The number of barriers can be increased, in series, to produce the multiple-well device. Many novel current-voltage characteristics can be obtained by inserting a resonant-tunneling structure into the device to transform it into a resonant-tunneling hot-electron transistor (RHET) or resonant-tunneling bioplar transistor (RTBT). These device can be used to form relatively complex circuit functions at high speed with reduced component counts.

The field of 1D (quantum wires) and 0D (quantum dots) devices is new, and progressing so fast that it is hard to predict future applications. Both systems require nanoscale lithographic techniques. In the electron device field, where ultrasmall structures are necessary for speed and integration, arrays of digital processors are being developed. This architecture may lead to new designs. In the optical field, with the concentration of electrons and holes over fewer k-states, the lower dimensions of injection lasers will enable higher gains [Wei91]. Electrooptic effects can have larger resonances due to the sharpening of the 1D and 0D densities of state compared to quantum wells.

Another major application of compound semiconductor devices is in optoelectronics. It is possible to integrate the field-effect, potential-effect, quantum-effect, and photonic devices to meet the future demands of electronic systems. Chapter 15 discusses the high-speed aspects of semiconductor photonic devices.

1.2 GaAs FOUNDRY

GaAs foundries have improved their production capacities over the years to meet the increasing demand for GaAs IC chips. *Microwave Journal* has done the survey of U.S. GaAs foundries [Ell91]. Of 14 responding foundries, 12 reported on their wafer-handling capabilities. There are plans that involve either expanding capacity to handle wafers of larger diameter or increasing the rate at which wafers of a given size can be processed.

1.2.1 Private/Commercial Requirements

The General Electric operation expects to remain a completely captive facility through 1994. Avantek has historically maintained its facilities exclusively for its own use, but it expects 30% of its capacity to be in the commercial market by 1994. Alpha and Anadigics were alone in relying completely on commercial business in 1991; Harris and TriQuint were not far behind with a 95% commercial business in 1991. Hughes, Litton, Raytheon, Texas Instruments, and TRW forecast rapidly increasing commercial production in 1994.

1.2.2 Analog/Digital Designs

More than 90% of the reported capacity is devoted to the production of analog devices or circuits. Eight of the facilities are now solely doing analog work. Anadigics, Texas Instruments, and TRW have small divisions for digital work. ITT maintains 15% digital capacity, and Raytheon expects to be using 20% of its capacity for digital circuits by 1994. TriQuint, with 70% of its capacity devoted to digital designs, is the firm most seriously involved in the digital market.

4 GALLIUM ARSENIDE DEVICES AND INTEGRATED CIRCUITS

1.2.3 Discrete Device/IC Production

Production of integrated circuits has dominated the work of the reporting foundries. IC production accounted for 72% of total capacity in 1985, and this figure is expected to rise to 90% by 1994. Except for Alpha and Litton, who foresee beginning some discrete device activity in 1994, overall the foundries indicate that their capabilities are either already completely dedicated to IC work or will be substantially increased for IC work.

1.2.4 Process Capability

In 1985 only one of the reporting foundries was working with a line width below 0.5 μm. Six others reached that level by 1988, six more by 1991, and all 14 expect to have sub-halfmicron capability by 1994. Average process capability for the reporting operations has progressed from 0.61 μm in 1985 to a projected 0.17 μm for 1994.

The reported progress of maximum device operating frequency does not precisely track reduced line width capabilities. Other process-related effects or interests are obviously at work there. However, the maximum operating frequency of the devices produced by the reporting foundries is expected to triple over the next decade.

1.2.5 Commercial Business

Forecasts of commercial business growth from 1985 through 1994 vary widely, as shown in Table 1.1 [Ell91]. Hughes has the most optimistic short-term view in this market. TriQuint's 70% growth forecast is a bit clouded by its concentration on digital work. Of the five forecasting growth of 30% or more from 1985 through 1994, only Litton expects a modest portion of that to occur between 1991 and 1994. Alpha, Hughes, ITT Defense, and TRW all expect their 1994 commercial business to grow

TABLE 1.1 Volume of Business (in dollars)

Years	1985	1988	1991	1994
Alpha	100	110	500	5000
Anadigics				
Avantek				
General Electric	0	0	0	0
Harris	100	172	326	500
Hughes	100	300	2000	10000
ITT Defense	100	300	1300	5000
Litton	100	3540	4400	6000
M/A-COM				
Raytheon		100	110	
Texas Instruments	100	160	460	530
TriQuint	100	500	3500	7000
TRW		100	500	3000
Westinghouse	100	150	150	300

Source: Ellowitz, H. I., *Microwave Journal*, **34**(8) (1991): 42–48.

from 4% to 10% of its 1991 level. All other respondents foresaw modest growth through 1994.

1.2.6 Unique Characteristics

The following list shows the comments that the *Microwave Journal* received in its survey on the operation of U.S. GaAs foundries in 1991:

Alpha   Works in the mm-wave range with 0.025-μm capability

Anadigics   Has high-volume, high-yield capabilities

Avantek   Is establishing a foundry for both high-power and low-noise, high-dynamic-range MMICs under a MIMIC Phase III contract

General Electric   Supplies to GE Aerospace; epitaxial material-based pilot line

Harris   Supplies mostly to military, with a 70/30% ratio of military/commercial product, has a standard product base and is totally manufacturing oriented

Hughes   Able to produce large numbers of high-power, high-efficiency MESFET MMICs and foundry service, using HEMT material and 0.25 μm gates from *E*-beam lithography

ITT   Offers a multifunction self-aligned gate (MSAG) process for uniformity and reproducibility and a process flow that permits microwave power and small signal ICs and digital ICs to be combined on a single chip

Litton   Offers a 0.25-μm HEMT-based process and a 0.4-μm MESFET process, both fully characterized for circuit designer use

M/A-COM (Corporate alliance of AT&T Microelectronics)   Provides access through M/A-COM to low-noise and high-power microwave processes, MBE and ion implant technologies and built-in second-source potential; low-noise and high power devices on a single wafer; on-wafer pulsed power, small signal and S-parameter testing

Raytheon   Has operated from inception on a foundry basis and has a broad range of available processes

Texas Instruments   Has, since 1986, 100% first pass success

TriQuint   Has a full-service foundry for digital, linear, and microwave needs with multiple processes to support the varied disciplines; offers 0.5-μm FETs with through-wafer Vias on 4-in. wafer

TRW   Offers HBT and HEMT technologies validated under DARPA's MIMIC program and fabricates wafers

Westinghouse   Processes MESFETs with a wide variety of options for microwave power and control applications to Ka-band and wafer-level RF small signal and/or power testing.

In 1991 the processing capacity was 933 in.^{2} per week of material, with 16,650 gross and 13,995 net circuits produced in a week (or 727,740 in a year). For 1994 per-weak processing capacity is planned at 17,192 in.^{2} to produce 26,820,000 usable circuits in a single shift. New commercial applications for circuits like DBS television service, computer chips, and new DOD equipment will further raise the demand for foundry products.

REFERENCES

[Ell91] Ellowitz, Howard I., *Microwave Journal*, **34**(8), August (1991): 42–48.
[Gei87] Geideman, W. A., et al., *Microwave Journal*, **30**, September (1987): 105–126.
[Ras86] Rasset, T. L., et al., *Computer*, **19**, October (1986): 71–81.
[Roc90] Rocchi, M., ed., *High-Speed Digital IC Technologies*, Artech House, Boston (1990).
[Sze90] Sze, S. M., *High-Speed Semiconductor Devices*, Wiley-Interscience, New York (1990).
[Wei91] Weisbuch, C. and B. Vinter, *Quantum Semiconductor Structures*, Academic Press, New York (1991).

2

GALLIUM ARSENIDE CRYSTAL STRUCTURE AND GROWTH

2.1 INTRODUCTION

The rapidly growing use of gallium arsenide—in discrete devices and integrated circuits for microwave, millimeter-wave, optoelectronic, and digital applications—is creating a huge demand for single-crystal substrates of this material. The attainment of low-cost, high-speed GaAs digital IC technology is critically dependent on the availability of large area (4-in.) wafers that exhibit high resistivity ($> 10^8$ ohm-cm) and low crystal defects and etch-pit densities. In this chapter we first describe the crystal and band structure, the transport properties of the crystal, and then address some of the different methods used to synthesize and grow bulk crystals, focusing on growth from stoichiometric melts in vertical and horizontal configurations. Crystal defects, impurities, and etch-pit densities will also be discussed.

2.2 CRYSTAL STRUCTURE

A crystal is an object of high symmetry. This symmetry can be used to obtain general information about the properties of crystals without complicated algebra. An ideally perfect single crystal consists in an infinite three-dimensional repetition of an identical building block. This building block, called a *basis*, can be an atom, a molecule, or a cluster of atoms or molecules [Ash76]. The basis is the quantity of matter contained in the *unit cell*, whose spatial volume in the shape of a three-dimensional parallelepiped may be translated discrete distances in three dimensions to fill all of its space.

The most obvious symmetry requirement of a crystalline solid is that of translational symmetry [Ash76]. This requires that three translational vectors **a**, **b**, and **c** be chosen such that the translational operation

$$\mathbf{T} = n_1 \mathbf{a} + n_2 \mathbf{b} + n_3 \mathbf{c}, \tag{2.1}$$

8 GALLIUM ARSENIDE CRYSTAL STRUCTURE AND GROWTH

where n_1, n_2, and n_3 are arbitrary integers, connects two locations in the crystal having identical atomic environments. The translation vectors **a**, **b**, and **c** lie along three adjacent edges of the parallelepiped unit cell. Any location in the crystal is designated as point **r**. The local arrangement of atoms must be the same about **r** as it is about any of the set of points:

$$\mathbf{r}' = \mathbf{r} + \mathbf{T}. \tag{2.2}$$

The set of operations **T** defines a space lattice or *Bravais lattice*. A real crystal lattice results when a basis is placed around each geometric point of the Bravais lattice.

The vectors **a, b, c**, and **r'** are called *primitive vectors* and are said to generate (or span) the lattice. The *primitive basis* is the minimum number of atoms or molecules that suffices to characterize the crystal structure, and is the amount of matter contained within the primitive unit cell.

A Bravais lattice is not an arrangement of atoms but a geometric arrangement of points in space. In describing a crystal structure, one must cite both the lattice and the symmetry of the *basis* of atoms associated with each lattice point. The basis consists of the atoms, their spacings, and bond angles, which recur in an identical fashion about each lattice point such that every atom in the crystal is accounted for.

From the standpoint of symmetry, there are 14 different kinds of Bravais lattices. The 14 lattices can be grouped into 7 crystal systems, as shown in Table 2.1. For the semiconductors considered in this text, the cubic Bravais lattice matters. The important semiconductors are characterized by the tetrahedral arrangement of their

TABLE 2.1 The 7 Crystal Systems and 14 Bravais Lattices in Three Dimensions

System	Restrictions on Conventional Unit Cell Dimensions and Angles	Bravais Lattices in the System
Triclinic	$a \neq b \neq c$ $\alpha \neq \beta \neq \gamma$	P (primitive)
Monoclinic	$a \neq b \neq c$ $\alpha = \gamma = 90° \neq \beta$	P (primitive) I (body-centred)
Orthorhombic	$a \neq b \neq c$ $\alpha = \beta = \gamma = 90°$	P (primitive) C (base-centered) I (body centered) F (face-centered)
Tetragonal	$a = b \neq c$ $\alpha = \beta = \gamma = 90°$	P (primitive) I (body-centered)
Cubic	$a = b = c$ $\alpha = \beta = \gamma = 90°$	P (primitive or simple cubic) I (body-centered) F (face-centered)
Trigonal	$a = b = c$ $120° > \alpha = \beta = \gamma = 90°$	R (rhombohedral primitive)
Hexagonal	$a = b \neq c$ $\alpha = \beta = 90°, \gamma = 120°$	P (primitive rhombohedral)

Source: Blakemore, J. S., *Solid State Physics*, 2d. ed., Cambridge University Press, 1974.

nearest-neighbor atoms. Each lattice has a cubic face with a basis of two atoms, as illustrated in Fig. 2.1. For silicon and germanium, these two atoms are the same; for GaAs and the III–V compounds, the two atoms are different. In GaAs that is crystallized in a cubic sphalerite (zincblende) structure, the fcc lattices of Ga and As, which are mutually penetrating, are shifted relative to each other by a quarter of the body diagonal.

The crystal properties along different planes are different, and the electrical and other device characteristics are dependent on a crystal's orientation. A convenient method of defining the planes in a crystal is to use *Miller indices*. These indices are obtained using the following steps [Sze81]:

1. Find the intercepts of each plane on the three cartesian coordinates in terms of the constant lattice.
2. Take the reciprocals of the numbers and reduce them to the smallest three integers having the same ratio.

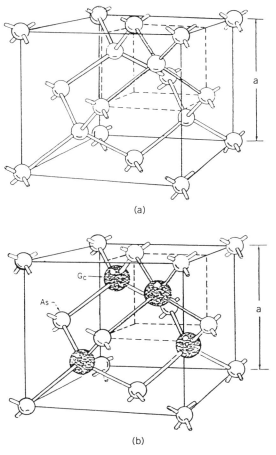

Fig. 2.1 (a) Diamond lattice; (b) zincblende lattice. [Sze, S. M., *Physics of Semiconductor Devicers*, 2d ed., Wiley-Interscience (1981)].

3. Enclose the result in parentheses (hkl) to show the Miller indices for a single plane.

Figure 2.2 gives the Miller indices of the main planes in a cubic crystal. Some other conventions are noted as follows:

(hkl) For a plane that intercepts the x-axis on the negative side of the origin, such as (100).

{hkl} For planes of equivalent symmetry, such as {100} for (100), (010), (001), (100), (010), and (001) in cubic symmetry.

[hkl] For a crystal direction, such as [100] for the x-axis. The [100] direction is perpendicular to the (100) plane, and the [111] direction is perpendicular to the (111) plane.

⟨hkl⟩ For a full set of equivalent directions, such as ⟨100⟩ for [100], [010], [001], [100], [010], and [001].

As can be seen from Fig. 2.3 [Shu87], the bonds between the nearest Ga and As atoms are in the ⟨111⟩ directions. The closest equivalent atoms are in the ⟨110⟩ directions. Owing to the lack of inversion symmetry in the zincblende structure, the directions [111] and [111] are not equivalent. The direction from Ga to the nearest As is usually denoted as [111], whereas [111] corresponds to the opposite direction.

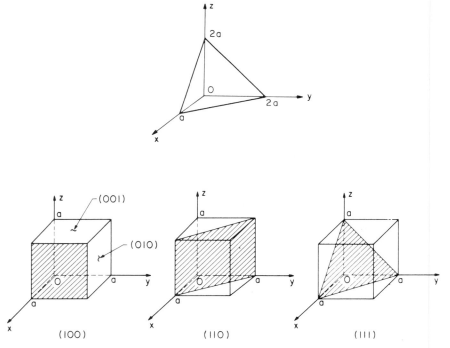

Fig. 2.2 Miller indices of some important planes in a cubic crystal. [Sze, S. M., *Physics of Semiconductor Devicers*, 3d ed., Wiley-Interscience (1981)]

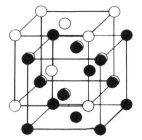

Fig. 2.3 GaAs structure: ○ Ga, ● As. [Shur, M., *GaAs Devices and Circuits*, pp. 1–10, Plenum Press (1987)]

The surface of the crystal cleaved along the (111) plane consists either of Ga atoms, which have three bonds with the crystal, or of As atoms, which have just one bond with the crystal. Of the eight planes in the {111} family for GaAs, four are (111A) planes containing only gallium atoms. The other four are (111B) planes comprised entirely of arsenic atoms. Thus a $\langle 111 \rangle$-oriented GaAs wafer with plane parallel faces has a (111A) plane for one face and a (111B) plane for the other. The difference between these two planes becomes apparent when the crystal is chemical by etched, subjected to the ion implantation, or covered by a passivating dielectric layer [Shu87].

Three basic crystal planes of the GaAs lattice are shown in Fig. 2.4. Each As atom on the (100) surface has two bonds with Ga atoms from the layer below. The two other bonds are free. The (110) plane contains the same number of Ga and As atoms. Each atom has one bond with the layer below. Atoms on the (111) surface have three bonds with the Ga atoms from the layer below. The fourth bond is free. The distance between the nearest neighbors is 2.44 Å. It is equal to the sum of the atomic radii of As (1.18 Å) and Ga (1.26 Å). The lattice constant is equal to 5.65 Å.

Figure 2.5 shows the first Brillouin zone (BZ) of reciprocal space for GaAs. Since sphalerite has fcc translational symmetry, the zone is the same as for an fcc solid [Bri78]. The zone comprises a truncated octahedron, lying within a cube with a wave-vector space (**k**-space) side length $(4\pi/A) = 2.223 \times 10^8 \, \text{cm}^{-1}$ [Bla82].

For the purpose of interpreting the vibrational spectrum of GaAs, or of describing the electronic energy band structure of this semiconductor, the most important paths through the Brillouin zone are those from the zone's center Γ to the high symmetry

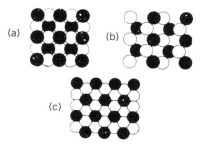

Fig. 2.4 GaAs crystal planes: (*a*) {100}; (*b*) {110}; (*c*) {111}; ○ Ga, ● As. [Shur, M., *GaAs Devices and Circuits*, pp. 1–10, Plenum Press (1987)]

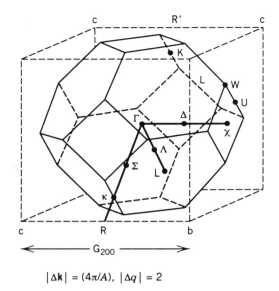

$|\Delta \mathbf{k}| = (4\pi/A),\ |\Delta q| = 2$

Fig. 2.5 First Brillouin zone for the GaAs lattice. [Blakemore, J. S., in *Key Papers in Physics—Gallium Arsenide*, ed. by J. S. Blakemore, American Institute of Physics, New York (1987)]

points X, L, and K on the zone's boundary. Figure 2.6 shows a (011) plane section through the zone's center. Note that in the figure the paths $\Gamma\Delta X$, $\Gamma\Lambda L$, and $\Gamma\Xi K$ are one quadrant of the BZ cross section in the (011) plane.

Partially heteropolar bonds in GaAs are stronger than homopolar bonds in Si or Ge. With the heteropolar bond, lattice vibrations are lower in amplitude, the melting point is higher, and the energy gap is wider.

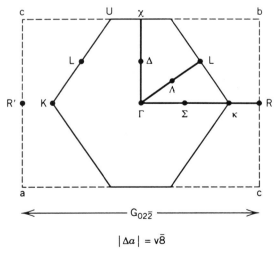

$|\Delta a| = \sqrt{8}$

Fig. 2.6 Hexagonal cross section of the GaAs first Brillouin zone as intersected by the {011} plane. [Blakemore, J. S., in *Key Papers in Physics—Gallium Arsenide*, ed. by J. S. Blakemore, American institute of Physics, New York (1987)]

2.3 ELECTRON ENERGY-BAND STRUCTURE

2.3.1 Energy-Band Calculations and Experimental Results

One of the earliest band calculations for GaAs was made by Herman [Her55], using the orthogonalized plane wave (OPW) method he had employed successfully for silicon and germanium [Her54a; Her54b]. Herman's approach benefited from the similarities of the valence-band and conduction-band systems for the diamond structure's group-IV elements and the sphalerite structure III–V compounds in general, and of isoelectronic Ge and GaAs in particular [Bla87]. In all of these solids the three upper valence bands have maxima at the zone's center, $\Gamma(000)$: the two uppermost bands (heavy holes and light holes) with a degenerate maximum, and the third separated by the spin-orbit splitting energy Δ_{so}. For GaAs, that splitting $\Delta_{so} = 341$ meV, only 18% larger than in germanium.

Early theory and experiment had shown that the lowest Si conduction minima are a set of six ellipsoids along [100], while for Ge the four $L(1/2\ 1/2\ 1/2)$ minima are slightly lower than the one at the center of the zone. Both Si and Ge are indirect gap solids. In contrast, the direct gap nature of GaAs was indicated quite early [Bla87], because of its intrinsic absorption [Osw54] and reflectance [Bar58] characteristics.

Figure 2.7 shows the general features of the $\varepsilon - k$ curves along high-symmetry directions in the zone for the four valence bands and the first several conduction bands. These curves were determined by Chelikowsky [Che76] using a nonlocal

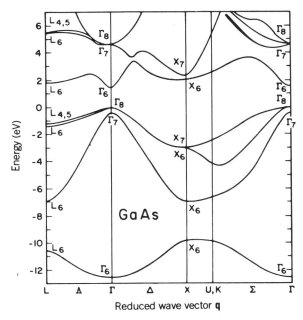

Fig. 2.7 Electron energy versus reduced wave vector, for the four GaAs valence bands, and the first several conduction bands. The top of the valence band ε_v is zero on this scale. [Blakemore, J. S., in *Key Papers in Physics–Gallium Arsenide*, ed. by J. S. Blakemore, American Institute of Physics, New York (1987)]

empirical pseudopotential model (EPM). Figure 2.8 shows another result of band calculation: In plotting $g(\varepsilon)$, the density of electron states is obtained with respect to energy. The solid line is $g(\varepsilon)$ as calculated by Chelikowsky et al. for the first 4.5 eV of the conduction-band system and for the full span of the valence bands. Chelikowsky and Cohen found that the valence band's total range is 12.5 eV, which can be compared with the 12.9 eV range reported in the experiments of Grobman et al. [Gro72]. The $g(E)$ curve was obtained from Wang's band calculation [Wan81]. This is similar to the solid line in Fig. 2.8 for the valence band—but it extends an extra 2 eV in the conduction range. The dashed line in Fig. 2.8 shows an experimentally obtained $g(\varepsilon)$ for the valence bands from X-ray photoemission spectroscopy data of Ley et al. [Ley74]. The resolution of their result was about 0.55 eV in the full-width half-maximum (FWHM).

Figures 2.7 and 2.8 present band features calculated over a 20-eV span of energy. However, the most important aspects of the band's structure are confined to a small fraction of an eV. The $\varepsilon - \mathbf{k}$ relationship at the Brillouin zone center of GaAs is good example. Figure 2.9 shows the degenerate extremum for both the heavy-hole (V1) and light-hole (V2) bands.

Another fine detail of GaAs bands to consider is the order of eigen energies of the three types of Brillouin zone locations where the lowest conduction band dips to a minimum. The order of increasing energy for Γ–L–X was definitively demonstrated by the 1976 electroreflectance experiments of Aspnes et al. [Asp76]. Figure 2.9 shows that sequence of Γ_6, L_6, and X_6 minima.

The compositional dependence of the energy gaps (E_g's) in $Al_xGa_{1-x}As$ has been studied [Lee80; Cas78]. Lee et al. [Lee80] also noted the dependence of E_g's on x at room temperature in the $Al_xGa_{1-x}As$ alloy as

$$E_g^\Gamma(x) = 1.425 + 1.155x + 0.37x^2, \qquad (2.3)$$

$$E_g^X(x) = 1.911 + 0.005x + 0.37x^2, \qquad (2.4)$$

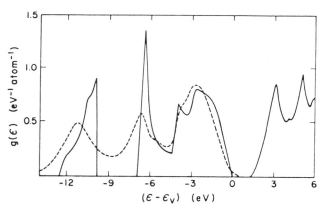

Fig. 2.8 Density of states with respect to energy $g(\varepsilon)$ for the lower part of the conduction-band system, and for the four valence bands. Solid curve is calculated by Chelikowsky and Cohen [Che76]. Dashed curve is the experimental result of Ley et al. [Ley74] for the valence bands obtained by X-ray photoelectron spectroscopy. [Blakemore, J. S., in *Key Papers in Physics—Gallium Arsenide*, ed. by J. S. Blakemore, American Institute of Physics, New York (1987)]

ELECTRON ENERGY-BAND STRUCTURE 19

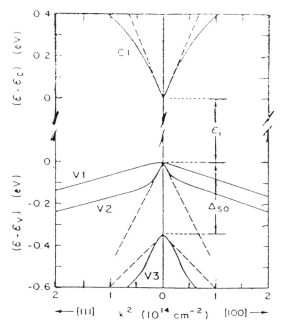

Fig. 2.11 An approximation for energy versus k^2 around the zone center for the lowest conduction minimum (C1) and the three highest valence bands (V1–V3). [100] and [111] directions are visualized. The range covered here is equivalent to $|\Delta \mathbf{k}|$ to about one-tenth of the Brillouin zone radius. [Blakemore, J. S., in *Key Papers in Physics—Gallium Arsenide*, ed. by. J. S. Blakemore, American Institute of Physics, New York (1987)]

One of the four hypothetical mini-maxima is presented at the left-hand side of Fig. 2.11, showing an energy just slightly higher than ε_v for the actual zone center. The nonparabolicity of $\varepsilon - \mathbf{k}$ for the C1, V2, and V3 extrema is also shown in the figure. Nonparabolicity of the light-hole band is a feature that has to be taken into account in describing the statistical weight of the V1–V2 combination for free holes. The need for that complication of V2 is mitigated, to some extent, by the higher density of states of the V1 band for energies slightly below ε_v.

For the lowest conduction band nonparabolicity cannot be ignored in evaluating the statistical weight for free electrons. The nonparabolicity for C1 (and also for V2 and V3) can be treated in a useful approximate manner by the $\mathbf{k} \cdot \mathbf{p}$ perturbation method.

A $\mathbf{k} \cdot \mathbf{p}$ perturbation approach toward making $\varepsilon - \mathbf{k}$ expansions about high-symmetry locations in the reduced zone was taken by Dresselhaus [Dre55], Parameter [Par55], Kane [Kan57], and Pollak et al. [Pol66]. The Kane approach for describing InSb also serves well for GaAs, once approximations relevant to $\varepsilon_i > \Delta_{so}$ are taken [Bla87].

The "three-level" Kane model, with ε_i and Δ_{so} considered as the two gaps, can describe features of the zone-center extrema for bands C1, V2, and V3 with respectable accuracy. This model describes the interactions among the four bands of Fig. 2.11 in terms of a momentum matrix element P, with dimensions of eV-cm. Those interactions can equivalently be parametrized by the quantity

$$\Omega = \frac{2m_0 P^2}{3\hbar^2} \tag{2.15}$$

with dimensions of energy [Her77].

Kane showed that the secular determinant has a solution for three of the bands (C1, V2, and V3) in the form of the cubic equation

$$\varepsilon(\varepsilon' + \varepsilon_i)(\varepsilon + \varepsilon_i + \Delta_{so}) = k^2 P^2 \frac{\varepsilon' + \varepsilon_i + 2\Delta_{so}}{3}. \tag{2.16}$$

The energy variable in Eq. (2.16) is

$$\varepsilon' = \frac{\varepsilon - \varepsilon_c - \hbar^2 k^2}{2m_0}, \tag{2.17}$$

where m_0 denotes the free electron mass. The three solutions of Eq. (2.17) can be identified with the bands C1, V2 or V3. Each solution reduces to a parabolic form, with energy varying by k^2 as a small-k asymptotic limit. These asymptotic forms are

$$\varepsilon = \varepsilon_c + \left(\frac{\hbar^2 k^2}{2m_0}\right) \times \left\{1 + \Omega\left[\left(\frac{2}{\varepsilon_i}\right) + (\varepsilon_i + \Delta_{so})^{-1}\right]\right\} \quad \text{(C1 band)}, \tag{2.18}$$

$$\varepsilon = \varepsilon_v - \left(\frac{\hbar^2 k^2}{2m_0}\right)\left[\left(\frac{2\Omega}{\varepsilon_i}\right) - 1\right] \quad \text{(V2 band)}, \tag{2.19}$$

$$\varepsilon = \varepsilon_v - \Delta_{so} - \left(\frac{\hbar^2 k^2}{2m_0}\right)\left[\frac{\Omega}{\varepsilon_i + \Delta_{so}} - 1\right], \quad \text{(V3 band)}. \tag{2.20}$$

The parabolic dispersion of Eqs. (2.18) through (2.20) ceases to be an acceptable approximation when $|\mathbf{k}|$ is more than a small fraction of the reciprocal lattice vector. This feature is demonstrated by the curves of Fig. 2.11, which depart markedly from the straight lines of the above-mentioned equations.

If Eq. (2.18) is construed as being the statement

$$\varepsilon = \varepsilon_c + \left(\frac{\hbar^2 k^2}{2m_{c_o}}\right) \quad (k \to 0) \tag{2.21}$$

in defining m_{c_o} to be the conduction band-edge effective mass, then this is related to χ_c for the band by

$$m_{c_o} = \frac{m_0}{1 + \chi_c[(2/\varepsilon_i) + (\varepsilon_i + \Delta_{so})^{-1}]}. \tag{2.22}$$

The low-temperature bands-edge effective mass $m_{c_o} = 0.067\, m_0$. When $|\mathbf{k}|$ is allowed to become a little larger, nonparabolicity must be allowed for, most innocuously through a k^4 term. This involves examining Eq. (2.18) for a quadratic for ε' and expanding the solution in powers of k^2 as far as k^4 terms. It can be written as

$$\varepsilon = \varepsilon_c + \left(\frac{\hbar^2 k^2}{2m_{co}}\right) + \left(\frac{\alpha}{\varepsilon_i}\right)\left(\frac{\hbar^2 k^2}{2m_{co}}\right)^2, \qquad (2.23)$$

where the nonparabolicity coefficient α turns out to be negative:

$$\begin{aligned}\alpha &= \frac{(1 - m_{co}/m_0)^2(3\varepsilon_i^2 + 4\varepsilon_i\Delta_{so} + 2\Delta_{so}^2)}{(\varepsilon_i + \Delta_{so})(3\varepsilon_i + 2\Delta_{so})} \\ &= \frac{-[1 - \varepsilon_i\Delta_{so}/(3\varepsilon_i + 2\Delta_{so})(\varepsilon_i + \Delta_{so})]}{[1 + \varepsilon_i(\varepsilon_i + \Delta_{so})/\chi_c(3\varepsilon_i + 2\Delta_{so})]^2}.\end{aligned} \qquad (2.24)$$

The low-temperature value $\alpha = -0.824$ is shown in Table 2.2.

2.3.4 Electron Effective Masses: Conduction-band System Statistical Weight

Effective Mass for $\Gamma_6(C1)$ Lowest Conduction Band

The definitions of effective mass lead to the same isotropic and parabolic numerical values for a band. However, these provisions are inapplicable to the lowest GaAs conduction band, whose nonparabolicity is more series oriented than anisotropic. One important definition of effective mass is band curvature:

$$m_c = \hbar^2 \left(\frac{d^2\varepsilon}{dk^2}\right)^{-1}. \qquad (2.25)$$

For phenomena such as Faraday rotation or magnetic susceptibility, the important

TABLE 2.2 Directionally Averaged (density of states) Band-edge Curvature Effective Masses and Related Parameters, for the Zone Center Extrema of Band C1, and Bands V1 through V3

Band Extremum	Band-Edge Density of States Effective Mass for $T \to 0$	$\mathbf{k \cdot p}$ Model Matrix Element, Expressed as $\chi = (2m_0 P^2/3\hbar^2)$	$\mathbf{k \cdot p}$ Model Nonparabolicity Parameter	Band-Edge Density of States' Effective Mass for $T = 300$ K
Conduction band C1	$m_{co} = 0.067 m_0$	$\chi_c = 7.51$ eV	$\alpha = -0.824$	$m_{co} = 0.063 m_0$
Heavy-hole band V1	$m_h = 0.51 m_0$	—	—	$m_h = 0.50 m_0$
Light-hole band V2	$m_l = 0.082 m_0$	$\chi_l = 10.0$ eV	$\beta = -3.80$	$m_l = 0.076 m_0$
Split-off band V3	$m_{co} = 0.154 m_0$	$\chi_{co} = 13.9$ eV	$\gamma = +10.8$	$m_{co} = 0.145 m_0$

Source: Blakemore, J. S., in *Key Papers in Physics—Gallium Arsenide*, ed. by J. S. Blakemore, American Institute of Physics (1987).

quantity is the "optical" effective mass:

$$m_{opt} = \hbar^2 k \left(\frac{d\varepsilon}{dk}\right)^{-1}. \quad (2.26)$$

If the $\varepsilon - k$ for the band can be approximated by Eq. (2.23), then both m_c and m_{opt} converge for the edge of the band. For energies slightly higher than ε_c, then

$$m_c = m_{c_o}\left[1 - \left(\frac{6\alpha}{\varepsilon_i}\right)(\varepsilon - \varepsilon_c)\right],$$

$$m_{opt} = m_{c_o}\left[1 - \left(\frac{2\alpha}{\varepsilon_i}\right)(\varepsilon - \varepsilon_c)\right]. \quad (\varepsilon - \varepsilon_c) \ll \varepsilon_i \quad (2.27)$$

Hence both versions of "effective mass" increase with energy (since $\alpha < 0$), and the rate of curvature increases three times faster than the optical mass.

Experimental values for the band-edge mass m_{c_o} have been deduced from a wide variety of experiments. Stillman et al. [Sti77] give a summary of 18 such experiments. Stillman et al. [Sti69] had obtained a quite accurate value $m_{c_o}/m_0 = 0.0665 \pm 0.0005$ from an analysis of Zeeman spectroscopy of shallow donor levels at liquid helium temperatures. Their consensus [Sti77] of all work reviewed is consistent with adoption of a low-temperature, band-edge value $m_{c_o} = 0.067 m_0$ and $m_{c_o} = 0.063 m_0$ at $t = 300$ K.

Other experiments indicate that m_{c_o} decreases by several percent on warming to ambient temperatures [Bla87]. Figure 2.12 shows how m_{c_o}/m_0 should decrease from

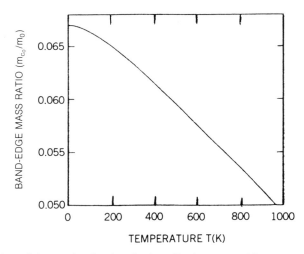

Fig. 2.12 Variation of the conduction band-edge effective mass with temperature, as described by the **k·p** model, Eq. (2.22). This assumes that $\Delta_{so} = 0.341$ eV independent of temperature and that $m_{c_o} = 0.067 m_0$ at low temperatures when $\varepsilon_i = 1.519$ eV, from which $\chi_c = (2m_0 P^2/3\hbar^2) = 7.51$ eV as the strength of the momentum matrix element, as noted in Table 2.2. [Blakemore, J. S., in *Key Papers in Physics–Gallium Arsenide*, ed. by J. S. Blakemore, American Institute of Physics, New York (1987)]

its low-temperature value when T rises and ε_i falls. As noted in Table 2.2, the low-temperature combination of knowledge of m_{co} and ε_i (and Δ_{so}) determines $\chi_c = 7.51$ eV.

Effective Densities of State for Γ_6 Conduction Band

For the parabolicity band the equilibrium concentration n_0 is given by

$$n_0 = N_{co} \int_0^\infty (2\sqrt{\pi})\zeta^{1/2} d\zeta = N_{co} F_{1/2}(\eta). \quad (2.28)$$

Here $F_{1/2}(\eta)$ is the Fermi-Dirac integral, and

$$N_{co} = 2\left(\frac{2\pi m_{co} kT}{h^2}\right)^{3/2} \quad (2.29)$$

is the effective density of state for a parabolic band. At room temperature for GaAs, $N_{co} = 3.99 \times 10^{17}$ cm^{-3}. Because of the nonparabolicity of the GaAs conduction band, the density of state N_{co} has to be rewritten as N'_c, where

$$N'_c = 8.63 \times 10^{13} T^{3/2}(1 - 1.93 \times 10^{-4} T - 4.19 \times 10^{-8} T^2)\,\text{cm}^{-3}$$

$$\text{for } (100 < T < 1200 \text{ K}). \quad (2.30)$$

N'_c and N_{co} are shown in Fig. 2.13 as a function of temperature [Bla87]. Table 2.3 provides a short list of degenerate Fermi energies that correspond to low-temperature n_0 values, the optical mass m_{opt}, and the curvature mass m_c.

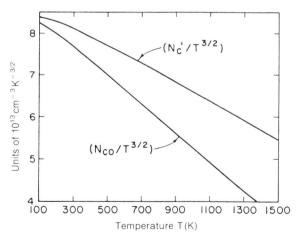

Fig. 2.13 Components of the temperature dependence for the effective density of states associated with the lowest GaAs conduction minimum. [Blakemore, J. S., in *Key Paper in Physics—Gallium Arsenide*, ed. by J. S. Blakemore, American Institute of Physics, New York (1987)]

TABLE 2.3 Low Temperature Degenerate Limit for the Lowest Conduction Band: Electron Density, Fermi Energy, and Two Different Definitions of Effective Mass

n_0 (cm^{-3})	$(\varepsilon_F - \varepsilon_c)$ (meV)	Optical Mass Ratio (m_{opt}/m_0)	Curvature Mass Ratio (m_c/m_0)
5×10^{16}	7.4	0.0675	0.0686
1×10^{17}	11.6	0.0679	0.0695
2×10^{17}	18.4	0.0683	0.0710
5×10^{17}	33.6	0.0694	0.0743
1×10^{18}	52.8	0.0708	0.0785
2×10^{18}	82.5	0.0730	0.0850
5×10^{18}	147	0.0770	0.0991
1×10^{19}	225	0.0834	0.1162

Source: Blakemore, J. S., in *Key Papers in Physics—Gallium Arsenide*, ed. by J. S. Blakemore, American Institute of Physics (1987).

Mass Parameters for L_6 and X_6

Conduction minima higher than Γ_6(C1) in the GaAs band structure have been attracted great interest. Aspnes et al. [Asp76] demonstrated the Γ–L–X conduction band order and gave the data concerning these two sets of upper minima, at L_6 and X_6. The values of the effective masses for the X and L are given as follows:

For the L_6 band,

$$m_L = 1.9 m_0,$$
$$m_t = 0.075 m_0,$$
$$m_L = (16 m_l m_t^2)^{1/3} = 0.056 m_0. \qquad (2.31)$$

For the X_6 band,

$$m_L = 1.9 m_0,$$
$$m_t = 0.19 m_0,$$
$$m_X = (9 m_l m_t^2)^{1/3} = 0.85 m_0. \qquad (2.32)$$

Here the m_L and m_X are density-of-state effective masses at the L and X bands, respectively.

Conduction Electron Thermal Distribution

Since $\Delta_{\Gamma L}$ of Eq. (2.12) is only a few kT for high temperatures, one can expect a major fraction of the thermal conduction electron population to make the $\Gamma \rightarrow L$ transition as temperature rises. While m_X is larger than m_L, the activation energy $\Delta_{\Gamma X}$ of Eq. (2.14) presents a more formidable barrier. The total conduction electron population

n_0 should be written as a sum of three thermal contributions:

$$n_0 = n_X + n_L + n_\Gamma. \tag{2.33}$$

For almost any N-type doping of GaAs, ε_F will remain low enough to permit an essentially Boltzmann distribution in the upper two bands. At any temperature T the three terms on the right of Eq. (2.33) should be

$$n_X = 2\left(\frac{2\pi m_X kT}{h^2}\right)^{3/2} \exp\left(\frac{\eta - \Delta_{\Gamma X}}{kT}\right),$$

$$n_L = 2\left(\frac{2\pi m_L kT}{h^2}\right)^{3/2} \exp\left(\frac{\eta - \Delta_{\Gamma L}}{kT}\right),$$

$$n_\Gamma = N_{co}[F_{1/2}(\eta) - \left(\frac{15\alpha kT}{4\varepsilon_i}\right) F_{3/2}(\eta). \tag{2.34}$$

Assume that $m_L = 0.52 m_0$ for all high temperatures and that $m_X = 0.85 m_0$ as in Eq. (2.32). The three quantities on the right-hand side of Eq. (2.34) can be added up to n_0 as

$$n_0 = N_c^* \exp\left[\frac{(\varepsilon_F - \varepsilon_c)}{kT}\right], \qquad n_0 < N_c^*, \tag{2.35}$$

where the total statistical weight of the conduction band system is

$$N_c^* = 8.63 \times 10^{13} T^{3/2} \bigg[(1 - 1.93 \times 10^{-4} T - 4.19 \times 10^{-8} T^2)$$
$$+ 21 \exp\left(\frac{-\Delta_{\Gamma L}}{kT}\right) + 44 \exp\left(\frac{-\Delta_{\Gamma X}}{kT}\right) \bigg] \text{cm}^{-3}. \tag{2.36}$$

Figure 2.14 presents the fractional contributions of n_Γ, n_L, and n_X toward the total n_0 for the same temprature range. In the figure that more than half of all conduction electrons are thermally in the L_6 band for $T > 900$ K. The proportion of electrons remaining in the $\Gamma_6(C1)$ conduction band has declined to less than 12% by the melting point.

2.3.5 Hole Effective Masses

Many observable properties of holes in GaAs can be interpreted fairly well in terms of scalar effective masses m_h and m_l for bands V1 and V2, respectively. The hole density-of-state effective mass m_v is given by

$$m_v = (m_h^{3/2} + m_l^{3/2})^{2/3}. \tag{2.37}$$

This is the quantity that one would like to use in describing the statistical weight of the valence band system. Various experiments suggest that low-temperature values

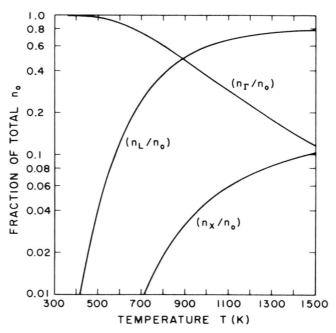

Fig. 2.14 Temperature dependence of the three-way split of conduction electrons among the $\Gamma_6(C1)$, L_6, and X_6 bands. [Blakemore, J. S., in *Key Papers in Physics—Gallium arsenide*, ed. by J. S. Blakemore, American Institute of Physics, New York (1987)]

are to be preferred [Bla87; Sko76]:

$$\frac{m_l}{m_0} = 0.082 \pm 0.004,$$

$$\frac{m_h}{m_0} = 0.51 \pm 0.02, \quad \text{for } T < 100 \text{ K}; \quad (2.38)$$

see Table 2.2. The two numbers lead to the density-of-state mass

$$m_v = 0.53 m_0, \quad \text{for } T < 100 \text{ K}, \quad (2.39)$$

in the V1–V2 combination of bands at low temperatures. Table 2.2 also shows the 300 K value of m_l and m_h for holes in GaAs.

The equilibrium concentration for holes in the valence band is given by

$$p_0 = N_v F_{1/2}(\xi), \quad (2.40)$$

where $\xi = [(\varepsilon_v - \varepsilon_F)/kT]$ is the dimensionless expression of ε_F relative to ε_v, and N_v is the effective density of states provided by the heavy-hole (V1) and light-hole (V2) bands combined.

Because of the absence of inversion symmetry, it should be clear that $\varepsilon - k$ is not parabolic for the heavy-hole band. The density of state depends on both the temperature and degeneracy:

$$N_v = 2\left(\frac{2\pi m_v kT}{h^2}\right)^{3/2} \left\{ m_h^{3/2} + m_l^{3/2}\left[1 - \left(\frac{15\beta kT}{4\varepsilon_i}\right) F_{3/2}(\xi)/F_{1/2}(\xi)\right]\right\}. \quad (2.41)$$

The $m_h^{3/2}$ decreases as temperature rises. The factor in the bracket $[\cdots]$ multiplying $m_l^{3/2}$ increases with temperature. The contribution of the factor $[\cdots]$ also depends on whether the free-hole density p_0 is larger or smaller than N_v, since $F_{3/2}(\xi) > F_{1/2}(\xi)$ for a degenerate P-type situation of $p_0 > N_v$, $\xi > 0$.

2.4 ELECTRON AND HOLE TRANSPORT

Carrier transport in GaAs and other III–V compounds concerns device physicists and engineers. This section is kept brief, with the intention of revealing the important aspects of carrier transport that the researcher commonly uses in an analysis of raw data. The boltzmann transport equation and its approximate solutions, scattering mechanisms, Monte Carlo methods, and the electron transport in ultra-small semiconductor devices will be covered in Chapter 6.

2.4.1 Hole Mobility

The mobility of holes in GaAs was studied early in the 1950s [Wel56]. The 1975 review by Wiley [Wil75] deals with hole transport in III–V compounds in considerably detail. D. C. Look has reviewed more recent data in undoped, doped, and ion-implanted GaAs [Loo90].

It will be supposed here, based on the discussions in Section 2.3.5, that heavy holes in the V1 valence band have an effective mass $m_h = 0.050 m_0$ averaged over all directions, and for all temperatures around and above the ambient temperature [Bla87]. The rms thermal speed of these heavy holes is

$$v_h(\text{rms}) = \left(\frac{3kT}{m_h}\right)^{1/2} = 1.65 \times 10^7 \left(\frac{T}{300}\right)^{1/2} \frac{\text{cm}}{\text{s}}. \quad (2.42)$$

For holes in the nonparabolic light-hole band, the speed $v_l(\text{rms})$ is

$$v_l(\text{rms}) = 3.4 \times 10^7 \left(\frac{T}{300}\right)^{1/2} \frac{\text{cm}}{\text{s}}. \quad (2.43)$$

The speed for the light and heavy holes combined is

$$\langle v_p \rangle (\text{rms}) = 1.77 \times 10^7 \left(\frac{T}{300}\right)^{1/2} \frac{\text{cm}}{\text{s}}. \quad (2.44)$$

Mobilities in GaAs are generally measured by means of the Hall effect method.

The measurement yield the Hall mobility

$$\mu_H = \sigma R_H = r_H \mu_p, \qquad (2.45)$$

where r_H is a Hall factor that depends on temperature, doping, magnetic induction strength, and other variables. In very large magnetic fields r_H tends to unity. Figure 2.15 shows the "high-purity" p-type GaAs Hall mobility as a function of temperature. Wiley [Wil75] suggests that the experimental results of three different groups [Hil70; Mea71; Zsc73] can be reasonably well fitted by

$$\mu_{H0}(T) = 400 \left(\frac{300}{T}\right)^{2.3} \frac{\text{cm}^2}{\text{V}-\text{s}}. \qquad (2.46)$$

Wiley's review [Wil75] goes into some detail on scattering mechanisms for holes. He points out that acoustic phonon and nonpolar optical mode scattering processes are of comparable importance for $T > 100$ K, with polar mode scattering being probably less important. Ionized impurity scattering dominates for the low-temperature region, as seen to a small extent in the mobility roll-off below 40 K for Hill's sample in Fig. 2.15.

The data of Fig. 2.16 [Bla87] came from experiments of Rosi et al. [Ros60], Hill

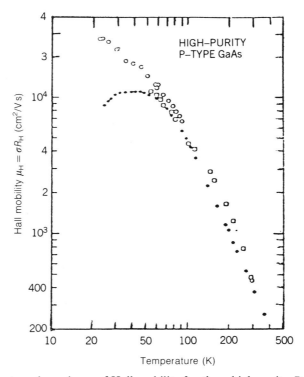

Fig. 2.15 Temperature dependence of Hall mobility for three high-purity P-type GaAs VPE samples, after Wiley [Wil75]. [Blakemore, J. S., in *Key Papers in Physics—Gallium Arsenide*, ed. by J. S. Blakemore, American Institute of Physics, New York (1987)]

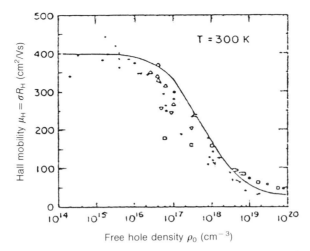

Fig. 2.16 Variation of 300 K Hall mobility with hole concentration, after Wiley [Wil75]. [Blakemore, J. S., in *Key Papers in Physics—Gallium Arsenide*, ed. by J. S. Blakemore, American Institute of Physics, New York (1987)]

[Hil64; Hil70], Vilms [Vil72], Rosztoczy et al. [Ros70], Emel'yanenko et al. [Eme60], and Gasanli et al. [Gas72]. Wiley noted that the curve in this figure could be calculated by using a combination of lattice scattering and Brooks-Herring-type ionized impurity scattering.

The simplest way to combine scattering contributions is by applying the Matthiessen's rule:

$$\mu_{H0} = \left[\left(\frac{1}{\mu_L}\right) + \left(\frac{1}{\mu_I}\right)\right]^{-1}. \tag{2.47}$$

A useful empirical equation for temperatures fairly close to room temperature was proposed by Blakemore [Bla87]:

$$\mu_{H0} = \left[2.5 \times 10^{-3}\left(\frac{T}{300}\right)^{2.3} + 4 \times 10^{-21} N_I \left(\frac{300}{T}\right)^{1.5}\right]^{-1} \frac{\text{cm}^2}{\text{V}-\text{s}}. \tag{2.48}$$

This equation could serve as a reasonable expectation for Hall mobility under weak field conditions.

2.4.2 Electron Velocity and Mobility

The rms speed of the conduction electrons for $(3/2)kT$ kinetic energy is

$$v_c(\text{rms}) = \left(\frac{3kT}{m_{c_o}}\right)^{1/2}\left[\frac{1 + 3\alpha kT}{\varepsilon_i}\right]$$

$$= 4.4 \times 10^7 \left(\frac{T}{300}\right)^{1/2} \frac{\text{cm}}{\text{s}}. \tag{2.49}$$

Here the second numerical approximation is suitable for temperatures fairly close to the ambient range.

A current pulse in n-type GaAs propagates at a speed called *electron drift velocity*, v_d. The drift velocity depends on the applied electric field **E**. For a small field $v_d = \mu_n \mathbf{E}$, where μ_n denotes the drift or conductivity mobility, $\mu_n = (\sigma/en_0)$. At room temperature $\mu_n(300) = 8000 \text{ cm}^2/\text{V} - \text{s}$ in lightly doped n-type GaAs. The low-field region at the left-hand side of Fig. 2.17 shows v_d rising with **E** at that slope. As the field increases, the electrons in the Γ_6 minimum start to warm up, and the slope decreases. Compilation of velocity-field characteristics for carriers in most heavily utilized semiconductors is shown in Fig. 2.17 [Sze90].

The electron mobility in GaAs for a small electric field can be written in terms of the Hall mobility

$$\mu_H = |\sigma R_H| = r_H \mu_n = \frac{\sigma r_H}{en_0}, \tag{2.50}$$

rather than as a conductivity or drift mobility μ_n. The Hall factor $r_H = 1$ for "strong-field" conditions ($B\mu_n \gg 1$ in the SI system of teslas and m^2/V-s).

Three figures from an important 1975 study by Rode [Rod75] of electron transport in III–V materials are reproduced here. Figure 2.18 shows μ_H as a function of temperature for various samples of "high-purity" n-type GaAs. The solid line in Fig. 2.18 of "lattice-scattering" mobility μ_L rises on cooling, but not in the power law form. The dashed line in Fig. 2.18 represents low-temperature falloff according to Dingle's formulation of ionized impurity scattering. The slightly better fit is based on the Brooks-Herring formulation [Cha81].

Figure 2.19 shows $\mu_H(T)$ for three samples, as reported by Stillman et al. [Sti70], with dashed lines indicating the expected contributions of (1) ionized impurity scattering, (2) deformation potential scattering by acoustic phonons, and (3) polar-mode scattering

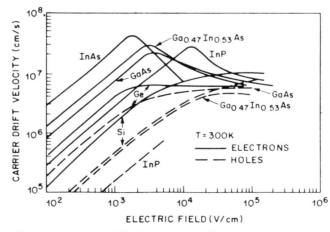

Fig. 2.17 Compilation of velocity-field characteristics for carriers in the most heavily utilized semiconductors. [Sze, S. M. ed., High-Speed Semiconductor Devices, John Wiley and Sons (1990)]

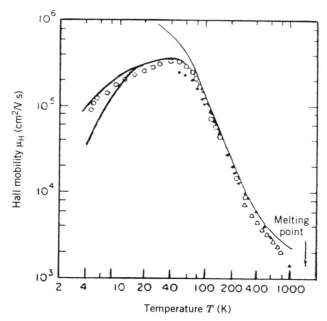

Fig. 2.18 Temperature dependence of the electron Hall mobility μ_H, as measured for $B \sim 5\,kG$ with rather pure N-type GaAs samples, after Rode [Rod75]. [Blakemore, J. S., in *Key Papers in Physics—Gallium Arsenide*, ed. by J. S. Blakemore, American Institute of Physics, New York (1987)]

by optical phonons. The third (polar optical scattering) dominates the course of the "lattice" mobility μ_L, while the first does an efficient job in controlling and modeling μ_H for low temperatures. In the figure the acoustic phonon scattering, which controls μ_L for semiconductors such as Ge and Si, appears to provide only about 10% of the lattice scattering for electrons in GaAs at room temperature.

The Hall factor r_H is greater than 1 for medium- or weak-field measurements with n-type GaAs. Figure 2.20 shows Rode's [Rod75] estimates for weak-field $r_{HO} = (\mu_{HO}/\mu_n)$ in changing temperatures. This is appropriate for weakly doped material in which impurity scattering plays a minor role above 100 K. Rode notes that the large lobe of $r_{HO} > 1$ for $T > 100$ K arises from the dominance of polar mode optical phonon scattering, whereas acoustic phonon scattering (deformation potential and piezoelectric) and impurity scattering determine the lower-temperature range.

A useful estimate for the electron mobility in low-doped samples at a low temperature close to room temperature was proposed by Blakemore [Bla87]:

$$\mu_n = 8000 \left(\frac{300}{T}\right)^{2.3} \frac{cm^2}{V-s}, \qquad (2.51)$$

$$\mu_{HO} = 9400 \left(\frac{300}{T}\right)^{2.3} \frac{cm^2}{V-s}. \qquad (2.52)$$

32 GALLIUM ARSENIDE CRYSTAL STRUCTURE AND GROWTH

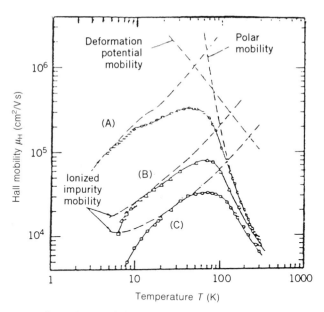

Fig. 2.19 Temperature dependence of the electron Hall mobility μ_H (for B = 5 kG) for three N-type GaAs samples, after Stillman et al. [Sti70] They estimated donor densities of (A) 5×10^{13} cm^{-3}, (B) 10^{15} cm^{-3}, and (C) 5×10^{15} cm^{-3} for the three samples so identified, with (N_c/N_e 0.3 to 0.4 in each case.) [Blakemore, J. S., in *Key Papers in Physics—Gallium Arsenide*, ed. by J. S. Blakemore, American Institute of Physics, New York (1987)]

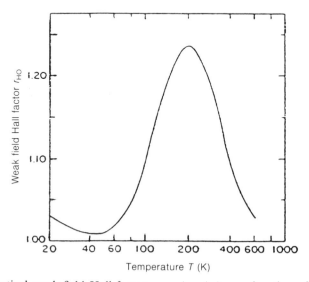

Fig. 2.20 Theoretical weak-field Hall factor $r_{H0} = (\mu_{H0}/\mu_n)$ as a function of temperature, as calculated by Rode [Rod75] for modestly doped N-type GaAs. [Blakemore, J. S., in *Key Papers in Physics—Gallium Arsenide*, ed. by J. S. Blakemore, American Institute of Physics, New York (1987)]

Figure 2.21 shows the Hall mobility for three samples [Nic80], compared with Rode's expressions (solid curve) for a lattice-scattering Hall mobility. At elevated temperatures the mobility decrease is quite substantial. Most of this decrease is caused by the intervalley transition into a higher minima of the conduction band (mostly to the L minima), where the mobility is smaller than in the central Γ minimum and decreases faster with temperature.

2.4.3 Electrical Conductivity

The ambipolar conductivity of a semiconductor can be written

$$\sigma = \sigma_n + \sigma_p = e(n_0\mu_n + p_0\mu_p) \tag{2.53}$$

where μ_n and μ_p are the drift mobilities with $\mu_n(300) = 8000 \, \text{cm}^2/\text{V} - \text{s}$ and $\mu_p(300) = 320 \, \text{cm}^2/\text{V} - \text{s}$. For GaAs doped predominantly with shallow donors or acceptors, the ratio n_0/p_0 is far from unity at ordinary temperatures that the conduction is unipolar. Doping conditions place the Fermi level ε_F near to its intrinsic location ψ, enforcing ambipolar conduction at ordinary temperatures. The room temperature resistivity of such semi-insulating GaAs in the range 10^5–10^9 Ω-cm. The achievement of the semi-insulating condition through creation of the EL2 centers will be discussed in Section 2.7. Chromium doping [Bla87] provides an alternative route to the near-intrinsic status for GaAs at room temperature.

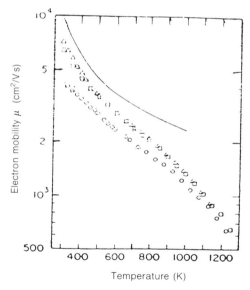

Fig. 2.21 Data of Nichols et al. [Nic80] for Hall mobility (at 3.5 kG) versus T for N-type GaAs epitaxial layers, above room temperature. The solid line shows Rode's expectation [Rod75] for high purity N-type GaAs. Doping of the three samples here is $\bigcirc$ $N_d \sim 4N_a \sim 1.2 \times 10^{17} \, \text{cm}^{-3}$; $\square$ $N_d \sim 4N_a \sim 10^{16} \, \text{cm}^{-3}$, and $\triangle$ $N_d \sim 3N_a \sim 2 \times 10^{15} \, \text{cm}^{-3}$. [Blakemore, J. S., in *Key Papers in Physics—Gallium Arsenide*, ed. by J. S. Blakemore, American Institute of Physics, New York (1987)]

Figure 2.22 shows n_0, p_0, and the resulting conductivity of Eq. (2.53) for $T = 300$ K as functions of the Fermi energy $(\varepsilon_F - \varepsilon_v)$. The intrinsic condition, $\varepsilon_F = \psi$, corresponds to $n_0 = p_0 = n_i$. Since the mobility ratio $b = (\mu_n/\mu_p) = (8000/320) = 25$ at room temperature for modestly doped material, the lowest conductivity $\sigma_{\min}$ occurs appreciably on the p-type side of ψ. Table 2.4 shows that $n_i(300) = 2.25 \times 10^6$ cm^{-3}, and hence that

$$\sigma_i(300) = en_i\mu_p(1+b) = 3.00 \times 10^{-9}\, \Omega^{-1}\text{cm}^{-1}. \qquad (2.54)$$

However, the conductivity is minimized when $p_0 = \sqrt{b}n_i = bn_0$, and then

$$\sigma_{\min} = 2en_i(\mu_n\mu_p)^{1/2} = \left[\frac{2\sigma_i}{b+1}\right]\sqrt{b}$$

$$= 0.385\sigma_i = 1.15 \times 10^{-9}\, \Omega^{-1}\text{cm}^{-1}. \qquad (2.55)$$

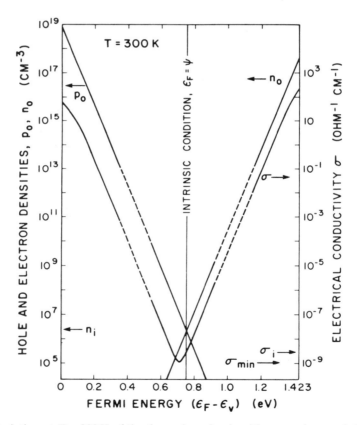

Fig. 2.22 Variation at $T = 300$ K of the thermal carrier densities n_0 and p_0, and the resulting electrical conductivity, with the Fermi energy location. Assume that $n_i = 2.25 \times 10^6$ cm^{-3} and that $\mu_n = 25\mu_p = 8000$ cm^2/V − s, with $\varepsilon_i = 1423$ meV, and $(\psi - \varepsilon_v) = 752$ meV. (Blakemore, J. S., in *Key Papers in Physics—Gallium Arsenide*, ed. by J. S. Blakemore, American Institute of Physics, New York (1987)]

TABLE 2.4 Intrinsic Carrier Pair Density n_i, and Consequent Intrinsic Fermi Energy ψ, for Gallium Arsenide

T (K)	ε_i (meV)	$\Delta_{\Gamma L}$ (meV)	$\Delta_{\Gamma X}$ (meV)	N_c^* (cm^{-3})	N_v' (cm^{-3})	n_i (cm^{-3})	$(\psi - \varepsilon_v)$ (meV)	$(\varepsilon_c - \psi)$ (meV)	$(\psi - \varepsilon_v - \frac{1}{2}\varepsilon_i)$ (meV)
250	1446.7	287.1	473.1	3.238 E17	7.234 E18	4.20 E3	756	689	34
300	1422.5	284.5	476.1	4.209 E17	9.509 E18	2.25 E6	752	671	40
350	1399.5	281.7	479.8	5.251 E17	1.198 E19	2.10 E8	747	653	47
400	1375.8	278.9	483.3	6.369 E17	1.464 E19	6.57 E9	742	634	54
450	1351.6	276.0	486.9	7.594 E17	1.747 E19	9.83 E10	737	615	61
500	1327.1	273.1	490.6	8.979 E17	2.046 E19	8.78 E11	731	596	67
600	1277.0	267.1	498.0	1.258 E18	2.690 E19	2.52 E13	718	559	79
700	1226.0	261.0	505.6	1.809 E18	3.389 E19	3.02 E14	701	525	88
800	1174.5	254.9	513.3	2.665 E18	4.141 E19	2.10 E15	682	493	95
900	1122.2	248.7	521.1	3.951 E18	4.941 E19	1.01 E16	659	463	98
1050	1043.8	239.3	532.8	6.943 E18	6.226 E19	6.50 E16	621	423	99
1200	964.6	229.8	544.6	1.152 E19	7.607 E19	2.79 E17	580	385	98
1350	885.1	220.4	556.4	1.793 E19	9.077 E19	8.99 E17	537	348	94
1500	805.3	210.8	568.3	2.642 E19	1.063 E20	2.35 F18	493	313	90

Source: Blakemore, J. S., in *Key Papers in Physics—Gallium Arsenide*, ed. by J. S. Blakemore, American Institute of Physics (1987).

36 GALLIUM ARSENIDE CRYSTAL STRUCTURE AND GROWTH

For 300 K the minimum conductivity situation occurs when $\varepsilon_F = [\psi - (1/2)kT\ln(b)] = (\varepsilon_v + 0.710\,\text{eV})$. GaAs near the $\sigma_{\min}$ condition will have a negative Hall coefficient even though $p_0 > n_0$. Not until $p_0 > bn_i > b^2 n_0$ does the Hall coefficient become positive.

2.5 CRYSTAL GROWTH OF SEMI-INSULATING GaAs FROM MELT

Development of high-performance GaAs microwave and millimeter-wave integrated-circuit technologies requires the semi-insulating substrate. Unlike-silicon, the semi-insulating GaAs substrate provides an electrically isolated active layer for high-speed IC devices after appropriate ion-implantation on epitaxy. Single crystals of GaAs have been grown by many techniques utilizing melt and solution approaches, including horizontal Bridgman (HB) and vertical Bridgman (VB), gradient freeze, Czochralski, liquid-encapsulated Czochralski (LEC), liquid-encapsulated Kyropoulous (LEK), float-zone, horizontal- and vertical-zone melting, and more recently magnetic LEC (MLEC). The Bridgman technique is dominant in terms of material quantity. However, its application is mainly in optoelectronic substrates, and it is the LEC-growth technique that is expected to dominate by the early 1990s.

2.5.1 Horizontal Bridgman Growth

The horizontal Bridgman technique, which has been employed for many years for the production of semi-insulating GaAs crystals [Bri25], is a variant of a method involving the progressive crystallization of a molten bar of the material passed through a temperature gradient. The process methodology for HB growth was recently reviewed by Rudolph and Kiessling [Rud86].

Figure 2.23 shows the basic components of a horizontal Bridgman reactor and growth system [AuC85]. The apparatus contains a GaAs charge formed separately in a quartz or pyrolytic boron nitride (PBN) boat from a seed crystal. Sufficient arsenic to maintain adequate vapor pressure during growth is placed in the quartz ampoule

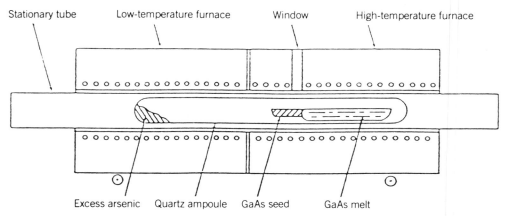

Fig. 2.23 Typical horizontal Bridgman apparatus for bulk growth of GaAs. [AuCoin, T. R., and R. O. Savage, in *Gallium Arsenide Technology*, ed. by D. K. Ferry, SAMS (1985)]

before it is sealed. A temperature difference is maintained along the tube such that at one end the Ga boat is at the melting point (1238°C) deg while at the other end arsenic is held at about 614°C. Stoichiometry is attained either by precise control of the temperature at the coldest point in the tube or by careful weighing of the Ga and As (taking account of the As in the vapor phase). The crystal is grown by moving the entire furnace so that the ampoule moves through a temperature gradient from the hotter to cooler section of the furance. With careful control of the seed-melt interface, the beginning of a single-crystal growth will be observed. A large system capable of synthesizing 5 kg of GaAs may have silica tubes that are 20 to 25 cm in diameter and 150 cm in length [Von74]. Parsey [Par81] has grown dislocation-free undoped GaAs by using a modified HB apparatus with four independent hot zones and a sodium heat pipe to control the arsenic vapor pressure. In this case a constant arsenic vapor in the growth ampoule was shown to be critically related to the perfection of the resulting GaAs crystals. The conventional viewing window used to observe the seeded growth was eliminated and replaced by careful thermal monitoring and precise control of thermal gradients. This eliminated the asymmetric thermal field associated with a window at the solid–liquid interface region. Thermal uniformity in the hot zone (melt) was found to be better than 0.5°C vertically and better than 0.1°C horizontally [AuC85].

In other experiments a modified 3T-HB furnace was developed [Aka87] [Nis82; Aka84]. This technique involves three temperature zones. Arsenic is not in the same temperature zone as the melt. The dislocation density of 2-in. and 3-in. wafer low Cr-doped HB GaAs was around 5000 cm^{-2} and showed fairly high electron mobility in implanted layers, similar to undoped SI LEC GaAs discussed later in this section. Very long GaAs single crystals of 1 meter in length can be grown. HB substrates are expected to be suitable for the substrates of HEMT-ICs, HBT-ICs, and especially for OEICs.

2.5.2 Horizontal Gradient Freeze

One of the earliest methods used for crystallization was the cooling of a static metal in a boat-shaped crucible along a thermal gradient [Bri25]. The horizontal gradient-freeze technique is in fact [Pla71] similar to the horizontal Bridgman technique. The conventional horizontal gradient-freeze technique induces the seed-oriented solidification of GaAs by programmed cooling of the temperature gradient over a static melt. Typically the gallium arsenide starting material is formed by a reaction between high-purity gallium metal and arsenic (99.9999%) inside a sealed quartz ampoule. The quartz boat containing the high-purity gallium is usually sandblasted to prevent wetting. Arsenic and the gallium-loaded quartz boat are placed inside a silica tube, which is sealed at one end. The evacuated tub is seald off and placed into a two-zone gradient-freeze furnace similar to that shown in Fig. 2.24 [Fer85]. The arsenic is located in the cooler region of the two-zone furnace, whose temperature is adjusted to yield approximately 1 atm of arsenic vapor pressure (630°C). The quartz boat containing the gallium metal is positioned at the high-temperature end of the furnace and is heated to approximately 1240°C. The arsenic is transported by vapor phase to the hot gallium where the two elements react to form GaAs. In conventional gradient-freeze growth the melt temperature is carefully adjusted to cause a slight meltback of the seed crystal. The seed–melt interface is visually monitored through

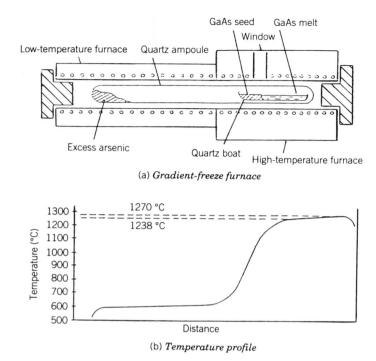

Fig. 2.24 Horizontal gradient freeze furnace and a typical temperature profile for the growth of GaAs. [AuCoin, T. R., and R. O. Savage, in *Gallium Arsenide Technology*, ed. by D. K. Ferry, SAMS (1985)]

a window located in the high-temperature region of the gradient-freeze furnace. The hotter zone temperature is then decreased at a controlled rate so that single-crystal growth can occur. After the melt has solidified, the crystal is slowed over an extended period of time in order to reduce lattice strain and associated dislocation.

Sources of impurities in the horizontal Bridgman gradient process are as follows [Par90]:

1. Starting materials (Ga, As).
2. Quartz C reacting with melt and evaporating during sealing of the tube (Si).
3. Backstreaming vacuum oil introduced during evacuation of the reactor tube (C).
4. PBN boats contributing B to the melt through dissociation in liquid GaAs.

The major impurity incorporated in the Bridgman technique is Si from reactions with quartz. The Si concentration in the crystal can be reduced significantly by controlled additions of oxygen.

2.5.3 Czochralski Crystal Growth

One of the most important techniques for the growth of large-diameter single crystals is the pulling method attributed to J. Czochralski [Czo18]. The system has a hot wall chamber to maintain the arsenic pressure at approximately 1 atm. However, arsenic loss and atmospheric contamination are a problem. The liquid-encapsulated Czochralski crystal growth (LEC) described in the following section can reduce most of the arsenic loss.

Liquid-Encapsulated Czochralski Technique

The conventional LEC technique utilizes a pressure puller in conjunction with a molten B_2O_3 encapsulating layer on the surface of the GaAs melt to prevent arsenic loss. The LEC technique with an encapsulant was first applied to the growth of PbTe by Metz [Met62]. J. B. Mullin [Mul75] reported on the use of LEC for the growth of InAs and GaAs from stoichiometric melts. In this method, as shown in Fig. 2.25, the vaporization of As from molten GaAs is inhibited by placing a layer of nonreactive B_2O_3 on the melt surface.

Boric oxide (B_2O_3) exhibits a number of properties ideal for application as an encapsulant material [Gra86]:

1. Low density (enabling it to float on the melt).
2. Not chemically reactive.
3. Low temperature melting point.
4. Transparency.

The boric oxide encapsulant typically weighs 500 g. The inert gas pressure, which is higher than the As partial pressure, is then maintained on the B_2O_3 encapsulating layer, and growth proceeds as in the standard Czochralski technique.

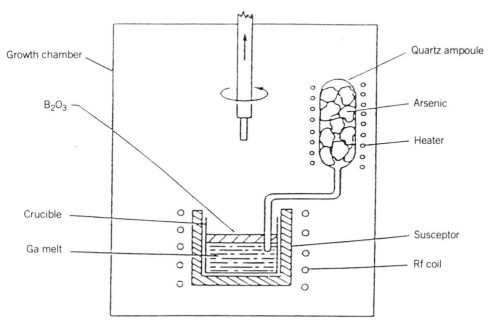

Fig. 2.25 Low-pressure injection compounding of GaAs using liquid encapsulation. [AuCoin, T. R., and R. O. Savage, in *Gallium Arsenide Technology*, ed. by D. K. Ferry, SAMS (1985)]

Low-Pressure in situ Compounding

Low-pressure in situ compounding was first reported by Pekarek [Pek70]. Pekarek described a process for synthesizing GaAs in a Czochralski puller at nitrogen pressures of 1.5 atm. Most current low-pressure in situ compounding schemes are a variation on this technique [Dun83; Wes83]. Gallium melt and the molten boric oxide encapsulant are heated together in a crucible to a temperature slightly higher than the melting point of GaAs. The fused silica arsenic cell is then positioned so that the injection quartz tube protrudes into the liquid gallium. Thermal radiation from the melt heats the cell, causing the solid arsenic to vaporize and react with the liquid GaAs under the boric oxide layer. After compounding is complete, the cell is moved out of the pulling path, the seed is lowered, and growth begins. This approach allows very precise control of the initial melt stoichiometry as well as an extra purification step for arsenic [Dun85].

High-Pressure in situ Compounding

AuCoin [AuC79] and Rumsby [Rum79] described a high-pressure in situ compounding growth process that yields semi-insulating GaAs on a reproducible basis. The high-pressure pullers are now available commercially. Stoichiometric quantities of high-purity gallium and arsenic were placed into PBN crucibles that were approximately 5 cm in diameter. The crucibles were cleaned prior to loading by etching in a 1:1 solution of electronic-grade HCl and deionized (DI) water. After etching, the crucibles were rinsed in DI water and vacuum dried overnight at 200°C [AuC85]. The loaded crucibles were placed in pyrolytic carbon-coated graphite susceptors. A dehydrated pallet of boric oxide encapsulant, weighing approximately 30 g, was placed on the charge, as shown in Fig. 2.26. The crucible was then heated in the puller chamber so that the boric oxide could melt and cover the arsenic with encapsulant before the arsenic could sublime.

After the charge and growth conditions were established in the puller, the system was evacuated and heated to 325°C to remove the volatile gallium/arsenic suboxides and residual moisture. The unit was then backfilled with high-purity nitrogen gas to about 4-atm pressure and heated to approximately 500°C to encapsulate the solid arsenic. The pressure was now increased to 60 atm and the temperature raised to 700°C, since arsenic reacts very exothermically with the gallium to form GaAs. When the reaction was completed, the temperature was raised above the melting point of GaAs and the pressure was reduced to 2 or 4 atm. Single-crystal growth was then conducted as in the conventional Czochralski technique.

Other LEC Growth Techniques

Undoped semi-insulating GaAs single crystals produced by the liquid-encapsulated Czochralski technique have received considerable attention because high-purity PBN crucibles are used. There are clear indications that large, round shaped homogeneous crystals with high quality (i.e., low dislocation density, low residual impurity, no thermal conversion, low cost, and high homogeneity) and high reproducibility are

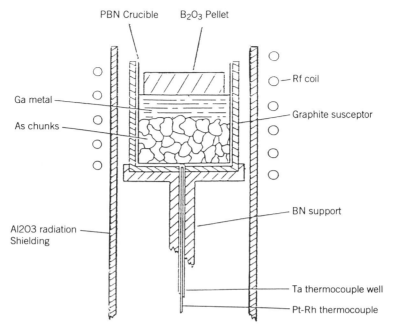

Fig. 2.26 Experimental arrangement for high-pressure in situ compounding of GaAs. [AuCoin, T. R., and R. O. Savage, in *Gallium Arsenide Technology*, ed. by D. K. Ferry, SAMS (1985)]

essential for fabrication of GaAs ICs [Fuk87]. To meet these requirements, GaAs crystals grown for ICs must have large diameters, high resistivity ($> 10^7 \Omega$-cm), no thermal conversion, and high reproducibility. As a result, the effort over the past few years has been largely concentrated on how to develop a controlled and reproducible growth technique for large-sized, high-quality GaAs crystals. Sumitomo has successfully grown a 5-in wafer. Optoelectronics Joint Research Lab (OJL) can grow very low dislocation density undoped SI substrates. In-doped SI substrates are virtually dislocation free.

There are other LEC growth techniques such as the arsenic injection technique, the liquid-phosphorus-encapsulated Czochralski technique, the liquid-encapsulated Kyropoulos technique, and the magnetic LEC technique. Next we will discuss the magnetic LEC (MLEC) techniques.

Magnetic LEC Techniques

In conventional LEC growth thermal convection currents are present. These currents affect the crystal's homogeneity and the shape of its interface. Utech [Ute66] has shown that magnetic fields can decrease thermal convection. Terashima et al. [Ter83; Ter84] investigated the effects of a magnetic field during LEC growth. Figure 2.27 shows the horizontal and vertical MLEC technique. It consists of an in-house modified Melbourne high-pressure puller and a superconducting coil with compact refrigerated systems directly set to the cryostat. This magnet can supply up to 5000 Oe to the

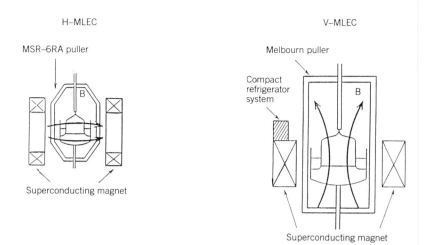

Fig. 2.27 Magnetic LEC technique. [Akai, S., in JAPAN–U.S. Perspective, U. Florida, (1987); Terashima, K., F. Orito, T. Katsumata, and T. Fukuda, *Jpn. J. Appl. Phys.* **23**, 7: L485–87 (1984)]

crucible center. It is designed to move vertically in order to apply the field effectively to the melt. A superconducting magnet is much smaller than a normal conducting magnet, about 1/5 its size and 1/10 its weight. Because of the compactness of the superconducting coil, the new MLEC apparatus can be constructed with a conventional high-pressure puller such as that widely used for volume production of GaAs.

2.5.4 Defect Densities in SI LEC GaAs

Undoped LEC GaAs contains crystal defects that affect both its semi-insulating behavior and the instability and nonuniformity of its substrates [Bro90]. In general, crystal defects can be divided into three groups:

1. Line defects (dislocations).
2. Volume defects (precipitates).
3. Point defects (EL2).

This section will briefly discuss these defects. Section 2.6 will cover EL2 defects in more detail.

Line Defects

Dislocations in commercial 3-in. and 4-in. diameter LEC GaAs crystals have recognizable characteristics. The dislocations result from polygonization and other slip interactions that are produced during or soon after solidification. The dislocation density D is not at all uniform and ranges from 10^4 to $10^5/cm^2$ [Jor80; Jor84; Rum80]. Such high values are unacceptable, so much work has been underway to reduce D.

Volume Defects

Volume defects are, in general, observed in association with dislocations. Those that are large enough to be studied have been analyzed by X-ray diffraction to be hexagonal arsenic precipitates [Cul80; Sti84]. Also the presence of microprecipitates consisting of GaAs particles has been reported [Cor85] and amorphous precipitates by others [Pon84]. It has been demonstrated that certain ingot heat treatments can cause a redistribution of the precipitates and hence an improvement in material homogeneity [Ina89; Faw90].

Point Defects

Point defects in GaAs can be divided into two categories [Bro90]: chemical impurities and native defects.

1. *Chemical impurities in LEC SI GaAs.* An excellent review of both the analytical techniques and results from various samples of GaAs has been given by Clegg [Cle82]. Typical results taken from Clegg are

S	mid-10^{14} to mid-10^{15} cm^{-3}
Se	low 10^{13} to low 10^{15} cm^{-3}
Te	10^{12} to 10^{14} cm^{-3}
Cr	10^{13} to mid-10^{13} cm^{-3}
Fe	around 10^{14} cm^{-3}
Mn	mid-10^{13} cm^{-3}
Mg	around 10^{14} cm^{-3}
Si	low 10^{14} to mid-10^{16} cm^{-3}
C	low 10^{15} to low 10^{16} cm^{-3}
B	mid-10^{14} to mid-10^{17} cm^{-3}

    The concentrations for Si, C, and B are a function of the crucible material and the H_2O content of the boric oxide encapsulant [Far82]. Carbon is observed at high concentration in crystals grown from either crucible type but only if "dry" B_2O_3 encapsulant is used. Boron is a troublesome impurity in crystals grown in a PBN crucible under dry B_2O_3.

2. *Native crystal defects and "deep levels."* Generally native defects can only be observed if they are associated with energy levels situated relatively deep in the band gap. They can be detected, and their concentrations measured in SI material using thermally stimulated current (TSC) spectroscopy [Bue72; Fan90]. This technique can be extended to detect deeper traps by using light excitation: optical transient current spectroscopy (OTCS) [Hur78]. Because of the difficulty in separating hole traps and electron traps using either TSC or OTCS, another method, deep level transient spectroscopy (DLTS), has been developed [Lan74].

2.6 EL2 CENTERS IN GaAs

High-speed integrated circuits made from GaAs require high-quality substrate materials. An understanding of the potential defects in the materials is crucial. One defect of considerable scientific and technological interest in semi-insulating GaAs grown by the LEC method is the deep donor level EL2. The free electron concentration, for example, is determined by the balance between EL2 deep donors and carbon acceptors. The electron concentration increases, and the resistivity decreases, as the EL2 concentration increases. It is important to evaluate the distribution of EL2 centers throughout the LEC GaAs crystals.

It has been shown that the average EL2 concentration along GaAs crystals is controlled by melt stoichiometry [Hol82; Lag82; Ta82] The average EL2 concentration increase, decreases, or remains constant depending on whether the melt is As rich, Ga rich, or near stoichiometric, respectively. It is preferable that the melt be slightly As rich [Hol82]. The origin of radial variations in EL2 is of special interest [Kir85]. A limited number of studies have been published concerning radial EL2 variations. In these studies linear EL2 profiles across 2-in. and 3-in. LEC GaAs crystals were obtained by the optical scanning technique [Mar80; Hol82]. The EL2 profiles were observed to follow a characteristic W pattern, and the dislocation density followed the same pattern. The possible correlation between EL2 and dislocations will be explored below. The electrical and optical characteristics and the annealing behavior of EL2 will also be discussed.

2.6.1 Electrical Characteristics of EL2

Mircea et al. [Mir76] demonstrated the donor nature of EL2. Their experiment consisted in studying the out diffusion of EL2 after a 600°–750°C annealing in VPE materials with an electron concentration of around 10^{15} cm^{-3}. Profiling of EL2 was carried out by a differential DLTS technique. The electron concentration profile observed at low temperature, with all the EL2 centers occupied by electrons, proved flat and unaffected by the annealing process. The high-temperature carrier profile (with all the EL2 centers ionized) closely followed the out-diffusion profile of EL2. This experiment clearly showed that the EL2 centers are electrically neutral when occupied by electrons and are positively charged when releasing these electrons [Mak84]. Kaminska et al. [Kam83] proposed that the deep donor is due to a defect center consisting of the As$_4$ antisite defect and an arsenic vacancy complex. Some recent observations [Von86; Miy86] obtained by electron paramagnetic resonance (EPR) and DLTS measurements on undoped and Si-doped LEC-grown GaAs suggest that EL2 is a complex formed by the trapping of an As interstitial by an As$_{Ga}$ antisite defect. The EPR-observed As$_{Ga}$ defect in as-grown, semi-insulating GaAs corresponds in concentration to that of EL2 and transforms into a metastable state for 140 K under low-temperature optical absorption in the EL2 absorption band; however, the EPR-observed As$_{Ga}$ defect formed in electron irradiated GaAs does not show the existence of a metastable state. Thermal annealing at 850°C transforms the photosensitive As$_{Ga}$ defect in the nonphotosensitive configuration and reduces simultaneously the EL2 concentration. A subsequent annealing in the 200°C range partly regenerates the EL2 concentration and renders the As$_{Ga}$ defect again photosensitive. The kinetics of this reaction have been studied by DLTS, which shows it to be of first order and to have an activation energy of 0.5 eV. From these observations we can deduce that the EL2 complex is in its two components: As$_{Ga}$ and As$_I$. The complex is restored by the thermal mobility of As$_{Ga}$ at 200°C. The theoretical models of EL2 will be discussed in the following section.

Evaluation of the electrical properties of undoped LEC GaAs reveals a strong dependence on melt stoichiometry, as shown in Figs. 2.28 and 2.29. As indicated in Fig. 2.28, the material is semi-insulating above and *p*-type below a critical As concentration in the melt—approximately 0.475 As atom fraction. The concentration of EL2 was determined by optical absorption on the melt stoichiometry.

Semi-insulating material is obtained only when EL2 is present, but *p*-type material is obtained in the presence of the double acceptor. The correlation of EL2 with the semi-insulating behavior suggests that the electrical activity of the material is determined by the deep donor EL2 and additional residual compensating acceptors. Carbon is the only acceptor identified in these measurements other than the deep double acceptor. The double acceptor is not observed in material simultaneously with EL2 [Kir85]. Thus the indicated mechanism for semi-insulating behavior is the compensating of EL2 by carbon acceptors.

The ionization of EL2 produces an ionized center plus an electron in the conduction band [Kir85]:

$$\text{Neutral EL2} \rightarrow \text{ionized EL2} + e^-. \tag{2.56}$$

According to the law of mass action, the concentration of ionized centers N_I, the concentration of electrons n, and the concentration of neutral centers N_U are related by the following equation:

$$\frac{N_I n}{N_U} = K, \tag{2.57}$$

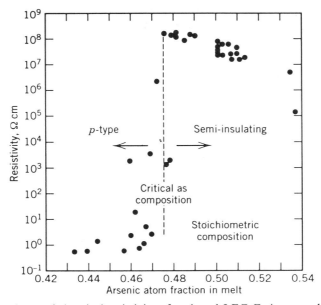

Fig. 2.28 Dependence of electrical resistivity of undoped LEC GaAs on melt stoichiometry. [Kirkpatrick, C. G., R. T. Chen, D. E. Holmes, and K. R. Elliott, in *Gallium Arsenide, Materials, Devices and Circuits*, pp. 39–94, ed. by M. J. Howes and D. V. Morgan, Wiley (1985)]

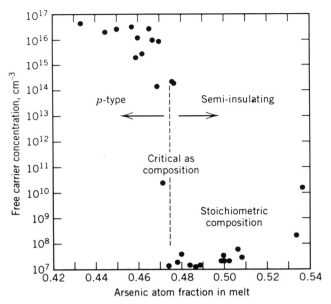

Fig. 2.29 Dependence of free-carrier concentration of undoped LEC GaAs on melt stoichiometry. [Kirkpatrick, C. G., R. T. Chen, D. E. Holmes, and K. R. Elliott, in *Gallium Arsenide, Materials, Devices and Circuits*, pp. 39–94, ed. by M. J. Howes and D. V. Morgan, Wiley (1985)]

where K is a constant determined by the thermodynamics of the system. N_I is equal to the net acceptor concentration, given as the difference in concentration between shallow acceptors N_A and shallow donors N_D:

$$N_I = N_D - N_A. \tag{2.58}$$

The concentration of acceptors is given as the sum of the concentrations of carbon and other residual acceptors N_A^R:

$$N_A = [\text{carbon}] + N_A^R. \tag{2.59}$$

The concentration of neutral centers is equal to the EL2 concentration as determined by optical absorption. That is, only EL2 centers that are occupied by electrons contribute to the optical absorption process:

$$N_U = [\text{EL2}]. \tag{2.60}$$

By substituting Eqs. (2.58) through (2.60) into Eq. (2.57), the following expression for the free electron concentration is obtained for the predominant centers in the material:

$$n = K \frac{[\text{EL2}]}{[\text{carbon}] + N_A^R - N_D}. \tag{2.61}$$

The expression can be rewritten in the following form:

$$[\text{Carbon}] = K \frac{[\text{EL2}]}{n} + N_D - N_A^R. \quad (2.62)$$

A plot of the carbon concentration as a function of the ratio of the EL2 concentration to the electron concentration, shown in Fig. 2.30, follows linear behavior, indicating that the electron concentration is indeed controlled by the balance between EL2 and carbon. Note that if some other impurity were the predominant acceptor, such as Mn, Fe, Cu, or Zn, the linearity predicted on the basis of equation (2.62) would still necessarily hold [Kir85]. The EL2 deep donors and carbon acceptors control and electrical compensation in semi-insulating LEC GaAs grown from melts range from 0.475 to 0.535 As atom fraction.

It is reported that in the magnetic field applied liquid-encapsulated Czochralski (MLEC) GaAs crystals, midgap level EL2 concentrations have been found to be strongly affected by the melt composition and crystal versus crucible rotational conditions [Ter86]. Two-inch diameter $\langle 100 \rangle$-oriented undoped GaAs singles were pulled from quartz crucibles with or without the vertical magnetic field present. Three ingots were pulled from melts with three different melt compositions $[\text{As}/(\text{Ga} + \text{As})] = 0.48, 0.50,$ and 0.51. Wafers were cut from the shoulders of the three ingots. The EL2

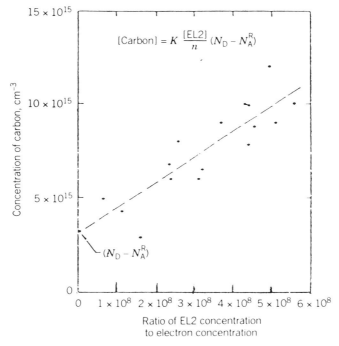

Fig. 2.30 Dependence of carbon concentration on ratio of EL2 concentration to electron concentration in undoped LEC GaAs. [Kirkpatrick, C. G., R. T. Chen, D. E. Holmes, and K. R. Elliott, in *Gallium Arsenide, Materials, Devices and Circuits*, pp. 39–94, ed. by M. J. Howes and D. V. Morgan, Wiley (1985)]

concentrations are shown versus the melt composition in Fig. 2.31. The EL2 concentrations in MLEC crystals ranged from 3×10^{14} in Ga-rich melt to $1 \times 10^{17}\,cm^3$ in As-rich melt. The EL2 concentrations in the MLEC crystals varied by more than two orders in magnitude. This phenomenon is assumed to be correlated with the suppression of convection in MLEC technique and/or the temperature stability of solid-liquid interface of GaAs during the crystal pulling. Figure 2.32 shows the variation among EL2 concentrations along the wafer's radius for the crystals grown from Ga-rich melts $[As/(Ga + As) = 0.48]$ with or without a magnetic field. The EL2 concentrations in the MLEC crystal was $2 \times 10^{16}\,cm^{-3}$ at the edge of the wafer. They decreased abruptly toward the wafer's center to 3×10^{14}.

2.6.2 Experimental Properties

There are several established experimental properties of the EL2 defect compiled by Baranowski [Bar90]:

1. The EL2 is a deep double donor. The level at 0.75 eV above the valence band corresponds to the neutral charge state (0/+ level). The level at 0.54 eV above the valence band corresponds to the positive charge state (+/2+ level).
2. In the neutral charge state EL2 is not paramagnetic. In the positive charge state the ESR quadruplet signal was identified as an As atom on a Ga position.
3. The EL2 defects are found in semi-insulating GaAs grown under As-rich conditions at a concentration level of a few times $10^{16}\,cm^{-3}$.

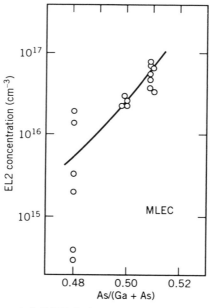

Fig. 2.31 EL2 concentrations in MLEC GaAs crystals versus the melt composition. [Terashima K., T. Yahata, and T. Fukuda, *J. Appl. Phys.* **59**, 3 (1986)]

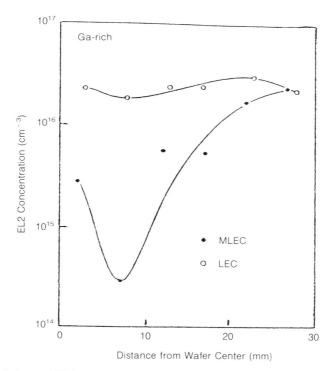

Fig. 2.32 Variations of EL2 concentrations along the wafer radius for the crystal grown from Ga-rich melts [As/(Ga + As) = 0.48] with or without a magnetic field. [Terashima, K., T. Yahata, and T. Fukuda, *J. Appl. Phys.* **59**, 3 (1986)]

4. The neutral EL2 exhibit an internal optical transition at $1.0\,\text{eV} < h\nu < 1.3\,\text{eV}$. This contributes to absorption but only little to conductivity. A zero-phonon line of this characteristic absorption has been observed at $1.037\,\text{eV}$.
5. Investigation of the zero-phonon line under uniaxial stress reveals the presence of an A1-T2 optical transition and therefore the tetrahedral symmetry of the EL2.
6. The optically detected electron nuclear double resonance (ODENDOR) done on the positive charge state of As_{Ga} has been interpreted within the frame of the As_{Ga}–As_I complex [Mey87; Mey87b].
7. For semi-insulating GaAs below 140 K EL2 can be optically bleached by photons from the internal transition range at $1.0\,\text{eV} < h\nu < 1.3\,\text{eV}$. The optical transition within the zero-phonon line contributes to bleaching as well [Sko85; Kus86]. For *n*-type GaAs optical bleaching takes place below 45 K [Tra87].
8. The bleached state corresponds to the metastable EL2-M center. The EL2-M is electrically, optically, and magnetically nonactive. The charge state of EL2-M is identical to the neutral EL2 center.
9. The EL2 ground state can thermally regenerated from the metastable configuration. In semi-insulating crystals regeneration takes place for $T > 140\,\text{K}$. The energy barrier for the regeneration is $0.34\,\text{eV}$ [Mit79]. For *n*-type GaAs thermal

regeneration takes place for T > 45 K. The barrier for this case is close to 0.1 eV [Tra87].

10. It is possible to regenerate the normal EL2 state at temperatures above 50 K by prolonged illumination with light of energy $0.6\,eV < h\nu < 1.0\,eV$ [von87; Par88]. It has been shown that hydrostatic pressure effectively increase the optical conversion from the metastable state even at liquid helium temperatures [Baj89].
11. It has been shown that in *n*-type material under hydrostatic pressure the metastable EL2 state acts as an electronic trap leading to a new, negatively charged state of EL2. The (EL2-M)-charge state appears under pressure only when EL2 is converted into the metastable state [Baj90].
12. EL2 can be destroyed by annealing at high temperature, about 1050°C, followed by rapid quenching. Regeneration of EL2 may be achieved by annealing at 850°C [Lag86].

2.6.3 Theoretical Models

Despite the wealth of experimental data there is no unified theory about the microscopy of the EL2 center [Bar90]. There are two microscopic models of EL2 that give reasonable experimental support for the intriguing properties of the EL2. One is that EL2 is simply As_{Ga}, the isolated arsenic antisite. The basic mechanism of the EL2 metastability has been proposed as connected with $As_{Ga} = V_{Ga}As_I$ structural transition [Kus86]. The other model is that EL2 is a loosely bound $As_{Ga}-As_I$ pair, where the As_I lies along the [111] antibonding direction at about one and a half bond lengths from the As_{Ga} [Mey87; Mey87b; von86b]. The experiment that most directly asserts the presence of the As_I on the [111] axis uses optically detected electron nuclear double resonance (ODENDOR) measurement. The binding energy and electronic structure of the $As_{Ga}-As_I$ defect pair have been calculated using the self-consistent Green's function technique [Bar87]. These calculations raise some questions about the model. In addition the presence of As_I at the location specified by the ODENDOR measurements is incompatible with the stress experiment [Bar89]. This implies there is no As_I near the As_{Ga}. More research is needed to reveal the nature of EL2.

REFERENCES

[Bla82] Blakemore, J. S., *J. Appl. Phys.* **53** (1982): R123–R181.
[Ash76] Ashcroft, N. W., and N. D. Mermin, *Solid State Physics* (1976).
[Lan69] Landsberg, P. T., *Solid State Theory* (1969).
[Sze81] Sze, S. M., *Physics of Semiconductor Devicers*, 2d ed., Wiley-Interscience (1981).
[Shu87] Shur, M., *GaAs Devices and Circuits*, Plenum Press (1987).
[Bri78] Brillouin, L., *Waves Propagation in Periodic Structures*, 2d ed., Dover (1978).
[Her55] Herman, F., *J. Electron.* **1** (1955): 103.
[Her54a] Herman, F., *Phys. Rev.* **93** (1954): 1214.
[Her54b] Herman, F., *Phys. Rev.* **95** (1954): 847.

[Bla87] Blakemore, J. S., ed., *Gallium Arsenide*, American Institute of Physics, New York (1987).
[Osw54] Oswald, F., and R. Schade, *Z. Naturforsch.* **A9** (1954): 611.
[Bar58] Barcus, L. C., A. Perlmutter, and J. Callaway *Phys. Rev.* **11** (1958): 167.
[Che76] Chelikowsky, J. R., and M. L. Cohen, *Phys. Rev.* B **14** (1976): 556.
[Gro72] Grobman, W. D., and D. E. Eastman, *Phys. Rev. Lett.* **29** (1972): 1508.
[Wan81] Wang, C. S., and B. M. Klein, *Phys. Rev.* B **24** (1981): 3392.
[Ley74] Ley, L., R. A. Pollak, F. R. McFeely, S. R. Kowalczyk, and D. A. Shirley, *Phys. Rev.* B **9** (1974): 600.
[Asp76] Aspenes, D. E., C. G. Olson, and D. W. Lynch, *Phys. Rev. Lett.* **37** (1976): 766.
[Lee80] Lee, H. J., L. Y. Juravel, and J. C. Woolley, *Phys. Rev.* B **21** (1980): 659.
[Cas78] Casey, H. C., Jr., and M. B. Panish, *Heterostructure Lasers* Academic Press (1978).
[Ada85] Adachi, S., *J. Appl. Phys.* **58** (1985): R1–R29.
[Mil72] Milnes, A. G., and D. L. Feucht, *Heterojunctions and Metal–Semiconductor Junctions*, Academic Press (1972).
[Bau83] Bauer, R. S., P. Zurcher, and H. W. Sang, Jr., *Appl. Phys. Lett.* **43** (1983): 663
[And62] Anderson, R. L., *Solid State Electron.* **5** (1962): 341.
[Sel74] Sell, D. D., H. C. Casey, and K. W. Wecht, *J. Appl. Phys.* **45** (1974): 2650.
[Stu62] Sturge, M. D., *Phys. Rev.* **127** (1962): 768.
[Afr68] Afromowitz, M. A., and D. Redfield, in *Proc. 9th Int. Conf. on Physics of Semiconductors*, Moscow, p. 98 (1968).
[Sha70] Shay, J. L., *Phys. Rev.* B **2**, p. 803 (1970).
[Cas75] Casey, H. C., D. D. Sell, and K. W. Wecht, *J. Appl. Phys.* **46** (1975): 250.
[Var67] Varshni, Y. P., *Physica* **34** (1967): 149.
[Pan69] Panish, M. B., and H. C. Casey, *J. Appl. Phys.* **40** (1969): 163.
[Thu75] Thurmond, C. D., *J. Electrochem. Soc.* **122** (1975): 1133.
[Asp76] Aspenes, D. E., *Phys. Rev.* B **14** (1976): 5331.
[Asp73] Aspenes, D. E., and A. A. Studna, *Phys. Rev.* B **7** (1973): 4605.
[Bal68] Balslev, I., *Phys. Rev.* **173** (1968): 762.
[Dre55] Dresselhaus, G., *Phys. Rev.* **100** (1955): 580.
[Par55] Parameter, R. H., *Phys. Rev.* **100** (1955): 573.
[Kan57] Kane, E. O., *J. Phys. Chem. Solids* **1** (1957): 249.
[Pol66] Pollak, F. H., C. W. Higginbotham, and M. Cardona, *J. Phys. Soc. Jap. suppl.* **21** (1966): 20.
[Her77] Herrmann, C., and C. Weisbuch, *Phys. Rev.* B **15** (1977): 823.
[Sti69] Stillman, G. E., C. M. Wolfe, and J. O. Dimmock, *Solid State Commun.* **7** (1969): 921.
[Sti77] Stillman, G. E., C. M. Wolfe, and J. O. Dimmock, in *Semiconductors and Semimetals*, ed. by R. K. Willardson and A. C. Beer, vol. 12. p. 169, Academic Press (1977).
[Sko76] Skolnick, M. S., A. K. Jain, R. A. Stradling, J. Leotin, J. C. Ousset, and S. Askenazy, *J. Phys.* C **9** (1976): 2809.
[Wel56] Welker, H., and H. Weiss, in *Solid State Physics*, vol. 3, p. 1, ed. by F. Seitz and D. Turnbull, Academic (1956).
[Wil175] Wiley, J. D., in *Semiconductors and Semimetals*, ed. by R. K. Willardson and A. Beer, vol. 10 p. 91, Academic, New York (1975).
[Loo90] Look, D. C., *Properties of GaAs*, 2d ed., IEENSPEC (1990).

[Hil70] Hill, D. E., *J. Appl. Phys.* **41** (1970): 1815.
[Mea71] Mears, A. L., and R. A. Stradling, *J. Phys. C* **4** (1971): L22.
[Zsc73] Szchauer, K. H., in *Gallium Arsenide and Related Compounds*, no. 17, p. 3, Institute of Physics, London (1973).
[Ros60] Rosi, F. D., D. Meyerhofer, and R. V. Jensen, *J. Appl. Phys.* **31** (1960): 1105.
[Hil64] Hill, D. E., *Phys. Rev.* **133** (1964): A866.
[Vil72] Vilms, J., and J. P. Garrett, *Solid State Electron.* **15** (1972): 443.
[Ros70] Rosztoczy, F. E., F. Ermanis, I. Hayashi, and B. Schwartz, *J. Appl. Phys.* **41** (1970): 264.
[Eme60] Emel'yanenko, O. V., T. S. Lagunova, and D. N. Nasledov *Sov. Phys. Solid State* **2** (1960): 176.
[Gas72] Gasanli, S. M., O. V. Emel'yanenko, V. K. Ergakov, F. P. Kesamanly, S. Lagunova, and D. N. Nasledov. *Sov. Phys. Semicond.* **5** (1972): 1641.
[Poz80] Pozela, J., and A. Reklaitis, *Solid State Electron.* **23** (1980): 927.
[Ruc68] Ruch, J. G., and G. S. Kino, *Phys. Rev.* **174** (1968): 921.
[Bra70] Braslau, N., and P. S. Hauge, *IEEE Trans. Electron Dev.* **ED-17** (1970): 616.
[Ash74] Ashida, K., M. Inoue, J. Shirafuji, and Y. Inuishi, *J. Phys. Soc. Jap.* **37** (1974): 408.
[Hou77] Houston, P. A., and A. G. R. Evans, *Solid State Electron.* 20, p. 197 (1977).
[Rod75] Rode, D. L., in *Semiconductors and Semimetals* **10** (1975): 1.
[Cha81] Chattopadhyay, D., and J. Queisser, *Rev. Mod. Phys.* **53** (1981): 745.
[Sti70] Stillman, C. E., C. M. Wolfe, and J. O. Dimmock, *J. Phys. Chem. Solids* **31** (1970): 1199.
[Nic80] Nichols, K. H., C. M. L. Yee, and C. M. Wolfe, *Solid State Electron.* **23** (1980): 109.
[Aka87] Akai, S. I., and S. K. Akai (Sumitomo Electric), in 1987 US–Japan Conference, University of Florida, Gainesville (1987).
[Bri25] Bridgman, P. W., *Proc. American Acad. Sci.* **60** (1925): 305.
[Rud86] Rudolph, P., and F. M. Kiessling, *Cryst. Res. Tech.* **23**, 10/11 (1986): 1209.
[AuC85] AuCoin, T. R., and R. O. Savage, in *Gallium Arsenide Technology*, ed. by D. K. Ferry, SAMS (1985).
[Von74] Von Neida, A. R., and J. W. Nielsen, J. W. *Solid State Tech.* **17** (1974): 90.
[Par81] Parsey, J. M., Y. Nanishi, J. Lagowski, and H. C. Gatos, *J. Electrochem. Soc.* **128**, (1981): 936.
[Nis82] Nishine, S., N. Kito, K. Fujita, M. Sekinobu, O. Shikatani, and S. Akai, *GaAs IC Symp. Tech. Digest* (1982): 58.
[Aka84] Akai, S., K. Fujita, S. Nishine, N. Kito, Y. Sato, Y. Yoshitake, and M. Sekinobu, *Proc. Symp. on III–V Opto-electronics Epitaxy and Device Related Processes*, p. 41, Electrochemical Society (1984).
[Pla71] Plaskett, T. S., J. M. Woodall, and A. Segmuller, *J. Electrochem. Soc.* **118** (1971): 115.
[Fer85] Ferry, D. K., ed., *Gallium Arsenide Technology*, SAMS (1985).
[Par90] Parsey, J. M., Jr., and E. M. Monberg, in *Properties of Gallium Arsenide*, INSPEC (1990).
[Czo18] Czochralski, J., *Phys. Chem.* **92** (1918): 219.
[Met62] Metz, E. P. A., R. C. Miller, and J. Mazelsky, *J. Appl. Phys.* **33** (1962): 2016.
[Mul65] Mullin, J. B., B. W. Stranghan, and W. S. Brickell, *J. Phys. Chem. Solids* **26** (1965): 782.
[Gra86] Grant, I. R., in *Galliun Arsenide for Devices and Integrated Circuits*, p. 22, ed. by H. Thomas and D. V. Morgan, IEE Press (1986).

[Pek70] Pekarek, L., *Czehoslovakian J. Phys.* B **20** (1970): 857.

[Dun83] Duncun, W. M., G. W. Westhal, and J. B. Sherer, *IEEE Electron Device Lett.*, **EDL-4** (1983): 199.

[Wes83] Westphal, W. M., W. M. Duncun, and J. B. Sherer, *Electron. Mater. Conf.*, Burlington, Vermont (1983).

[Dun85] Duncun, W. M., and W. M. Westphal, *VLSI Electronics: Microstructure Science*, vol. 11, ed. by N. G. Einspruch, Academic Press (1985).

[AuC79] Aucoin, T. R., R. L. Ross, M. J. Wade, and R. O. Savage, *Solid State Tech.* **22** (1979): 59.

[Rum79] Rumsby, D. A., IEEE Workshop on Compound Semiconductor Mater. and Dev., Atlanta (1979).

[Fuk87] Fukuda, T., Florida US–Japan Conf. Notes (1987).

[Ute66] Utech, H. P., and M. C. Flemings, *J. Appl. Phys.* **37** (1966): 2021.

[Ter83] Terashima, K., T. Katsumata, F. Orito, T. Kikuta, and T. Fukuda, *Jap. J. Appl. Phys.* **22**, 6 (1983): L325–L327.

[Ter84] Terashima, K., F. Orito, T. Katsumata, and T. Fukuda, *Jap. J. Appl. Phys.* **23**, 7 (1984): L485–L487.

[Koh85] Kohda, H., *J. Cryst. Growth* **71** (1985): 813.

[Bro90] Brozel, M. R., in *Properties of GaAs*, 2d ed., INSPEC (1990).

[Jor80] Jordan, A. S., R. Caruso, and A. R. von Neida, *Bell Syst. Tech. J.* **59** (1980): 593.

[Jor84] Jordan, A. S., A. R. von Neida, and R. Caruso, *J. Cryst. Growth* **70** (1984): 555.

[Rum80] Rumsby D., R. M. Ware, and M. Whittaker, *Proc. Conf. on Semi-insulating III–V Materials*, p. 59, Nottingham UK (1980).

[Cul80] Cullis, A. G., P. D. Augustus, M. R. Brozel, and E. J. Foulkes, *J. Appl. Phys.* **51**, 5 (1980): 2556–2560.

[Sti84] Stirland, D. J., P. D. Augustus, M. R. Brozel, and E. J. Foulkes, *Proc. Conf. on Semi-insulating III–V Materials*, p. 91, Kah-nee-ta, OR, Shiva Publishing (1984).

[Cor85] Cornier, J. P., M. Duseaux, and J. P. Chevalier, *Inst. Phys. Conf. Ser.*, no. 74, p. 95 (1985).

[Pon84] Ponce, F. A., F. C. Wang, and R. Hiskes, *Proc. Conf. on Semi-insulating III–V Materials*, p. 91, Kah-nee-ta, OR, Shiva Publishing (1984).

[Ina89] Inada, T., Y. Otoki, K. Ohata, S. Taharasako, and S. Kuma, *J. Cryst. Growth* **96**, 2 (1989): 327.

[Faw90] Fawcett, T. J., M. R. Brozel, and D. J. Stirland, *Inst. Phys. Conf. Ser.*, no. 106, p. 19 (1990).

[Cle82] Clegg, J. B., *Proc. Conf. on Semi-insulating III–V Materials*, p. 80, Shiva Publishing (1982).

[Far82] Farges, J. P., G. Jacob, C. Schemali, G. M. Martin, A. Mircea-Roussel, and J. Halais, *Proc. Conf. on Semi-insulating III–V Materials*, p. 80, Shiva Publishing (1982).

[Bue72] Buehler, M. G., *Solid State Electron.* **15** (1972): 69.

[Fan90] Fang, Z. Q., L. Shan, T. E. Schlesinger, and A. G. Milnes, *Mater. Sci. Eng.* B **5**, 3 (1990): 397.

[Hur78] Hurtes, C., M. Boulou, A. Mittonneau, and D. Bois, *Appl. Phys. Lett.* **32**, 12 (1978): 821.

[Lan74] Lang, D. V., *J. Appl. Phys.* **45**, 7 (1974): 3023.

[Che83] Chen, R. T., and D. E. Holmes, *J. Cryst. Growth* **61** (1983): 111.

[Hol82] Holmes, D. E., R. T. Chen, K. R. Elliot, and C. G. Kirkpatrick, *Appl. Phys. Lett.* **40** (1982): 46.

[Lag82] Lagowski, J., H. C. Gatos, J. M. Parsey, K. Woda, M. Kaminski, and W. Walakeiwicz, *Appl. Phys. Lett.* **40** (1982): 342.

[Ta82] Ta, L. B., H. M. Hobgood, A. Rohatgi, and R. N. Thomas, *J. Appl. Phys.* **53** (1982): 5771.

[Kir85] Kirkpatrick, C. G., R. T. Chen, D. E. Holmes, and K. R. Elliott, in *Gallium Arsenide, Materials, Devices and Circuits*, pp. 39–94, ed. by M. J. Howes and D. V. Morgan, Wiley (1985).

[Mar80] Martin, G. M., G. Jacob, G. Poiland, A. Goltzene, and C. Schwab, in "Defects and Radiation Effects in Semiconductors", Oiso (1980).

[Hol82] Holmes, D. E., K. R. Elliott, R. T. Chen, and C. G. Kirkpatrick, in *Semi-insulating III–V Compounds*, p. 19–27, ed. by S. Makram-Ebeid and B. Tuck, Shiva (1982).

[Mir76] Mircea, A., A. Mitonneau, L. Hollan, and A. Briere, *Appl. Phys.* **11** (1976): 153.

[Mak84] Makram-Ebeid, S., P. Langlade, and G. M. Martin, in *Semi-Insulating III–V Materials*, p. 184, ed. by D. C. Look and J. S. Blakemore, Shiva Publishing (1984).

[Kam83] Kaminska, K., M. Skrowronski, J. Lagowski, J. M. Parsey, and H. C. Gatos, *Appl. Phys. Lett.* **43** (1983): 302.

[Von86] Von Bardeleben, H. J., D. Stievenard, J. C. Bourgoin, and A. Huber, in *Conf. on Semi-insulating III–V Materials*, May 18–21, Hakone, Japan (1986).

[Miy86] Miyazawa, S., in *Conf. on Semi-insulating III–V Materials*, May 18–21, Hakone, Japan (1986).

[Kir85] Kirkpatrick, C. G., in *Gallium Arsenide*, p. 67, ed. by M. J. Howes and D. V. Morgan, Wiley (1985).

[Ter86] Terashima, K., T. Yahata, and T. Fukuda, *J. Appl. Phys.* **59**, 3 (1986): 982.

[Bar90] Baranowski, J. M., in *Properties of Gallium Arsenide*, 2d ed., IEE INSPEC (1990).

[Mey87] Meyer, B. K., D. M. Hofman, J. R. Niklas, and J. M. Spaeth, *Phys. Rev.* B **36**, 2 (1987): 1332–1335.

[Mey87b] Meyer, B. K., D. M. Hofmann, and J. M. Spaeth, *J. Phys.* C **20**, 16 (1987b): 2445–2452.

[Sko85] Skowronski, M., J. Lagowski, and H. C. Gatos, *Phys. Rev.* B **3**, 6 (1985): 4264–4267.

[Kus86] Kuszko, W., and M. Kaminska, *Acta Phys. Pol.* A **69**, 3 (1986): 427–30.

[Tra87] Trautman, P., M. Kaminska, and J. M. Baranowski, *Acta Phys. Pol.* A **71**, (1987): 269–71.

[Mit79] Mittonneau, A., and A. Mircea, *Solid State Commun.* **30**, 3 (1979): 157–162.

[Von87] von Bardeleben, H. J., N. T. Bagraev, and J. C. Bourgoin, *Appl. Phys. Lett.* **51**, 18 (1987): 1451–1453.

[Par88] Parker, J. C., and R. Bray, *Phys. Rev.* B **37**, 11 (1988): 6368–6376.

[Baj89] Baj, M., and P. Dreszer, *Phys. Rev.* B **30**, 14, (1989): 10470–10472.

[Baj90] Baj, M., P. Dreszer, and A. Babinski, submitted to 20th Int. Conf. on the Phys. of Semiconductors, Thessaloniki, Greece, Aug. 6–10 (1990).

[Lag86] Lagowski, J., H. C. Gatos, C. H. Kang, M. Skowronski, K. Y. Ko, and D. G. Liu, *Appl. Phys. Lett.* **49**, 14 (1986): 892–894.

[Kus86] Kuszko, W., P. J. Walczak, P. Trautman, M. Kaminska, and J. M. Baranowski, *Mater. Sci. Forum* (Switzerland) **10–12**, pt. 1 (1986): 317–322.

[von86b] von Bardeleben, H. J., D. Stievenard, D. Deresmes, A. Huber, and J. C. Bourgoin, *Phys. Rev.* B **34**, 10 (1986): 7192–7202.

[Bar87] Baraff, G. A., and M. Schluter, *Phys. Rev.* B **34** (12), (1987): 6154–6164.

[Bar89] Baraff, G. A., *Phys. Rev. Lett.* **62**, 18 (1989): 2156–2159.

3

EPITAXIAL GROWTH PROCESSES

3.1 INTRODUCTION TO EPITAXIAL GROWTH PROCESSES

There are four main techniques by which GaAs and AlGaAs epitaxial films are grown: chloride transport vapor phase epitaxy (VPE) [Jai70], liquid phase epitaxy (LPE) [Hei80], molecular beam epitaxy (MBE) [Pan83], and metalorganic chemical vapor deposition (MOCVD) [Man81]. These epitaxial growth techniques are compared in Table 3.1 [Che87].

In the chloride transport VPE growth system, any silicon contamination is serious for it creates unfavorable thermodynamics in the compound and alloy containing Al, making it very difficult for growth to occur. The LPE technique, through it has been successfully used in compound semiconductors, is not suitable for mass production by substrate limitation. The abruptness of the interface in LPE growth is unsatisfactory for GaAs high-speed device fabrication.

MBE is considered to be the most promising future growth technique because it allows for precise thickness control, dopant control, and pattern drawing. Growth is performed under an ultra-high vacuum chamber with a low growth rate (0.1 to 10 μm/h). This permits one to accurately control the impinging atoms or molecules and thus the thickness of the film. The system has multiple sources that allow accurate stoichiometric growth. The straight-line beam impinging on the substrate also prevents collision, scattering, or diffusion during beam flight.

The MOCVD technique has demonstrated its effectiveness for growing the widest variety of III–V materials. As pointed out in Table 3.1, the MOCVD technique has several advantages over other growth techniques [Joh84; Dap84]: (1) The formation of the desired compound occurs via the pyrolysis of the metalorganics and hydrides, and the subsequent recombination of the atomic or molecular species occur at or near the substrate surface. (2) The composition and impurity concentration can be controlled well by fixing the flow rates of the various reactants with electronic mass flow controllers. (3) Complex multilayer epitaxial structures are readily formed by

TABLE 3.1 Comparison of Epitaxial Growth Techniques

	LPE	Hydride VPE	MOCVD	MBE
Al alloys	Capable	Difficult	Capable	Capable
Range of growth rate (μ/min)	$0.1 \sim 10$	$0.01 \sim 0.5$	$0.005 \sim 1.5$	few ~ 0.05
Minimum thickness (Å)	500	250	20	5
Homogeniety	good	good	good	good
Surface morphology	bad	good	good	good
Abruptness of interface	bad	good	good	excellent
Doping level (cm^{-3})	$10^{14} \sim 10^{19}$	$10^{14} \sim 10^{19}$	$10^{14} \sim 10^{19}$	$10^{14} \sim 10^{19}$
Number of heating point	1	2	1	3
Productivity	low	high	high	very low

Source: Chen, L. P., Ph.D. Dissertation, National Cheng Kung University, Taiwan (1987).

exchanging one gas composition for another gas using automatic gas mixing system. (4) This technique is suitable for mass production based on its similarity to silicon CVD process.

Chemical beam epitaxy (CBE) is an alternate epitaxial growth technique for GaAs and InP that is similar to MBE. This technique uses gaseous group-III and group-V alkyls. The In and Ga were derived by the pyrolysis of either trimethylindium or triethylindium and trimethylgallium or triethylgallium at the heated substrate surface, respectively. The As or P were obtained by thermal decomposition of trimethylarsine or arsine and triethylphosphine or phosphine in contact with heated Ta or Mo at 950°–1200°C, respectively [Ta85].

Two other newly developed growth techniques are atomic layer epitaxy (ALE) and vapor transport epitaxy (VTE). Atomic layer epitaxy has a self-limiting growth mechanism that has weak dependence on growth parameters such as partial pressures of source gases, beam flux, growth temperature, and growth time. The operating principle of vapor transport epitaxy is used to direct vapor transport from uncoupled sources (elemental or metalorganic) into a common flux distribution manifold and then to the substrate. Films grown are free from oval defects. In this chapter we will discuss the growth techniques and mechanisms of MBE, MOCVD, CBE, ALE, and VTE.

3.2 MOLECULAR BEAM EPITAXY

Molecular beam epitaxy (MBE) is a versatile technique for epitaxial growth of semiconductor, metal, and insulator thin films [Cho75; Cho79; Plo80; Plo81; Cha75; Fox80; Far77; Bea81]. It is a highly refined form of vacuum deposition with precise control of the beam fluxes and deposition conditions. Compounds or elements are evaporated from heated crucibles, called *Knudsen cells*, onto clean ordered substrates. The deposition rate of the components and the temperature of the substrate must be carefully chosen and controlled, and the substrate surface must be clean and as free of defects as possible. The need for extreme cleanliness means that the substrates and the Knudsen cells must be enclosed in an ultra-high-vacuum chamber, so the MBE process is generally carried out in large stainless steel high-vacuum systems [Her84]. However, provided that all the relevant parameters are carefully controlled, it is possible to grow good quality single crystal materials of large area in this way (>50 cm^2). Growth rates of about 1 μm per hour to 10 μm per hour are typical, and the growth is generally carried out on single crystal gallium arsenide substrates. If a third effusion cell containing, for example, aluminum is included in the system, it is possible to grow alloys such as $Al_xGa_{1-x}As$. The technique thus offers the flexibility to prepare elemental semiconductors, compound semiconductors, semiconductor alloys, and heterostructures that involve alternating layers of different materials. The thickness of these layers can be accurately controlled, and the interfaces between them may be made remarkably abrupt [Cho69; Cho71; Art74].

The III–V compound is a new class of semiconductors for microwave devices, high-speed digital integrated circuits, and highly efficient optoelectronic devices. These compound semiconductors usually consist of the group-III elements, Ga, Al, and In, and the group-V elements, As, P, and Sb. Several compounds such as GaAs [Cho69; Cho70], GaP [Cho70b], $Al_xGa_{1-x}As$ [Cho70b], $GaAs_xSb_{1-x}$ [Cho77], and $Ga_xIn_{1-x}As_yP_{1-y}$ [Cho79] were first studied. The potential for excellent thickness control of MBE was first demonstrated by the growth of $GaAs/Al_xGa_{1-x}As$ periodic structures [Cho71b]. In the process of evaluating device performance, it was found that the photoluminescent intensity increased more than an order of magnitude when the substrate temperature was increased from 540° to 650°C during growth [Cas75]. Excellent results with double-heterostructure lasers [Tsa79; Tsa79b; Tsa79c; Col82; Yam82], microwave field-effect transistors [Mor82; Fen82; O'Co82; Duh86; Hwa82; Che82], pseudomorphic HEMT [Cha85], heterojunction bipolar transistors, hot-electron transistors, and resonant-tunneling transistors, coupled with the high throughput and highly uniform growth with rotating sample holders [Cho81; Che81] made MBE an important thin film technology.

3.2.1 MBE Growth Systems and Deposition Sources

The rapid development of MBE system in a relatively short period has changed from custom-designed special ultra-high-vacuum (UHV) evaporators to dedicate high-throughput complete MBE instruments with proven ability to fabricate high-quality material.

System Configuration

Nowadays ultra-thin films and superlattice structure with arbitrary number of layers and layer thicknesses are grown by MBE. The defect density during crystal growth should be reduced to a minimum. The study of monoatomic layer control in MBE becomes important. The multiple sources should be kept clean.

The art of UHV technology usually consists of two aspects. The first is about pumping, and the second is about materials used in a chamber and their outgassing characteristics. In an MBE chamber the vacuum condition is more determined by the outgassing characteristics of various contributing materials than by the pumping capability with which the chamber is equipped. An MBE chamber is also less forgiving in terms of background contaminants.

A typical commercial MBE system is shown in Fig. 3.1. It has three vacuum chambers: a growth chamber (left), an analysis chamber (center), and a small chamber for sample load-lock (right). The three chambers are vacuum-isolated from each other

58 EPITAXIAL GROWTH PROCESSES

Fig. 3.1 Commercial gas source MBE machine manufactured by the Riber Division of Instruments SA, France. [Permission by Riber Division of Instruments SA]

by either metal-sealed or viton-sealed gate valves. The function of the sample load-lock is to facilitate transfer of wafers in and out of the MBE system with minimum disturbance of the vaccum.

The analysis chamber is where most of the surface-analysis instruments are housed. Auger electron spectrometer (AES) and a secondary ion mass spectrometer (SIMS) are used to analyze the sample.

The growth chamber is the most important component in an MBE system. A growth chamber usually consists of source furnaces, shutters, a substrate manipulator, cryoshrouds, and some surface-analysis equipment. The cryoshrouds, high-quality source furnaces, and sample load-locks have been most instrumental to the development of MBE as a material preparative technique. A schematic of the MBE system viewed from the top is shown in Fig. 3.2. The sample exchange load-lock permits the maintenance of an ultra-high vacuum while changing substrates. The cryoshroud is used to enclose the entire growth area in order to minimize the residual water vapor and carbon-containing gases in the vacuum chamber during epitaxy. The modern commercial MBE growth chamber is often equipped with a rotary substrate manipulator capable of turning the substrate azimuthally during growth at a speed of about 3–5 rpm. With this feature the substrate can be heated more uniformly, resulting in epilayers of very good thickness and doping uniformity [Cho81; Che81]. To increase the throughput and yield of GaAs ICs, commercial MBE systems also have the capability of handling 4-in. diameter substrates, for Indium-free mounting of substrates and for loading a large number of substrates (10 to 20) in a cassette [Eri85].

The effusion cells are generally 2.5 cm in diameter and 7.5–12 cm in length, and they are made of pyrolytic boron nitride (PBN). The PBN crucible is usually selected for work with reactive materials at high temperatures. These large-capacity crucibles are offered in two configurations: an "upward-looking" version and a "downward-looking" reverse-insert version. The upward-looking crucible kit is designed for furnaces mounted on the lower half of the MBE GeN II source flange. The downward-

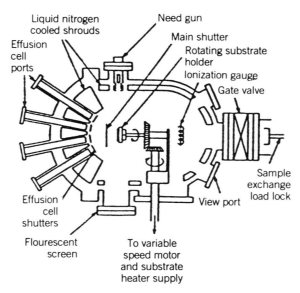

Fig. 3.2 Schematic of the MBE system viewed from the top. The rotating sample holder has a variable speed from 0.1 to 5 rpm. [Cho, A. Y., in *Molecular Beam Epitaxy and Heterostructures*, p. 191, ed. by L. L. Chang and K. Ploog, Kluwer, Academic Press (1985)]

looking crucible kit is for furnaces mounted on the upper half of the source flange. The volume of the PBN reverse-insert crucible is 16 cc, and the volume of the standard PBN crucible is 40 cc. The growth of uniform epitaxial films from multiple effusion cells requires special effusion cell geometry and continuous rotation of the substrate around an axis normal to the substrate surface. The substrate holder can feature rotation speeds up to 125 rpm. The control unit remotely orients the sample holder into any of hour positions: growth, transfer, E-beam, and auxiliary. The controller also allows remote continual adjustment of rotation speed.

Deposition Sources

The molecular beams are usually generated in Knudsen cells. These cells may be made of graphite or boron nitride. A water-cooled enclosure usually surrounds each Knudsen cell, and the cell arrangement is surrounded by surfaces cooled to liquid nitrogen temperatures, since it is most important to avoid cross contamination of the cells [Her85]. The mean free path of the atoms or molecules is much greater than the cell orifice, so the molecular rather than the hydrodynamic flow pattern of pressure is a concern.

In practice the conditions are rather far removed from those of the ideal Knudsen because large orifices are usually employed to obtain faster growth rates, as well as to improve the uniformity of films. The flux emanating from the Knudsen cells is controlled by accurate control of the temperature, and the flux arriving at the sample may be regulated by shutters in front of the Knudsen cells.

The flux arriving in the sample position is monitored by using an ion gauge attached to the reverse side of the sample holder. To calibrate the flux, the ion gauge is turned into the molecular beam by rotating the sample holder through 180°. Several Knudsen cells may be incorporated in the growth chamber, in order to dope the semiconductor or to grow multicomponent compounds with alloys.

Sometimes additional cells called *cracker cells* are inserted between a Knudsen cell and the substrate. The effusion beams are directed from a conventional Knudsen crucible enclosure via a high-temperature (cracker) region onto the substrate. At an elevated temperature this region will provide a multiple collision path that will dissociate the molecular species emanating from the Knudsen cell. Thus the heating of solid arsenic and phosphorous in a Knudsen cell generates As_4 and P_4 tetramers. By allowing these molecules through the cracker cell at a temperature between 800° and 1,000°C, it is possible to generate a beam of dimers, As_2 and P_2. The dimer sources offer the same advantages as the group-V MBE source and are likely to be accepted as the standard form of the source. From a practical point of view the sticking coefficient of As_2 has been shown to be twice that of As_4 [Cha76], so only half the arsenic flux should be needed for each growth run. It is also found that use of the dimer source reduces deep levels in the resulting gallium arsenide thin films [Nea80].

3.2.2 In situ Analysis

In the initial development of MBE, surface analysis performed during deposition played a major role in the understanding of the growth process. It has a reflective high-energy electron diffraction (RHEED) apparatus and an ion gauge in the growth chamber. A modern MBE system often includes other surface diagnostic facilities such as Auger electron spectroscopy (AES), secondary ion mass spectroscopy (SIMS), X-ray photoelectron spectroscopy (XPS), and scanning electron microscopy (SEM).

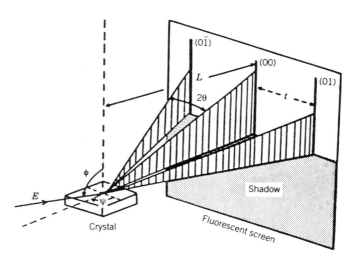

Fig. 3.3 Experimental geometry of the RHEED technique. [Herrenden-Harker, W. G., and R. H. Williams, Epitaxial growth of GaAs: MBE and MOCVD in *Gallium Arsenide for Devices and Integrated Circuits*, p. 57, ed. by H. Thomas, D. V. Morgan, B. Thomas, J. E. Aubrey, and G. B. Morgan, IEEE UWIST (1986)]

Reflection High-Energy Electron Diffraction

RHEED apparatus is essential for the growth chamber because it gives the information about substrate cleanliness and proper growth conditions. The basis of the RHEED technique is illustrated in Fig. 3.3. Electrons are generated by an electron gun, and are incident on the sample at grazing angles with energies typically in the 5 to 15 keV range. Because of the grazing angle conditions the electrons sample the surface two-dimensional unit mesh, and this is indicated by streaks on the opposite fluorescent. If the surface of the sample happens to be rough, electrons may penetrate through asperities on the surface, producing three-dimensional diffraction or spots. The same may be seen if growth in MBE is in the form of an island or a cluster. The electron beam current must be kept low; otherwise, it may polymerize the residual hydrocarbon gases, resulting in carbon contamination on the substrate.

In analyzing the RHEED pattern, we can imagine a crystal made up of sets of parallel net planes in which the atoms are located. Bragg's law relates the wavelength to the angle through which the electron beam is diffracted, and it can be written

$$2d_h \sin \theta = \lambda, \tag{3.1}$$

where d_h is the plane spacing, θ is the incident angle, and λ is the electron beam wavelength,

$$\lambda = \left(\frac{150}{V(1 + 10^{-6} V)} \right)^{1/2} \text{Å}. \tag{3.2}$$

An electron-accelerating voltage range from 3 to 100 kV would correspond to wavelengths from 0.2 to 0.04 Å.

The net plane spacing d_h in cubic structures can be expressed in terms of the Miller indices h_1, h_2, and h_3:

$$d_h^2 = \frac{1}{\left(\dfrac{h_1}{a}\right)^2 + \left(\dfrac{h_2}{a}\right)^2 + \left(\dfrac{h_3}{a}\right)^2}, \tag{3.3}$$

where a is the lattice constant.

The relation between the atom arrangements on the substrate and the observed diffraction pattern on the fluorescent screen can best be described in Fig. 3.4 and may be expressed as

$$d_h = \frac{2\lambda L}{D}, \tag{3.4}$$

where L is the distance between the substrate and the fluorescent screen and D is twice the distance between the diffraction spots measured on the fluorescent screen. One can therefore deduce the atom periodicities by measuring the diffraction spot spacings.

The RHEED technique is very useful and convenient for determining the right substrate surface condition for starting the epitaxial growth. The intensity of the

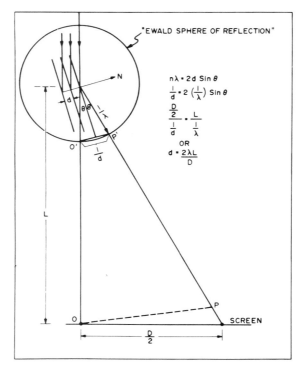

Fig. 3.4 The high-energy electron diffraction patterns. [Cho, A. Y., in *Molecular Beam Epitaxy and Heterostructures*, p. 191, ed. by L. L. Chang and K. Ploog, Kluwer, Academic Press (1985)]

oscillation period is equal to the time to grow one monolayer. It measures precisely growth rate over 30-s intervals. As the substrate is heated up from room temperature, the surface oxide gradually desorbs. With the surface oxide present, no RHEED pattern but diffused reflection is observed.

The RHEED patterns from a clean crystal surface also contain information about the surface reconstructions. On an ideal (100) GaAs surface each surface atom exhibits two dangling bonds, giving rise to a large total surface energy. This energy can be reduced by a rearrangement of the surface atoms often resulting in a surface structure that has a periodicity that is different from that of the bulk lattice; the surface is said to be *reconstructed*.

A particularly useful feature of the RHEED technique is that the intensities associated with the diffraction beams can be used to yield a calibration of the thickness of the material grown. Figure 3.5 shows the intensity associated with the specular, integral, and half-order beams for GaAs (001) growth. Growth commences when the shutter in front of the gallium cell is open, and the intensity oscillates with time thereafter as shown. The reason for this remarkable effect is illustrated in Fig. 3.6 [Her85]. Initially the surface is smooth and the corresponding diffraction conditions are well satisfied, enabling sharp RHEED features. For a coverage of one complete monolayer the situation is similar, but for a coverage of about half a monolayer there is a great deal of surface disorder. This in turn leads to a reduction in the maximum diffraction of intensities. The amplitude of the oscillations decreases with the number

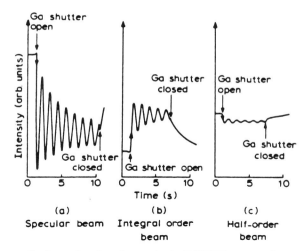

Fig. 3.5 Intensity oscillations of various beams in the RHEED pattern from a GaAs (001)–2 × 4 reconstructed surface in the (110) azimuth: (*a*) specular (00) beam, (*b*) integral order beam, (*c*) half-order beam. [Herrenden-Harker, W. G., and R. H. Williams, Epitaxial growth of GaAs: MBE and MOCVD, in *Gallium Arsenide for Devices and Integrated Circuits*, p. 57, ed. by H. Thomas, D. V. Morgan, B. Thomas, J. E. Aubrey, and G. B. Morgan, IEE UWIST (1986)]

of layers grown. But, if growth is interrupted, by holding the substrate at the growth temperature for a few minutes, the diffraction intensity recovers as growth is recommenced.

Auger Electron Spectroscopy

Auger electron spectroscopy (AES) is a nondestructive analytical method for determining the chemical composition of the outermost atomic layers of surfaces with a detection sensitivity ranging from a 0.1 to 0.01 monolayer. AES typically uses 1 to 5 keV electrons to generate Auger electrons [Wil90]. It is not sensitive enough to study the dopant concentration used in the semiconductor devices, which is on the order of one part per million. It may be used for several different purposes such as [Tsa85]

1. characterization of the initial substrate surface,
2. experimental verification of surface accumulation of dopant elements during epitaxial growth,
3. determination of the relative change in the relative ratio of the constituent elements of reconstructed surface structures.

Secondary Ion Mass Spectroscopy

The mass-spectrometer was first used to study the adsorption–desorption kinetics of atoms on solid surfaces. The mean adsorption lifetime, sticking coefficient, and the activation energy of certain elements on the substrate can be determined. The intensity of the detected peak is a strong function of the resolution setting, ionization cross section, ion-accelerated voltage, ionization energy, and the geometry of the spectrom-

64 EPITAXIAL GROWTH PROCESSES

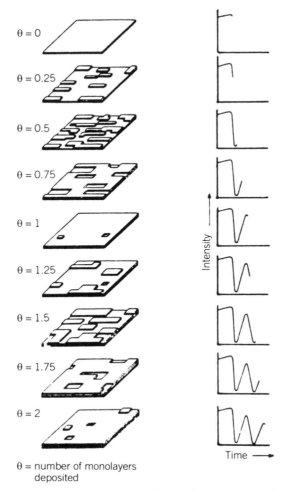

θ = number of monolayers deposited

Fig. 3.6 First-order growth model (two monolayers) in relation to the RHEED intensity behavior. [Herrenden-Harker, W. G., and R. H. Williams, Epitaxial growth of GaAs: MBE and MOCVD, in *Gallium Arsenide for Devices and Integrated Circuits*, p. 57, ed. by H. Thomas, D. V. Morgan, B. Thomas, J. E. Aubrey, and G. B. Morgan, IEE UWIST (1986)]

eter. The lower limit of the detection can be 0.1 Å/s. Secondary ion mass spectroscopy is mainly for determination of the chemical composition of the outermost atomic layers of the grown structure. The quadrupole mass spectrometer included in the SIMS system can be used for diagnostics of the ambient gas in the growth chamber [Dav85]. It measures the charge/mass ratio for molecules impinging on ionizer.

X-ray Photoelectron Spectroscopy

X-ray photoelectron spectroscopy (XPS) and angle-resolved ultraviolet photoelectron spectroscopy (ARUPS) are mainly for studying the electron structure of the epilayer surface, or the energy band distribution at heterointerface [Her89].

Scanning Electron Microscopy

Scanning electron spectroscopy is used to display the structure of the film deposited on the substrate surface.

3.2.3 Growth of III–V Compounds

Epitaxial growth of III–V compounds by MBE involves a series of events: (1) adsorption of the constituent atoms and molecules, (2) surface migration and dissociation of the adsorbed molecules, and (3) incorporation of the atoms to the substrate resulting in nucleation and growth. An important understanding of the epitaxial growth process for GaAs was obtained by the kinetic studies of gallium and arsenic atoms on GaAs surfaces [Art68, Art68b]. Arsenic has a very low sticking coefficient above 500°C unless it is combined with a gallium atom to form GaAs. Stoichiometric GaAs is formed as long as an excess of arsenic is supplied at the growing surface. The arsenic that does not form a bond to gallium will reevaporate from the surface. However, in the case of growing ternary and quaternary compounds such as $Ga_xIn_{1-x}As$ and $Al_xIn_{1-x}As_yP_{1-y}$, precise ratios of the beam fluxes are required for the compounds to grow with the desired mole fractions.

Processes of Thin Film Formation

The chemical reactions involved in the molecular beam epitaxial growth processes are governed by thermodynamic considerations. However, the rate at which a system moves toward thermodynamic equilibrium is generally governed by kinetic considerations. In this section we will consider growth from $Ga + As_4$ and $Ga + As_2$ and their reactions on GaAs (001) surfaces [Fox74; Joy85].

A beam of neutral atoms or molecules, having thermal velocities and with an intensity in the range 10^{11}–10^{16} atoms (or molecules) $cm^{-2}s^{-1}$, is directed at a substrate surface and the desorbing flux detected mass spectrometrically. The experiment is performed under UHV conditions ($<10^{-10}$ torr) to minimize surface impurities effects. Some provision may be made for structural and compositional analysis of the surface, most frequently by RHEED and AES.

Consider the interaction of arsenic and arsenic + gallium of a GaAs surface. Gallium is always monatomic, but the arsenic flux comprises either As_2 or As_4 molecules. The reaction processes of As_2 and As_4 on GaAs substrate are quite different.

The Reaction Process of As_4   By measuring the desorption rate of As_4 as a function of the incident As_4 flux at a fixed Ga flux, we obtained the result shown in Fig. 3.7 [Joy85]. At low fluxes (low surface concentrations) the desorption rate is second order with respect to the incident rate but becomes first order as the incident flux is increased. This does not imply a change of mechanism; it is the rate controlling step of the reaction that changes as the As_4 surface population increases.

The model for the growth mechanism of GaAs from Ga and As_4 beams that we have constructed from these results is shown in Fig. 3.8 [Joy85]. The critical feature is the pairwise dissociation of As_4 molecules adsorbed on adjacent Ga atoms. From any two As_4 molecules four As atoms are incorporated in the GaAs lattice and the other four desorbed as an As_4 molecule. This is consistent with the observed relationship $S_{As_4} = J_{As_4}/4J_{Ga}$ when $J_{Ga} \ll J_{As_4}$, that is, when the surface population of Ga is

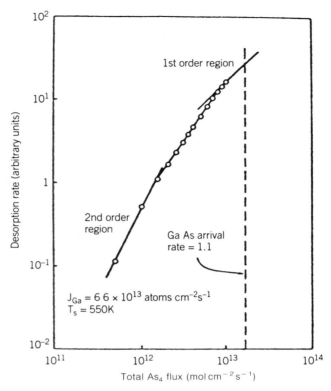

Fig. 3.7 Desorption rate of As_4 as a function of As_4 adsorption rate under Ga-rich surface conditions. The desorption rate is second order with respect to the incident flux at low fluxes, gradually becoming first order as the flux increases. [Joyce, B. A., Kinetic and surface aspects of MBE, in *Molecular Beam Epitaxy and Heterostructures*, ed. by J. L. Chang and K. Ploog, Martinus Nijhoff Publishers (1985)]

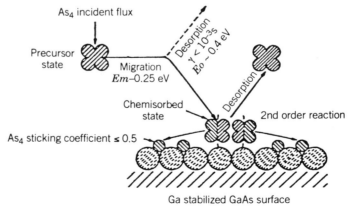

Fig. 3.8 Model of the growth of GaAs from molecular beams of Ga and As_4. [Joyce, B. A., Kinetic and surface aspects of MBE, in *Molecular Beam Epitaxy and Heterostructures*, ed. by L. L. Chang and K. Ploog, Martinus Nijhoff Publishers (1985)]

large. The model also explains the second-order As$_4$ desorption rate at low relative incident rates, that is, low surface concentration of As$_4$ molecules, since under these conditions the desorption rate will be determined by probability of pairs of As$_4$ molecules being adsorbed on adjacent sites, which is simply proportional to the square of the adsorption rate. The change to first order at higher coverages reflects the change to the supply rate limited desorption rate.

The Reaction Process of As$_2$ As–Ga–GaAs interactions Arthur [Art69] and Foxon and Joyce [Fox77] have shown that the sticking coefficient of As$_2$ (S_{As_2}) increases linearly with the Ga adatom population of one monolayer. Above ~600 K the additional surface processes become significant [Joy85]. Below 600 K there is no measurable dissociation of GaAs, but some incident As$_2$ molecules may associate on the surface to form As$_4$ before desorbing [Fox77], as shown in Fig. 3.9. The desorbing As$_2$ flux decreases monotonically with decreasing temperature, but the desorption rate of As$_4$ reaches a maximum at 450 K. The decrease at lower temperatures is an artefact arising from the use of a GaAs source to produce the As$_2$. It leads to the nondissociative adsorption of some As$_4$ molecules on the Ga atoms that arrive with the As$_2$ flux.

The Ga–As$_2$ interactions on GaAs are summarized in the growth model shown in Fig. 3.10. According to this model As$_2$ molecules are first adsorbed into a mobile, weakly bound precursor state. Dissociation of adsorbed As$_2$ can only occur when the molecules encounter paired Ga lattice sites while migrating on the surface. In the absence of free surface Ga adatoms, As$_2$ has a measurable surface lifetime, but no permanent condensation occurs. Thus As has a very low sticking coefficient above

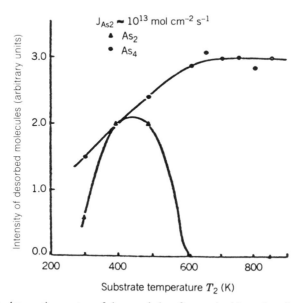

Fig. 3.9 Relative desorption rates of As$_2$ and As$_4$ for an incident As$_2$ flux as a function of temperature. Note the surface association of As$_2$ to form As$_4$ below 600°K. [Joyce, B. A., Kinetic and surface aspects of MBE, in *Molecular Beam Epitaxy and Heterostructures*, ed. by L. L. Chang and K. Ploog, Martinus Nijhoff Publishers (1985)]

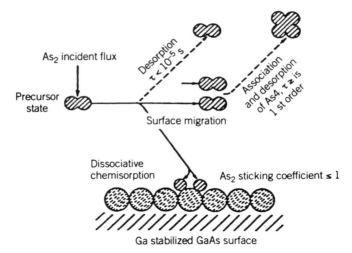

Fig. 3.10 Model of the growth of GaAs from molecular beams of Ga and As_2. [Joyce, B. A., Kinetic and surface aspects of MBE, in *Molecular Beam Epitaxy and Heterostructures*, ed. by L. L. Chang and K. Ploog, Martinus Nijhoff Publishers (1985)]

773 K unless it is combined with a Ga atom to form GaAs. Stoichiometric GaAs is formed as long as an excess As is supplied at the growing surface. At lower substrate temperatures (< 600 K), a pairwise association of adsorbed As molecules, followed by desorption as As_4 molecules, commences and becomes increasing dominant.

Effects of Arsenic Species on Film Properties The different growth mechanisms of GaAs films prepared from As_2 or As_4 might be expected to influence film properties. The crucial difference between the two reaction systems is that As_2 chemisorption involves a single Ga surface atom, while with As_4 there is an interaction involving adjacent pairs of Ga surface atoms. The steady state arsenic surface population should therefore be higher with As_2; with As_4 the maximum coverage will be 100%. Some proportion of single sites (10%) will always remain unoccupied. The nonoccupied surface sites leads to (As) vacancy introduction, which will influence the deep level concentrations. Neave et al. [Nea80] have conclusively demonstrated that the concentrations of three characteristics deep states (M1, M3, and M4) in MBE-grown GaAs are substantially lower in films grown from As_2 than from As_4, all other conditions being identical, and this has been confirmed by Künzel et al. [Kün82]. The states involved are all electron traps in n-type material, and they cannot unequivocally be related to native defects.

We now consider the influence of the arsenic species, and the associated surface chemistry, on the incorporation of an amphoteric dopant, germanium, in GaAs films. On the basis of these models the relatively higher arsenic surface population obtained with As_2 should favor the incorporation of Ge as a donor, and the degree of autocompensation should consequently be less with As_2 than As_4-grown films. This was tentatively suggested by Neave et al. [Nea80] and confirmed by Künzel et al. [Kün82], who observed almost a decade difference in the ratio at a growth temperature of 820 K. If the system is grown under gallium-stabilized conditions, a stoichiometric

GaAs composition cannot be maintained, and the surface morphology degrades rapidly.

Growth Conditions

Nucleus Formation For III–V compounds the passivated oxide layer serves as a protection for the freshly chemical-etched substrate from atmospheric contamination before epitaxial growth. After the MBE system is pumped down, the liquid nitrogen shroud cooled, and the effusion cells brought up to the desired temperatures, one begins to heat the substrate. In the case of GaAs, the oxide on the substrate desorbed between 580° and 600°C, and for InP, the oxide desorbed at about 520°C [Che81b]. At this point the substrate is almost atomically clean and ready for epitaxial growth. Assuming that the substrate is properly prepared and atomically clean, the epitaxial layer will be mirror shiny if the group-V to group-III ratio in the molecular beam is above a certain value, giving an As-stabilized surface structure [Cho71].

This value is also a function of the substrate temperature. The approximate relationship, also referred as the "MBE phase diagram," is shown in Fig. 3.11. In commercial MBE systems a GaAs growth rate of 10 µm/h may be achieved with a substrate temperature of 620°C.

The construction of this phase diagram is made possible from the knowledge

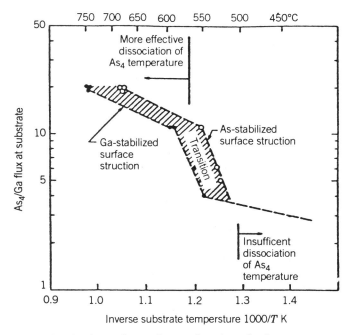

Fig. 3.11 As_4/Ga molecular beam flux ratio as a function of substrate temperature when the transition between As-stabilized and Ga-stabilized structures occurs on the (001) GaAs surface. The beam flux was measured by an ion gauge at the substrate position with Ga flux equal to 8×10^{-7} torr, giving a growth rate of about 1 µm/h. [Cho, A. Y., by in *Molecular Beam Epitaxy and Heterostructures*, p. 191, ed. by L. L. Chang and K. Ploog, Kluwer, Academic Press (1985)]

gained about surface atom structures by the use of RHEED [Cho70; Cho71]. In the case of GaAs in the $\langle 100 \rangle$ and $\langle 111 \rangle$ directions, the crystal is formed with alternate layers of Ga and As atoms. The terms Ga-rich and As-rich surface structures are used to describe the growth conditions where the top layer is terminated with Ga or As, respectively [Cho70; Cho71]. On the (100) surface the Ga-stabilized surface structure is C(8 × 2), and the As-stabilized surface structure is C(2 × 8). These results were later confirmed with excellent mass spectrometry [Art74; Nea78] and Auger studies [Cho76; Plo77]. There are many more surface structures reported on (100) [Cho76] and (111) [Cho70] surfaces. The ratio of As/Ga in the molecular beam to form an As-stabilized surface on a (100) surface is different from that on a (111) surface for a given substrate temperature; the latter requires a higher As/Ga ratio.

Higher growth temperature resulted in higher-quality epitaxial layers, and it was related to the efficiency of photoluminescence [Cas75]. This luminescence result has been proved to be directly related to the performance of double-heterostructure (DH) lasers [Cas75].

The upper limit of the growth temperature is controlled by the availability of group V over pressure, or the group-V arrival rate that prevents the noncongruent evaporation of the compound. A higher growth temperature requires a larger As consumption in growing GaAs or $Al_xGa_{1-x}As$. Furthermore above 640°C the Ga adsorption lifetime becomes sufficiently short to affect the growth rate [Fis83]. The growth rates of GaAs, AlAs, and the GaAs fraction in $Al_xGa_{1-x}As$ as a function of the substrate temperature are shown in Fig. 3.12. Below 640°C the growth rates are nearly independent of the substrate temperature, implying that the sticking coefficient of Ga is nearly unity. Note that only a small percent of Al in the beam can significantly increase the Ga sticking coefficient at high temperatures [Art68].

Morphological Defects In the MBE growth system there are several surface-morphological problems related to growth conditions [Chai81; Woo81; Bac81; Baf82; Mor82; Suz84; Wan85; Baf83; Ito84]. Among the reported defects on the MBE-grown layers are oval defects [Chai81; Woo81; Bac81; Baf82], polycrystallites [Chai81; Woo81; Bac81], whiskers [Woo81; Bac81; Baf82], stacking faults [Suz84], and dislocations

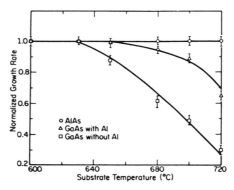

Fig. 3.12 Growth rates normalized to low temperature ($\leqslant 620°C$) values as a function of substrate temperature. The curve labeled "GaAs with Al" represents the Ga fraction of the growth rate. [Cho, A. Y., in *Molecular Beam Epitaxy and Heterostructures*, p. 191, ed. by L. L. Chang and K. Ploog, Kluwer, Academic Press (1985)]

[Shi84]. Some of the observed defects are easily removed by preparing the substrate properly. Others like oval defects, however, are difficult to eliminate. These are oval-shaped hillocks, oriented along the [110] direction on a (001) substrate [Din85]. The size is about 100 μm in length, 4 to 5 μm in width and 0.1 to 0.2 μm in height. At present the typical oval defect density is 500 to 2000 per square centimeter.

Wang et al. [Wan86] have studied the morphologically defects in MBE-grown layers. Figure 3.13 deals with the oval defect density as a function of substrate temperature. The measured samples were consecutively grown under the same growth conditions. The average oval defect density measured by Nomarski phase contrast microscopy was obtained from five regions of 100-μm diameter. Two sets of epilayers were investigated; one with a growth rate of 1.7 μm/h and the other with approximately 1 μm/h. The oval defect density decreases monotonically with increasing substrate temperature.

The origins of these oval defects are not fully understood. Several attempts were

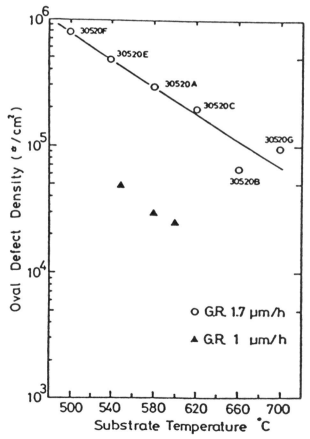

Fig. 3.13 Oval defect density as a function of substrate temperature. The growth conditions are identical except for the sample 30520G growth with higher As pressure to maintain a mirrorlike surface. [Wang, Y. H., W. C. Liu, C. Y. Chang, and S. A. Laio, *J. Vac. Sci., Tech.* B **4**, 1 (1986): 30]

made to uncover their origins. Gallium oxide in the Ga flux is the major contaminant. Chai et al. [Chai81] have shown conclusively that the oval defect density can be reduced from 6×10^5 to 2×10^3 by careful elimination of gallium oxide. Wang et al. [Wan86] proposed that the reaction of Ga and As fluxes can be described as

$$4GaAs(s) + Ga_2O_3(s) \longrightarrow 3Ga_2O(v) + As_4(v), \tag{3.19}$$

$$4Ga(s) + Ga_2O_3(s) \longrightarrow 3Ga_2O(v). \tag{3.20}$$

Since the Ga_2O_3 is a nonvolatile oxide on the GaAs substrate, it becomes a nucleation center for the oval defect in the epilayers. Raising the substrate temperature will increase the maximum allowable Ga_2O pressure, leading the reaction Eq. (3.19) at the right-hand side. Therefore the amount of Ga_2O_3 on the surface decreases, and the number of nucleation centers for the oval defects is reduced. The observed Ga droplets on the growth layers can be interpreted by Eq. (3.20). The increase of Ga on the surface arises from the gallium oxide. It has also been reported [Pet84] that no oval defects were observed when growth took place under the influence a strong Mg flux, which is a getter of oxides.

Figure 3.14 shows the relation between RHEED patterns as functions of substrate temperature and As pressure. The growth rate is 1 μm/h. The transition temperature for the change from the As-stabilized condition to the Ga-stabilized condition at the same As pressure is about 50°C in the study [Wan86]. Figure 3.15 illustrates the effect of growth rates on the oval defect density. The epilayers were grown at 580°C. Oval defect density decreases with decreasing growth rate. The surface morphology of the high growth rate is inferior to those of the low growth rate, even at the same thickness.

Figure 3.16 demonstrates the dependence of the oval defect density on the epilayer thickness. The oval defect density monotonically increases with increasing epilayer thickness. It means that not all of the oval defects are formed in the initial starting

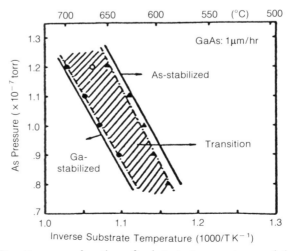

Fig. 3.14 RHEED pattern as a function of substrate temperature and As pressure. [Wang, Y. H., W. C. Liu, C. Y. Chang, and S. A. Laio, *J. Vac. Sci., Tech.* B **4**, 1 (1986): 30]

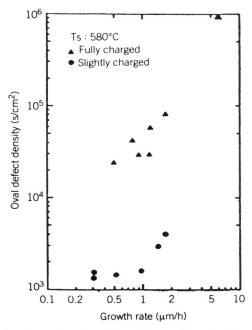

Fig. 3.15 Oval defect density as a function of growth rate. Two sets of samples were investigated: one with almost fully charged Ga sources and the other with only a little Ga melt in the crucible. [Wang, Y. H., W. C. Liu, C. Y. Chang, and S. A. Laio, *J. Vac. Sci., Tech.* B **4**, 1 (1986): 30]

growth (i.e., epilayer-substrate interface), although most defects arise from here. The relation between the oval defect density and background As pressure, or As/Ga ratios, is shown in Fig. 3.17 [Wan86]. The thickness of inspected undoped GaAs layers is approximately 2 µm. It indicates that the oval defect density slightly increases with increasing As/Ga ratios consistent with the trend of Eq. (3.19). As the As/Ga ratio increases, the nonvolatile oxide for nucleated sites is enhanced to form the oval defects. It is also found that very large As/Ga ratios show large whisker density which is believed to be vapor-liquid-solid (VLS) mechanism, on account of Ga droplets [Woo81; Mor82].

The conclusions on the observed oval defects are drawn as follows:

1. The oval defect density increases with increasing growth rates and epitaxial thickness. Most of the observed oval defects are formed during growth. The low growth rate at 0.3 µm/h can reduce the oval defect density from 10^6 to 10^3 cm^{-2} or less.

2. The effects of substrate types, doping concentrations, and the compositions of etching solution are minor.

3. The oval defect density decreases with increasing substrate temperature and decreasing As pressure, as is the case with the model proposed by Kirchner et al. [Kir81].

Shinohara et al. [Shi84] suggested that oval defects have a higher resistivity. An increase in oval defect area inside the source-to-drain region will decrease the source-

74 EPITAXAL GROWTH PROCESSES

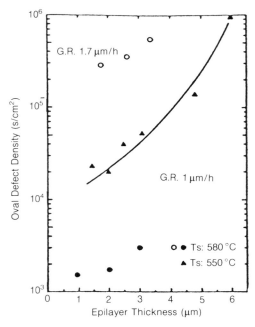

Fig. 3.16 Oval defect density as a function of epitaxal thickness. Two sets of samples at various growth rates of 1 and 1.7 μm/h were considered; ● shows the samples grown at little Ga melt in the crucible. [Wang, Y. H., W. C. Liu, C. Y. Chang, and S. A. Laio, *J. Vac. Sci., Tech.* B **4**, 1 (1986): 30]

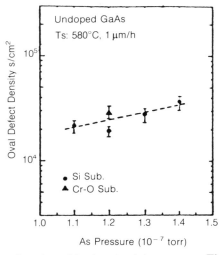

Fig. 3.17 Oval defect as a function of background As pressure. The growth rate was kept at 1 μm/h. [Wang, Y. H., W. C. Liu, C. Y. Chang, and S. A. Laio, *J. Vac. Sci., Tech.* B **4**, 1 (1986): 30]

to-drain channel conductance. The breakdown voltage of the MESFET will degrade when the core of the oval defects is at or near the gate edge [Shi84].

A higher growth temperature will increase the surface mobility of Ga atoms, and a lower growth rate [Met83] will increase the residence time for Ga atoms to find a favorite site. Both of these conditions will minimize Ga segregation. Lowering the Ga flux while maintaining the same growth rate [Abr84], decreasing the distance between the substrate and the Ga cell, or increasing the Ga crucible diameter will minimize Ga spitting. Careful preparation and choice of the substrates, as well as proper outgassing of a fresh Ga source after the growth chamber is exposed to air, will also minimize the oval defect density from Ga_2O. Since MOCVD-grown high-quality selectively doped heterostructures have been reported without oval defects, MOMBE (CBE) may be a feasible approach to grow high-quality films. Defect density low enough for MSI and LSI circuits will be achieved in the near future, and this will open the door to GaAs VLSI circuit fabrication.

3.3 METAL–ORGANIC CHEMICAL VAPOR DEPOSITION

Metal–organic chemical vapor deposition (MOCVD) technology represents the fastest growing and most promising technology for the new compound semiconductor industry. The technique involves the reaction, at a temperature well below the melting point of the resultant solid, of two or more chemically reactive gases at atmospheric or reduced pressure.

Here we consider mainly pyrolysis reactions where one or more of the sources involved is a metal alkyl. In a typical reaction of this type, a lower-order metal alkyl, such as trimethylgallium (TMGa), is mixed in the vapor phase with a hydride, such as arsine. As a result of pyrolysis, atomic or molecular species combine at or near a heated substrate. The deposition results in epitaxy on a single-crystal substrate such as GaAs or InP. The emphasis is on epitaxial growth, and on the fact that the metal alkyls in typical use belong to a broader class of compounds known as metal organic chemical vapor deposition (MOCVD) [Str85]. There are several reasons for MOCVD to become an important epitaxial growth technology in a wide variety of III–V and II–VI materials. For example, all constituents are in the vapor phase, which allows for accurate electronic control of such important system parameters as gas flow rates and hence partial pressures. The pyrolysis reaction is relatively insensitive to growth temperature, allowing for efficient and reproducible deposition of thin layers and abrupt interfaces between deposited layers. Complex multiple layer heterostructures can be grown utilizing computer-controlled automatic gas-exchange systems.

Since all of the source chemicals involved in MOCVD are either extremely toxic or spontaneously inflammable in air, rigorous safety precautions are necessary for their use. Safety training of all personnel associated with operations and maintenance is crucial.

3.3.1 Metal–Organic Sources

The result metal–organic sources normally have two basic characteristics: (1) They must have suitable vapor pressures (~ 10 torr) at reasonable temperature ($-20°$ to $+20°C$), and (2) they must thermally decompose at typical growth temperatures to yield the desired group-III or group-V element for the growth process [Str85]. The vapor pressure of several OM sources used for III–V growth are plotted versus temperature in Figs. 3.18 and 3.19 [Str89].

In general, alkyls with the highest vapor pressure, usually the lowest molecular weight, are preferred. Thus TMGa and TMAl are used whenever possible for GaAs and AlGaAs growth. TMIn is a solid at room temperature due to its unique tetrameric nature; hence it is either sublimed or heated to above its melting point. This necessitates heating all lines between the TMIn bubbler and the reactor, or dilution with large quantity of H_2 before leaving the heated part of the bubbler. TMIn has the highest vapor pressure of all the indium alkyls. The triethyl-III (TE-III) compounds are also used, but their lower vapor pressures usually yield lower growth rates. They are even less stable, tending to form polymers in atmospheric pressure reactors, which further reduces the growth rate [Bha82; Kuo83] and may lead to inhomogeneous growth. However, Kuech et al. [Kue85] and Chang et al. [Cha81] have reported that less carbon contamination may be expected in the GaAs epilayer when it is deposited by the TEGa method.

Triethylindium (TEIn) is a liquid source. Due to its loose polymer structure, it has a very low vapor pressure. It is also much less thermally stable than the trimethyl

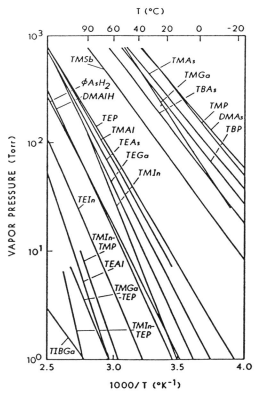

Fig. 3.18 Temperature dependence of vapor pressure for common group III and group V organometallic sources. [Stringfellow, G. B., *Organometallic Vapor Phase Epitaxy*, Academic Press (1989)]

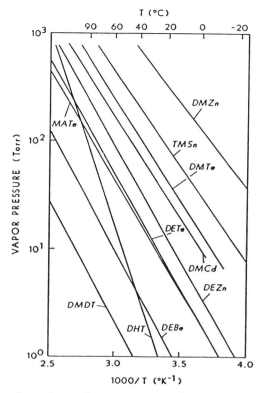

Fig. 3.19 Temperature dependence of vapor pressures for common group II and group VI organometallic sources. [Stringfellow, G. B., *Organometallic Vapor Phase Epitaxy*, Academic Press (1989)]

and decomposes appreciably above 40°C, as well as on exposure to hydrogen carrier gas. To minimize these problems, TEIn is used in reduced pressure deposition systems held at 35° to 40°C, and nitrogen is used as a carrier gas.

The group-III alkyls listed pyrolize efficiently in normal low pressure, or 1-atm, reactors. The group-V alkyls and hydrides are known to pyrolize more slowly. TEP and TMP will decompose very little, even less than PH_3, and are thus an ineffective P source [Mos81]. TMAs is known to pyrolize less rapidly than AsH_3[Che84], but it is still a useful source of As. Trimethyl and triethyl antimony are both very effective sources of Sb and are more convenient to use than SbH_3 [Coo80; Clo69].

3.3.2 Nonhydride Group-V Sources for MOCVD

Historically the hydrides (e.g., AsH_3 and PH_3) have been used for MOCVD because of their ready availability in relatively high pure form. A major obstacle to the use of MOCVD in large-scale production operations is in the use of large quantities of highly toxic AsH_3 and PH_3. The threshold limit values (TLVs) the maximum permissable exposure limit based on a time weighted average for an eight hour day [Str88], for AsH_3 and PH_3 are 0.05 and 0.3 ppm, respectively [CRC67]. The value of LC_{50} (rats), defined as the lethal concentration for 50% of the population in rat testing, necessarily involves both concentration level and length of exposure. Typically only the concentration is given. The exposure time, which is assumed to be on the order of four hours, is reported by Sax [Sax79] to be 11 ppm for PH_3. No such value is listed for arsine. In recent years the general public has become increasingly aware and alert to the dangers associated with the use of toxic materials near residential neighborhoods. A large fraction of the expense of both purchasing and operating a reactor deemed safe by today's standards is devoted to safety features. Fortunately a number of organometallic group-V sources are much less toxic than the hydrides. The ideal group-V source would be a nontoxic liquid with a moderate vapor pressure (50 to 500 torr). In this section recent research on newly developed sources will be discussed [Str88].

Requirements for Group-V Sources

The requirements for group-V sources for MOCVD are stringent. They include high vapor pressure, low-temperature stability, pyrolysis at temperatures at and above 400°C, no inherent purity limitations such as excess carbon contamination, and no interactions with the group-III sources producing parasitic reactions.

The commonly used nonhydride sources are low vapor pressure liquids or solids at room temperature, which is advantageous from a safety viewpoint. However, to be useful for MOCVD, at room temperature the vapor pressures should be greater than 50 torr, in order to avoid using extremely high carrier gas flow rates through the bubblers in which the sources are contained or heating of the bubbler and downstream lines. The group-V sources that have been successfully used for MOCVD are listed in Table 3.2 [Str88] along with their vapor pressures. Clearly the elemental sources must be heated to temperatures well above 300 K. Di- and triethylarsine also have vapor pressures too low to be conveniently used without heating the bubblers.

TABLE 3.2 Vapor Pressure of Group V Sources for OMVPE.

Compound	Vapor Pressure (torr)	Temperature (°C)
Phosphorus		
P	1	260
PH_3	760	−87.8
TMP	381	20
TEP	46.5	50
IBP	122	23
TBP	286	23
Arsenic		
As	1	370
AsH_3	760	−55
TMAs	238	20
TEAs	15.5	37
DEAs	0.6	18
TBAs	96	−10

Source: Stringfellow, G. B., *J. Electron. Mat.*, **17**, 1 (1988): 67.

Two factors are important when considering the stability of the group V sources. First, the materials must be relatively easy to synthesize and purify. Second, the materials suitable for sources should have shelf lifetimes at room temperature measured in years.

The group-V source molecules must pyrolyze at the relatively low temperatures used for MOCVD growth. At present, GaAs growth temperatures as low as 550°C are common. For smaller band-gap materials, with lower bond strengths giving lower melting points, the optimum growth temperature may be considerably lower. For example, InSb melts at 530°C, so the growth temperatures must be somewhat lower. Only the slower pyrolysis of common sources such as AsH_3 and TMSb prevents the use of even lower growth temperatures. AsH_3 is 50% pyrolyzed only at temperatures of greater than 757°C [Den86]. Even higher temperatures of greater than 850°C are required for PH_3 pyrolysis [Lar86].

Another practical requirement is the absence of reactions of the group V sources with the group-III organometallic sources, leading to depletion of the source materials from the vapor phase upstream from the substrate. Naturally a reaction that yields the III–V semiconductor at high temperatures is desired. However, parasitic reactions frequently lead to decreased growth rates due to deposition of undesirable nonvolatile material on the reactor walls. An example is the interaction of both TEGa and TEIn with AsH_3 and PH_3 to form adducts that subsequently eliminate methane forming nonvolatile polymers [Str85]. Fortunately the more commonly used TMGa and TMIn do not interact with the group-V hydrides in parasitic reactions at ordinary growth temperatures [Str85].

Results for Nonhydride Group-V Sources

Elemental Sources Table 3.2 illustrates a practical problem with the elemental group V sources. Their low vapor pressures require the use of high-temperature containers and, in some cases, heated lines to the reactor. Bhat [Bha85] was the first to show that the obstacle could be overcome in a slightly modified MOCVD apparatus. Passing hydrogen over the elemental As held in a boat at 425° to 475°C upstream from the substrate, the growth of GaAs using TMGa was demonstrated. Unfortunately, the layers were always p-type for $T \leqslant 700°C$, due to incorporation of carbon. Tokumitsu et al. [Tok84] and Ishikawa [Ish86] also used an elemental As source with TMGa for GaAs growth through metalorganic molecular beam epitaxy (MOMBE). The MOMBE techniques will be discussed in Section 3.4. In this vacuum deposition technique the layers were again always p-type with carbon levels as high as 10^{20} cm^{-3}.

Kuck et al. have reported success with GaAs growth using an elemental As source [Str88]. They used a high-temperature stainless steel As "bubbler" in a configuration similar to that used for the organometallic bubblers. The bubbler temperature was in the range 400°–460°C, and the stainless steel lines leading to the reactor, as well as the reactor walls, was held at 400°C. The growth of n-type GaAs layers occurred under normal growth temperatures and a normal As/Ga ratio. The carbon concentrations were, however, orders of magnitude below the values obtained by MOMBE. This was attributed to the effect of ambient H_2, which has been reported to react with TMGa producing CH_4 [Yos85].

Kuch et al. [Str88] have also reported the growth of InP using TEIn and an elemental P source, in a Pyrex bubbler containing white phosphorus at a temperature in the range 80°–130°C. They found that no epitaxial films could be produced with the P_4 source, The P_4 produced had to be cracked to P_2 using a remote plasma before the phosphorus became an effective MOCVD source. Improved electrical characteristics were obtained with carrier concentrations in the mid-10^{16} cm^{-3} range and mobilities as high as 3950 cm^2/V-s [Nai87].

Trimethyl Sources Trimethylphosphine is not an effective P source for MOCVD. It is so stable that no P is produced at ordinary growth temperatures. Benz et al. [Ben81] formed TMIn–TMP adducts used for the growth of InP. Without the addition of PCl_3 to the system, the growth was unsuccessful.

Trimethylarsenic presents similar problems. The pyrolysis temperature is significantly higher than that for AsH_3 [Che84]. The resulting epitaxial layers using TMAs are contaminated with significant impurity levels. Using ^{13}C tagged TMAs, Lum et al. [Lum88] have shown that C from the TMAs is incorporated at a concentration of 5×10^{16} cm^{-3}. This is consistent with the arguments advanced by Kuech and Veuhoff [Kue84]. An advantage of TMAs is low toxicity. The LC_{50} is reported to be 20,000 ppm [Org86].

Triethyl Sources Triethylphosphine (TEP) is not an effective P source for MOCVD because its high stability prevents pyrolysis at normal growth temperatures. Moss and Evans [Mos81] reported that P from TEP is not incorporated into the solid; GaInAs can be grown from a TMIn–TEP adduct, TMGa, and AsH_3 with no trace of P incorporation.

Triethylarsenic (TEAs) does pyrolyze at reasonable temperatures, but, as seen in Table 3.2, the vapor pressure is inconveniently low. Speckman and Wendt [Spe87] report the successful growth of GaAs using TEAs and TMGa. In the temperature range 540°–650°C and at values of the V/III ratio varying from 6.7 to 11, specular surface morphologies were obtained. These layers have high carbon and Si impurities present in the mid-10^{17} cm^{-3} range. The carbon probably comes from CH_3 radicals produced during the TMGa pyrolysis. With no atomic H produced during the TEAs decomposition, this results in carbon incorporation into the solid. The toxicity is significantly lower than for AsH_3, with an LC_{50} of 1060 ppm for female rats [Hat87].

Diethylarsine The use of diethylarsine (DEAs) is an attempt to replace only two of the H atoms in AsH_3 with organic radicals. The vapor pressure of DEAs [Bha87] is lower than that of TEAs, as seen in Table 3.2. Its toxicity is unknown. The best layers, grown at 500° to 580°C, were *n*-type with background free-carrier concentrations of 0.3 to 5×10^{15} cm^{-3} and liquid nitrogen mobilities as high as 64,600 cm^3/V-s. The low-temperature PL spectra indicates that diethylarsine can grow purest GaAs layers within the non-AsH_3 group-V sources [Bha87].

Tertiarybutylarsine Tertiarybutylarsine (TBAs) has a vapor pressure of approximately 96 torr at 10°C [Che87]. It pyrolyzes at much lower temperatures than for AsH_3. For example, pyrolysis is 50% complete at 425°C compared to 575°C for AsH_3 [Den86]. The pyrolysis products are mainly C_4H_{10} and AsH, with some C_4H_8 and AsH_3 [Lar86] produced by unimolecular reactions analogous to those for TBP pyrolysis.

The MOCVD growth of GaAs using TBAs and TMGa has been investigated by several groups. Due to the low pyrolysis temperature, good morphology layers could be grown at V/III ratios of approximately unity, more than a factor of ten lower than for AsH$_3$ [Miz84]. The initial data indicate that the carbon concentration determined from PL measurements was reduced when TBAs was substituted for AsH$_3$ during growth at 640°C at V/III ratio of ten. However, the carrier concentration and mobility were found to be nearly independent of growth temperature, as shown in Fig. 3.20, indicating the donor impurity to be a nonvolatile group-IV impurity [Hsu86], probably Ge. The best room temperature electron mobilities were approximately 4500 cm^2/V-s at carrier concentrations in the mid-10^{16} cm^{-3} range. Lum et al. [Lum87] reported similar results, with background doping levels as low as 10^{16} cm^{-3} and room temperature electron mobilities of approximately 4000 cm^2/V-s. They found a slight increase in doping at increased growth temperatures in the range 625°–800°C, as shown in Fig. 3.20. The lowest background doping levels were obtained at the lowest values of V/III ratio, as seen in Fig. 3.21.

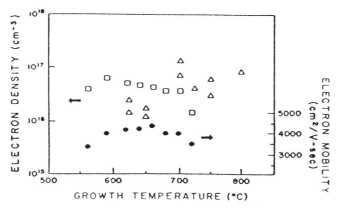

Fig. 3.20 Room temperature carrier concentration and mobility versus growth temperature for GaAs grown using TMGa and TBAs. [Stringfellow, G. B., *J. Electron. Mat.* **17** (1988)]

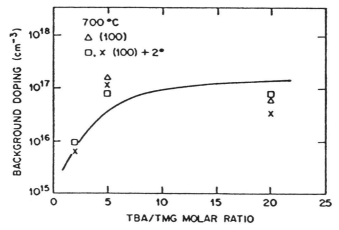

Fig. 3.21 Dependence of background carrier concentration on V/III ratio. [Stringfellow, G. B., *J. Electron. Mat.* **17** (1988)]

The toxicity tests indicate the LC_{50} to be 70 ppm for TBAs [acute inhalation of LC_{50} (rats) equals approximately 70 ppm]. This value is still considered to be dangerous. The leading candidate for replacement for AsH_3 is probably TBAs. However, the toxicity and possible difficulties in large-scale production based on instability of the molecule suggest that a better substitute must be found in the future.

3.3.3 Basic Reactions

During growth by CVD a number of processes take place, partly in series and partly in parallel. Their relative importance depends on both the chemical nature of the species and the design of the reactor. Reactor chamber designs are divided into between horizontal types (e.g., [Bas75]) and vertical types (e.g., [Man68]). The growth rate is determined by the slowest process in series of events needed to come to deposition. In the case of MOCVD, it has been mentioned by several authors [Reed83; Duc78] that the growth rate is epitaxial growth is controlled by the diffusion of the metalorganic components through the boundary layers. The epitaxial layers of GaAs can be grown at atmospheric pressure or under reduced pressure (LP-MOCVD) system.

The fundamental aspects of the CVD process have been investigated by many workers [Gha83; Sze81], and the models developed for the process are based on the assumption that the arrival of reactant species on the growth surface is limited by diffusion through a boundary layer. In this section the basic considerations of CVD process will be described.

Basic Chemical Reactions in MOCVD

In the discussion of basic MOCVD reactions for the growth of compound semiconductor epitaxial layers, we will limit ourselves to those reactions involving organometallic compounds or mixed organometallic compounds and hydrides. The exact chemical decomposition pathways in MOCVD are not yet clearly understood. The nature of the reactions are in part determined by the dynamics of the gases, that is, by the velocity and temperature profiles in the vicinity of the susceptor and the subsequent concentration and thermal gradients that are established [Str84; Kop84; Kus85].

The reaction pathways of course are also strongly influenced by the choice of precursor chemicals. Alkyls of the group-III metals, and hydrides of the group-V elements are usually used as precursor species in MOCVD. Dilute vapors of these chemicals are transported at or near room temperature to a hot zone where a pyrolysis reaction occurs. The most commonly used reactions for the growth of compound semiconductor MOCVD layers is given in the general form

$$R_n M + XH_n \longrightarrow MX + nRH, \qquad (3.21)$$

where

R = an organic radical,
M = one component of the resulting semiconductor layers,
X = the other component,
n = an integer.

In the reaction of Eq. (3.21) the simple organometallic species represented by R_n–M reacts with X–H_n to generate the compound semiconductor MX and a residual organic compound RH. Important examples of these reactions are

$$(CH_3)_3Ga + AsH_3 \longrightarrow GaAs + 3CH_3, \qquad (3.22)$$

$$x(CH_3)_3Al + (1-x)(CH_3)_3Ga + AsH_3 \longrightarrow Al_xGa_{1-x}As + 3CH_4. \qquad (3.23)$$

Although relatively simple to describe in an equation, the actual kinetics and mechanism associated with he heterogeneous reaction between the materials at the surface of the growing interface and the gas phase reactions are quite complex. The right-hand side of Eq. (3.23) is particularly interesting because it reflects impact of alloy layers and heterostructures on compound semiconductor devices. In all of these reactions, the metal alkyl is stored as a liquid at low temperature ($-10°C$ for TMGa) and the hydrides are gaseous. The metal alkyl is transported to the quartz reaction cell (termed a *Bass cell*) by hydrogen carrier gas where it is intimately mixed with arsine and brought into the vicinity of a heated susceptor. The susceptor can be heated using an RF coil that surrounds the cell. The reactants are pyrolyzed by the heat of the susceptor and fragment into atomic or molecular forms of the component species, which then combine to form the deposited semiconductor.

Another alternative is the substitution of a related organometallic compound for the hydride source. This is described by a reaction of the general form [Col85]

$$R_nM + R'_mX \longrightarrow MX + nRH + mR'H, \qquad (3.24)$$

where

R, R' = organic radicals,

M, X = compound semiconductor components.

The H_2 indicates that these reactions typically take place in a reducing atmosphere. Some typical examples of these type of reaction are

$$(CH_3)_3Ga + (CH_3)_3As \longrightarrow GaAs + 6CH_4 \qquad (3.25)$$

and

$$(1-x)(CH_3)_3In + x(CH_3)_3Ga + (CH_3)_3Sb \longrightarrow In_{1-x}Ga_xSb + 6CH_4. \qquad (3.26)$$

Reactions involving only organometallic sources have not been studied to the extent that the metal alkyl-hydride systems have been studies. There is the possibility that organometallic compounds are easier to purify than are gaseous hydrides [Dap81]. Background doping levels are also lower in epitaxial layers grown only with organometallic compounds.

Some metal alkyls, particularly indium compounds, react rapidly at room temperature with hydrides, such as PH_3, to form adducts of the form

$$R_nM:XH_n. \qquad (3.27)$$

An example of this is

$$(CH_3)_3In:PH_3, \qquad (3.28)$$

which is thought to be polymeric at room temperature and a simple 1:1 adduct at lower temperature [Ois82]. The group III metal alkyls are strong Lewis acids, and hydrides are Lewis bases. It is thus not surprising that these adducts are formed [Mos81]. No such intermediate reactions are observed in the TMGa–AsH_3 system.

Several empirical approaches have been used to minimize the formation of intermediate reaction products or parasitic polymers that can deleteriously affect the growth rate or composition of an MOCVD-grown epitaxial layer. The first approach is the physical solution. It is based on the recognition that the rate of formation of the adduct and the rates of the elimination reactions may be rather slow, possibly on the order of a fraction of a second [Fro77]. Through the use of low pressure, the residence time of the source molecules, together in the gas phase before reacting with the hot substrate, is reduced by at least one order of magnitude. This has allowed Duchemin et al. [Duc81] and Yoshino et al. [Yos81; Yos81b] to grow layers of InP and GaInAsP alloys using TEIn and PH_3. Duchemin et al. [Duc78b] prepyrolyzed the PH_3 prior to mixing it with TEIn in a low pressure (0.1 atm) MOCVD system. Oishi et al. [Ois82] produced GaInAs layers using TEIn, TEGa, and AsH_3 but observed no intermediate reaction product or parasitic polymer formation problems. Duchemin et al. [Duc81] tried the additional step of passing the PH_3 through a 760°C furnace to partially crack the PH_3, using a carrier gas of 50% N_2 and 50% H_2 [Ois82].

The second approach might be termed the *chemical solution*. The adduct $(C_2H_5)_3In-PH_3$ probably decomposes by the spontaneous elimination of C_2H_6 molecules formed from one C_2H_5 of In and one H of P. Simply cracking the PH_3 before H enters the reactor could eliminate this process, especially at low H_2 pressure [Ley81]. Alternatively, we could substitute TEP or TMP for PH_3. The elimination reaction is clearly less favorable in this case, partly because no stable organic by-product is produced and partly because of the tight binding of the ethyl radical to the phosphorus. Moss et al. [Mos81] report that the P from TEP is not incorporated into the solid; GaInAs can be grown from TMIn–TEP, TMGa, and AsH_3. Benz et al. [Ben81] have used a similar approach using a TMIn–TMP adduct as the In source and PCl_3 as the P source. Cooper et al. [Coo80] have substituted TMAs for AsH_3 in the growth of In-.and As-containing alloys.

Basic Growth Mechanisms

The basic growth mechanisms as a function of growth temperature is shown in Fig. 3.22 [Hun91]. At lower temperatures (<550°C) the growth mechanism is surface kinetic limited. Betewen 550°C and 750°C it is mass transport limited, which is nearly temperature independent. At higher temperature (>800°C) it is thermodynamics limited, where temperature increase leads to growth rate decrease. With the onset of kinetic growth, the region shifts to higher temperatures and the bond strength of the used As compound (AsH_3 < TEAs < TMAs) increases [Bra91].

3.3.4 Purity and Dopants

The purity of GaAs grown by MOCVD is controlled by the growth temperature, the purity of the starting materials, and the arsenic-to-gallium ratio in the reactor. The purity of GaAs grown by TMGa and AsH_3 increases monotonically with decreasing

METAL–ORGANIC CHEMICAL VAPOR DEPOSITION 85

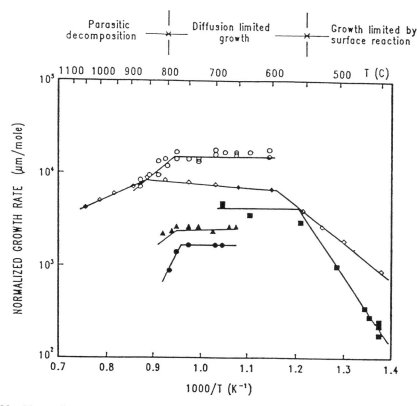

Fig. 3.22 Normalized growth rate as a function of inverse of temperature. [Danny Hung, in *Chemical Vapor Deposition*, p. 245, ed. by M. L. Hitchman and J. E. Jensen, Academic Press (1992)]

growth temperature [Dap81; Nak77]. Dapkus et al. [Dap81] indicate that this increase in purity is due to a monotonic decrease in the incorporation of carbon and silicon in GaAs grown by MOCVD. Similar studies by Nakanisi [Nak77] indicate a less strong, but important, dependence of the total, electrically activated, impurity concentration in GaAs on the arsenic-to-gallium ratio in the reactor. The purity of both the TMGa and AsH_3 is vitally important to obtaining high-purity GaAs.

Since the initial work on MOCVD of GaAs, researchers have been concerned with the incorporation of carbon as a background impurity [Rai69; Sek75; Kue85]. This concern was accentuated by early studies that showed high background C levels [Rai69]. The acceptor behavior of C in GaAs is well known, and C acceptors can be readily distinguished by using low-temperature PL techniques [Kob85; Yos85]. Carbon acceptors are seen in virtually all GaAs grown using TMGa [Püt86]. Stringfellow showed that an increase in growth temperature results in large increase in carbon incorporation [Str89; Kue87].

In the MOCVD process the carbon (the alkyl radicals resulting from the surface decomposition reaction) is removed from the growing surface much more easily [För89]. The interpretation for this effect is the possibility of β-elimination processes in the decomposition of TEGa [Wil82]:

$$(C_2H_5)_3Ga \longrightarrow (C_2H_5)_2GaH + C_2H_4 \quad (\beta\text{-elimination}). \quad (3.29)$$

The $Ga(C_2H_5)_3$ (TEGa) decomposes stepwise into components $GaH_n(C_2H_5)_{3-n}$ with simultaneous formation of C_2H_4. The stable C_2H_4 molecules formed from TEGa does not lead to the carbon incorporation [Str79; 'tHO81]. The C_2H_4 molecule has a very low probability of sticking on the GaAs surface even at room temperature [För89].

β-elimination is not possible in the TMGa decomposition; the split-off of Ga leaves a very reactive CH_3 radical, which sticks to the surface quite readily. The CH_3 can be transformed into the little-reactive CH_4 by adding atomic H, the result of a AsH_3 decomposition.

Carbon contamination in $Al_xGa_{1-x}As$ is a more severe problem. Electron mobility is dramatically reduced with increasing x. This is believed to be due to C being incorporated into the strong Al–C bond, which decreases the activity coefficient of C in the solid. The C concentration determined from the temperature dependence of electron mobility is approximately $10^{17}\,cm^{-3}$ for $Al_xGa_{1-x}As$ grown with TMGa, TMAl, and AsH_3.

Another major residual impurity is oxygen. For GaAs a few parts per million (ppm) of O_2 and/or H_2O in the gas stream has little effect on the properties. In fact ^{18}O-doping studies indicate that when a few ppm of O_2 are added to the gas stream, less than $10^{16}\,cm^{-3}$ is incorporated into the solid [Kis82]. For $Al_xGa_{1-x}As$, on the other hand, ^{18}O-doping studies clearly show that 1 ppm of $^{18}O_2$ in the vapor gives rise to $10^{20}\,cm^{-3}$ of ^{18}O in the solid. Assuming that the $P(O_2)$ at the growing interface is zero, this is exactly what would be predicted for a total flow rate of 2 liters/min if the reactor is about 10% efficient.

In the solid $10^{19}\,cm^{-3}$ of oxygen is sufficient to affect adversely the PL efficiency. This is a serious problem. It is very difficult to maintain a gas stream with <0.1 ppm O_2 from leaks, desorption from walls, outgassing from the susceptor, and perhaps most important, an impurity in the AsH_3 [Str85]. Studies by several groups have shown that PL efficiency [Wag81] and mobility [Dap81] are strongly affected by the AsH_3 source. Several techniques have been described for purification of the gas stream just above the substrate.

The use of graphite baffles to catalyze the reaction between TMAl and any O_2 or H_2O in the gas stream is now well established. The effect was first observed as an increase in PL efficiency of MOCVD AlGaAs when graphite baffles were introduced into the gas stream [Nis61]. If the SiC-coated graphite baffles are replaced by simple graphite baffles, there is a reduction of ^{18}O in the epitaxial layer by a factor greater than 10^3.

We next consider the n- and p-type dopants. A number of donor impurities have been used in MOCVD, including the group VI elements S, Se, and Te and the group-IV elements Si and Ge. There are fewer acceptor impurities to choose from, since the group-IV elements are not amphoteric. Thus we are left with Zn, the major p-type dopant, and Cd, Be, and Mg. These same dopants have been used for the entire range of III–V compounds and alloys.

Selenium is probably the most widely used n-type dopant in MOCVD material. Hydrogen selenide (H_2Se) is used for the Se source. It has a distribution coefficient of somewhat less than unity, (the distribution coefficient can be defined by the fraction of the group-III or V lattice sites filled by the dopant atoms, divided by the ratio of input partial pressure) depending on the growth temperature and the III/V ratio. It is expected to have a lower distribution coefficient with increasing temperature because of its increased volatility [Str85], as is generally seen to be the case for the volatile dopants Se, S, and Zn. There is a decrease in electron carrier concentration when the growth temperature increases, which is due to surface desorption of selenium. The use of Se apparently produces good PL efficiencies and minority carrier lifetimes in GaAs and $Al_xGa_{1-x}As$ [Nis61].

Tellurium is obtained from DETe, which may be purchased diluted in H_2 in high-pressure cylinders. In the MOCVD growth of GaAs, Te can be used to produce doping levels as high as 10^{18} cm^{-3} with only the expected increase in the PL efficiency [Mor81]. At high doping levels, all impurities produce nonradiative recombination centers either through complex formation or precipitation. However, low-threshold current density lasers with very good reliability can be fabricated using Te as the n-type dopant.

Silicon acts as donor in MOCVD GaAs and AlGaAs. The effective distribution coefficient and PL efficiency of the resulting material both increase strongly with increasing substrate temperature. The electron carrier concentration increases with growth temperature for a fixed silane molefraction. Silicon is the n-type dopant used at several laboratories for injection-laser devices. In summary, the best donor impurities for MOCVD growth appear to be Se, from an H_2Se source, Te using DETe, and Si from SiH_4.

The standard p-type dopant for MOCVD is Zn from either DEZn or DMZn. Although DMZn and DEZn sources are suitable for the p-type dopant over the entire doping range anticipated for most usages, DEZn appears to be the most controllable and provides the widest range of doping for all alloy compositions. Zinc has a small distribution coefficient especially at high substrate temperatures. The effect of the V/III ratio is the same as that expected for a group-II impurity substituting on the group-III sublattices. That is, as the V/III ratio increases, so does the Zn concentration in the solid.

3.4 CHEMICAL BEAM EPITAXY

Chemical beam epitaxy (CBE) is an alternate epitaxial growth technique for GaAs and InP that is similar to MBE. It is particularly useful for the growth of precision layers of III–V material containing both As and P, such as GaInAsP, where control of the As/P ratio is difficult in conventional MBE. In this technique all the sources were gaseous group-III and group-V alkyls. The In and Ga were derived by the pyrolysis of trimethylindium, triethylindium and trimethylgallium, or triethylgallium at the heated substrate surface. The As_2 and P_2 were obtained by thermal decomposition of trimethylarsine or arsine and triethylphosphine or phosphine in contact with heated Ta or Mo at 950° to 1200°C [Tsa86]. The AsH_3 and PH_3 are thermally decomposed in an unconventional high-pressure gas source, which is a cracker that emits As_2 and/or P_2 and H_2. Deposition could only be achieved if the AsH_3 and PH_3 are partially cracked before injecting it into the apparatus [Tsa89]. The gas cracker replaces one of the conventional effusion ovens, and ion pumps are replaced with diffusion, cryo-, or turbomolecular pumps so that H_2 generated by the cracking of the hydrides can be effectively removed from the system.

Unlike conventional vapor phase epitaxy, in which the chemicals reach the substrate surface by diffusion through a stagnant carrier gas boundary layer above the substrate, the chemicals in CBE are admitted into a high-vacuum growth chamber and impinged directly onto the heated substrate surface in the form of molecular beams. The InP and GaAs epilayers grown have smooth surfaces. It has also been demonstrated that the epilayers grown by CBE are free of oval defects over an entire substrate surface of 8 cm in diameter [Tsa85].

3.4.1 Chemical Beam Epitaxy System

The CBE growth system is shown in Fig. 3.23 [Tsa89]. In this CBE system all the sources were gaseous and derived from group-III and group-V alkyls. Triethylgallium (TEGa) maintained at 30°C, trimethylindium (TMIn) at 37°C, and trimethylaluminum (TMAl) at 25°C were used. The In and Ga were derived by the pyrolysis of either TMIn or TEIn, and TMGa or TEGa, at the heated substrate surface, respectively. The As_2 and P_2 were obtained by thermal decomposition of the TMAs and TEP passing through a Ta- or Mo-buffered heated alumina tube, respectively. As in MOCVD, and shown in Fig. 3.23, the flow rates of the various gases (and hence the growth rates and chemical compositions) were controlled by precision electronic mass flow controllers. Separate effusion ovens were used for group-III and group-V alkyls. No cracking was used for the group-III alkyls, while the group-V alkyls were cracked at

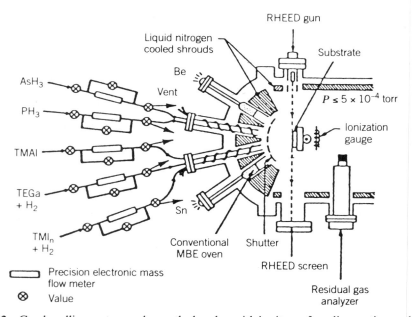

Fig. 3.23 Gas-handling system and growth chamber with in situ surface diagnostic capabilities incorporated in a CBE system. [Tsang, W. T., Chemical beam epitaxy, in *Beam Processing Technologies*, vol. 21, ed. by N. G. Einspruch, S. S. Cohen, and R. N. Singh, VLSI Electronics (1989)]

950° to 200°C after mixing with palladium-diffused H_2. This separate oven arrangement also prevented the formation of polymer products between the In alkyls and As alkyls, which can result in significantly reduced and irreproducible growth rates. Both Ar and H_2 have been used as the carrier gases for the group-III alkyls, which were maintained at constant temperatures with thermal baths. The chemicals in CBE that were admitted into the high-vacuum chamber impinge line-of-sight onto the heated substrate surface in the form of molecular beams. Typical growth rates were 2 to 3 µm/h for GaAs, 4 to 6 µm/h for AlGaAs, 4.0 µm/h for GaInAs and GaInAsP, and 1.5 to 2.5 µm/h for InP [Tsa86]. Such growth rates are higher than those typically used in MBE and low-pressures MOCVD. The beam nature of CBE allowed for mechanical shutters to be used in reducing the transient flow effect during valving, as well as for very abrupt composition, doping changes, and ultrathin layers.

We now briefly discuss the cracker cell of group-V and group-III systems. Figure 3.24 gives a schematic view of the high-temperature, low-pressure cracker cell [Tsa89] for AsH_3 and PH_3. The high-temperature portion of the cell, including the baffles, is constructed of PBN. No tantalum or tungsten is used in the high-temperature zone. A cracking efficiency of essentially 100% for both AsH_3 and PH_3 is routinely and reproducibly achieved at 900° to 1000°C. The two-zone temperature heating is also used with the end section at a higher temperature than the rest of the PBN oven. This is important because the end of the oven tends to be cooler due to heat loss by radiation, and a higher temperature at this end section is needed to reduce the process of dimer recombination to form tetramers.

Figure 3.25 shows a schematic diagram of the group-III cell. The cell is constructed

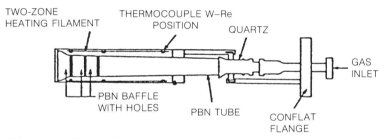

Fig. 3.24 Schematic of the high-temperature, low-pressure cracker cell for AsH_3 and PH_3. [Tsang, W. T., Chemical beam epitaxy, in *Beam Processing Technologies*, vol. 21, ed. by N. G. Einspruch, S. S. Cohen, and R. N. Singh, VLSI Electronics (1989)]

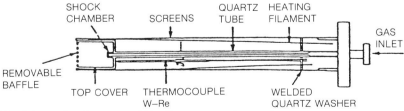

Fig. 3.25 Schematic of a group III cell constructed out of quartz. [Tsang, W. T., Chemical beam epitaxy, in *Beam Processing Technologies*, vol. 21, ed. by N. G. Einspruch, S. S. Cohen, and R. N. Singh, VLSI Electronics (1989)]

entirely of quartz. This is because the cell is always at low temperature and quartz is easy to work into different geometries. All the different metalorganics used are combined and introduced into the cell by passing through a narrow tube ending in a shock chamber. The end of the shock chamber is covered with an interchangeable baffle that has an array of holes of different sizes. This allows detailed tailoring of the flux distribution to achieve a high degree of thickness uniformity. The shock chamber helps to randomize the flux effused out from a set of small holes in the sidewall of the narrow tube. A narrow tube is employed to reduce dead volume.

3.4.2 The Advantages of CBE

The chemical beam epitaxial growth technique combines many potential advantages of MBE and MOCVD [Tsa86]. Compared with MBE, these include

1. the use of room-temperature gaseous group-III metal–organic sources that simplifies multiwafer scale-up,
2. semi-infinite source supply and precision electronic flow control,
3. a single group-III beam that guarantees material composition uniformity,
4. no oval defects even at high-growth rates (important for LSI applications),
5. high-growth rates of desired.

Compared with MOCVD, the presence of a beam in CBE eliminates the flow pattern problem in MOCVD reactors. With flow patterns there can be thickness and compositional nonuniformity. The various flow patterns created depend on the reactor geometry, gas flows, temperature, and rotation speed of the substrate holder. The beam aspect and the separation of group-III and group-V beams in CBE also prevent chemical reactions in the gas phase before impinging on the substrate surface. Thus the advantages include (1) no flow pattern problem encountered in multiwafer scale-up, (2) a beam that produces very abrupt heterointerfaces and ultrathin layers conveniently, and (3) a clean growth environment.

3.4.3 Growth Kinetics of CBE

The growth kinetics of CBE is completely different from that of conventional MBE and in some aspects different from MOCVD, as schematically depicted in Fig. 3.26 [Tsa86b]. In conventional MBE the atomic group-III beams impinge on the heated substrate surface, migrate into the appropriate lattice sites, and deposit epitaxially in the presence of excess impinging group-V molecular beams. Since the sticking coefficient of the group-III atoms on the substrate surface at the usual growth temperature is practically unity, the growth rate is determined by the arrival rate of the group-III atomic beams. No chemical reaction is involved in deriving the group-III atoms at the substrate surface because they are generated by thermal evaporation from solid elemental sources.

In MOCVD the group-III alkyls in the gas stream of H_2 and N_2, Ar or they diffuse through a stagnant boundary layer above the heated substrate, thermally dissociate the three alkyl radicals at the substrate surface to yield the atomic group-III elements. These elements migrate into the appropriate lattice sites and deposit epitaxially by capturing a group-V atom derived as a result of thermal dissociation

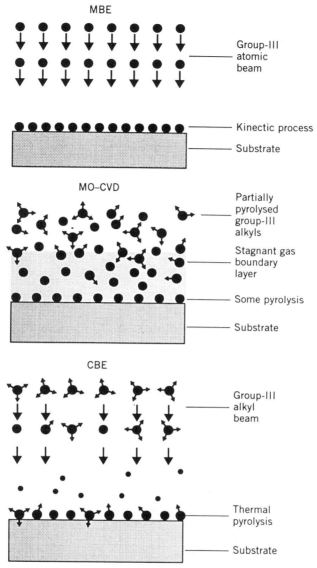

Fig. 3.26 (a), (b), and (c) The growth kinetics involved in conventional MBE, MOCVD, and CBE, respectively. [Tsang, W. T., Chemical beam epitaxy, in *Beam Processing Technologies*, vol. 21, ed. by N. G. Einspruch, S. S. Cohen, and R. N. Singh, VLSI Electronics (1989)]

of the hydrides. For the usual growth temperature employed, the growth rate is limited by the diffusion rate of the group-III alkyls through the boundary layer. The group-III alkyl molecules are adsorbed on the heated substrate surface and are dissociated into elemental group-III atoms. In CBE the beam of group-III alkyl molecules impinges directly into the heated substrate surface as in the conventional MBE process. There is no boundary layer in front of the substrate surface nor are there space-flight molecular collisions on the path because of the long mean free path of the molecules at the pressure of $<5 \times 10^{-4}$ torr. The group-III molecule and acquire enough thermal energy from the heated substrate and dissociate all its three alkyl radicals, leaving the elemental group-III atom on the surface after it strikes at the substrate surface. Figure 3.27 shows the growth rates of InP from TMIn and GaAs from TEGa as a function of substrate temperature for different absolute flow rates of the group-III alkyls. A flow of -3cc/min occurred, resulting in a constant growth rate with the substrate temperature increasing up to 700°C. When the flow rate was halved, the growth rate for the substrate temperature above 550°C was approximately halved, indicating that in this temperature regime the growth rate was limited by the arrival rate of the TEGa. Between 490° and 550°C the substrate temperature was high enough to completely dissociate all the adsorbed TEGa molecules at the lower rate of 1.5 cc/min.

3.4.4 Epilayer Morphologies of GaAs-Grown CBE

One of the key problems for epitaxial layers grown by MBE is the presence of oval defects [Woo81; Cha81; Baf83; Cal83; Kir81; Pet84] in gallium-containing compound semiconductor epilayers. These oval defects are likely to present a serious obstacle for large-scale integration work. The oval defects were eliminated in epilayers grown by CBE [Tsa85; Tsa84]. In the CBE process all gaseous TMGa or TEGa and TMAs sources were used. The Ga atoms are derived by thermal pyrolysis of the metal alkyls at the heated substrate surface, while the As atoms are believed to derive from the As_2 dimers thermally cracked by passing a mixture of TMAs and H_2 through heated Ta at 950° to 1200°C. No precracking of the group-III alkyls was carried out. No oval defect was observed for GaAs epilayers grown by CBE. The use of CBE reproducibly yielded GaAs epilayers free of oval defects over the entire substrate surface of 8 cm in diameter. Recently high-quality GaAs/Al_xGa_{1-x}As quantum well heterostructures have been prepared by CBE [Tsa86]. Such oval defect-free epilayers may prove to be important for LSI applications.

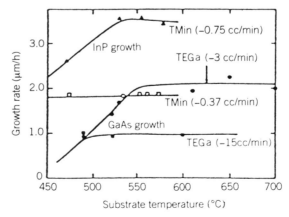

Fig. 3.27 Growth rates of InP from TMIn and GaAs from TEGa as a function of substrate temperature at different absolute flow rates of the group III alkyls. [Tsang, W. T., *Appl. Phys. Lett.* **45**, 11 (1984): 1234]

3.4.5 Photo-CBE

One of the attractive features of CBE is that photochemical reaction can be used. Tokumitsu et al. [Tok89] has used Triisobutylaluminum (TIBA) in the photo-CBE system instead of triethylaluminum (TEA) to enhance the photodecomposition.

The molar extinction coefficient ε is defined as

$$I_t = I_0 \times 10^{-\varepsilon Cl},$$

where I_t indicates the transmitted light intensity, I_0 the incident light intensity, C the molar concentration of reactant gas (mol/liter) and l the light path length (in centimeters). Figure 3.28 [Tok89] shows the measured optical density, $-\log(I_t/I_0)$ as a function of the partial pressure of the metalorganic compounds. An ArF excimer laser that radiates 193-nm ultraviolet light was used as a light source. The molar extinction coefficient for TIBA was found to be three larger than those of TEG and TEA.

The photo-CBE system consists of a growth chamber and an ArF excimer laser. The ultraviolet laser pulse from the ArF excimer was incident onto the substrate through the quartz window. A mask was positioned in the growth chamber in the path of the UV light to make a pattern on the substrate. The distance between the mask and the substrate was ~ 10 cm. TIBA was introduced into the growth chamber through a leak valve and directed toward the substrate. AlGaAs layers were grown at 350°C. The partial pressure of TEG, TIBA, and As_4 were 1×10^{-6}, 1×10^{-5}, and 5×10^{-5} torr, respectively. The large enhancement of the AlGaAs layer thickness was observed by the laser irradiation. The thickness of the AlGaAs layer with the laser irradiation was 0.2 μm, whereas that of the film not exposed to the laser irradiation was 200 to 300 Å.

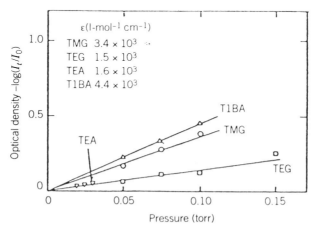

Fig. 3.28 Optical density, $-\log(I_t/I_0)$, of metalorganic compounds as a function of the partial pressure. [Tokumitsu, E., T. Yamada, M. Konagai, and K. Takahashi, *J. Vac. Sci. Tech.* A 7, 3 (1989): 706]

94 EPITAXIAL GROWTH PROCESSES

3.5 ATOMIC LAYER EPITAXY AND MOLECULAR LAYER EPITAXY

To obtain the superlattice, the quantum well, and the various modulation structures, very thin uniform layers must be deposited. In conventional growth methods such as MBE, MOCVD, and VPE there are a lot of parameters to be controlled, such as partial pressures of source gases, beam flux, growth temperature, and growth time. In contrast, atomic layer epitaxy (ALE), which was originally proposed by Suntola [Sun77], has a self-limiting growth mechanism. The growth in ALE has very weak dependence on growth parameters, as mentioned earlier [Nis85; Tis86; Doi86; Wat87].

Several approaches have been applied to achieve the ALE of III-V compounds. In this section we will describe mainly the work developed at NEC in Japan by Usui and his coworkers [Usu86b] involving chloride transport and the photo-assisted single molecular layer epitaxy by Nishizawa [Nis87].

The ALE of III–V compound semiconductors using chloride source gases such as GaCl and InCl [Usu86b] and metalorganic chloride gas such as diethylgalliumchloride (DEGaCl) [Mor87] has been studied. The chloride source gases are quite suitable for the self-limiting growth. The chloride ALE growth proceeds approximately in a monolayer unit under a wide range of growth conditions. In contrast to other epitaxial growth mechanisms, ALE growth is digital epitaxy in which the thicknesses of deposited films are determined by the number of cycles rather than the growth time.

The growth process of chloride atomic layer epitaxy using GaCl can be studied by the temperature-programmed desorption (TPD) method, by the surface photo-absorption (SPA) method, or by *ab initio* molecular orbital calculation [Usu92]. From TPD measurements the GaCl adsorption energy is obtained as 32 Kcal/mol and 38 Kcal/mol for the As-terminated surface and the Ga-terminated surface, respectively. Without H_2, the Ga–Cl bond is very stable in the adsorption state. The process simulation using a simple cluster model of (AsH_2)–GaCl indicates that the adsorption is completed by forming σ and π covalent bonds and that the self-dissociation of the Ga–Cl bond hardly occurs. In this section we will briefly describe the TPD and SPA methods.

3.5.1 Growth Process and Growth Apparatus

The ALE growth process for GaAs, using GaCl and As_4, is shown in Fig. 3.29. The growth process consists of four steps.

Step 1: Expose the substrate to GaCl or DEGaCl source gas. GaCl is supplied with H_2, as shown in Fig. 3.29(a). The hydrogen immediately extracts Cl from adsorbed GaCl on the surface and that Cl is released as HCl. Figure 3.30 shows the growth thickness uniformity over a 3-in. diameter substrate grown by DEGaCl–ALE. When the exposure time of DEGaCl to the substrate in 9 s, the uniformity is excellent compared with the results obtained in conventional MOCVD. The results suggest the monolayer adsorption of GaCl in step 1.

Step 2: Purge the source gas out of the substrate region. Figure 3.29(b) shows the GaCl purge where a monolayer of Ga is deposited on the surface. The purging time dependence of the growth rate at the growth temperature of 450°C was investigated, as shown in Fig. 3.31. The result indicates that the GaCl desorption from the substrate surface is negligibly small at this temperature.

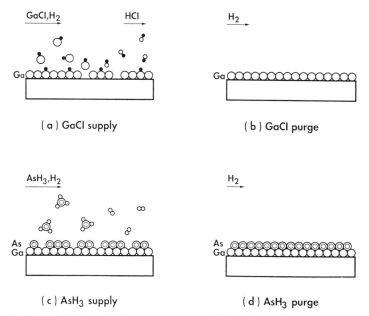

Fig. 3.29 GaAs ALE growth process. [Permission from A. Usui]

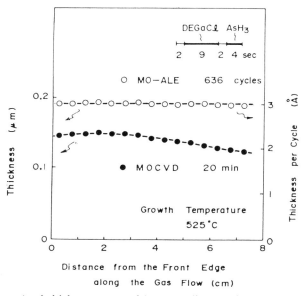

Fig. 3.30 Thickness (and thickness per cycle) versus distance from the front edge along the gas flow. Chloride partial pressure dependence of growth rate for various III–V compounds. [Permission from A. Usui]

96 EPITAXIAL GROWTH PROCESSES

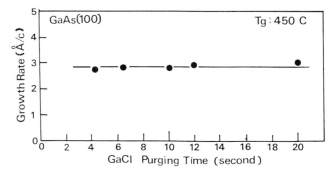

Fig. 3.31 Growth rate dependence of GaCl purging time at growth temperature (T_g) equals to 450°C. [Permission from A. Usui]

Step 3: Supply AsH$_3$ to the substrate surface. Figure 3.31(c) shows the AsH$_3$ supplied with H$_2$.

Step 4: Purge unreacted AsH$_3$ and by-product gases from the substrate region. A monolayer of As is deposited when AsH$_3$ (and any by-product gases) is purged, as shown in Fig. 3.29(d).

The ALE system was developed to use chloride source gases of group-III elements, such as GaCl and InCl, for III–V compound materials. Chloride gases have strong thermal stability, which is a necessary property for obtaining the wide ALE window. Since GaCl exists at high temperatures and cannot be transported by carrier gas from outside the reactor, a special growth apparatus has to be designed. The first chloride ALE of GaAs was carried out in a dual-chamber reactor with a substrate transfer mechanism for the hydride vapor phase epitaxy (VPE) [Usu86]. GaCl was formed by the reaction between Ga metal and HCl gas upstream of the reactor. Mori et al. [Mor88] developed a diethylgalliumchloride (DEGaCl) ALE system using a horizontal MOCVD reactor in which the thermal cracking of DEGaCl at 400° to 500°C generated the GaCl. This was a convenient method because DEGaCl easily responded. However, carbon contamination of the grown layers was not avoided. Recently Usui et al. developed a reactor system that uses TEGa and HCl gases. The gases were mixed in the low-temperature region ($\sim 200°C$) of the reactors and they generated GaCl in the high temperature region ($\sim 700°C$). In this method the thermal cracking of TEGa is considered to be enhanced by the hot wall of the reactor, in which way it differs from the MOCVD reactor. The hot wall reactor is thought to reduce carbon contamination by suppressing the generation of hydrocarbon radicals at the substrate surface region. With the chloride ALE apparatus, growth rate saturation at one monolayer/cycle was obtained in the wide temperature ranges of 400–600°C for GaAs and 320°–425° for InP [Usu91].

3.5.2 Experiment

The temperature-programmed desorption (TPD) method was used to investigate the adsorption/desorption process [Sas91]. Experiments were conducted in a vacuum chamber. A GaCl gas cell was developed to produce a pure GaCl molecular beam from Ga metal and Cl_2 gas. At the cell temperature of 800°C, the beam contained only GaCl molecules. The 2×4 As-stabilized and 4×6 Ga-stabilized surfaces of GaAs (100) substrate were exposed to the GaCl beam of 4 Langmuir at about 100°C. The substrate temperature was raised at a constant rate of 0.8°C/s. Figure 3.32 shows the typical TPD spectra obtained by the desorption. Desorbing species were detected by using QMS. Since the peak temperature (T_p) of the desorption was found to be independent of the surface coverage, the desorption rate can be described by the first-order reaction. Using the equation $r = \exp(-E_{ad}/RT)$, where r is the desorption rate, v the vibration frequency (2.19×10^{13}/s), E_{ad} the adsorption energy, the adsorption energy of GaCl E_{ad} was calculated to be 38 Kcal/mol and 32 Kcal/mol, for the 2×4 As-stabilized and for 4×6 Ga-stabilized surfaces, respectively.

The optical reflection observation by the SPA method was also performed in chloride ALE reactor designed by Usui et al. The growth was carried out by placing DEGaCl and AsH_4 under the H_2 carrier gas in a hot wall reactor. The upstream of the reactor was heated to about 680°C. DEGaCl is considered to completely decompose to GaCl and hydrocarbons [Sas88]. The reactor pressure was 760 torr. The flow rates of DEGaCl, AsH_4, and total H_2 carrier gas were 1, 4, and 8000 cc/min, respectively. A chopped and monochromatized 488 nm p-polarized light was irradiated on the substrate at an angle of 75°C, which is close to the Brewster angle.

The reflected light intensity changed by alternating the supply of GaCl and AsH_3. When the (100) GaAs substrate was used and the beam incidence azimuth was [011], the reflected intensity dropped steeply on exposure to GaCl and saturated immediately. The intensity remained unchanged after the GaCl stopped. When AsH_3 was supplied, the intensity recovered to a level that would indicate an As-stabilized surface. This oscillation continued during the ALE growth and was observed in the wide temperature

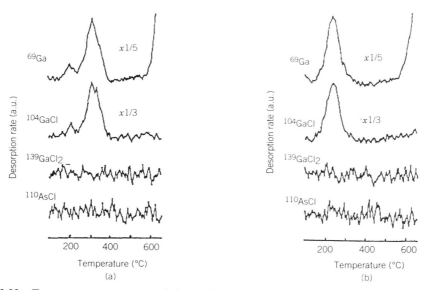

Fig. 3.32 Temperature programmed desorption (TPD) spectra of GaCl adsorbed on GaAs (100) after 4 Langmuir GaCl exposure: (a) 2×4 As-stabilized surface; (b) 4×6 Ga-stabilized surface. [Permission from A. Usui]

range of 400°C to 600°C. Figure 3.33 shows the spectra obtained at 520°C from the GaCl-supplied and the Ga-supplied surfaces. On the vertical axis is the normalized spectra ($\Delta R (= R - R_{As})/R_{As}$) due to the AsH$_3$; the Ga surface was prepared by supplying TEGa. The duration of the TEGa supply was carefully controlled to cover the surface with one monolayer of Ga from the decomposition of TEGa. There is a clear similarity between the two spectra in their peak positions and their magnitudes. This indicates that the surface under the GaCl exposure was identical to a Ga-stabilized surface.

3.5.3 Characteristics of Chloride ALE

1. Self-limiting growth using chloride source gases, self-limiting growth, or digital epitaxy are obtained under a wide range of growth conditions [Usu87]. Figures 3.34 and 3.35 show the chloride partial pressure and temperature dependence of the growth rate.
2. Defect-free As-grown surface InP, or InAs and GaP layers are grown besides GaAs, at 450°C. The ALE-grown layers are nearly defect free.
3. ALE-grown layers have excellent uniformity of growth thickness. The DEGaCl system is particularly suitable for the growth on a large-diameter substrate because there is no need for substrate transfer in the growth process. Mori et al. [Mor87] demonstrated digital epitaxy of GaAs on 3-in. diameter GaAs substrates.
4. The chloride ALE exhibits complete selectivity and digital selective growth. A thin film is sometimes deposited selectively on a substrate that is partly covered with SiO$_2$. Two important advantages of using chloride ALE in selective growth are there is no deposition on the SiO$_2$ mask area, and digital epitaxy occurs in the selective growth as well as in the growth on normal substrate.

3.5.4 Atomic-Plane Doping

Atomic-plane doping into GaAs has been carried out using H$_2$Se as an impurity source gas. H$_2$Se rather than AsH$_3$ gas was supplied to the GaCl-adsorbed surface.

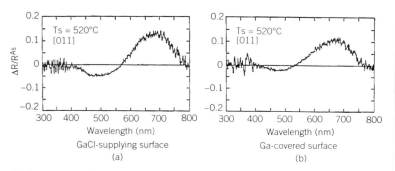

Fig. 3.33 SPA spectra obtained (a) from the GaCl-supplying surface and (b) from the Ga-covered surface at 520°C with [011] incidence azimuth. The Ga surface was prepared by supplying TEGa. [Permission from A. Usui]

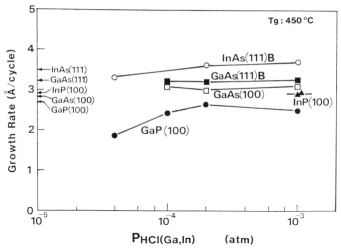

Fig. 3.34 Chloride partial pressure dependence of the growth rate at T_g equal to 450°C. [Permission from A. Usui]

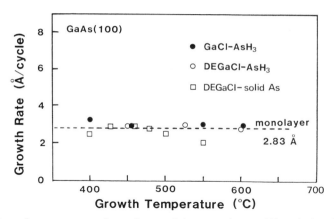

Fig. 3.35 Growth temperature dependence of the growth rate. [Permission from A. Usui]

The growth temperature dependence of the incorporated Se concentration and of the carrier concentration are shown in Fig. 3.36 and 3.37, respectively. It was found that the quantity of selenium that will form an impurity monolayer can be incorporated at lower growth temperatures.

Figure 3.38 shows the relationship between mobility and sheet carrier concentration. Data from MBE-grown delta-doped (Se) samples are shown. This figure suggests that the ALE layers are comparable in quality with those of MBE.

3.5.5 Photo-Assisted Molecular Layer Epitaxy

Photo-assisted molecular layer epitaxy (MLE) is a combination of photoepitaxy and molecular layer epitaxy developed by Nishizawa et al. [Nis86]. This photochemical

100   EPITAXIAL GROWTH PROCESSES

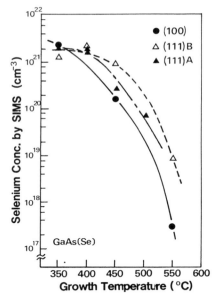

Fig. 3.36 Growth temperature dependence of incorporated Se concentration. Values were obtained from SIMS measurements. [Permission from A. Usui]

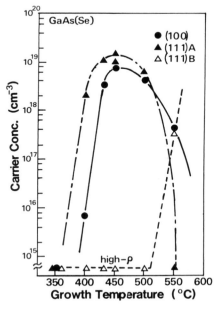

Fig. 3.37 Growth temperature dependence of free-carrier concentration. [Permission from A. Usui]

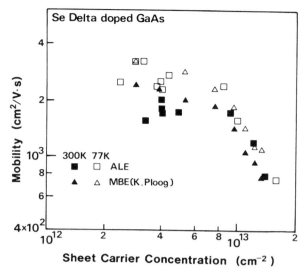

Fig. 3.38 Hall mobility versus sheet carrier concentration in (Se) delta-doped GaAs. [Permission from A. Usui]

technique allows each elementary process to be individually controlled by a monochromatic light of whichever wavelength is suited to the activation energy of the process (i.e., the selective generation of it specified radical or ion). The advantage of using the photochemical technique is that it can localize the reaction region by a sharply focused beam.

Molecular layer epitaxy is a crystal growth method that responds to the chemical reaction of absorbates on the semiconductor surface, where gas molecules containing one element of the compound semiconductor are introduced alternately into the growth chamber. In GaAs MLE, TEG or TMG and arsine are introduced into the growth chamber, resulting in a higher-quality epitaxial layer at the lower crystal growth temperature. An excimer laser, which is operated with ArF (193 nm), KrCl (222 nm), KrF (249 nm), XeCl (308 nm), or XeF (350 nm), is mainly used as a light source for irradiation during crystal growth. A diagram of the experimental setup for photo-MLE is shown in Fig. 3.39. After cleaning, the GaAs substrate is loaded inside the growth chamber, and the chamber is evacuated to $p < 5 \times 10^{-8}$ Pa. The substrate is then heated up over 500°C to remove the contaminants on the surface before crystal growth is started.

AsH_3 is introduced for 20 s and evacuated for 3 s. Then MO gas is introduced for 4 s and evacuated for 3 s. The admittance pressure is set at 6.7×10^{-3} Pa for TMG and AsH_3, and 4×10^{-4} and 6.7×10^{-2} Pa for TEG and TEG and AsH_3, which was the standard gas admittance pressure in the experiment. Light beams from an excimer laser are reflected by the mirrors incident onto the substrate through the window of the lamp housing. The light beams are uniformly directed on a substrate area of 1 cm². Light beam power is monitored at the exit of the laser housing. A repetition rate of 60 Hz was mainly used. In the case of conditioned photoirradiation, the light beam is mechanically chopped; the chopping modes may be synchronized with gas admittance modes.

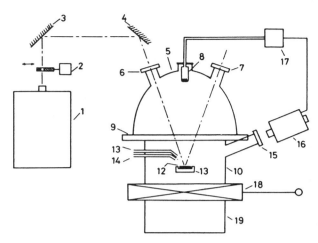

Fig. 3.39 Schematic of experimental setup: (*1*) excimer laser, (*2*) light beam chopper, (*3*), (*4*) mirror, (*5*) lamp housing, (*6*), (*7*) quartz window, (*8*) lamp, (*9*) quartz plate, (*10*) growth chamber, (*11*) quartz sample holder, (*12*) substrate, (*13*), (*14*) gas admittance nozzle, (*15*) sapphire window, (*16*) pyroscope, (*17*) temperature controller, (*18*) gate valve, and (*19*) pumping unit. [Nishizawa, J., H. Abe, T. Kurabayashi, and N. Sakurai, *J. Vac. Sci. Technol.* A **4**, 3 (1986): 707]

The substrate's temperature is measured by a pyroscope. The heating lamp's power is controlled by an automatic device connected to the output signal of the pyroscope. The substrate's temperature can be stabilized within $\pm 1°C$ during the experimental growth. The intensity of the irradiation at the substrate is about 0.3 W if it is 2.1 W at the exit of the laser housing. The substrate's temperature will rise a few degrees at room temperature when measuring by a thermocouple. This temperature rise is controlled by an automatic device during MLE.

Monolayer growth takes place in the TEG–AsH_3 system at a lower temperature ($\sim 300°C$). The surface morphology of epitaxial films is observed to have a little roughness. The substrate's temperature rise due to the Hg lamp irradiation is almost negligible, but the surface morphology of epitaxial films is changed by irradiation. Photo irradiation enhances surface migration during epitaxy.

The model proposed by Nishizawa for the surface desorption species with TMG is shown in Fig. 3.40 [Nis89]. It was estimated that the surface-adsorbed species was Ga in the high-temperature range 500°C, MMG (monomethylgallium) in the range 350°C, and DMG (dimethylgallium) in the range >350°C. With Ga it is easy to form multilayer adsorption. MMG will migrate until it becomes trapped by the crystal surface and combines with As atoms, and then nearly all dangling bonds of the surface become saturated with the adsorbate in a rather broad range of TMG pressure. When AsH_3 was applied to the MMG adsorbate, the desorption of CH_4 and other hydrocarbons was detected. Figure 3.41 gives a diagram of the reaction mechanism of the Ga compound adsorbate with AsH_3 [Nis87]. As can be seen in the figure, the desorption of CH_4 seems to be caused by the surface reaction of MMG with AsH_3.

A simplified model for the reaction of MLE on the (100) surface of GaAs is as follows: First, when TMG is supplied to the GaAs substrate, the reaction is

$$Ga(CH_3)_3(gas) \longrightarrow Ga(CH_3)_2(ad) + CH_3 \tag{3.30}$$

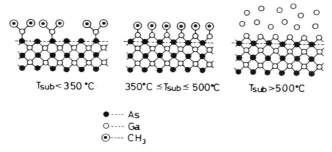

Fig. 3.40 Schematic of the adsorbates formed by TMG supply on the (100) surface of GaAs. [Nishizawa, J., T. Kurabayashi, and Y. Iwasaki, in *Colloids and Surfaces*, vol. 38, pp. 103–112, Elsevier Science (1989); permission by Prof. J. Nishizawa]

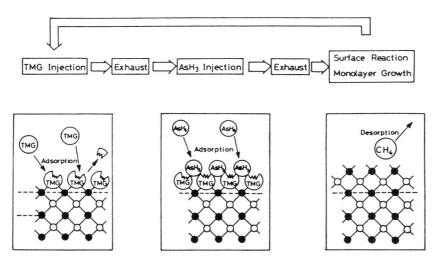

Fig. 3.41 Schematic of the reaction mechanism of Ga compound adsorbate with AsH_3. [Nishizawa, J., and T. Kurabayashi, IEEE Tokyo Section, *Denshi Tokyo* No. 26, pp. 120–124 (1987); permission by Prof. J. Nishizawa]

or

$$Ga(CH_3)_2(ad) \qquad (T_{sub} < 350°C),$$

since there is the possibility that TMG is the adsorbace. Then

$$Ga(CH_3)_3 \text{(gas)} \longrightarrow GaCH_3(ad) + 2CH_3 \qquad (350°C \leqslant T_{sub} \leqslant 500°C), \quad (3.31)$$

$$Ga(CH_3)_3 \text{(gas)} \longrightarrow Ga(ad) + 3CH_3 \qquad (T_{sub} > 500°C). \quad (3.32)$$

Next, when AsH_3 is supplied, the reaction with these adsorbates (ad) is

$$Ga(CH_3)_2(ad)$$

or

$$Ga(CH_3)_3(ad) + AsH_3 \longrightarrow \text{no reaction} \qquad (\text{at least } T_{sub} < 350°C), \qquad (3.33)$$

$$GaCH_3(ad) + AsH_3 \longrightarrow GaAs + CH_4 + H_2 \qquad (350°C \leqslant T_{sub} \leqslant 500°C). \qquad (3.34)$$

Monomolecular layer growth obeys the reaction

$$Ga(ad) + AsH_3 \longrightarrow GaAs + \tfrac{3}{2}H_2 \qquad (T_{sub} > 500°C). \qquad (3.35)$$

$Ga(CH_2)_2(ad)$ and $Ga(CH_3)_2(ad)$ do not react with AsH_3 in the (TMG–AsH_3) system. On the other hand, when TEG is used instead of TMG, the Ga or Ga compound species react with AsH_3, and GaAs epitaxial growth occurs even below 300°C. The reaction of AsH_3 may be limited by the form of the Ga compound adsorbate. The surface reaction of AsH_3 occurs at a considerably lower temperature than the vapor phase reaction of AsH_3.

3.6 VAPOR TRANSPORT EPITAXY

The resolution of several problems common to MBE (vapor flux control, low deposition rates, oval defects, etc.) and MOCVD (complex flow/chemistry dynamics, carbon contamination, high toxicity and costly materials) is the motivation for the development of vapor transport epitaxy (VTE). The operating principle of VTE is to direct vapor transport from uncoupled sources (elemental, gas, or metalorganic), which are controlled by flux-regulating valves, into a common flux distribution manifold and then to the substrate.

3.6.1 Principle of Vapor Transport Epitaxy

The VTE growth region is shown in Fig. 3.42 [Gur92]. Vapor from the sources are delivered through transport lines. Each source's pressure is controlled by its temperature. The flux is regulated by valves, which can also generate a pressure drop resulting in viscous flow in the sources and transport lines and molecular flow downstream of the valves. The pressure drop at the valve minimizes back diffusion and cross contamination. The regulatory valves maintain flux stability while allowing for rapid (seconds) flux changes over a wide range. The fluxes next enter a common flux distribution manifold located beneath the inverted substrate. The transport lines, valves, and common flux distribution manifold are heated to a high enough temperature to evaporate in a flash any atomic clusters and to prevent condensation. The collisionless and cluster-free transport environment eliminates the source of oval defects in the GaAs film's growth.

The VTE system is presented in Fig. 3.43. The reactor is made of stainless steel. The substrate was inserted into the reactor through the load-lock. Diffusion and mechanical pumps provide a base vacuum in the reactor of $\sim 2 \times 10^{-8}$ torr. The load-lock uses turbomolecular and mechanical pumps. The pumped gases are exhausted to the atmosphere through a toxic gas adsorber. An electron cyclotron resonance (ECR) plasma source and residual gas analyzer (RGA) are mounted on the reactor.

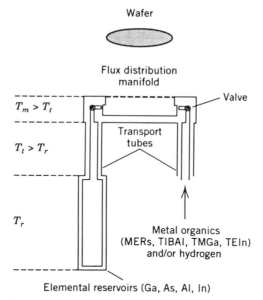

Fig. 3.42 Schematic of vapor transport epitaxy growth region. T_r is the reservoir temperature, T_t is the transport-line temperature, and T_m is the flux distribution manifold temperature. [Gurary, A., G. S. Tompa, C. R. Nelson, and R. A. Stall, *J. Vac. Sci. Technol.* A **10**, 4 (1992): 1453–1457]

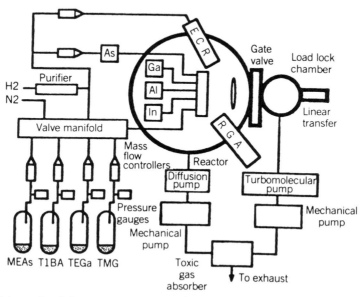

Fig. 3.43 Schematic of the vapor transport epitaxy system. [Gurary, A., G. S. Tompa, C. R. Nelson, and R. A. Stall, *J. Vac. Sci. Technol.* A **10**, 4 (1992): 1453–1457]

3.6.2 Experiment and Results

There are three types of vapor sources used in the VTE system; elemental (As, Ga, Al, and In), metaloraganic monoethylarsine (MEAs), triethylgallium (TEGa), triisobutylaluminum (TIBAl), and trimethylindium)], and gas (H_2). Two different VTE systems are the elemental source and vapor sources VTEs. The elemental fluxes are controlled by a regulating valve in the vapor transport line and by carrier gas flow through the elemental vapor sources. GaAs films are deposited under hydrogen flow that varies from 0 to 6.0 sccm. The temperatures of the Ga source, As source, flux distribution manifold, and substrate will be 1215°, 415°, 1235°, and 550°C, respectively, the growth rate is not strongly affected by H_2 at flows up to 6 sccm, as shown in Fig. 3.44. However, highest flow of H_2 (6 sccm) could degrade the surface morphology.

For the metalorganic sources (MOVTE) system, the source temperatures for the MEAs and the TEGa are $-20°C$ and $25°C$, respectively. The substrate temperature varies from 400° to 600°C. For MOVTE the transport tubes and flux distribution manifold are maintained at a temperature sufficiently low to prevent precracking of the metalorganics.

Figure 3.45 shows growth rates under different wafer temperatures of MOVTE. The flows of MEAs and TEGa are 2.0 and 0.5 sccm, respectively. The maximum growth rate is $\sim 3.5\,\mu m/h$ for the substrate temperatures between 450° and 550°C. In this temperature regions growth appears to be enhanced by the mass-transport-limited process. In lower temperature regions the growth rate is limited by the thermal decomposition of the metalorganics, while in higher temperature regions it is limited by the desorption of the metalorganics on the substrate surface. The best surface morphology is obtained at a substrate temperature of 500°C. Above this temperature the surface becomes hazy. Some of the TEGa may precrack and deposit in the flux box. This will effectively decrease the Ga flux, since Ga cannot re-evaporate from the box.

The relation between the GaAs growth rate and TEGa flow at a fixed MEAs flow of 2.0 sccm is shown in Fig. 3.46. The selected wafer temperature is 500°C. The TEGa flow is varied from 0.25 to 1.0 sccm. The process pressure inside the reactor, which

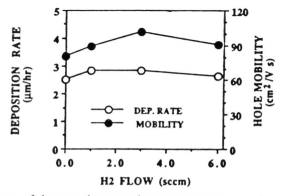

Fig. 3.44 Dependence of the growth rate and room temperature Hall mobility of the GaAs film grown using elemental sources from hydrogen flow. [Gurary, A., G. S. Tompa, C. R. Nelson, and R. A. Stall, *J. Vac. Sci. Technol.* A **10**, 4 (1992): 1453–1457]

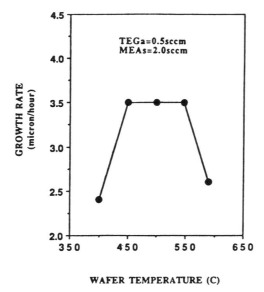

Fig. 3.45 The growth rate versus wafer temperature of MOVTE growth. [Gurary, A., G. S. Tompa, C. R. Nelson, and R. A. Stall, *J. Vac. Sci. Technol.* A **10**, 4 (1992): 1453–1457]

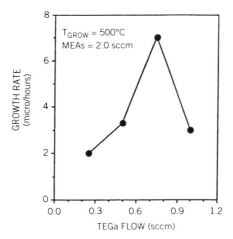

Fig. 3.46 The growth rate versus TEGa flow at $T_g = 500°C$. [Gurary, A., G. S. Tompa, C. R. Nelson, and R. A. Stall, *J. Vac. Sci. Technol.* A **10**, 4 (1992): 1453–1457]

is limited by pumping system, ranges from 0.01 to 0.5 torr for these runs. The highest growth rate is 7.0 µm/h at a TEGa flow of 0.75 sccm.

The VTE techmique has been demonstrated to grow epitaxial films of GaAs from elemental and metalorganic sources. The cracking temperature of the metalorganic is generally lower and sufficient for elemental transportation. Films grown from the elements have been shown to be free of oval defects [Tom92].

The MEAs has been used to replace highly toxic arsine. The successful use of the VTE technique with elemental and metalorganic vapor sources suggests that it may be possible to build a universal deposition system can employ metalorganic or elemental vapor sources.

REFERENCES

[Jai70] Jain, V. K., and S. K. Sharma *Solid State Electron.* **13** (1970): 1145.

[Hei80] Heish, J. J., Liquid phase epitaxy, in *Handbook on Semiconductor*, vol. 3, pp. 415–497, ed. by S. P. Keller, North-Holland (1980).

[Pan83] Panish, M. B., and A. Y. Cho, *IEEE Spectrum* (April 1983): 18.

[Man81] Manasevit, H. M., *J. Cryst. Growth* **55** (1981): 1.

[Che87] Chen, L. P., Ph.D. dissertation, National Cheng Kung University, Taiwan (1987).

[Joh84] Johnson, E., R. Tsui, D. Convey, N. Meller, and J. Curless, *J. Cryst. Growth* **69** (1984): 497.

[Dap84] Dapkus, P. D., *J. Cryst. Growth* **68** (1984): 345.

[Tsa85] Tsang, W. T., *Appl. Phys. Lett.* **46**(11) (1985): 1086.

[Cho75] Cho, A. Y., and J. R. Arthur, *Progress in Solid-State Chemistry*, ed. by G. Somorjai and J. McCaldin, vol. 10, p. 157, Pergamon (1975).

[Cho79] Cho, A. Y., *J. Vac. Sci. Tech.* **16** (1979): 275.

[Plo80] Ploog, K., *Crystal Growth, Properties, and Applications*, ed. by H. C. Freyhardt, vol. 3, p. 73, Springer-Verlag (1980).

[Plo81] Ploog, K., *Am. Rev. Mater. Sci.* **11** (1981): 171.

[Cha75] Chang, L. L., and R. Ludeke, *Epitaxial Growth*, p. 37, ed. by J. W. Mathews, Academic Press (1975).

[Fox80] Foxon, C. T. and B. A. Joyce, *Current Topics in Materials Science*, vol. 7, ed. by E. Kaldis, North-Holland (1980).

[Far77] Farrow, R. F. C., *Crystal Growth and Materials*, vol. 1, p. 237, ed. by E. Kaldis and H. J. Schul, North-Holland (1977).

[Bea81] Beam, J. C., *Growth of Doped Silicon by Molecular Beam Epitaxy*, p. 177, ed. by F. F. Y. Wang, North-Holland (1981).

[Her85] Herrenden-Harker, W. G., and R. H. Williams, *Epitaxial Growth of GaAs: MBE and MOCVD*, p. 57, ed. by H. Thomas, D. V. Morgan, B. Thomas, J. E. Aubrey, and G. B. Morgan, IEE UWIST (1985).

[Cho69] Cho, A. Y., *Surface Sci.* **17** (1969): 494.

[Cho71] Cho, A. Y., *J. Appl. Phys.* **42** (1971): 2074.

[Art74] Arthur, J. R., *Surface Sci.* **43** (1974): 449.

[Cho70] Cho, A. Y., M. B. Panish, and I. Hayashi, *Proc. 3rd Int. Symp. on GaAs*, p. 18, London (1970)

[Cho70b] Cho, A. Y., *J. Appl. Phys.* **41** (1970): 2780.

[Cho77] Cho, A. Y., H. C. Casey, Jr., and P. W. Foy, *Appl. Phys. Lett.* **30** (1977): 397.
[Cho79] Cho, A. Y., *J. Vac. Sci. Tech.* **16** (1979): 275.
[Cho71b] Cho, A. Y., *Appl. Phys. Lett.* **19** (1971): 467.
[Cass75] Casey, H. C., A. Y. Cho, and P. A. Barnes, *IEEE J. Quantum Elect.* **QE-11** (1975): 467.
[Tsa79] Tsang, W. T., *Appl. Phys. Lett.* **34** (1979): 473.
[Tsa79b] Tsang, W. T., C. Weisbuch, and R. C. Miller, *Appl. Phys. Lett.* **35** (1979): 673.
[Tsa79c] Tsang, W. T. and R. A. Logan, *IEEE Quantum Elect.* **QE-15** (1979): 451.
[Col82] Collins, D. M., 1982 MBE Workshop, Urbana, IL, Oct. 21–22 (1982).
[Yam82] Yamakoshi, S., O. Wada, T. Fujii, S. Hiyamizu, and T. Sakurai, IEEE IEDM, San Francisco (1982).
[Mor82] Morkoc, H., T. J. Drummond, and M. Omori, *IEEE Trans. Elect. Dev.* **ED-29** (1982): 222.
[Fen82] Feng, M., V. K. Eu, I. J. D'Haenens, and M. Braunstein, *Appl. Phys. Lett.* **41** (1982): 633.
[O'Co82] O'Connor, P., T. P. Pearsall, K. Y. Cheng, A. Y. Cho, J. C. M. Hwang, and K. Alavi, *IEEE Elect. Dev. Lett.* **EDL-3** (1982): 64.
[Duh86] Duh, K. H. G., P. C. Chao, P. M. Smith, L. F. Lester, B. R. Lee, and J. C. M. Hwang, 44th Annual Device Research Conf., June 23–25, Amherst, MA (1986).
[Hwa82] Hwang, J. C. M., D. G. Flahive, and S. H. Wemple, *IEEE Elect. Dev. Lett.* **EDL-3** (1982): 320.
[Che82] Chen, C. Y., A. Y. Cho, K. Y. Cheng, T. P. Pearsall, and P. O'Connor, *IEEE Elect. Dev. Lett.* **EDL-3** (1982): 152.
[Cho81] Cho, A. Y., and K. Y. Cheng, *Appl. Phys. Lett.* **38** (1981): 360.
[Che81] Cheng, K. Y., A. Y. Cho, and W. R. Wagner, *Appl. Phys. Lett.* **39** (1981): 607.
[Eri85] Erickson, L. P., G. L. Carpenter, D. D. Siebel, P. W. Palmberg, P. Pearah, W. Kopp, and H. Morkoc, *J. Vac. Sci. Tech.* B **3** (1985): 536–537.
[Cha76] Chang, L. L., A. Segmulle, and L. Esaki, *Appl. Phys. Lett.* **28** (1976): 39.
[Nea80] Neave, J. H., P. Blood, and B. A. Joyce, *Appl. Phys. Lett.* **36** (1980): 311.
[Wil90] Williams, R., *Modern GaAs Processing Methods*, Artech House (1990).
[Dav85] Davies, G. J., and D. Williams, in *The Technology and Physics of Molecular Beam Epitaxy*, ed. by E. H. C. Parker Plenum Press (1985).
[Her89] Herman, M. A., and H. Sitter, *Molecular Beam Epitaxy*, Springer-Verlag (1988).
[Art86] Arthur, J. R., *J. Appl. Phys.* **39** (1968): 4032.
[Fox74] Foxon, C. T., J. W. Boudry, and B. A. Joyce, *Surf. Sci.* **44** (1974): 69.
[Joy85] Joyce, B. A., Kinetic and surface aspects of MBE, in *Molecular Beam Epitaxy and Heterostructures*, ed. by L. L. Chang and K. Ploog, Martinus Nijhoff (1985).
[Art69] Arthur, J. R., *Structure and Chemistry of Solid Surface*, ed. by G. A. Somorjai, 46-1, Wiley (1969).
[Fox77] Foxon, C. T., and B. A. Joyce, *Surf. Sci.* **64** (1977): 293.
[Kün82] Künzel, H., J. Knecht, H. Jung, K. Wünstel, and K. Ploog, *Appl. Phys.* **A28**, p. 167 (1982).
[Che81b] Cheng, K. Y., A. Y. Cho, W. R. Wagner, and W. A. Bonner, *J. Appl. Phys.* **52** (1981): 1015.
[Cho71] Cho, A. Y., *J. Appl. Phys.* **42** (1971): 2074.
[Cho70] Cho, A. Y., *J. Appl. Phys.* **41** (1970): 2780.
[Art74] Arthur, J. R., *Surf. Sci.*, **43** (1974): 449.

[Nea78] Neave, J. H., and B. A. Joyce, *J. Cryst. Growth* **44** (1978): 387.
[Cho76] Cho, A. Y., *J. Appl. Phys.* **47** (1976): 2841.
[Plo77] Ploog, K., and A. Fischer, *Appl. Phys.* **13** (1977): 111.
[Fis83] Fischer, R., J. Klem, T. J. Drummond, R. E. Thorn, W. Kopp, H. Morkoc, and A. Y. Cho, *J. Appl. Phys.* **54**, (1983): 2508.
[Cas75] Casey, H. C., A. Y. Cho, and P. A. Barnes, *IEEE J. Quantum Elect.* **QE-11** (1975): 467.
[Tsa80] Tsang, W. T., I. K. Reinhart, and J. A. Ditzenberger, *Appl. Phys. Lett.* **36** (1980): 118.
[Art69] Arthur, J. R., *Structure and Chemistry of Solid Surface*, ed. by G. A. Somorjai, Wiley (1969).
[Fox77] Foxon, C. T., and B. A. Joyce, *Surf. Sci.* **64** (1977): 293.
[Kün82] Künzel, H., J. Knecht, H. Jung, K. Wünstel, and K. Ploog, *Appl. Phys.* A **28** (1982): 167.
[Chai81] Chai, Y. G., and R. Chow, *Appl. Phys. Lett.* **38** (1981): 796.
[Woo81] Wood, C. E. C., L. Rathbaum, H. Ohno, D. Desimone, *J. Cryst. Growth* **51** (1981): 299.
[Bac81] Bachrach, Z., and B. S. Krusor, *J. Vac. Sci. Tech.* **18** (1981): 754.
[Baf82] Bafleur, M., A. Munoz-Yague, and A. Rocher, *J. Cryst. Growth* **59** (1982): 531.
[Mor82] Morkoc, H., R. Stamberg, and E. Krikorinar, *Jap. J. Appl. Phys.* **21**, (1982): L234.
[Suz84] Suzuki, Y., M. Seiki, Y. Horikoshi, and H. Okamoto, *Jap. J. Appl. Phys.* **23** (1984): 1641.
[Wan85] Wang, Y. H., W. C. Liu, S. A. Liao, K. Y. Cheng, and C. Y. Chang, *Jap. J. Appl. Phys.* **24** (1985): 514.
[Baf83] Bafleur, M., and A. Munoz-Yague, *Thin Solid Films* **101** (1983): 299.
[Ito84] Ito, T., M. Shinohara, and Y. Imamura, *Jap. J. Appl. Phys.* **23** (1984): L524.
[Din85] Dingle, R., M. D. Feuer, and C. W. Tu, in *VLSI Electronics*, vol. 11, ed. by N. G. Einspruch and W. Wisseman, Academic Press (1985).
[Pet84] Petit, G. D., J. A. Woodall, S. L. Wright, P. D. Kirchner and J. L. Freeout, *J. Vac. Sci. Tech.* B **2** (1984): 241.
[Wan86] Wang, Y. H., W. C. Liu, C. Y. Chang, and S. A. Laio, *J. Vac. Sci. Tech.* B **4**, 1 (1986): 30.
[Kir81] Kirchner, P. D., J. M. Woodall, J. F. Freeouf, and G. D. Pettit, *Appl. Phys. Lett.* **38** (1981): 427.
[Met83] Metze, G. M., A. R. Calawa, and J. G. Mavroides, *J. Vac. Sci. Tech.* B **1** (1983): 166.
[Abr84] Abrokwah, J. K. N. C. Cirillo, Jr., M. J. Helix, and M. Longerbone, *J. Vac. Sci. Tech.* B **2** (1984): 252.
[Str85] Stringfellow, G. B., in *Semiconductor and Semimetals*, vol. 22, p. 209, ed. by W. Tsang Academic Press (1985).
[Str89] Stringfellow, G. B., *Organometallic Vapor-Phase Epitaxy*, Academic Press (1989).
[Bha82] Bhat, R., P. O'Connor, H. Temkin, R. Dingle, and V. G. Keramidas, *Inst. Phys. Conf. Ser.* **63** (1982): 101.
[Kuo83] Kuo, C. P., R. M. Cohen, and G. B. Stringfellow, *J. Cryst. Growth* **64** (1983): 461.
[Kue85] Kuech, T. F., and R. Potemski, *Appl. Phys. Lett.* **477** (1985): 821.
[Cha81] Chang, C. Y., Y. K. Su, M. K. Lee, L. G. Chen, and M. P. Houng, *J. Cryst. Growth* **55** (1981): 24.
[Mos81] Moss, R. H., and J. S. Evans, *J. Cryst. Growth* **55** (1981): 129.

[Che84] Cherng, M. J., R. M. Cohen, and G. B. Stringfellow, *J. Electron. Mater.* **13** (1984): 799.

[Coo80] Cooper, C. B., M. J. Ludowise, V. Aebi, and R. L. Moon, *J. Electron. Mat.* **9** (1980): 299.

[Clo69] Clough, R. B., and J. J. Tietjen, *Trans. Metall. Soc. AIME* **245** (1969): 583.

[Str88] Stringfellow, G. B., *J. Electron. Mat.* **17**, 4 (1988): 327–335.

[CRC67] *CRC Handbook of Laboratory Safety*, ed. by N. V. Steere, Chemical Rubber Co., Cleveland, OH (1967).

[Sax79] Sax, N. I., in *Dangerous Properties of Industrial Materials*, Van Nostrand Reinhold (1979).

[Den86] DenBaars, S. P., B. Y. Maa, P. D. Dapkus, A. D. Danner, and H. C. Lee, *J. Cryst. Growth* **77** (1986): 188.

[Lar86] Larsen, C. A., and G. B. Stringfellow, *J. Cryst. Growth* **75** (1986): 247.

[Str85] Stringfellow, G. B., in *Semiconductors and Semimetals*, vol. 22, p. 209, ed. by W. Tsang, Academic Press (1985).

[Bha85] Bhat, R., *J. Electron. Mater.* **14** (1985): 433.

[Tok84] Tokumitsu, E., Y. Kudou, M. Konagai, and K. Takahashi, *J. Appl. Phys.* **55** (1984): 3163.

[Ish86] Ishikawa, H., K. Kondo, S. Sasa, H. Tanaka, and S. Kiyamizu, *J. Cryst. Growth* **76**, (1986): 521.

[Kuc88] Kuck, M., H. S. Marek, and W. Wiener, Paper E9.26 presented at Materials Research Society Meeting, Boston (1988).

[Yos85] Yoshida, M. H., Watanabe, and F. Uesugi, *J. Electrochem. Soc.* **132** (1985): 677.

[Nai87] Naitoh, M., and M. Umeno, *Jap. J. Appl. Phys.* **26** (1987): L1538.

[Ben81] Benz, K. W., H. Renz, J. Weidlein, and M. H. Pilkuhn, *J. Electron. Mater.* **10** (1981): 185.

[Che84] Cherng, M. J., R. M. Cohen, and G. B. Stringfellow, *J. Electron. Mater.* **13** (1984): 799.

[Lum88] Lum, R. M., J. K. Klinger, D. W. Kisker, D. M. Tennant, M. D. Morris, D. L. Malm, J. Kovalchick, and L. A. Heimbrook, *J. Electron. Mater.* **17** (1988): 101.

[Kue84] Kuech, T. F., and E. Veuhoff, *J. Cryst. Growth* **68** (1984): 148.

[Org86] *Organometallics for Vapor Phase Epitaxy, Literature and Product Review*, Morton Thiokol Inc., Alfa Products, Danvers, MA (1986).

[Spe87] Speckman, D. M., and J. P. Wendt, *Appl. Phys. Lett.* **50** (1987): 676.

[Hat87] Hata, M., Y. Zempo, N. Fukuhara, K. Sawara, and T. Maeda, Paper E-4, Electronic Materials Conference, Santa Barbara, CA (1987).

[Bha87] Bhat, R., M. A. Koza, and B. J. Skromme, *Appl. Phys. Lett.* **50** (1987): 1194.

[Che87] Chen, C. H., C. A. Larsen, and G. B. Stringfellow, *Appl. Phys. Lett.* **50** (1987): 218.

[Den86] DenBaars, S. P., B. Y. Maa, P. D. Dapkus, A. D. Danner, and H. C. Lee, *J. Cryst. Growth* **77** (1986): 188.

[Lar86] Larsen, C. A., N. I. Buchan, D. S. Li, and G. B. Stringfellow, *J. Cryst. Growth* **75** (1986): 247.

[Miz84] Mizuta, M., S. Kawata, T. Iwamoto, H. Kukimoto, *Jap. J. Appl. Phys.* **23** (1984): L283.

[Hsu86] Hsu, C. C., J. S. Yuan, R. M. Cohen, and G. B. Stringfellow, *J. Cryst. Growth* **74** (1986): 535.

[Lum87] Lum, R. J., J. K. Klingert, and M. G. Lamont, *Appl. Phys. Lett.* **50** (1987): 284.

[Bas85] Bass, S. J., *J. Cryst. Growth* **31** (1975): 172.
[Man68] Manasevit, H. M., *Appl. Phys. Lett.* **12** (1968): 156.
[Ree83] Reep, D. H., and S. K. Ghandhi, *J. Electrochem. Soc.* **130** (1983): 675.
[Duc78] Duchemin, J. P., M. Bonnet, F. Koelsch, and Huggi, D. *J. Cryst. Growth* **45** (1978): 181.
[Gha83] Ghandhi, S. K., *VLSI Fabrication Principles*, Wiley (1983).
[Sze81] Sze, S. M., *Physics of Semiconductor Devices*, Wiley-Interscience (1981).
[Str84] Stringfellow, G. B., *J. Cryst. Growth* **68** (1984): 111.
[Kop84] Koppitz, M., O. Vestavik, W. Plestschen, A. Mircea, M. Heyen, and W. Richter, *J. Cryst. Growth* **68** (1984): 136.
[Kus85] Kusumoto, Y., T. Hayashi, and S. Komiya, *Jap. J. Appl. Phys.* **24** (1985): 620.
[Col85] Coleman, J. J., and P. D. Dapkus, in *Gallium Arsenide Technology*, ed. by D. K. Ferry, Howard W. Sams & Co. (1985).
[Dap81] Dapkus, P. D., H. M. Manesevit, K. L. Hess, T. S. Low, and Stillman, G. E. *J. Crystal Growth* **55** (1981): 10.
[Ois82] Oishi, M., and K. Kuroiwa, *Jap. J. Appl. Phys.* **21** (1982): 203.
[Mos81] Moss, R. H., and J. S. Evans, *J. Cryst. Growth* **55** (1981): 129.
[Fro77] Frolor, I. A., J. Beldyrevskii, B. L. Druz, and E. B. Sokolov, *Inorg. Mater.* **13** (1977): 632.
[Duc81] Duchemin, J. P., J. P. Hirtz, M. Razeghi, M. Bonnet, and S. D. Hersee, *J. Cryst. Growth* **55** (1981): 64.
[Yos81] Yoshino, J., T. Iwamoto, and H. Kukimoto, *J. Cryst. Growth* **55** (1981): 74.
[Yos81b] Yoshino, J., T. Iwamoto, and H. Kukimoto, *Jap. J. Appl. Phys.* **20** (1981): L290.
[Duc78b] Duchemin, J. P., M. Bonnet, G. Beuchet, and F. Koelsch, *Gallium Arsenide and Related Compounds*, Institute of Physics Conference, series no. 45, pp. 10–18 (1978).
[Ley81] Leys, M. R., and H. Veenvliet, *J. Cryst. Growth* **55** (1981): 145.
[Ben81] Benz, K. W., H. Renz, J. Widlein, and M. H. Pilkuhn, *J. Electron. Mater.* **10** (1981): 185.
[Coo80] Cooper, C. B., M. J. Ludowise, V. Aebi, and R. L. Moon, *J. Electron. Mater.* **9** (1980): 299.
[Hun92] Hung, D., in *Chemical Vapor Deposition*, ch. 6, p. 245, ed. by M. L. Hitchman and J. E. Jensen, Academic Press, (1992).
[Bra91] Brauers, A., *J. Cryst. Growth*, **107** (1991): 281.
[Dap81] Dapkus, P. D., H. M. Manasevit, K. L. Hess, T. S. Low, and G. E. Stillman, *J. Crystal Growth* **55** (1981): 10.
[Nak77] Nakanishi, T., T. Udagawa, A. Tanaka, K. Kamei, *J. Crystal Growth* **38** (1977): 23.
[Rai69] Rai-Chudhury, P., *J. Electrochem. Soc.* **116** (1969): 1745.
[Sek75] Seki, Y., K. Tanno, K. Iida, and E. Ichiki, *J. Electrochem. Soc.* **122** (1975): 1108.
[Kue85] Kuech, T. F., and R. Potemski, *Appl. Phys. Lett.* **477** (1985): 821.
[Kob85] Kobayashi, N., and T. Makimoto, *Jap. J. Appl. Phys.* **24** (1985): L824.
[Yos85] Yoshida, M., H. Watanabe, and F. Uesugi, *J. Electrochem. Soc.* **132** (1985): 677.
[Püt86] Pütz, N., H. Heinecke, M. Heyen, P. Balk, M. Weyers, and H. Lüth, *J. Cryst. Growth* **74** (1986): 292.
[Kue87] Kuech, T. F., *Materials Science Reports* 2, p. 3 (1987).
[För89] Förster, A., and H. Lüth, *J. Vac. Sci. Tech. B*, **7**(4) (1988): 720–724.

[Wil82] Wilkinson, G., G. A. Stone, and E. W. Abel, ed., *Comprehensive Organomatallic Chemistry*, Pergamon Press (1982).

[Str79] Stringfellow, G. B., and G. Hom, *Appl. Phys. Lett.* **34** (1979): 794.

['tHo81] 'tHooft, G. W., C. Van Opdorp, H. Veenvliet, and A. T. Vink, *J. Cryst. Growth* **55** (1981): 173.

[Kis82] Kisker, D., J. N. Miller, and G. B. Stringfellow, *Appl. Phys. Lett.* **40** (1982): 614.

[Wag81] Wagner, E. E., G. Hom, and G. B. stringfellow, *J. Electron. Mater.* **10** (1981): 239.

[Dap81] Dapkus, P. D., H. M. Manasevit, K. Hess, T. S. Low, and G. E. Stillman, *J. Cryst. Growth* **55** (1981): 10.

[Nis61] Nishizawa, J., *J. Metals* **25** (1961): 149.

[Mor81] Mori, Y., and J. Watanabe, *Appl. Phys.* **52** (1981): 2792.

[Tsa86] Tsang, W. T., and R. C. Miller, *Appl. Phys. Lett.* **48**, 9 (1986): 1288.

[Tsa89] Tsang, W. T., Chemical beam epitaxy, in *Beam Processing Technologies*, VLSI Electronics, vol. 21, ed. by N. G. Einspruch, S. S. Cohen, and R. N. Singh, Academic Press (1989).

[Tsa85] Tsang, W. T., *Appl. Phys. Lett.* **46**, 11 (1985): 1086.

[Tsa86b] Tsang, W. T., *J. Electron Mater.* vol. 15, p. 235 (1986b).

[Woo81] Wood, C. E. C., L. Rathbun, H. Ohmo and D. De Simone, *J. Cryst. Growth* **51** (1981): 299.

[Cha81] Chai, Y. G., and R. Chow, *Appl. Phys. Lett.* **38** (1981): 796.

[Baf82] Bafleur, M., A. Munoz-Yague, and A. Rocher, *J. Cryst. Growth* **59** (1982): 531.

[Cal83] Calawa, A. R., G. M. Metze, and M. J. Manfa, Electron. Mater. Conf., Burlington, Vermont (1983).

[Kir81] Kirchner, P. D., J. M. Woodall, J. F. Freeouf, and G. D. Pettit, *Appl. Phys. Lett.* **38** (1981): 427.

[Pet84] Pettit, G. D., J. M. Woodall, S. L. Wright, P. D. Kirchner, and J. L. Feeeouf, *J. Vac. Sci. Tech.* **B2** (1984): 241.

[Tsa84] Tsang, W. T., *Appl. Phys. Lett.* **45**, 11 (1984): 1234.

[Tok89] Tokumitsu, E., T. Yamada, M. Konagai, and K. Takahashi, *J. Vac. Sci. Tech.* A **7**, 3 (1989): 706.

[Sun77] Suntola, T., and M. J. Anton, U.S. Patent no. 4-058-430 (1977).

[Nis85] Nishizawa, J., H. Abe, and T. Kurabayashi, *J. Electrochem. Soc.* **132** (1985): 1197.

[Tis86] Tischler, M. A., and S. M. Bedair, *J. Cryst. Growth* **77** (1986): 89.

[Doi86] Doi, A., Y. Aoyagi, S. Namba, *Appl. Phys. Lett.* **49** (1986): 785.

[Wat87] Watanabe, H., and A. Usui, *Proc. 13th Int. Symp. on GaAs and Related Compounds 1986*, pp. 1–8, Bristol, Institute of Physics (1987).

[Usu86] Usui, A., and H. Snakawa, *Jap. J. Appl. Phys.* **25** (1986): L212.

[Usu86b] Usui, A., and H. Sunakawa, *Proc. 13th Int. Symp. on GaAs and Related Compounds 1986*, pp. 129–134, Bristol, Institute of Physics (1987).

[Mor87] Mori, K., M. Yoshida, A. Usui, and H. Terao, *Abstracts of 1987 Spring Meeting of Jap. Soc. Appl. Phys.*, p. 125 (1987).

[Usu92] Usui, A., Presnted at 2d ALE Symp., Releigh, NC (1982).

[Mor88] Mori, K., M. Yoshida, A. Usui, and H. Terao, *Appl. Phys. Lett.* **52** (1988): 27.

[Usu91] Usui, A., and H. Watanabe, *An. Rev. Sci.* **21**, (1991): 185.

[Sas91] Sasaoka, C., Y. Kato, and A. Usui, *Jap. J. Appl. Phys.* **30** (1991): L1756.

[Sas88] Sasaoka, C., M. Yoshida, and A. Usui, *Jap. J. Appl. Phys.* **27** (1988): L490.

[Nis86] Nishizawa, J., H. Abe, T. Kurabayashi, and N. Sakurai, *J. Vac. Sci. Tech.* A **4**, 3 (1986). pp. 706–710.

[Nis89] Nishizawa, J., T. Kurabayashi, and Y. Iwasaki, *Colloids and Surfaces*, **38** (1989): 103–112.

[Nis87] Nishizawa, J., and T. Kurabayashi, IEEE Tokyo Section, Denshi Tokyo no. 26, pp. 120–124 (1987).

[Gur92] Gurary, A., G. S. Tompa, C. R. Nelson, and R. A. Stall, *J. Vac. Sci. Tech.* A **10**, 4 (1992): 1453–1457.

[Tom92] Tompa, G. S., A. Gurary, C. R. Nelson, R. A. Stall, S. Liang, and Y. Lu, *J. Vac. Sci. Tech.* B **10** (1992): 975.

4

PROCESS TECHNIQUES

4.1 INTRODUCTION TO PROCESS TECHNIQUES

The demands of modern commercial and military industries have stimulated the research and development in process techniques. In this chapter we will describe the cleaning of GaAs wafers for IC applications, wet and dry etching methods, plasma-enhanced chemical vapor deposition (PECVD), and rapid thermal processing for GaAs ICs.

Wet chemical etching has been the technique most widely employed in device fabrication. Wet etching processes include pattern formation, polishing, and defect or damaging visualization. However, the need to interconnect a number of separate elements to form a circuit and to shrink this down to the dimensions of a device has led to dry processing techniques.

Reactive plasma-etching techniques have rapidly been developed to manufacture silicon-based integrated circuits, gallium arsenide MESFETs, and heterostructure high-speed circuits. Dry etching offers substantial directionality and etch anisotropy. In this chapter we will discuss the reactive ion etching techniques and laser-assisted dry etching techniques.

Plasma-assisted techniques are especially important to GaAs and other III–V compound processing because they can be accomplished at relatively low temperatures. Deposition of thin film materials is an excellent example of this low-temperature advantage. In the case of GaAs processing, PECVD can grow thin films at a lower substrate temperature (300°C) than CVD. An electron-cyclotron-resonance (ECR) plasma CVD has been developed, and it offers a method of silicon nitride film formation.

As device dimensions are reduced, junction depths shrink, so dopant diffusion must be more precisely controlled. In a conventional furnace it is difficult to reduce the time accurately because of thermal stresses and other factors. Rapid thermal annealing covers a range of time durations from nanoseconds to about 100 seconds.

116   PROCESS TECHNIQUES

Annealing with times above one second give nearly isothermal wafer heating. Finally, in this chapter we will discuss the thermal treatment on III–V compound materials.

4.2 CLEANING AND CLEANLINESS

Cleaning and cleanliness are important matters to the semiconductor industry. By cleaning, we mean removing undesired material from the wafer before the subsequent process steps. Cleanliness refers to preventing contamination, to maintaining the level of cleanliness already present [Wil84]. Environmental contaminants and handling are other considerations to address in discussing cleanliness.

Cleanliness is crucial to achieving high yields and reproducible processes in the production of semiconductor devices. Undesired particles, chemical and environmental, can degrade or destroy virtually any aspect of the fabrication process: metal adhesion, resist application and patterning, wet etching, dry etching, and plating, for example [Wil84].

Most chemicals and gases can be reliably obtained in a highly pure form. Many chemicals are intended precisely for semiconductor use and are designated *semiconductor grade*. However, minor traces of contaminants on breakers, stirring rods, thermometers, tweezers, wafer holders, or other items can degrade the purity of the chemicals.

Production of semiconductor devices places a high premium on uniformity and yield. Uniformity in performance follows directly from high-quality wafer, uniformity in material, and process. Yield is tied to cost because affordable products demand good yield. Elimination of particles becomes more important as the number of gates increases. Particles on the wafer can cause pattern defects, resulting in electrical failures such as shorts in capacitors or opens in gate fingers. Particles that cause openings in dielectric films can in turn cause reliability problems. Particles that become included in resist films interfere with proper thickness uniformity and exposure. Particles can also the damage masks used in contact photography. In general, resist operations must be free from particle contamination.

4.2.1 Environment and Handling

Cleanliness is maintained by attention to environmental conditions and wafer-handling techniques. Environmental factors include control of the number and size of particles present in the fabrication area, the temperature, and the humidity. The fabrication standards require that most operations occur in environmentally controlled areas called "clean rooms." Air in the clean room is monitored and classified with respect to particulates. A "class 100" environment has a maximum of 100 particles per cubic foot ($\sim$ 3500 particles per cubic foot with particle size larger than 5.0 µm [Dou81]. In critical lithography areas particle densities are typically maintained below 10 particles per cubic foot. The filtered air flows from ceiling to floor at more than 85 linear feet per minute ($\sim$ 26 linear meters per minute), which is approximately the threshold for laminar flow (i.e., for uniform velocity of air following parellel flow lines without turbulence). Particulates can emanate from process equipment as well as from human beings. Protective clothing for personnel may include smocks, hoods, face masks, gloves, safety glasses, and boots. Pads of sticky sheets may be placed near entrance areas to remove loose dirt from shoe bottoms. Gloves not only restrict particles, skin oils, and salts from being spread from hands but also protect hands from dangerous chemicals. Safety glasses protect eyes from chemical splashes, flying pieces due to breakages or explosions, and other mishaps. Plastic shields are sometimes placed over microscopes to prevent the breath of the viewer from contaminating the wafer [Wil90]. An even cleaner environment can be achieved when operations are performed in a laminar flow hood.

Paper is another source of particles. Lint-free paper is available for use in the clean room. Use of lead pencils and/or erasers within clean rooms can generate large amounts of particles and should be prohibited.

Temperature and humidity need to be controlled principally for resist work. When *relative humidity* (RH) is too low (below 20% RH), static electricity becomes a problem. Extreme dryness can undermine the photoresist. When the humidity is too high (over 50% RH), the resist may lose adhesion or exhibit cracking. Clearly poor control of humidity in resist areas can cause big problems. Relative humidity should be controlled to within 5% RH and should be continually monitored.

4.2.2 Cleaning Techniques

Cleaning operations are performed before all major processing steps. Since organic solvents, vapor degreasing, acids, bases, and/or plasma etchers may be used in any of these step, maintaining chemical purity is obviously important. Many solvents, acids, and bases can be obtained in semiconductor grade (SC grade).

Organic solvents are effective in removing oils, greases, waxes, and organic material such as photoresist. Organic solvents are also innocuous to almost all materials intended to be permanently present on GaAs devices, including GaAs, metals, and dielectrics. Some properties of solvents used in GaAs processing are listed in Table 4.1 [Wil84]. Cleaning with solvents is usually done at elevated temperatures, raised even to the boiling point of the solvent. A typical sequence of treatments is 1,1,1-trichloroethane, acetone, methanol, and iso-propyl alcohol. Cleaning may also take place in a vapor degreaser, which uses the solvent's vapors to clean the wafer. As the solvent is distilled, only its vapors condense on the wafer. The more common practice is to immerse the wafer in the heated solvent. Agitation during these steps removes any inorganic particles.

Waxes are sometimes used in GaAs processing to mount wafers on stronger substrates (for lapping, handling after lapping, etc.). They are somewhat soluble in solvents. For example, polyglycol is soluble in acetone; paraffin is soluble in tetrachloroethylene.

4.2.3 Acids, Bases, and Pure Water Systems

Acids are used to remove metal contaminants, to remove oxides, or as part of wet etchants to remove GaAs material and provide a fresh surface. GaAs etchants generally contain an acid whose function is to remove a small amount of surface material and expose "clean" surface. Such etchants usually also contain oxidizing agents that leave an oxide layer on the etched surface. Ordinary exposure to air will leave an oxide on GaAs surfaces that is generally between 10 and 50 Å thick. An acid can be used

TABLE 4.1 Solvents and Properties

Solvents	Boiling Point (°C)	Flash Point (°C)[a]	Water Sol.	Density (g/ml)[b]	Safety[c]
Acetone	56.2	−16	100%	.784	F
Benzene	80.1	−11	<1%	.874	F[C,N]
n-Butyl acetate	126.	22	<1%	.876	F
Carbon tetrachloride (tetrachloroethylene)	76.8	None	<1%	1.58	[C,N,T]
Chlorobenzene	132.	29	<1%	1.10	F
ortho-Dichlorobenzene	180.	74	<1%	1.3	
Ethanol	78.5	60	100%	.789	
Ethylene dichloride	83.5	13	<1%	1.25	F[C]
Isobutyl alcohol	108.	35	8.5%	.798	F[C]
Methanol	64.7	12	100%	.787	F
Methyl ethyl ketone	79.6	−1	24%	.800	F[T]
Methylene chloride	39.8	None	1.6%	1.32	
Propanol-1	97.2	25	100%	.800	F[C]
Propanol-2 (isopropyl alcohol)	82.3	22	100%	.781	F
Tetrachloroethylene (perchloroethane)	121.	None	<1%	1.62	
Trichloroethylene	87.2	None	<1%	1.45	

[a] Flash point temperatures depend on the conditions of measurement; different references may give slightly different values.
[b] At 25°C.
[c] F = Flammable; [C] = possible carcinogen; [N] = possible neoplastic agent (cause nonmalignant tumor); [T] = possible teratogen (cause physical defects in developing embryo).
Source: Williams, R. E., Gallium Arsenide Processing Techniques, Artech House (1990): p. 163.

to remove this oxide but not etch the GaAs. Acids used for oxide removal or cleaning are usually used in highly diluted concentrations. Diluted HCl is a common choice.

Base or alkaline solutions can be used as cleaners in GaAs processing. Because oxides formed on GaAs tend to be amphoteric, bases (e.g., ammonium hydroxide NH_4OH) as well as acids can be used to dissolve the oxide. These are usually diluted before use.

Water is essential for IC fabrication, but ultra-pure, electronic-grade water must be used. Tap water contains particulates and contaminants such as sodium, copper, and iron that, when deposited on the wafers, lead to device degradation. Water with a specific resistivity of 18 MΩ-cm is considered to have a low ionic content.

A pure water system consists of several sections [Hel81]. In-flowing water is passed through charcoal filters and into electrodialysis units that filter and demineralize with water. The water then passes through resin tanks, followed by mixed-bed ion-exchange resin units—to remove more minerals—and nuclear-grade resin canisters to minimize sodium ion content. Undissolved solids, bacteria, and other organic matter are removed by a series of filters ranging in pore size from 10 μm down to 0.01 μm. Bacterial content is minimized by continually circulating the water through an ultraviolet sterilization unit. The research on improving pure water systems is still being carried on.

4.3 ETCHING

4.3.1 Wet Etching

Wet etching has three main purposes: to form patterns to polish, and to enable visualization of defects or damages [Ash86]. Some specific characteristics of wet etches are

1. preferential etching based on crystallographic orientation, with the (111) Ga face generally etching a factor of 2 to 5 slower than the (100), (110), or (111) As faces;
2. greater tendency for etch pits to form on (111) faces;
3. etching reactions generally based on oxidation of Ga.

There are two categories of wet etches: nonelectrolytic [Ker78; Gan74; Iid71; Mer76] and electrolytic [Ker78; Nue70; Har67; Sch76; Ash86]. Nonelectrolytic etching rates can be either diffusion limited or chemical-reaction limited. For diffusion-limited etches the rate is controlled by the mass transport of reactants to the surface or of products from the surface. Those etches tend to be isotropic and relatively insensitives to temperature, though highly sensitive to changes in the nature and degree of agitation. Diffusion-limited etches are especially good for polishing wafers. For chemical-reaction-limited etches the rate is controlled by the chemical reactions at the GaAs surface. These etches tend to be anisotropic with respect to certain crystallographic orientations, quite sensitive to temperature, and relatively insensitive to agitation. They are well suited for etching geometric shapes along crystallographic planes. Whether a given etching rate is diffusion or chemical-reaction controlled is often determined by the relative proportions of the constituents of the etching solution.

Electrolytic etches can provide accurate depth control but require ohmic contact to the wafer. They are based on anodic oxidation of GaAs followed by dissolution of products. Etching of n-type material is much slower than p-type unless it is illuminated to photo-generate holes in the surface region or unless sufficient voltage is applied to break down the surface barrier. First we will discuss the difference between preferential etching and selective etching.

In our discussion of specific etching processes we will use the word "preferential" to describe the etching of certain crystallographic planes that occurs faster than that of some other planes. Etching is then anisotropic with respect to crystal directions. However, we will use the word "selective" for etches that remove one material significantly faster than another. The differences in etch rates could be used by either different etch mechanisms or differences in contact potentials with respect to the etch solution. The layer of material in contact with GaAs can be a metal, a dielectric, a photoresist, or an electron beam resist.

Most etchants for GaAs contain an oxidizing agent, a complexing agent, and a dilutant, such as water. The oxidizers usually are Br_2 [Tar73], H_2O [Ots76; Yen83; Sha81], $AgNO_3/CrO_3$ [Abr65], HNO_3 [Tuc75], and $NaOCl$ [Sti76]. The oxidized layer is usually insoluble in water. It is made soluble by complex agent such as NH_4OH [Gan74; Gre77; Ken82], $NaOH$ [Shi77], H_2SO_4 [Koh80; Iid71; Sha81; Shi77], HCl [Sha81], HF [Abr65; Nis79], H_3PO_4 [Mor78], and critic acid [Ots76]. In the bromine-methanol system the dilutant (methanol) dissolves the reaction products for GaAs–Br_2 reactions (mainly bromides) [Tar73].

In the next section we will discuss the diffusion-limited and reaction-rate-limited etching. Diffusion-limited reactions often occur in very viscous etch solutions containing high concentrations of the complexing agent and low concentrations of the oxidizing agent, such as H_2O_2. However, an etch solution need not be highly viscous in order to have diffusion-controlled properties.

Diffusion-Limited and Reaction-Rate-Limited Etching

Agitation Dependence In reaction-rate-limited etching the diffusion boundary layer between the etch solution and the GaAs surface oxide layer is negligible [Muk85]. Agitation does not change the surface adsorption rate significantly and does not affect the etch rate. In diffusion-limited etching, on the other hand, agitation-induced turbulence does decrease the diffusion boundary layer thickness, causing the diffusion flux to rise. As a result both the surface adsorption rate and etch rate increase.

Preferential Etching In the GaAs system a free {111} Ga surface has Ga atoms attached firmly to three As atoms underneath it so that the valency of 3 is completely satisfied. The {111} As plane contains As atoms that have two extra unbounded electrons per atom owing to its valency of 5 [Muk85]. Since oxidation involves loss of electrons, the As atoms present on a {111} As surface react much more readily with the oxidizer than do Ga atoms present on a {111} Ga surface. Once an As atom from the {111} As surface is removed by oxidation, the Ga atoms in the plane underneath are dislodged easily by the oxidation process. Consequently the {111} As etch rate is found to be the highest in GaAs for reaction-rate-limited etching processes. Figure 4.1 shows etch rate as a function of GaAs crystal orientation. For the H_2SO_4–H_2O_2–H_2O etchant a reaction-rate-limited solution (1:8:1) (curve B) is compared with a diffusion-limited solution (8:1:1) (curve A). The diffusion-limited solution etches {111} Ga planes at a slower rate than all the others but does not etch {111} As any faster than any other direction. For diffusion-limited processes the etch rate dependence on orientation almost disappears [Mor78; Iid71].

Temperature Dependence In wet etching the etch rates always increase with etchant temperature because the rates of all the participating physical and chemical phenomena increase as $\exp(-E/kT)$ with temperature T, E being the relevant activation energy. Most diffusion phenomena in common water-based etchants have activation energies ranging from 5 to 8.5 kcal/mol (0.22 to 0.36 eV/mol). The surface adsorption process of the oxidizing molecules on partially oxidized GaAs surface has activation energies between 8 and 16 kcal/mol (0.35 to 0.70 eV/mol). As a result the reaction-rate-limited etching process generally shows a stronger temperature dependence than the diffusion-limited etching process [Muk85].

Time Dependence During diffusion-limited etching in a stagnant solution, the layers of solution close to the etched surface slowly become depleted of the molecules. This creates the etching, and the etch rate drops as $E \propto t^{1/2}$. Depending on the reaction and diffusion rates, a steady state is reached after prolonged exposure to the etchant. In contrast, reaction-rate-controlled etching remains constant over time [Mor78], as shown in Fig. 4.2(a). Vigorous agitation can change diffusion-limited etching to

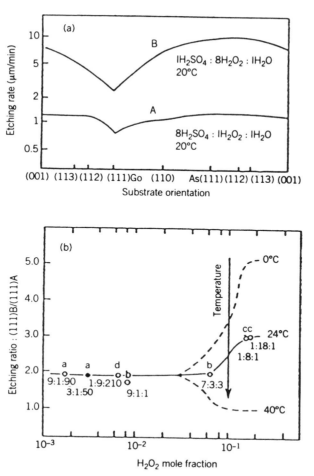

Fig. 4.1 Etch rate as a function of GaAs crystal orientation: (*a*) For the H_2SO_4–H_2O_2–H_2O etchant, a reaction-rate-limited mixture (1:8:1) (curve *B*) is compared with a diffusion-limited one (8:1:1) (curve *A*). (*b*) The ratio of the etch rate for {111} As to that for {111} Ga plotted against H_2O_2 mole fraction for the H_3PO_4–H_2O_2–H_2O etchant at three different temperatures: 0, 24, and 40°C. [Mukerjee, S. D., and D. W. Woodard, in *Gallium Arsenide*, ed. by M. J. Howes and D. V. Morgan, Wiley (1985)]

reaction-rate-limited etching, causing the etch rate to become constant over time [Log73], as shown in Fig. 4.2(*b*).

Laser-Assisted Wet Etchant Method

Laser-assisted wet etching usually involves direct participation of photo-generated carriers in the etching reaction. Photoelectrolytic etching requires ohmic contacts to the GaAs and an external power supply. Etch rates are controlled by the applied current density, while photons are used to create a plentiful supply of holes in *n*-type GaAs. This allows anodic oxidation and oxide dissolution to occur readily [Ell80]. The dependences of the dissolution potential on carrier concentration and surface

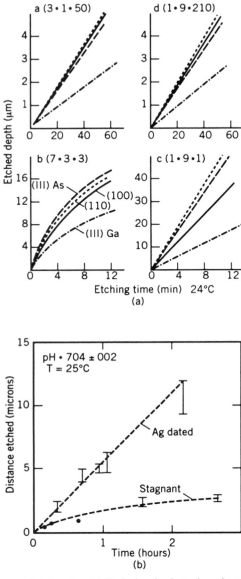

Fig. 4.2 Time dependence of etch rates. (*a*) Etch depth plotted against time for the four major directions in GaAs for the H_3PO_4–H_2O etchant with different ratios. The 7:3:3 solution is diffusion limited, the rest are reaction-rate limited. (*b*) In a diffusion-limited etchant, 30 wt% H_2O_2 is brought up to pH 7.04 ± 0.02 with the addition of a few drops of NH_4OH solution; the etch rate decreases with time owing to the depletion of NH_4OH from regions near the surface. [Mukerjee, S. D., and D. W. Woodard, in *Gallium Arsenide*, ed. by M. J. Howes and D. V. Morgan, Wiley (1985)]

orientation seen in dark electrolytic etching diminish as incident photon flux is increased [Yam75]. Uniform thicknesses of n-GaAs on semi-insulating GaAs can be achieved by selecting the appropriate depletion layer thickness [Hof81] and light intensity [Shi79].

4.3.2 Dry Etching

In recent years reactive plasma-etching techniques have been increasingly applied in the manufacture of silicon-based integrated circuits, gallium arsenide MESFETs, and heterostructure high-speed circuits. A plasma is a fully or partially ionized gas composed of ions, electrons, and neutrons. Dry etching techniques are those that use plasma-driven chemical reactions and/or energetic ion beams to remove material.

Dry etching has several advantages over wet etching. It offers substantial directionality and etch anisotropy, since etching can proceed more rapidly in the vertical direction than in the horizontal. The lateral etch rates in dry etching are close to zero, so undercutting of masking patterns can be greatly reduced. This is essential when etching geometries have lateral dimensions the thickness of the film. Etch conditions can also be adjusted to yield smoothly sloped edge profiles when needed for metal crossovers.

Dry etching is very important to GaAs device processing. There is no liquid etchants that can etch with little or no undercutting of the masking pattern. To reduce source inductances, many discrete GaAs devices and microwave monolithic integrated circuits (MMIC) connect, via holes through the GaAs substrate, the front-side metallization to the back-side ground plane. These narrow, deep holes require the dry etching techniques. The high resolution property of dry etching has been utilized to fabricate quantum dots, zero- and one-dimensional structures etched into superlattice materials.

All dry etching techniques use a gas for etching, and almost all use that gas in the form of a plasma. Obviously a variety of gases, pressure ranges, bias voltages, and electrode configurations are possible. Consequently every combination of gas, pressure range, and electrode configuration used for etching could conceivably be given a distinct name. There are at least 3 different names applied to dry etching techniques [Fon85]. Fonash classified four basic dry etching techniques according to the mechanism being used to achieve the etching: physical, chemical, chemical-physical, and photochemical. We will discuss the dry etching techniques commonly used in GaAs device processing. This includes plasma etching, reactive ion etching (RIE), reactive ion beam etching (RIBE), and ion milling. All belong to the above-mentioned four basic dry etching techniques, and their use depends on operating conditions.

Plasma Etching

The composition of the etching gas in plasma etching is chosen to efficiently volatilize the layer to be etched and to provide good selectivity with respect to each mask and underlying substrate. The equipment of a parallel-plate, planar plasma etching system is shown in Fig. 4.3. It usually consists of two parallel plates, with the power being supplied to the upper plate and the substrates loaded onto a cooled lower plate. The pressure in the chamber is typically 0.01 to 1 torr. The two electrodes have the same

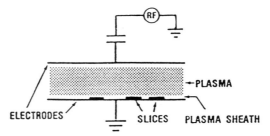

Fig. 4.3 Configuration of parallel plate, planar plasma etching machine. [Williams, R. E., *Gallium Arsenide Processing Techniques*, p. 190, Artech House (1985)]

area. The plasma does not extend completely up to the electrodes. There is a zone, known as the plasma sheath, that separates the plasma from the electrode. The electrodes become charged by electrons which move from the plasma on the surfaces. The electrons have greater mobility than the positive ions, and so the electrodes become negative with respect to the plasma. This results in an electric field across the plasma sheath, between the plasma and the electrodes. This field causes ions at the edge of the plasma to be accelerated across the plasma sheath. Because of the general geometry, they move perpendicular to the electrodes except near the outer radius when distortions in etch profile occur.

The perpendicular arrival of energetic ions opens the possibility of anisotropic etching. As shown in Fig. 4.4, any masking substance will prevent these ions from initially hitting the sidewalls of etching patterns. The undercutting depends on the chemical etch rate of the neutral species. If etching can occur by purely chemical processes, it will tend to be isotropic. However, if additional energy from ionic bombardment is needed to help drive the total reaction, the process will be anisotropic. Figure 4.5 shows the isotropic etching with no preferred direction (*a*), anisotropic etching with the ion impinging vertically, giving straight-walled profiles (*b*), and crystallographic chemical etching that performs straight or angled smooth profiles [Ibb88]. The directionality is sensitive to the operating parameters. Anisotropy is aided by lower pressure and/or increased power.

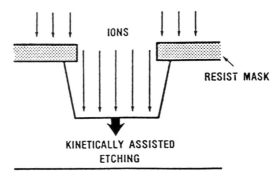

Fig. 4.4 Anisotropic etching driven by kinetically assisted chemical reactions. [Williams, R. E., *Gallium Arsenide Processing Techniques*, p. 191, Artech House (1985)]

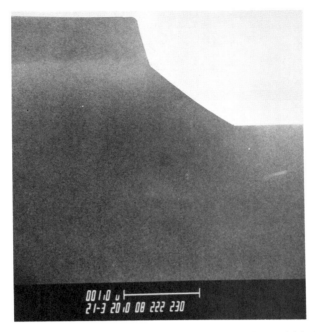

Fig. 4.5 (a) Isotropic chemical etching has no preferred direction, which leads to circular profiles that undercut a mask. (b) In anisotropic etching, ions impinge vertically, forming straight-walled profiles. (c) Crystallographic chemical etching can form straight- or angled smooth-sidewall profiles, which can preserve mask linewidth if aligned with slow-etching planes. [Ibbotson, D. E., and D. L. Flamm, *Solid State Tech.* Oct. (1988): 77–79]

Chemical etching of III–V compounds tends to be crystallographic, and critical dimensions can be maintained if their features are strategically aligned along certain lattice planes (Figure 4.6) [Ibb88]. Preferential crystallographic etching is observed with (111) As greater than (100), greater than (110), greater than (111) Ga [Ibb83; Ash90]. Chlorine atoms are the most common active species, but Br_2 and H_2 have also been used. Sources of chlorine includes Cl_2, chlorocarbons, PCl_3, HCl, and $COCl_2$ [Smo81]. The etch rate with Cl_2 increases with temperature with an Arrhenius-type dependence at a given power density [Don82]. In contrast, etch rates with chlorocarbons such as CCl_4 decrease with increasing temperature; this has been attributed to a temperature-dependent morphology of the deposited chlorocarbon film, which is believed to control rates by controlling reactant or product diffusion [Got82].

Plasma etching has some disadvantages. The disadvantages are smaller wafer capacity, more radiation damage effects, and, in some cases, residues left by the etching.

The Equivalent Circuit and Matching Network Koeing and Maissel [Koe70] have provided a simple equivalent circuit for RF discharges. The basic concepts are shown in Fig. 4.7. The dark space sheaths are primarily capacitive due to the negative surface charge on the electrode or grounded adjacent surface. The presence of the blocking capacitor results in charge storage, and although the ion and electron currents differ,

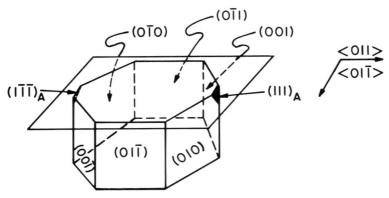

Fig. 4.6 Schematic diagram of the crystallographic planes that are exposed when (100) surface of GaAs is etched in Br_2 plasma under a mask aligned along the $\langle 011 \rangle$ directions. The relative etch rates are $111_B \gg 100 > 110 > 111_A$ where the B planes are As-rich and the A planes are Ga-rich. [Ibbotson, D. E., and D. L. Flamm, *Solid State Tech.* Oct. (1988): 77–79]

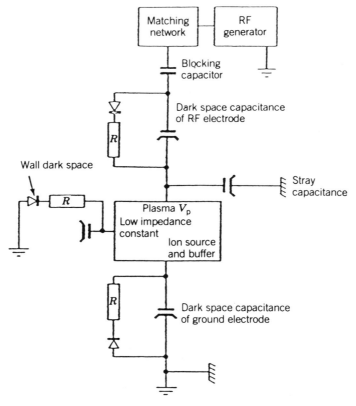

Fig. 4.7 A simple diagram for the equivalent circuit of an RF discharge. [Morgan, R. A., *Plasma Etching in Semiconductor Fabrication*, Elsevier (1985)]

their total charge transfer of each half cycle is equal and balanced. The ion sheath characteristics can change considerably with gas pressure due to recombinations and collisions. These are a function of number density and each collision's cross section (i.e. mean free path).

Matching networks are used in high-frequency discharges (above 1 MHz) to match the impedance of the load to the RF generator output impedance. The purpose of this network is to increase the power dissipation in the discharge and to protect the generator. When the impedances of supply and load are matched (i.e., are equal), the power transferred by the generator to the discharge is at a peak value.

We can understand this better by considering the dc counterpart (Fig. 4.8) in which a cell of emf E and internal resistance r is supplying power to an external load of variable resistance R. The current in the circuit is $E/(r + R)$, and therefore the power P dissipated in R is given by

$$P = \frac{E^2 R}{(r + R)^2}. \tag{4.1}$$

The power P will vary with the value of R. The maximum value of P is obtained by differentiation:

$$\frac{dP}{dR} = \frac{E^2(r + R)^2 - 2(r + R)E^2 R}{(r + R)^2}. \tag{4.2}$$

This differential has a zero value for $R = r$, that is, for maximum P. Therefore to dissipate maximum power in a load, the resistance of the load has to be matched to the resistance of the power supply.

Etching Chemistry Commercial etching of silicon and its compounds is almost exclusively done by forming the volatile fluoride or chloride. The fluorides of group-V elements are involatile, but fortunately both group-III and group-V chlorides, bromides, and iodides have usable vapor pressures, especially at elevated temperatures [Ibb88b]. GaAs, InP, and other III–V semiconductors have usually been patterned in chloride- and bromine-containing plasmas.

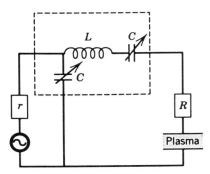

Fig. 4.8 Simplified diagram of a typical matching network. [Morgan, R. A., *Plasma Etching in Semiconductor Fabrication*, Elsevier (1985)]

Hydrogen plasma III–V etching was reported some years ago [Cha81]. However, a new hydrocarbon chemistry, in which 15% to 25% CH_4 is added to H_2 plasma, is reported to offer high etch rates, anisotropic etching, and less surface damage. CH_4/H_2 is an unsaturated polymer-forming mixture (similar to the CH_xH_y/H_2 which feeds sometimes with O_2 or N_2O added) used for selective anisotropic etching of SiO_2. It is likely that a hydrocarbon film coats GaAs and InP in the mixture, where ion bombardment stimulates CH_x/substrate interfacial reactions. In this case the volatile products could be organometallic compounds, such as $Ga(CH_3)_x$ and $P(CH_3)_x$, or hydrides, such as GaH_x and PH_x.

In halocarbon-oxidant plasma feed mixtures, there are reactions between halogen atoms and unsaturated halocarbon species formed from dissociation reactions. Unsaturated species, such as CCl_2 and CF_2, are formed by electron-impact dissociation of halocarbons like CCl_4 and CF_2Cl_2. These dissociation reactions also produce etchant species (Cl), though in halocarbon plasmas where the additional etchant (Cl-atoms) is not limited by rapid combination reactions with unsaturated CCl_xF_x species. One of the reasons why O_2 and Cl_2 are added to halocarbon etchant feed mixtures is that both gases increase the etchant atom concentrations: Cl_2 dissociation adds directly to the supply of Cl atoms, while O-atoms form the additional etchant (Cl, Br) through reactions with CF_yCl_x or CF_yBr_x species [Don84]. However, large amounts of suppressing oxygen unsaturate completely and form enough free oxygen atoms to promote surface oxidation, thus suppressing etch rates. Unsaturated radicals can form thin sidewall films, which stimulate anisotropic etching with ion bombardment. Added oxidants serve to adjust the ratios and concentrations of etchant and unsaturated species for acceptable anisotropy and etch rates.

Since polymer growth increases rapidly with the partial pressure of unsaturated species, lower system pressure or feed dilution with a noble gas (which lowers the partial pressure of these species) is another way to suppress film growth relative to etching. Inert gases (usually Ar or He) are often added to help stabilize a plasma, enhance anisotropy, or lower the etching rate (by dilution). Since helium has high thermal conductivity, it also improves heat transfer between wafers and supporting electrodes.

The feed gas combinations have been used in several circumstances to etch III–V compounds. Tables 4.2 and 4.3 give low and high pressure etching of III–V compounds using the discharge mixtures [Ibb88b].

Reactive Ion Etching

Reactive ion etching (RIE) uses ions whose energy is greater than the positive space-charge potential of the plasma, which assists the etching. RIE is similar to plasma etching in that reactive species generated from a plasma are used to etch the material. RIE is principally distinguished from plasma etching by its emphasis on directionality, leading to changes in pressure and equipment configuration. In this section we will review some general principles of reactive ion etching, process parameters, equipment, and applications with respect to various materials and structures.

General Principles RIE is generally performed in a reactor that is somewhat different from the type used in plasma etching. There are three distinguishing characteristics of RIE's operating conditions:

ETCHING   129

TABLE 4.2 Low Pressure Plasma Etching of III–V Compounds

Material	Gasses/Conditions	Profile
GaAs, InP	$CCl_2F_2/O_2/Ar$ 2.10 mtorr, 13 MHz, −600 V self-bias	More vertical at lower P (Fig. a)
InP, GaInAsP	Cl_2/O_2 1–50 mtorr, 13 MHz, 100–400 V self-bias sample rotated in Faraday cage at angles 0° and 30°	More vertical and rough at lower P (Fig. b)
GaAs	Cl_2/Ar 15 mtorr, 13 MHz, 200–350 V self-bias	80% Cl_2: orientation dependence 20% Cl_2: vertical walls
GaAs	$CCl_2/F_2/He$ 40 mtorr, 3 kV DC discharge	No undercutting
GaAs, AlGaAs	$CCl_2/F_2/He$ 4–40 mtorr, 13 MHz, 100–500 V self-bias	(Fig. c)
GaAs	$SiCl_4$ 2–110 mtorr, 13 MHz, 20–130 V self-bias	< 20 mtorr vertical walls > 20 mtorr orientation dependence
GaAs	CCl_4/O_2, CCl_4/H_2 10 mtorr, 13 MHz	Isotropic, rough Vertical walls, smooth
InP, GaAs	CH_4/H_2 1–15 mtorr, 13 MHz	Anisotropic

Source: Jbbotson, D. E., and D. L. Flamm, *Solid State Technology*, Oct. (1988): 77–79

1. *Asymmetric electrodes.* The electrode that holds the substrates is significantly smaller in area than the second electrode which usually is the chamber.
2. *Substrates placed on the powered electrodes.* The substrate electrode is powered and the other electrode is ground. The applied voltage being dropped at the electrode is generally on the order of a few volts.
3. *Relatively low operating pressure ranging from about 10^{-3} to 10^{-1} torr.*

This operating condition lengthens the mean free path of the reactive ions generated in the plasma and permits ions to impinge at near-normal incidence upon the substrate, resulting in a considerable improvement in the anisotropy of the etch. RIE can be used for via hole etching.

Figure 4.9 illustrates a typical RIE system [Gor84]. A plasma is produced and

TABLE 4.3 High Pressure Plasma Etching of III–V Compounds

Material	Gasses/Conditions	Profile
GaAs, GaAs-oxide	$CCl_2F_2, CCl_4PCl_3, HCl, Cl_2, COCl_2$ 75–500 mtorr, 13 MHz	Partially anisotropic
GaAs	BCl_3/Cl_2 (6%) 150 mtorr, 13 MHz	Anisotropic
GaAs, GaP, InP, GaInAsP	$CCl_4/O_2, CCl_2F_2, Cl_2/O_2$ 50–150 mtorr, 55 kHz, 300°C, 165–425°C	Some orientation effects for GaAs (Fig. a)
GaAs, InP	Cl_2, 0.1–1 torr, 0.01–20 MHz, 25–300°C	Orientation dependence (Fig. b)
GaAs	Br_2, Cl_2 0.5 torr, 13 MHz Very low power or no plasma	Crystallographic, smooth surfaces (Fig. c)

Source: Ibbotson, D. E. and D. L. Flamm, *Solid State Technology*, pp. 77–79, Oct. (1988).

sustained in a glow discharge between two electrodes, one of which is capacitively coupled to a high-frequency (13.56 MHz) power supply through an impedance matching network while the other is grounded. The wafers are placed on the powered electrode. A flat electrode holds the substrates. The area of the other electrode is much larger, and usually the entire chamber is used as the second electrode. The substrate electrode is usually the RF-powered electrode, and the chamber the RF ground. The large ratio of electrode areas causes most of the voltage drop between electrodes to appear across the plasma sheath, and the voltage drop is divided between the two sheaths. The expression usually quoted for this division is

$$\frac{V_c}{V_a} \sim \left(\frac{A_a}{A_c}\right) \tag{4.3}$$

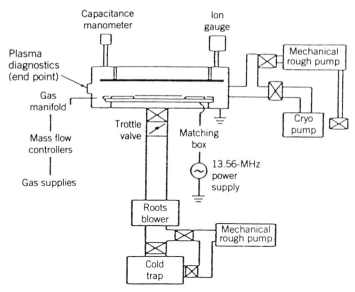

Fig. 4.9 Outline of typical RIE system components. [Gorowitz, B., and R. J. Saia, in *VLSI Electronics*, vol. 8, pp. 69–90, Academic Press (1984)]

where A_a and A_c are the anode and cathode areas, respectively, and V_a and V_c are the voltage drop associated with the anode and cathode. This expression is valid only at low operating pressures. The plasma model used in this analysis treats the sheaths as pure capacitors so that larger areas have larger capacitances. The electrode/plasma/electrode circuit then acts as a capacitance voltage divider with the larger voltage across the larger impedance (smaller area).

Process Parameter Effects There are a number of readily accessible process parameters that can be used to control key criteria such as etch rates, selectivity, resolution accuracy, feature profiles, and etch uniformity. The most obvious factors are gas composition, flow pressure, and input power or voltage [Gor84]. The effects of these and other control parameters on microscopic phenomena such as particle energies, gas-phase chemistry, surface chemistry, and electrical factors are difficult to separate because of the complex nature of their interactions. In this section we will first examine the overall effects of major process parameters and later review them with respect to specific materials.

1. *Flow.* The etch rate is dependent on the rate of generation of active species, the rate of consumption of these species, and the rate of removal of reaction products. Excessive flow rate at constant pressure results in decreased residence times and can consequently result in a decrease in the active species generation rate if the power is kept constant. At low flow rates the depletion of species during the etch process also leads to a decrease in the etch rate. The etch rate of a given surface can be used to calculate the flow rate of the major effluent species (i.e., of SiF_4 for Si etching). Chapman and Minkiewicz [Cha78] refer to this effluent/input gas flow ratio as a utilization factor.

132 PROCESS TECHNIQUES

Since gas flow utilization is related to the surface area to be etched, the etch rate R can also be examined in terms of what Mogab [Mog77] has defined as a loading effect:

$$R = \frac{B\tau G}{(1 + KB\tau A)}, \quad (4.4)$$

where G is the generation rate of active species, B the reaction rate constant, τ the lifetime of active species in the absence of etchable material, A the surface area of etchable material, and K a constant for a given material and reactor geometry. This effect is clearly important in a batch reactor.

2. *Power.* The applied power affects the electron energy distribution. Etch rates generally increase monotonically with applied power [Mel78], as shown in Figs. 4.10 through 4.12. Compensation for excessive flow rates and decreased residence times may be made by increasing the RF power. In RIE this may result in excessive ion

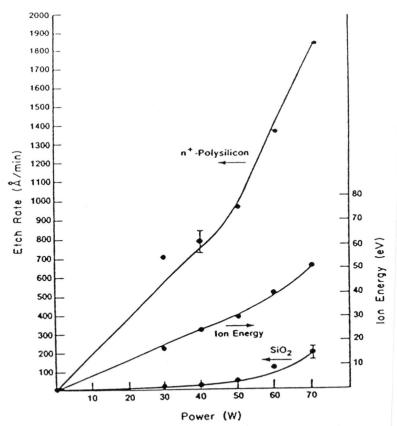

Fig. 4.10 Etch rate of n^+ polysilicon in CCl_4/N_2 plasmas as a function of RF power. The etching parameters are flow, 15 sccm CCl_4/5 sccm N_2; pressure, 60 mtorr; power density, 60 mtorr; power density, 60 W, 0.47 W/cm². [Melliar-Smith, C. M., and C. J. Mogab, in *Thin Film Processes*, p. 497, ed. by J. L. Vossen and W. Kerns, Academic Press (1978)]

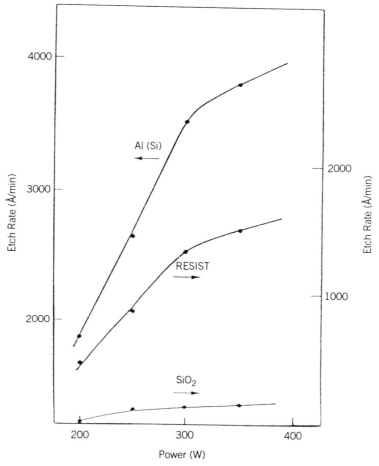

Fig. 4.11 Etch rates of Al/1% Si as a function of power in batch RIE system. The etching parameters are flow, 40 sccm pressure, 70 nm torr; $BCl_3/CCl_4/O_2$ is 30 sccm/8 sccm/2 sccm. [Melliar-Smith, C. M., and C. J. Mogab, in *Thin Film Processes*, p. 497, ed. by J. L. Vossen and W. Kerns, Academic Press (1978)]

bombardment and lower selectivity, photoresist degradation, and an aggravation of the loading effect and lead to etching nonuniformities. An increase in the RF power cannot compensate for reactant depletion, since it may increase the reactant generation rate but not necessarily affect the relative consumption rates. Therefore it appears desirable to cope with the loading effect by using high flow rates and high pumping speeds to maintain low pressures (e.g., < 100 mtorr) and to use sufficiently low power to prevent deleterious power-related effects.

3. *Pressure.* RIE is usually operated in the 5.0- to 100.0-mtorr range, since these define the range of ion energies available. As pressure is increased, the residence time and particle collision rate will increase, and the average electron energy will decrease [Mel78]. Since the electron energy determines the generation rate of active species, the etch rate will decrease with increasing pressure. Similarly the ratio of the flux of

134 PROCESS TECHNIQUES

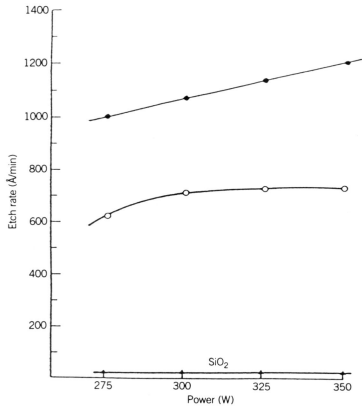

Fig. 4.12 Etch rates of molybdenum as a function of power. The etching parameters are flow, 25 sccm CCl_4/75 sccm O_2; pressure, 250 mtorr; direct-current-magnetron-sputtered Mo, ○; S-gun-sputtered Mo, ●. [Melliar-Smith, C. M., and C. J. Mogab, in *Thin Film Processes*, p. 497, ed. by J. L. Vossen and W. Kerns, Academic Press (1978)]

atomic species to ions will increase, and directionality of the etch may be diminished, increasing the possibility of undercutting.

4. *Temperature.* The temperature of the surface being etched also affects the reaction rate and the rate at which the reaction products leave the surface. Table 4.4 lists the boiling points of some typical etch products. The vapor pressure curves can be useful in the selection of etchant gases. Reaction rates generally might be expected to exhibit an Arrhenius-type temperature dependence, with activation energies lying in the range of 0.05 to 0.5 eV [Mel78]. The ion bombardment contributes a significant portion of the energy of activation, and hence the calculations based on the thermal response of etch rates can be misleading.

Typical RF power inputs, etch rates, and heats of reaction can produce a temperature rise of 100° to 200°C in a 75-mm wafer [Gor84]. The power input and the etch rate are consequently limited by the thermal stability of the resist. The wafer-carrying electrode is electrically and somewhat thermally isolated from the rest of the chamber. It can be maintained at a lower temperature during the etch cycle with an independent

TABLE 4.4 Boiling Points of Typical Etch Products

Element	Chlorides	Boiling Point (°C)	Fluorides	Boiling Point (°C)
Al	$AlCl_3$	177.8 (subl.)	AlF_3	1291 (subl.)
As	$AsCl_3$	130.2	AsF_5	−53
Cr	CrO_2Cl_2	117	CrF_2	>1300
	$CrCl_3$	1300 (subl.)		
Cu	$CuCl$	1490	CuF	1100 (subl.)
Ga	$GaCl_3$	201.3	GaF_3	1000
In	$InCl_3$	300 (subl.)	InF_3	>1200
Mo	$MoCl_5$	268	MoF_6	35
	$MoOCl_3$	100 (subl.)	MoF_5	213.6
			MoO_2F_2	270 (subl.)
			$MoOF_4$	180
Nb	$NbCl_5$	254	NbF_4	236
	$NbOCl_3$	335 (subl.)		
Si	$SiCl_4$	57.57	SiF_4	−86
Ta	$TaCl_5$	242	TaF_5	229.5
Ti	$TiCl_4$	136.4	TiF_4	284 (subl.)
W	WCl_4	346.7	WF_6	17.5
	WCl_5	275.6	WOF_4	187.5
	$WOCl_4$	227.5		

Source: Gorowitz, B., and R. J. Saia, in *VLSI Electronics*, vol. 8, Academic Press (1984): pp. 69–90.

heat-exchange system. Typical chamber temperatures range from 40° to 80°C when an auxiliary heating system is used.

In RIE, GaAs is etched through the combined effect of chemical reaction and physical sputtering. Ashby [Ash85] lists etch rates for RIE under a variety of conditions. In general, slower etch rates give greater anisotropy and better surface morphology. For a fixed power density, the rate increases with pressure, reaches a maximum between 5 and 20 mtorr, and then decreases with any further increase in pressure [Ash85]. This is due to the interplay between ion energy, which decreases with increasing pressure due to decreasing self-bias voltage, and reactant concentrations, which increase with increasing pressure.

At a constant bias voltage, rates increase linearly with pressure [Ash85]. When the self-bias voltage is increased, etch rates increase with increasing RF power [Ash85]. Pronounced dependence on self-bias voltage indicates the importance of ion-assisted mechanisms. Rates first increase and then reach a plateau with increasing flow rates. More vertical walls are achieved by increasing the importance of ion-assisted process by increasing bias voltage or RF power density and by decreasing pressure. Highest rates are achieved with Cl_2 as the Cl source, but Cl produces surface roughness and exhibits low reproducibility in etch rates from run to run [Ash90]. Reproducibility can be improved by an initial short etch in H_2 or CCl_3F [Hu84]. Chlorocarbons such as CCl_4 and CCl_3F produce smoother surfaces but can deposit chloropolymer films. Etching eith BCl_3 is slower but highly reproducible [Tam84]. Addition of O_2 to CCl_4 or CCl_3F increases etch rates by increasing Cl atom densities [Hu80].

Addition of Ar to Cl, CCl_4, or CCl_3F increases anisotropy [Yam85; Cha83]. Addition of H_2 to CCl_4, CCl_3F, or CCl_2F_2 increases anisotropy, decreases damage, and gives a residue-free surface [Sem84a; Sem84b].

Reactive Ion-Beam Etching

Reactive ion-beam etching (RIBE) is a further extension of reactive ion etching. It describes the technique of removing material from a surface by impinging on it a beam of chemically reactive ions. Etching occurs by a combination of ion bombardment and chemical reaction. A comparison made of the etching parameters of RIBE with RIE and ion beam etching is given in Table 4.5 [Hea84]. Since RIBE is essentially a derivative of ion milling produced by the use of a reactive gas instead of argon, the only differences between these two techniques relate to the chemical reactivity of the species emanating from the ion source. This results in very dissimilar etch characteristics. Most important, more specific selectivity of the etch for one material over another can be achieved in RIBE than in ion milling. Selectivity is necessary to complete the etch without erosion of the thin underlying material. In RIBE the material to be etched is for the most part chemically converted to volatile reaction products that are then removed by the vacuum system. Redeposition of the eroded material does not occur as in ion milling.

Both RIBE and RIE are techniques that rely on a combination of ion bombardment and chemically reaction as their etching mechanisms. The pressure used in RIBE is much lower than that used in RIE, while the ion current is somewhat higher. In the RIBE system the chemical and physical etching mechanisms can be independently controlled. The sample is isolated from the plasma in RIBE—in contrast to RIE where it is immersed in the plasma—and this leads to a major difference in the ratios of ion flux to neutral and radical flux in the two techniques. The lack of short-lived

TABLE 4.5 Characteristics of Some Dry-Etching Processes

Characteristic	Reactive Ion Etching	Reactive Ion-beam Etching	Ion-beam Milling
Location of substrate	In plasma on powered electrode	Remote from plasma, in beam	Remote from plasma, in beam
Pressure (torr)	$n \times 10^{-2}$	1×10^{-4}	1×10^{-4}
Flow rate (sccm)	$n \times 10$	$n \times 1$	$n \times 1$
Ion energy (eV)	$10 - n \times 10^2$	$n \times 10^2$	$n \times 10^2$
Ion Current (mA/cm^2)	0.01–0.2	0.03–1.0	0.2–1.0
Neutral flux (per cm^2 sec)	10^{18}–10^{20}	$<10^{17}$	
Atom or free-radical flux (per cm^2 sec)	$<10^{16}$		
Active species	Reactive ions, radicals	Reactive ions, neutrals	Ar^+ ions
Products	Volatile	Volatile	Nonvolatile
Mechanism	Chemical/physical	Chemical/physical	Physical

Source: Gorowitz, B., and R. J. Saia, in *VLSI Electronics*, vol. 8, Academic Press (1984): pp. 69–90.

ETCHING   137

radical etching in RIBE results in an extremely anisotropic etch. The ion energy and current density can be independently set to maximize the etch rate. In addition the high ion energy and ion current yield a relatively high sputtering rate that can be advantageous when etching a material with a component that does not form volatile etch products. Next we will describe the apparatus used for RIBE, as well as some of its applications to etch materials in device and circuit fabrication.

Ion Sources and Operations   The most common method of producing large-diameter ion beams for etching applications has been electron bombardment of a low-pressure gas in a magnetically confined dc discharge. An ion source of this type is called a *Kaufman source*. A typical Kaufman ion source is shown in Fig. 4.13. Electrons are thermally emitted from a heated cathode, usually a resistively heated tungsten filament, and accelerated toward an anode held typically at $+40$ to $100\,\text{V}$ from the cathode. Magnetic fields serve to confine energetic electrons and produce a glow discharge in the source volume at pressures of 10^{-4} to 10^{-3} torr.

The magnetically confined discharge is the source of positive ions that are extracted to form a beam. Discharge operating parameters that can be adjusted to control the beam current or profile include anode potential, discharge current, and magnetic field strength. The plasma itself is nearly a field-free region with a potential close to the anode potential. Primary electrons from the cathode have energy that is roughly the cathode-to-anode potential difference.

An ion beam is formed by extracting ions from the plasma through one or more

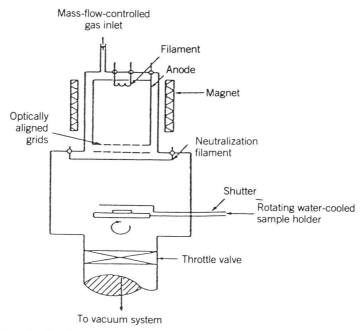

Fig. 4.13   Reactive ion-beam etching apparatus. [Heath, B. A., and T. M. Mayer, in *VLSI Electronics*, vol. 8, pp. 365–410, Academic Press (1984); reprinted by Permission of Electrochemical Society]

perforated grids. In a two-grid system, such as the one shown in Fig. 4.14, the first or cathode grid is held at the cathode potential while the second, or accelerating grid, is biased negatively with respect to the cathode to accelerate ions out of the plasma. The maximum current that can be extracted from a source such as this is determined by the space-charge limitations of extraction optics. This limitation can be approximated by Child's law [Chi11], which gives the maximum current density between two parallel planes as

$$i = \frac{(4\varepsilon_0/9)(2e/m)^{1/2} V^{3/2}}{l^2}, \qquad (4.7)$$

where V is the potential difference, l the separation between the planes, and e/m the charge/mass ratio of the ions.

Early RIBE machines used hot cathodes to emit electrons, which then generate reactive ions. The chlorinated and fluorinated hydrocarbons attack each cathode, thus decreasing its life. A variety of alternatives to the basic hot cathode, gridded optics ion sources have been proposed for RIBE applications. A microwave discharge (at 2.45 GHz) using electron cyclotron resonance (ECR) to produce a dense plasma [Mat82; Miy83] was demonstrated to eliminate the hydrocarbon attachment problem. Figure 4.15 shows the basic design. The microwave power is coupled to the discharge chamber through a rectangular wave guide with a fused quartz window. The discharge chamber functions as a microwave cavity resonator. ECR uses a magnetic field (about 875 G) to change the normal straight-line trajectories of electrons into circular or helical orbits. This increases the average lifetimes of electrons by increasing the average time of flight before a collision with the chamber wall, which is a major cause of recombination.

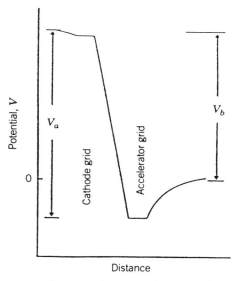

Fig. 4.14 Potential parameters for two-grid ion optics. [Heath, B. A., and T. M. Mayer, in *VLSI Electronics*, vol. 8, pp. 365–410, Academic Press (1984)]

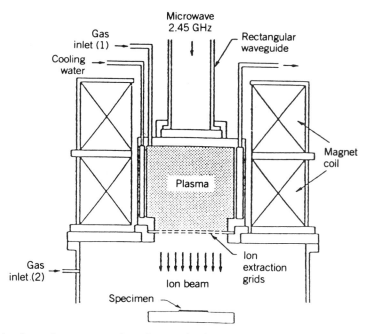

Fig. 4.15 A schematic representation of a reactive ion-beam etcher with an electron cyclotron resonance chamber as the ion source. [Matsuo, S., and Y. Adachi, *Jap. J. Appl. Phys.* **21**, L4 (1982); Miyamura, M., T. Uchiyama, O. Tsukakoshi, and S. Komiya, in *Proc. Int. Ion Eng. Congress*, p. 1623, Kyoto (1983)]

Ion-Beam Etching (Milling)

Ion-beam etching, also known as ion milling, is a purely physical process in which energetic inert ions (usually 500 to 1000 eV) are extracted from a broad-beam ion source and delivered to the substrates as a uniform directional beam. The process is one of momentum transfer between the impinging ions and the surface atoms by which the latter gain momentum. The physics of ion etching will be briefly discussed here.

There are several incentives for using ion-beam etching. First, it is a universal etchant. It can pattern any material (e.g., permalloy) or combinations of materials (e.g., multilayer Ti, Pt, Au metalizations) that are difficult to etch by using wet-chemical or plasma methods. Chemical etch rates are often highly selective, whereas ion etch rates usually differ by less than an order of magnitude. Hence complex multilayer structures can be etched in a continuous ion-beam process. Second, ion-beam etching has demonstrated high resolution of the patterning technique and is able to define structures below 100 Å [Bro76]. Finally, ion etching is highly anisotropic, relying on shadow masking to expose various surfaces to the beam but not requiring superior mask adhesion or suffering from beneath-the-mask etchant penetration that leads to undercutting and loss of pattern fidelity (which frequently occurs in wet-chemical etching). Since the substrate can be positioned at different angles to the incident beam, a wide variety of sidewall etch profiles can be achieved.

The major advantages of ion milling are high resolution, good uniformity, high anisotropy, residue-free etching, good control over process parameters, and relative freedom from radiation damage effects resulting from the physical isolation of the plasma and the source. The disadvantages are lack of sensitivity, possible trenching and redeposition, faceting of the photoresist, damage from energetic ion bombardment, and low etch rates resulting in low wafer throughput [Lee84].

Focused Ion-Beam Etching (FIBE)

In three-dimensional quantum-size structures, a focused ion beam is used for clean maskless, selective etching. An enclosed through processing system has been built [Asa87]. The advantages for clean maskless etching are (1) no surface residue, (2) chemically high enhancement, (3) submicron resolution, and (4) direct drawing [Har87].

Laser-Assisted Dry Etching Methods

Laser-assisted dry etching methods are based on heating effects (photothermal or pyrolytic), photochemical generation of reactants (gas-phase photochemical), or photogeneration of carriers (carrier-driven photochemical) [Ash90]. The highest etch rates are achieved at high pulsed power densities (exceeding 10^5 W/cm^2) [Tuc83; Tak83; Tak84; Bre85] when samples are heated to the melt [Tuc83]. These relatively high power (> 5000 W/cm^2) etching processes using CW lasers can produce high rates and narrow lines if the beam is scanned [Tak84; Tak83; Tak84b], but the high temperatures involved may adversely affect GaAs properties [Tak84]. Patterns can be etched into GaAs using substrate contact masking, projection etching, and direct writing with scanned beams [Bre85]. Laser-assisted etching with CF_3Br proceeds at 100 times the physical ablation rate [Bre85b]. Ablation becomes significant at power densities above 2×10^6 to 4×10^6 W/cm^2 [Bre85, 85b].

Etching of GaAs can occur when the laser beam passes through the reactant gas parallel to the surface, but it is slower and less localized than when perpendicular beams are used [Bre85c]. With HBr or CF_3Br orientation-dependent etch rates are in the sequence (111) As, (100), (110), (111) Ga [Bre85c].

Carrier-driven photochemical etching is based on the reaction of GaAs with Cl atoms. The reaction is driven by photo-generation of electrons and holes in the semiconductor surface, using protons of band-gap or greater energy [Ash84; Ash85a; Ash85b]. The requirement for photo-generated carriers at the surface permits very selective wavelength-controlled etching of GaAs and GaAs-related ternary materials [Ash85a] and bias-voltage-controlled selective etching based on dopant concentrations [Ash85b]. Materials of nearly equal composition can be selectively etched.

4.4 PLASMA-ENHANCED DEPOSITION

Plasma-assisted techniques are especially important in GaAs and other III–V compound processing because they can be accomplished at relatively low temperatures. Deposition of thin film materials is an excellent example of the low-temperature requirement. Such materials can be deposited on wafers by the chemical vapor deposition method. The temperatures required to drive the chemical reactions are often between 700° and 1000°C. GaAs cannot stand up to such temperatures: Not only does arsenic evolve in this temperature range, but metals commonly present on the wafer cannot be exposed to these extremes [Wil84]. The low-temperature plasma-assisted deposition process plays an important role in GaAs processing. The major advantage of PECVD for GaAs processing is the ability to grow thin films at relatively low substrate temperatures ($\sim 300°C$) as compared to CVD. However, the compositional control of the thin film material is difficult. PECVD yields films that are amorphous in nature. The film may be randomly bonded, highly crosslinked, and of variable composition. The chemical species other than the desired ones are often included in the film. In PECVD films a range of stoichiometry is possible depending on the plasma and operating parameters. The etch rate is a function of composition.

The dc and ac glow discharges will be reviewed in this section. Frequency effects on RF plasma reactor behavior will be discussed. Different plasma-enhanced CVD reactors will be compared. The electron cyclotron resonance (ECR) CVD reactor will be included in this section.

4.4.1 dc/ac Glow Discharges

A glow discharge in a low pressure gas (~ 1 torr) created by a dc-applied voltage is nonuniform in appearance [She87]. A typical discharge is shown in Fig. 4.16 [Bro59]. Since the cathode is cold, the discharge is maintained by secondary electrons produced there by positive ion impacts. The ions experience a strong electric field near the cathode, which causes them to accelerate toward it. The sheath is the region next to the cathode in which charge neutrality is not obeyed and relatively few collisions occur. This encompasses the Aston, Crookes, and Faraday dark spaces, and the cathode and negative glow regions. There is an excess of ions in this region, hence the net positive charge there. The positive column has no net space charge. This is a region of high electrical conductivity. Ions and electrons in this region can be lost by gas-phase recombination, or diffusion to the tube walls. They can be regenerated by electron impact ionization in the positive column, or the secondary electron emission from the cathode. A smaller sheath appears at the anode.

If the glow discharge of Fig. 4.16 is operated under alternating voltage conditions, we observe a discharge with two dark spaces. Up to about 10 kHz the frequency is low enough so that the discharge lights and extinguishes on each cycle. There is sufficient time between cycles for most electrons to leave the positive column and be lost to the tube walls. The loss of electrons extinguishes the glow discharge. Above 10 kHz there is no enough time for the electrons in the positive column to be lost to the walls, so the discharge remains lit continuously. Depending on the discharge geometry and gas involved, the starting voltage for an ac discharge can depend on the RF frequency and pressure [Bro56].

When the AC discharge is operated with a blocking capacitor between the power supply and one of the electrodes, that electrode will assume negative self-bias. Such an average negative on this electrode can serve to accelerate ions toward it with considerable energies.

The forming of self-bias has been described by Butler and Kino [But63]. Figure 4.17 shows the creation of negative self-bias in ac discharge [But63]. In Fig. 4.17(a) an alternating potential is applied to a conducting probe in a plasma. When the probe sees a positive voltage, a large electron current flows. Reversing the voltage produces

142 PROCESS TECHNIQUES

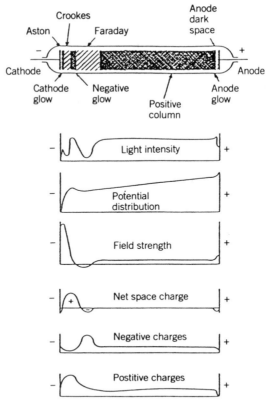

Fig. 4.16 A dc glow discharge at low pressure. [Brown, S. C., *Basic Data of Plasma Physics*, Wiley (1959); Sherman, A., in *Chemical Vapor Deposition for Microelectronics, Principles, Technology, and applications*, Noyes Publication (1987)]

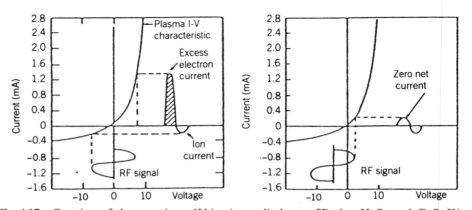

Fig. 4.17. Creation of the negative self-bias in ac discharge. [Butler, H. S., and G. S. Kino, *Phys. Fluids* **6** (1963): 1346; Sherman, A., in *Chemical Vapor Deposition for Microelectronics, Principles, Technology, and Applications*, Noyes Publication (1987)]

only a small current flow due to the immobility of heavy ions. When the probe is conducting, the self-bias is zero and large net current flows on average. On the other hand, when the probe is nonconducting (i.e., electrode attached to the blocking capacitor), the behavior is shown in Fig. 4.17(b). For the average current to be zero, the average applied voltage (i.e., RF signal) must be negative.

4.4.2 Frequency Effects on RF Plasma Reactor Behavior

When the RF discharges are used to create plasmas in PECVD reactors, the influence of the frequency at which the discharge is operated is another question that must be explored. A lower frequency discharge (50 to 100 kHz) will produce a CVD film with greater compressive stress than a film created in a higher frequency discharge (13.56 MHz) [She87]. The speculation has been that the ion bombardment is more intense at lower frequencies, and this bombardment causes the film stress to be compressive.

There are two potential explanations of why the ion bombardment is more intense at low frequencies. First, the sheath potential drop, on average, will be higher at lower frequencies. The electrons are lighter than ions, so they tend to preferentially diffuse out of the plasma, and the electrode assumes a negative bias. As frequency is increased, there is increasingly less time available for charged particles to diffuse to the reactor walls between cycles. Therefore, if there is less opportunity for electrons to diffuse out at higher frequencies, there will be less need for a strong bias to form. If we recognize that the negative bias formed has the effect of accelerating ions toward the surface, then more bias (lower frequencies) means more ion bombardment on the forming film.

Second, the plasma potential will vary with time, depending on the cycle of the applied potential [Bru81]. If the ion can transit the sheath before the applied electric field reverses, it can experience the maximum sheath potential. As the frequency is raised, the ion cannot cross the sheath before the field reverses, so it experiences the average sheath potential, which is approximately one-third the maximum. Therefore ion bombardment will be less intense at the highest frequencies.

4.4.3 Equipment for PECVD

Plasma processing occurs in one of two general types of plasma reactors: a barrel (or tube) reactor or a planar reactor. These are illustrated in Fig. 4.18 [She87]. In barrel systems the plasma is excited with inductive coils or capacitive electrodes outside of the quartz or glass tube. The substrates are generally held in the vertical position by a wafer holder and are immersed in the plasma with no electrical bias applied. The wafer surfaces are at a floating potential near that of the glow discharge. Uniformity of growth, or etch rate across a wafer, is almost impossible to obtain using tube reactors. This follows from nonuniformity in the plasma, the gas flow pattern, and the wafer temperature.

Good uniformity can be achieved using the radial flow planar reactor shown in Fig. 4.19. The reactant gases are introduced at the outer radius and flow radially between the electrodes. Substrates are placed flat on the lower electrode which is used to heat them. The basic radial flow planar reactor was announced in 1973 [Rei73] and has become enormously popular.

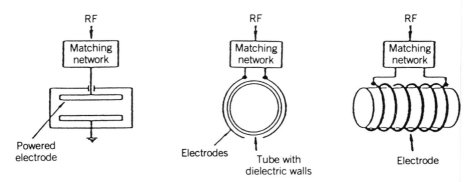

(a) Planar electrode system   (b) Glass shell electrode system   (c) Coil type electrode system

Fig. 4.18 Geometries of plasma-assisted CVD reactors: (*a*) parallel-plate discharge, (*b*) tube with capacitive coupling, (*c*) tube with inductive coupling. [Sherman, A., in *Chemical Vapor Deposition for Microelectronics, Principles, Technology, and Applications*, Noyes Publication (1987)]

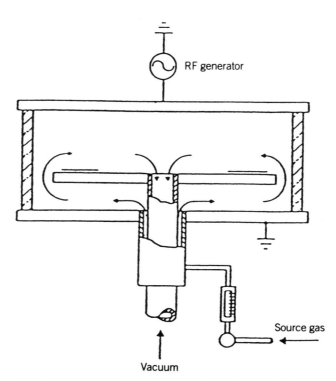

Fig. 4.19 Radial-flow, plasma-enhanced CVD reactor after Reinberg. [Reinberg, A. R., Radial flow reactor, U.S. Patent 3,757,733, Sept. 11, (1973); Sherman, A., in *Chemical Vapor Deposition for Microelectronics, Principles, Technology, and Applications*, Noyes Publication (1987)]

Plasma processing may employ or generate highly reactive species, including oxygen, chlorine, or fluorine, and/or explosive gases, such as hydrogen. Special pumps, pump fluids, or cold traps used to handle these species may be required for safety precautions and better performances. Good vacuum techniques is needed for seals and pumps. A minor leak in the system may allow contaminants into the plasma reactor, and such contaminants can play havoc with processes.

4.4.4 Electron-Cyclotron-Resonance (ECR) Plasma CVD

The plasma CVD method has an advantage of low substrate temperature; however, the method has a disadvantage of degradation caused by hydrogen contamination of source materials such as silane (SiH_4) and ammonia (NH_3). Pure nitrogen (N_2) has been used in place of NH_3 to eliminate the hydrogen contamination, but the film's quality is insufficient comparing to films with NH_3. It has been considered that the nitrogen may not be sufficiently activated. Electron-cyclotron-resonance (ECR) plasma CVD has been developed [Mat82] and become attractive as a method of silicon nitride film formation.

The ECR plasma CVD apparatus used for the deposition of silicon nitride films is schematically shown in Fig. 4.20. Microwave power with frequency at 2.45 GHz is introduced into the plasma chamber through a rectangular wave guide and a window made of a fused quartz plate. The interior of the plasma chamber is 20 cm in diameter and 20 cm in height, and it operates as a microwave cavity resonator (TE_{113}). Magnetic coils are arranged around the periphery of the chamber for ECR plasma excitation. The electron cyclotron frequency is controlled by the magnet coils so as to coincide with the microwave frequency (magnetic flux density, 875G). The ECR condition enables the plasma to effectively absorb the microwave energy. Highly activated

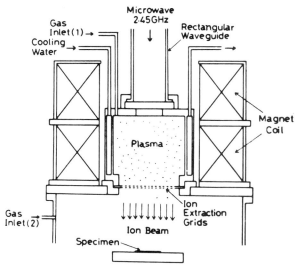

Fig. 4.20 ECR plasma deposition apparatus. Deposition area, 20 cm in diameter. Gas pressure, about 10^{-4} torr. Substrate without heating. [Matsuo, S., and Y. Adachi, *Jap. J. Appl. Phys.* **21**, L4 (1982)]

plasma is easily obtained at low gas pressures in the range from 10^{-5} to 10^{-3} torr.

Deposition gases are introduced through two respective inlet systems into the plasma chamber and the specimen chamber. The plasma chamber and the magnetic coils are water cooled. The vacuum system consists of an oil diffusion pump (2400 liter/min) and a mechanical rotary pump (500 liter/min).

The divergent magnetic field that has been developed for plasma extraction from the plasma chamber to the specimen chamber is gradually weakened from the plasma chamber to the specimen table, as shown in Fig. 4.21. High-energy electrons in circular motions peculiar to ECR plasma are accelerated by the interaction between their magnetic moments and the magnetic field gradient, and they bring about a negative potential toward the specimen table, which is electrically isolated from the plasma chamber. Therefore a static electric field, which accelerates ions and decelerates electrons, is generated along the plasma stream between the plasma chamber and the specimen table, and the effective ion transport and ion bombardment to the specimen surface are enhanced during deposition.

The negative potential generated by the divergent magnetic field method is shown in Fig. 4.22 as a function of the distance from the plasma extraction window. The negative potential increases corresponding to the degree of decrease of the magnetic field intensity. The energy of the ion bombardment is of the order of 20 eV at the position of the specimen table, adding the effect of the plasma sheath generated at the specimen surface.

The plasma extraction window is 10 cm in diameter, and the plasma stream at the specimen table in the deposition area is 20 cm in diameter. Deposition uniformity is within ± 5% in the 10-cm-diameter middle area. The experiments on film deposition were carried out without substrate heating. The specimen temperature during deposi-

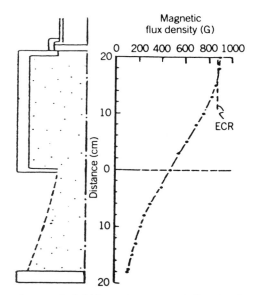

Fig. 4.21 Distribution of magnetic field intensity, magnetic flux density, from the top of plasma chamber to specimen table. [Matsuo, S., and M. Kiuchi, *Jap. J. Appl. Phys.* **22**, 4 (1983): L210–212]

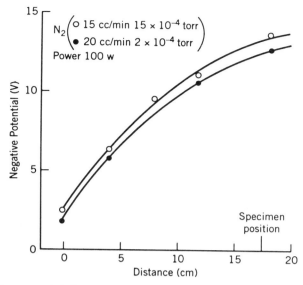

Fig. 4.22 Negative potential distribution with respect to distance, with an N_2 flow rate as a parameter. [Matsuo, S., and M. Kiuchi, *Jap. J. Appl. Phys.* **22**, 4 (1983): L210–212]

tion was in the range 50° to 150°C, due to some heating effect by the plasma. The ECR plasma deposition apparatus can realize a deposition of dense and high-quality films, such as Si_3N_4 and SiO_2, without the need for substrate heating [Mat83].

4.4.5 Thin Film Deposition

The major materials that have been grown using PECVD are silicon nitride and silicon dioxide. These materials are used primarily for interlevel dielectrics, capacitor dielectrics, and encapsulation layers, where they provide scratch protection. Thin film materials are usually grown to thicknesses of 0.1 to 1.0 μm. Growth rate is on the order of 100 to 500 Å/min. The thin films will usually have pinholes in them. Pinhole density can be determined by etching the substrate and film in an etchant that does not attack the film but rather the underlying material. Silicon nitride (Si_3N_4), silicon dioxide (SiO_2), silicon oxynitride (SiO_xN_y), and aluminium nitride (AlN) are treated in the following sections.

Silicon Oxide

Silicon dioxide (SiO_2) is used in GaAs processing for many of the same purposes as silicon nitride. Its low dielectric constant (compared to most other stable dielectrics including Si_3N_4) makes it a popular choice for use as an interlevel spacer to separate metal crossovers with minimum parasitic capacitance. The plasma-deposited silicon oxide may have a Si/O ratio that departs from 1/2, and a certain amount of hydrogen will exist in the film. Nitrogen from either the input gases or from background sources is very difficult to exclude completely. PECVD of silicon oxide is generally done using silane as the silicon source, and O_2, N_2O, CO, or another similar gas as the

148 PROCESS TECHNIQUES

oxygen source. The desired reaction, for example, using N_2O, would be [Wil84]

$$SiH_4 + 2N_2O \longrightarrow SiO_2 + 2N_2 + 2H_2. \tag{4.10}$$

Oxygen has a much greater affinity for reacting with silane than does nitrogen. Hence silicon oxide formation dominates over silicon nitride formation. The same type of radial flow planar reactors described for deposition of silicon nitride are also suitable for silicon oxide deposition, often without major changes in operating parameters other than gas composition and flow rates.

Silicon Nitride

Silicon nitride (Si_3N_4) is the dielectric widely used in GaAs processing. It is a better diffusion barrier than silicon dioxide [Hol78], so it is superior for encapsulation or glassification. It is preferred for capacitor dielectrics, since it has a higher dielectric constant then silicon dioxide. General information about PECVD Si_3N_4 may be found in several references [Hol78; Ros76; Swa67].

Silane (SiH_4) is generally used for silicon source and nitrogen or ammonia for the nitrogen source. Ammonia is preferred because it has a lower ionization potential then N_2. If nitrogen is used, it is difficult to control the refractive index to be in the desired range (~ 2.0) and still flow enough silane to get practical deposition rates [Ros76]. Silane is generally mixed with an insert carrier gas such as argon. The desired, overall reaction would be [Wil84]

$$3SiH_4 + 4NH_3 \longrightarrow Si_3N_4 + 12H_2, \tag{4.11}$$

or

$$3SiH_4 + 2N_2 \longrightarrow Si_3N_4 + 6H_2. \tag{4.12}$$

The plasma conditions make it likely that SiH_4, SiH_3, and SiH_2 are present as well as NH_3, NH_2, and ionized hydrogen. Oxygen may be present from water vapor. Carbon may be present from background hydrocarbons, such as pump oil. The resulting "silicon nitride" may include impurities like Si, N, O, H, and C. Plasma Si_3N_4 usually contains 20–25% hydrogen [Lan78]. Greater amounts of hydrogen are incorporated at lower growth temperatures. Some of the hydrogen can be removed by annealing [Hol78]. The Si/N ratio will be a function of the operating parameters and will not necessary be 3/4. The departure of PECVD silicon nitride from true, stoichiometric silicon nitride grown by high-temperature CVD is indicated by the etch rate. PECVD silicon nitride will etch five to ten times faster in acid etches than will the CVD nitride [Hol78].

Figure 4.23 illustrates the increase in the index of refraction as the SiH_4/N_2 ratio increases, forming silicon-rich films [Ger72]. This ratio must be kept low to obtain the desired (~ 2.0) index of refraction. The dependence of the index of refraction and the growth rate on pressure and temperature is shown in Figs. 4.24 and 4.25. The growth rate decreases with increasing pressure or temperature, although the pressure dependence is minimal. The relationship among density, composition (Si/N ratio), and refractive index of plasma nitride films is indicated by the Lorentz-Lorenz relationship show in Fig. 4.26. The depositioh rate increases with the concentration of silane

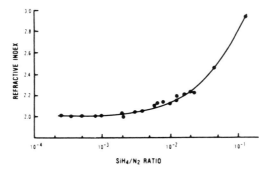

Fig. 4.23 Typical data illustrating the dependence of the refractive index of a PECVD silicon nitride film on the SiH_4/N_2 ratio. [Williams, R. E., *Gallium Arsenide Processing Techniques*, p. 175, Artech House (1985); Gereth, R., and W. Scherber, *J. Electrochem. Soc.* **119** (1972): 1248]

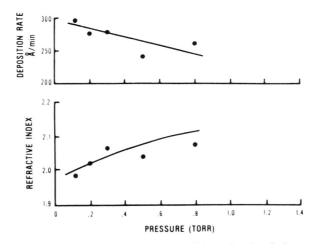

Fig. 4.24 Typical data illustrating the dependence of the refractive index and the deposition rate of a PECVD silicon nitride film on total pressure. [Williams, R. E., *Gallium Arsenide Processing Techniques*, Artech House (1985); Rosler, R. S., W. C. Benzing, and J. Baldo, *Solid State Tech.* **19**, 6 (1976): 45]

[Ros76; Swa67]. Any nonuniformity in the film thickness can be caused by differential depletion of reactant gases as they flow across the wafers. The surface can be improved by higher flow rates. For a given temperature, the etch rate of the silicon nitride film in HF is a function of the refractive index, as shown in Fig. 4.27. Figure 4.28 shows the silicon nitride etch rate in HF as a function of deposition temperature [Ros76].

The dielectric constant increases with the concentration of silane [Swa67]. It also increases with decreased film thickness, especially below 1500 Å, as shown in Fig. 4.29. The dependence on film thickness has been attributed to stress [Swa67] or to nonuniform composition in the vertical direction resulting from nonuniformities during initiation of growth.

Silicon nitride films produced by ECR have proved to be comparable to those

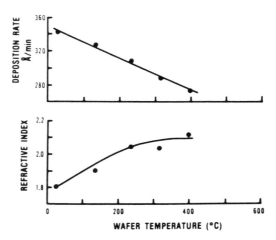

Fig. 4.25 Typical data illustrating the dependence of the refractive index and the deposition rate of a PECVD silicon nitride film on the deposition temperature. [Williams, R. E., *Gallium Arsenide Processing Techniques*, Artech House (1985); Rosler, R. S., W. C. Benzing, and J. Baldo, *Solid State Tech.* **19**, 6 (1976): 45]

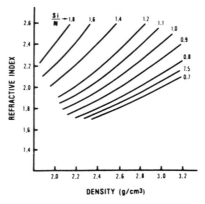

Fig. 4.26 Typical Lorentz-Lorenz correlation curves for PECVD silicon nitride. [Williams, R. E., *Gallium Arsenide Processing Techniques*, Artech House (1985); Sinha, A. K., *Electrochem. Soc. Extended Abstr.* 76-2 (1976): 625]

created in parallel-plate reactors in terms of stoichiometry and hydrogen content, as can be seen from the data in Figs. 4.30 and 4.31. Silicon nitride films deposited in parallel-plate, plasma-enhanced CVD reactors typically have a refractive index on the order of 2.0, partly because of hydrogen incorporated into the layer. The ECR films appear similar. The buffer oxide etch (BOE) rate for such films gives some indication of their hydrogen content, with high rates indicating very high atomic percentages of hydrogen. Etch rates for silicon nitride films produced in parallel-plate reactors will frequently range up to 100 Å/min, so the ECR films appear to be of good bulk quality [She87].

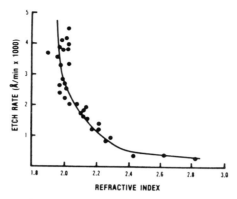

Fig. 4.27 PECVD silicon nitride etch rate in 48% HF as a function of refractive index. [Williams, R. E., *Gallium Arsenide Processing Techniques*, Artech House (1985); Rosler, R. S., W. C. Benzing, and J. Baldo, *Solid State Tech.* **19**, 6 (1976): 45]

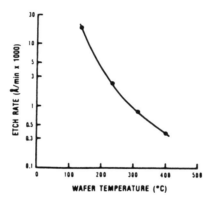

Fig. 4.28 PECVD silicon nitride etch rate in 48% HF as a function of deposition temperature. [Williams, R. E., *Gallium Arsenide Processing Techniques*, Artech House (1985); Rosler, R. S., W. C. Benzing, and J. Baldo, *Solid State Tech.* **19**, 6 (1976): 45]

Silicon Oxynitride

The silicon oxynitride films on GaAs can be formed by the plasma-enhanced chemical vapor deposition. The formation of high-quality dielectric film on GaAs and on related compound semiconductors has roused considerable interest because of its potential use in the fabrication of high-speed metal-insulator-semiconductor (MIS) devices, surface passivation for ion implantation annealing, and various device-processing techniques [Wu86]. Native oxide on GaAs grown by anodic oxidation or plasma oxidation and dielectric films such as SiO_x, SiN_y, and Al_2O_3, formed on GaAs by chemical vapor deposition or sputtering have been studied by earlier workers [Saw79; Ito74; Qua72; Miy75; Mim78; Yok79; Yok79b; Mei77]. Previous studies have shown that both the anodic oxide/GaAs and the plasma oxide/GaAs interface move into the crystal during oxidation and that the stoichiometry at the interface

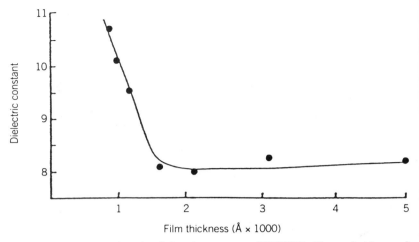

Fig. 4.29 Typical data showing the dielectric constant of PECVD silicon nitride as a function of final thickness. [Williams, R. E., *Gallium Arsenide Processing Techniques*, Artech House (1985); Swann, R. C. G., R. R. Mehta, and T. P. Cauge, *J. Electrochem. Soc.* **114** (1967): 713]

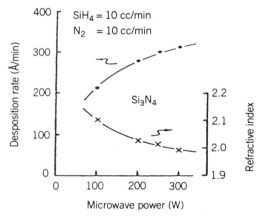

Fig. 4.30 Deposition rate and refraction index for silicon nitride film as a function of microwave power for ECR reactor. [Sherman, A., in *Chemical Vapor Deposition for Microelectronics, Principles, Technology, and Applications*, Noyes Publication (1987); Matsuo, S., and M. Kiuchi, *Jap. J. Appl. Phys.* **22**, 4 (1983): L210–212]

may be uncontrollable due to different oxidation rates of gallium and arsenic atoms [Yok79].

SiO_x grown at low temperature ($< 400°C$) exhibits several notable properties such as low hydrogen content, low dielectric constant, scratch resistance [Hes84], large barrier height, poor passivation in water and ion contamination, and tensile stress. SiN_y films, in contrast, have better passivation, a higher dielectric constant, good adhesion, high conformity, crack resistance, no pinholes, higher hydrogen content, lower barrier height, and compressive stress. Because of these complementary properties, Wu and

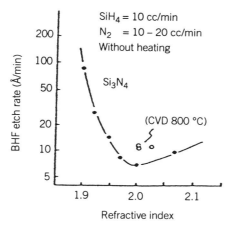

Fig. 4.31 BOE etch rate versus refractive index for silicon nitride films for the ECR reactor. [Sherman, A., in *Chemical Vapor Deposition for Microelectronics, Principles, Technology, and Applications*, Noyes Publication (1987); Matsuo, S., and M. Kiuchi, *Jap. J. Appl. Phys.* **22**, 4 (1983): L210–212]

Lin [Wu86] have grown a SiO_xN_y film that is a mixture of SiO_x and SiN_y in order to obtain a better-quality dielectric film that possesses the best characteristics of both materials for GaAs MIS application.

Aluminum Nitride

Aluminum nitride (AlN) is a wide band-gap material ($\sim 6.2\,eV$) that can be used in an insulator or in passivation film. The AlN has been grown by several methods such as vapor phase growth [Man71; Yim73; Cal74], sputtering [Duc71], and reactive MBE [Yos77; Yam79]. Most of these are single crystalline films on sapphire or Si substrates, and the growth temperature ranges from 1100° to 1200°C. However, to use AlN film as an insulator of MISFETs or a passivation films, the growth temperature must be less than 500°C. Amorphous AlN film can be deposited at reasonably low temperatures by a plasma CVD if Al is supplied by TMAl and N is supplied by NH_3 via separate gas lines [Has86].

4.5 RAPID THERMAL PROCESS

In recent years there have been two major trends in semiconductor processing: (1) a reduction in the dimensions of devices to increase speed and packing density, and (2) an introduction of larger wafer sizes in order to increase the number of circuits per wafer and thus reduce processing costs [Wils86]. Both trends put constraints on the thermal processing to which a wafer is subjected.

As the dimensions of devices are reduced, junction depths must also be reduced and dopant diffusion precisely controlled. In particular, in thermal processing dopant diffusion must be minimized during and subsequent to the doping steps. Two approaches have been tried to minimize such diffusion: (1) high pressure oxidation and lower reflow temperature of CVD glasses in *low-temperature processing*, and (2) shortened time in *high-temperature processing* [Wol86]. In a conventional furnace, however, it is difficult to shorten the time accurately because of thermal stresses, the effect of radial heating and cooling, and the large thermal mass of a boat load of wafers.

To overcome these problems many scientists and engineers have been working in the area of rapid thermal processing (RTP). In RTP a single wafer is heated to temperatures of 400° to 1400°C (depending on the process desired) for times on the order of 1 to 60 s. The wafers are thermally isolated, and the heating and cooling are by radiation. The temperature and time can be precisely controlled on a given wafer and from wafer to wafer.

RTP potentially covers a range of time durations from nanoseconds to about 100 seconds. The shortest time regimes are implemented with pulsed lasers and electron beams. Step thermal gradients exist for the shortest time regimes, and both the spatial and time temperature dependencies follow the laws of thermal diffusivity. Annealing with times above 1 second give nearly isothermal wafer heating, and the case may have more widespread practical applications [Sei87].

RTP has been applied to a variety of thermal processing steps [Wils86; Wils85], among these annealing of ion implants [Wils84], formation of silicides [Pra85], reflow of phosphosilicate and borophosphosilicate glass [Hir84], contact formation [Pai85; Kat89], dopant activation and grain growth in polycrystalline silicon [Wils85b], growth of thin oxides and nitrides [Gat85], epitaxial silicon deposition [Gib85], and GaAs device processing. In this section we discuss recent activities in rapid thermal processing on III–V compound materials. Our primary focus will be on its application to high-speed GaAs-integrated circuits [Kuz87]. Thermal treatments of GaAs include ion implantation for FET channels, contacts, and isolations.

4.5.1 n-Type Channel Implants

GaAs-integrated circuit technology requires uniformly activated, high-quality, shallow n-type channel layers with controlled doping levels. Furnace annealing involves long heating times and may have various undesirable effects, such as substrate quality degradation and implanted dopant diffusion in both vertical and lateral directions. Detailed activation conditions for channel implants were reported by Kuzuhara et al. [Kuz82]. They annealed $5 \times 10^{12} \, \text{cm}^{-2}$ Si implants under various conditions using a PID-controlled RTA apparatus. Sheet carrier concentration and sheet mobility dependence on annealing temperature are shown in Fig. 4.32. All samples were annealed for 2 s under unencapsulated conditions with an implanted surface facing a Si support wafer. Electrical activation reaches a maximum at 950° to 1000°C, beyond which it begins to decrease drastically. The sheet mobility remains unchanged even at high temperatures, suggesting that the observed activation degradation is not associated with As loss caused by surface dissociation.

4.5.2 n^+ Contact Implants

Unlike the case of low-dose implants, it is rather difficult to obtain sufficient electrical activation for high-dose donor implants. Achievable doping levels have been limited to below $2 \times 10^{18} \, \text{cm}^{-2}$, when samples were processed using conventional furnace annealing [Kuz87]. Davies et al. [Dav82] reported the first successful RTA application

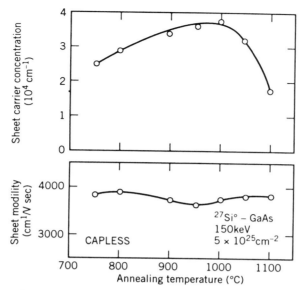

Fig. 4.32 Sheet carrier concentration and sheet mobility dependence on RTA temperature (capless). [Kuzuhara, M., in *Mat. Res. Soc. Symp. Proc.*, vol. 92: *Rapid Thermal Processing of Electronic Materials*, pp. 401–410, ed. by S. R. Wilson, R. Powell, and D. E. Davies, MRS (1987)]

to high-dose (200 keV, 4×10^{14} cm^{-2}) Si implant activation in which a peak electron concentration as high as 6.5×10^{-3} was realized by 1000°C, 2.5 s RTA using an Si$_3$N$_4$ encapsulant.

Figure 4.33 illustrates the electrical activation for 150 keV, 7×10^{13} cm^{-2} Si implants as a function of RTA temperature ranging from 800° to 1100°C. All samples were annealed for 4 s using three kinds of encapsulants: SiO$_2$, Si$_3$N$_4$, and SiO$_x$N$_y$. For each encapsulating material, electrical activation increases with the increase in RTA temperature. Figure 4.34 shows a depth profile of the carrier concentration for 150 keV, 4×10^{14} cm^{-2} Si implants after SiO$_x$N$_y$ encapsulated RTA at 1120°C in 5 s. A peak electron concentration of 9×10^{18} cm^{-3} is obtained with negligible dopant diffusion. An attractive application of heavily doped layers is the formation of non-alloyed ohmic contacts [Kuz87].

4.5.3 Slip Lines and Wafer Distortion

One of the striking differences between RTA and furnace annealing is that the annealed wafer is thermally isolated from the surrounding ambient in RTA. This situation tends to induce temperature gradients across the wafer during annealing, resulting in formation of some undesirable defects, such as crystallographic slip lines and wafer distortion.

The slip lines are generally evident around the periphery of the wafer [Pea86], extending 5 to 15 mm toward the center [Blu85]. Blunt et al. [Blu85] have analyzed the possible mechanism responsible for inducing the temperature gradient across a rapidly annealed 2-in.-diameter GaAs wafer, placed on a 4-in.-diameter Si support

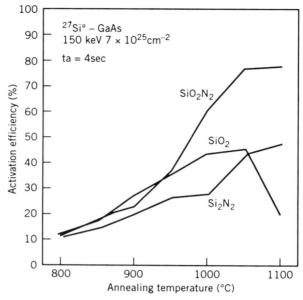

Fig. 4.33 Activation dependence on RTA temperature. [Kuzuhara, M., in *Mat. Res. Soc. Symp. Proc.*, vol. 92: *Rapid Thermal Processing of Electronic Materials*, pp. 401–410, ed. by S. R. Wilson, R. Powell, and D. E. Davies, MRS (1987)]

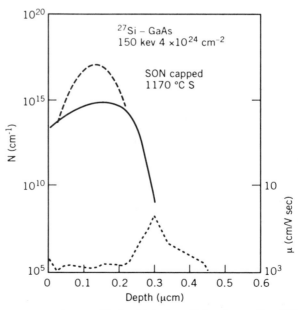

Fig. 4.34 Carrier concentration profile for high-dose Si implants after SiO_xN_y capped RTA. [Kuzuhara, M., in *Mat. Res. Soc. Sym. Proc.*, vol. 92: *Rapid Thermal Processing of Electronic Materials*, pp. 401–410, ed. by S. R. Wilson, R. Powell, and D. E. Davies, MRS (1987)]

wafer. They have suggested that at least two mechanisms give rise to the temperature nonuniformities across the GaAs wafer under uniform irradiation conditions over the wafer's surface. The first mechanism originates from radiative heat losses at the wafer edges. The second mechanism is due to difference in thermal mass across the large-diameter Si wafer supporting the GaAs wafer to be annealed. The former is a well-recognized effect in the Si RTA processing [Kom84]; the latter relates to GaAs RTA processing which employs Si support wafers. The heat transfer by these mechanisms during RTA is illustrated in Fig. 4.35 [Kuz87].

A guard ring structure placed around the wafer has been shown to be an efficient in reducing the temperature nonuniformities [Blu85; Tam86; Cum86]. The temperature gradients are largely consumed in the guard ring region. Thus the thermoplastic stresses, which cause the crystallographic slip lines near the peripheral region of the wafer, are considerably reduced.

4.5.4 Rapid Thermal Processing Equipment

A schematic of an ideal RTP unit is shown in Fig. 4.36 [Wils86]. The setup consists of one or more heat sources that cause the wafer to heat and cool quickly and uniformly. The wafer holder is of low thermal mass and very low thermal conductivity, so the contact of the wafer and holder does not produce a temperature gradient across the wafer. This setup is necessary to prevent slip or a nonuniform process. The wafer temperature is precisely measured with a temperature monitor, and this information is fed back to a microprocessor that controls the radiation on the wafer. The ambient in the process chamber may be vacuum or a process gas depending on the thermal process being performed. The walls of the process chamber may be highly reflective or highly absorbing of radiation. The process chamber temperature gradually rises with time and thus changes the process conditions between the first and last wafer in the group.

Wafer transport is a critical issue in RTP. The RTP unit must move wafers through the system without breaking them and at a throughput that is compatible with the rest of the process line. The ideal rapid thermal processor has a computer for programming and storing process recipes. The computer then operates the equipment in a fully automatic mode.

RTP systems use radiation sources to heat the wafers. An ideal radiation source has low initial and operating costs as well as high reliability. The radiation source has a fast response time so that a feedback loop is established between the wafer temperature and the source output.

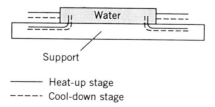

Fig. 4.35 Heat transfer caused by edge radiation and thermal mass effects during RTA. [Kuzuhara, M., in *Mat. Res. Soc. Sym. Proc.*, vol. 92: *Rapid Thermal Processing of Electronic Materials*, pp. 401–410, ed. by S. R. Wilson, R. Powell, and D. E. Davies, MRS (1987)]

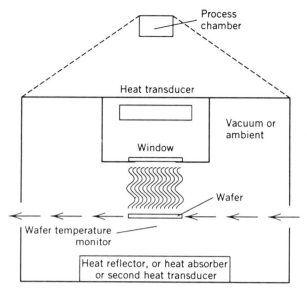

Fig. 4.36 Schematic of an ideal RTP unit. [Wilson, S. R., Gregory, R. B., and Paulson, W. M., *Mat. Res. Soc. Symp. Proc.*, vol. 52, p. 181, MRS (1986)]

The three most commonly used sources are tungsten-halogen lamps, water-walled arc lamps, and a planar sheet of resistively heated graphite. The tungsten-halogen lamps have been around for more than 30 years. These lamps have a tungsten filament inside a quartz tube, and inside the tube is a small amount of one of the halogens (usually bromine). The tungsten filament is resistively heated, and the quartz tube reaches an operating temperature of 300°–400°C. At these temperatures the tungsten evaporated from the filament reacts with the bromine to form a volatile tungsten bromide compound. As these gas molecules reach the filament the tungsten will deposit on the filament and release the bromine for another cycle. This reduces the darkening of the quartz tube and allows the filament to be operated at higher temperatures. Since each tungsten-halogen lamp is rated for only a few kilowatts, an array of lamps is required to heat an entire wafer.

The water-walled arc lamp is essentially a dc arc lamp with an improved cooling and stabilization system [Wils86]. This lamp uses water cooling which is injected in such a way that a wall of water spirals around the inside of the quartz tube at high speed. This forms a thin layer with an empty region in the center. An argon gas mixture is injected such that it spirals in the center region. A discharge is induced in the spiraling gas column, producing a stable and tightly positioned arc. The flowing water also traps any sputtered material from the electrodes, preventing it from depositing on the tube walls and darkening the tube. The flowing water provides very efficient cooling and prevents the tube or end seals from heating significantly. Due to this efficient cooling the lamp has a power capability of 100 kW. The spectral irradiance of this lamp is essentially that of a 5500 K blackbody radiator [Piv61], with some line structure in the 600- to 1000-nm region [Gep83]. The spectral distribution extends from the quartz cutoff in the ultraviolet at about 330 nm into the infrared. The arc represents a dimensionally stable line, which makes it possible to design a reflector to produce a uniform, large-area radiation field.

Another heater source commercially available is the planar graphite heater. It consists of a slotted sheet of graphite which is resistively heated to temperatures of 1000° to 1400°C. The plate heater is dished out in the center to provide a more uniform radiation across a large wafer. The plate heater is kept in a vacuum to prevent oxidation of the heater when it is at high temperature. The heater must be vacuum fired after construction to eliminate mobile ion impurities. Because of the thermal mass of this source, the response time of this heater is much slower than the lamp systems. Therefore, instead of controlling the output power of this source, the exposure time of the wafer to the source is controlled by a shutter between the source and the wafer. The feedback loop is from the wafer temperature sensor to the shutter rather than the source power supply. This source acts exactly like a blackbody radiator with the spectral distribution determined by the graphite temperature. At the normal operating temperatures (1000° to 1400°C), most of the spectral distribution is in the infrared.

There are more than five different RTP vendors in the United States and a few other vendors overseas. There are small systems or large units with serial processing cassette-to-cassette capabilities to heat wafers from room temperature to process temperature in the range 400°–1400°C. A variety of process ambients are available.

REFERENCES

[Wil84] Williams, R. E., *Gallium Arsenide Processing Techniques*, Artech House (1984).

[Dou81] Douglas, E. C., *Solid State Tech.* (May 1981): 65.

[Hel81] Helmke, G. E., *Semiconductor Int.* 119 (Aug. 1981): 119.

[Wil90] Williams, R. E., *Gallium Arsenide Processing Techniques*, 2d ed., Artech House (1990).

[Ash86] Ashby, C. I. H., in *Properties of Gallium Arsenide*, sec. 16.1, IEE INSPEC Publication (1986).

[Ker78] Kern, W., RCA Rev. **39** (1978): 278.

[Gan74] Gannon, J. J., and C. J. Nuese, *J. Electrochem. Soc.* **121** (1974): 1215.

[Iid71] Iida, S., and K. Ito, *J. Electrochem. Soc.* **118** (1971): 118.

[Mer76] Mertz, J. L., and R. A. Logan, *J. Appl. Phys.* **47** (1976): 3503.

[Nue70] Nuese, C. J., and J. J. Gannon, *J. Electrochem. Soc.* **117** (1970): 1094.

[Har67] Harvey, W. W., *J. Electrochem. Soc.* **114** (1967): 472.

[Sch76] Schwartz, B., F. Ermanis, and M. H. Brastad, *J. Electrochem. Soc.* **123** (1976): 1089.

[Tar73] Tarui, Y., Y. Komiya, and T. Yamaguchi, *J. Jap. Soc. Appl. Phys.*, suppl. **42** (1973): 78.

[Ots76] Otsubo, M., T. Oda, H. Kumabe, and H. Miiki, *J. Electrochem. Soc.* **123** (1976): 676.

[Yen83] Yenigalla, S. P., and C. L. Ghosh, *J. Electrochem. Soc.* **130** (1983): 1377.

[Sha81] Shaw, D. W., *J. Electrochem. Soc.* **128** (1981): 847.

[Abr65] Abrahams, M. S., and C. J. Buiocchi, *J. Appl. Phys.* **36** (1965): 2855.

[Tuc75] Tuck, B., *J. Mat. Sci.* **10** (1975): 321.

[Sti76] Stirland, D. J., and B. W. Straughan, *Thin Solid Films* **31** (1976): 139.

[Gan74] Gannon, J. J., and C. J. Nuese, *J. Electrochem. Soc.* **121** (1974): 1510.

[Gre77] Greene, L. I., *J. Appl. Phys.* **48** (1977): 3739.

[Ken82] Kenefick, K., *J. Electrochem. Soc.* **129** (1982): 2380.

[Shi77] Shiota, I., K. Motoya, T. Ohmi, N. Miyamoto, and J. Nishizawa, *J. Electrochem. Soc.* **124** (1977): 155.

[Koh80] Kohn, E., *J. Electrochem. Soc.* **127** (1980): 505.

[Nis79] Nishizawa, J., Y. Oyama, H. Tadano, K. Inokuchi, and Y. Okuno, *J. Cryst. Growth* **47** (1979): 434.

[Mor78] Mori, Y., and N. Watanabe, *J. Electrochem. Soc.* **125** (1978): 1510.

[Muk85] Mukerjee, S. D., and D. W. Woodard, in *Gallium Arsenide*, ed. by M. J. Howes and D. V. Morgan, Wiley (1985).

[Log73] Logan, R. A., and F. K. Reinhart, *J. Appl. Phys.* **44** (1973): 4172.

[Ash85] Ashby, C. I. H., in *Properties of Gallium Arsenide*, IEE INSPEC (1985).

[Ell80] Elliott, C. R., and J. C. Regnault, *J. Electrochem. Soc.* **127** (1980): 1557.

[Yam75] Yamamoto, A., and S. Yano, *J. Electrochem. Soc.* **122** (1975): 260.

[Hof81] Hoffmann, H. J., J. M. Woodall, and T. I. Chappell, *Appl. Phys. Lett.* **38** (1981): 564.

[Shi79] Shimano, A., H. Takagi, and G. Kano, *IEEE Trans. Electron Dev.* **ED-26** (1979): 1690.

[Fon85] Fonash, S. J., *Solid State Tech.* **28**, 1 (Jan. 1985): 150–158.

[Ibb88] Ibbotson, D. E., and D. L. Flamm, *Solid State Tech.* **31** (Oct. 1988): 77–79.

[Ibb83] Ibbotson, D. E., D. L. Flamm, and V. M. Donnelly, *J. Appl. Phys.* **54** (1983): 5974.

[Ash90] Ashby, C. I. H., in *Properties of GaAs*, 2d ed. Sec. 20.4, IEE INSPEC (1990).

[Sch84] Schwartz, G. P., G. J. Gualtieri, L. H. Dubois, and W. A. Bonner, *J. Electrochem. Soc.* **131** (1984): 1716.

[Smo81] Smolinsky, G., R. P. Chang, and T. M. Mayer, *J. Vac. Sci. Tech.* **18** (1981): 12.

[Don82] Donnelly, V. M., D. L. Flamm, C. W. Tu, and D. E. Ibbotson, *J. Electrochem. Soc.* **129** (1982): 2533.

[Got82] Gottscho, R. A., G. Smolinsky, and R. H. Burton, *J. Appl. Phys.* **53** (1982): 5908.

[Koe70] Koeing, H. R., and L. I. Maissel, *IBM J. Res. Dev.* **14** (1970): 172.

[Ibb88b] Ibbotson, D. E., and D. L. Flamm, *Solid State Tech.* **31**, 11 (1988): 105–108.

[Cha81] Chang, R. P. H., and S. Darack, *Appl. Phys. Lett.* **38** (1981): 898.

[Don84] Donnelly, V. M., D. L. Flamm, and D. E. Ibbotson, in *Ion Bombardment Modification of Surfaces*, ch. 8, p. 323, ed. by O. Auciello, R. Kelly, Elsevier Science (1984).

[Gor84] Gorowitz, B., and R. J. Saia, in *VLSI Electronics*, vol. 8, pp. 69–90, Academic Press (1984).

[Sch88] Schultz, R. J., in *VLSI Technology*, ed. by S. M. Sze, 2d ed. (1988).

[Cha78] Chapman, B. N., and V. J. Minkiewicz, *J. Vac. Sci. Tech.* **15** (1978): 329.

[Mog77] Mogab, C. J., *J. Electrochem. Soc.* **124** (1977): 437.

[Mel78] Melliar-Smith, C. M., and C. J. Mogab, in *Thin Film Processes*, p. 497, ed. by J. L. Vossen and W. Kerns, Academic Press, New York (1978).

[Sch88] Schultz, R. J., in *VLSI Technology*, ed. by S. M. Sze ed., 2d ed. (1988).

[Ash85] Ashby, C. I. H., in *Properties of GaAs*, sec. 16.4, IEE Press (1985).

[Hu84] Hu, E. L., and R. E. Howard, *J. Vac. Sci. Tech.* B **2** (1984): 85.

[Hu80] Hu, E. L., and R. E. Howard, *Appl. Phys. Lett.* **37** (1980): 1022.

[Tam84] Tamura, H., and H. Kurihara, *Jap. J. Appl. Phys.* **23** (1984): L731.

[Yam85] Yamada, H. Ito, and H. Inaba, *J. Vac. Sci. Tech.* B **3** (1985): 884.

[Sem84a] Semura, S., H. Saitoh, and K. Asakawa, *J. Appl. Phys.* **55** (1984): 3131.

[Sem84b] Semura, S., and H. Saitoh, *J. Vac. Sci. Tech.* A **2** (1984): 474.

REFERENCES

[Hea84] Heath, B. A., and T. M. Mayer, in *VLSI Electronics*, vol. 11, pp. 365–410, Academic Press (1984).

[Chi11] Child, C. D., *Phys. Rev.* **32** (1911): 492.

[Mat82] Matsuo, S., and Y. Adachi, *Jap. J. Appl. Phys.* **21** (1982): L4.

[Miy83] Miyamura, M., T. Uchiyama, O. Tsukakoshi, and S. Komiya, in *Proc. Int. Ion Eng. Congress*, p. 1623, Kyoto, Japan (1983).

[Bro76] Broers, A. N., W. W. Molzen, J. J. Cumno, and N. D. Wittels, *Appl. Phys. Lett.* **29** (1976): 596.

[Lee84] Lee, R. H., in *VLSI Electronics: Microstructure Science*, vol. 8, Academic Press (1984).

[Asa87] Asakawa, K., NEC Corp., Japan/US Perspectives, University of Florida, October 26–28, (1987).

[Har87] Harriott, L. R., R. E. Scotti, K. D. Cummings, and A. F. Ambrose, *J. Vac. Sci. Tech.* B **5**, 1 (Jan./Feb. 1987): 207–210.

[Tuc83] Tucker, A. W., and M. Birnbaum, *Proc. Laser Processing Semiconductor Devices*, vol. 385, p. 131, SPIE (1983).

[Tak83] Takai, M., J. Tokuda, H. Nakai, K. Gamo, and S. Namba, *Jap. J. Appl. Phys.* **22** (1983): L757.

[Tak84] Takai, M., J. Tsuchimoto, H. Nakai, K. Gamo, and S. Namba, *Jap. J. Appl. Phys.* **23** (1984): L852.

[Bre85] Brewer, P. D., D. McClure, and R. M. Osgood, Jr., *Appl. Phys. Lett.* **47** (1985): 310.

[Tak84b] Takai, M., J. Tokuda, H. Nakai, K. Gamo, and S. Namba, *Mat. Res. Soc. Symp. Proc.* **29** (1984b): 211–216.

[Bre85] Brewer, P. D., G. M. Reksten, and R. M. Osgood, Jr., *Solid State Tech.* **28**, 4 (1985): 273.

[Bre85b] Brewer, P. D., S. Halle, and R. M. Osgood, Jr., *Appl. Phys. Lett.* **47** (1985): 310.

[Bre85c] Brewer, P. D., D. McClure, and R. M. Osgood, Jr., *Appl. Phys. Lett.* **47**, 3 (1985c): 310.

[Ash84] Ashby, C. I. H., *Appl. Phys. Lett.* **45** (1984): 892.

[Ash85a] Ashby, C. I. H., and R. M. Biefeld, *Appl. Phys. Lett.* **47** (1985): 62.

[Ash85b] Ashby, C. I. H., *Appl. Phys. Lett.* **46** (1985): 752.

[She87] Sherman, A., in *Chemical Vapor Deposition for Microelectronics, Principles, Technology, and Applications*, Noyes Publication (1987).

[Bro59] Brown, S. C., *Basic Data of Plasma Physics*, Wiley (1959).

[Bro56] Brown, S. C., Breakdown in gases: Alternating and high-frequency fields, in *Handbuch der Physik*, vol. 22, ed. by S. Flugge, Springer-Verlag (1956).

[But63] Butler, H. S., and G. S. Kino, *Phys. Fluids* **6** (1963): 1346.

[Bru81] Bruce, R. H., *Proc. Symp. on Plasma Etching and Deposition*, 81-1, p. 243, Electrochemical Society, Pennington, NJ (1981).

[Rei73] Reinberg, A. R., Radial Flow Reactor, U.S. Patent 3,757,733, Sept. 11, 1973.

[Mat83] Matsuo, S., and M. Kiuchi, *Jap. J. Appl. Phys.* **22**, 4 (April 1983): L210–L212.

[Hol78] Hollahan, J. R., and R. S. Rosler, in *Thin Film Process*, p. 335, ed. by J. L. Vossen and W. Kern, Academic Press (1978).

[Ros76] Rosler, R. S., W. C. Benzing, and J. A. Baldo, *Solid State Tech.* **19**, 6 (1976): 45.

[Swa67] Swann, R. C. G., R. R. Mehta, and T. P. Cauge, *J. Electrochem. Soc.* **114** (1967): 713.

[Lan78] Lanford, W. A., and M. J. Rand, *J. Appl. Phys.* **49** (1978): 2473.

[Ger72] Gereth, R., and W. Scherber, *J. Electrochem. Soc.* **119** (1972): 1248.

[Wu86] Wu, C. Y., and M. S. Lin, *1986 EDMS*, p. 224, Tainan (1986).

[Saw79] Sawada, T., and H. Hasegawa, *Thin Solid Film* **56** (1979): 183.

[Ito74] Itoh, T., and Y. Sokai, *Solid State Electron.* **17** (1974): 751.
[Qua72] Quast, W., *Electron Lett.* **8** (1972): 419.
[Miy75] Miyazaki, T., N. Nakamura, A. Doi, and T. Tokuyama, *Jap. J. Appl. Phys.*, suppl., **14**, 2, pt. 2 (1975).
[Mim78] Mimura, T., N. Yokoyama, Y. Nakayama, and M. Fukuta, *Jap. J. Appl. Phys.*, suppl., **17**, 1 (1978): 153.
[Yok79] Yokoyama, S., K. Yokitomo, M. Hirose, and Y. Osaka, *Thin Solid Films* **56** (1979): 81.
[Yok79b] Yokoyama, N., T. Mimura, K. Odani, and M. Fukuta, *Appl. Phys. Lett.* **32** (1979): 58.
[Mei77] Meiners, L. G., R. P. Pan, and J. R. Sites, *J. Vac. Sci. Tech.* **14** (1977): 961.
[Hes84] Hess, D. W., *J. Vac. Sci. Tech.* A **2**, 2 (1984): 244.
[Man71] Manesevit, H. M., F. M. Erdmann, and W. I. Simpson, *J. Electrochem. Soc.* **118** (1971): 1804.
[Yim73] Yim, W. M., *J. Appl. Phys.* **44** (1973): 292.
[Cal74] Callagham, M. P., *J. Cryst. Growth* **22** (1974): 85.
[Duc71] Duchene, J., *Thin Solid Film* **8** (1971): 69.
[Yos77] Yoshida, S., S. Misawa, and S. Itoh, *Proc. 7th Int. Vacuum Congress*, p. 1797, Vienna (1977).
[Yam79] Yamashita, H., S. Misawa, and S. Yoshida, *J. Appl. Phys.* **50** (1979): 896.
[Has86] Hasegawa, F., T. Takahashi, I. Onodera, and Y. Nannichi, *Extended Abstracts of the 18th Conf. on Solid State Devices and Materials*, p. 663, Japanese Society of Physics (1986).
[Wils86] Wilson, S. R., R. B. Gregory, and W. M. Paulson, *Mat. Res. Soc. Symp. Proc.* **52** (1986): 181.
[Wol86] Wolf, S., and R. N. Tauber, in *Silicon Processing for the VLSI Era*, vol. 1: *Process Technology*, Lattice Press, Sunset Beach, CA (1986).
[Sei87] Seidel, T. E., *The Impact of Rapid Thermal Processing*, p. 237, New England Combined Chapter of the American Vacuum Society (1987).
[Wils85] Wilson, S. R., W. M. Paulson, and R. B. Gregory, *Solid State Tech.* **28** (1985): 185.
[Wils84] Wilson, S. R., W. M. Paulson, R. B. Gregory, A. H. Hamdi, and F. D. McDaniel, *J. Appl. Phys.* **55** (1984): 4162.
[Pra85] Pramanik, D., M. Deal, A. N. Saxena, and K. T. Owen, *Semiconductional Int.* **8** (1985): 94.
[Hir84] Hiroshi, M., and M. Yoshimaru, *Proc. SEMI Technology Symp. 1984*, p. 1 (1984).
[Pai85] Pai, C. S., E. Cabreros, S. S. Lau, T. E. Seidel, and I. Suni, *Appl. Phys. Lett.* **46** (1985): 652.
[Kat89] Katz, A., P. M. Thomas, S. N. G. Chu, J. W. Lee, and W. C. Dautremont-Smith, *J. Appl. Phys.* **66** (1989): 2056.
[Wils85b] Wilson, S. R., R. B. Gregory, W. M. Paulson, S. J. Krause, J. D. Gressett, A. H. Hamdi, F. D. McDaniel, and R. G. Downing, *J. Electrochem. Soc.* **132** (1985): 922.
[Gat85] Gat, A., and J. Nulman, *Semiconductor Int.* **8** (1985): 120.
[Gib85] Gibbons, J. F., C. M. Gronet, and K. E. Williams, *Appl. Phys. Lett.* **47** (1985): 721.
[Kuz87] Kuzuhara, M., in *Materials Research Society Symp. Proc.*, vol. 92: *Rapid Thermal Processing of Electronic Materials*, pp. 401–410, ed. by S. R. Wilson, R. Powell, and D. E. Davies, MRS (1987).
[Kuz82] Kuzuhara, H. Kohzu, and Y. Takayama, *Appl. Phys. Lett.* **41** (1982): 755.
[Dav82] Davies, D. E., P. J. McNally, J. P. Lorenzo, and M. Julian, *IEEE Electron Device Lett.* **EDL-3** (1982): 102.

[Pea86] Pearton, S. J., J. M. Gibson, D. C. Jacobson, and J. M. Poate, *Mat. Res. Soc. Symp. Proc.* vol. 52, p. 351, MRS (1986).

[Blu85] Blunt, R. T., M. S. M. Lamb, and R. Szweda, *Appl. Phys. Lett.* **47** (1985): 304.

[Kom84] Komatsu, R., and K. Kajiyama, *J. Appl. Phys.* **56** (1984): 486.

[Blu85] Blunt, R. T., M. S. M. Lamb, and R. Szweda, *Appl. Phys. Lett.* **47** (1985): 304.

[Tam86] Tamura, A., T. Uenoyama, K. Inoue, and T. Onuma, *13th Int. Symp. GaAs and Related Compounds*, p. 533, Las Vegas (1986).

[Cum86] Cummings, K. D., S. J. Pearton, and G. P. Vella-Coleiro, *J. Appl. Phys.* **60** (1986): 163.

[Gib87] Gibbons, J. F., S. Reynolds, C. Gronet, D. Vook, C. King, W. Opyd, S. Wilson, C. Nauka, G. Reid, and R. Hull, *Mat. Res. Soc.* **92** (1987): 281.

[Piv61] Pivovonsky, M., and M. R. Nagel, *Tables of Blackbody Functions*, Macmillan (1961).

[Gep83] Gepley, J. C., and P. O. Stump, *Microelectronic Manufacturing and Testing* **6**, 9 (1983): 22.

5

LITHOGRAPHY

5.1 INTRODUCTION TO LITHOGRAPHY

In semiconductor processing area-patterning techniques are very important. Lithography is the process of transferring patterns of geometric shapes on a mask to a thin layer of radiation-sensitive material called *resist covering* the surface of a semiconductor wafer [Wil84]. These patterns define the various regions in an integrated circuit such as the implantation regions, the contact windows, and the bonding pad areas. The resist film is then subjected to a development process that selectively removes either the exposed or unexposed resist. The remaining pattern will be replicated in other materials by evaporation or etching. The exposure may be accomplished using light, ultraviolet light, electron beams, X rays, or ion beams. Photoresists can be classified as positive or negative, depending on how they respond to radiation. The positive type removes the exposed resist. The net result is that the patterns formed in the positive resist are the same as those on the mask. The negative type forms patterns that are the reverse of the mask patterns because the exposed regions are made less soluble. The image reversal techniques, which give better resolution, contrast, and process latitude will be discussed in this chapter. Photolithography is the dominant technique using in GaAs processing.

GaAs processing requires different lithography techniques than silicon processing [Wil90]. The basic silicon process uses resist as an etch mask. Consider a metal is being patterned such as that in Fig. 5.1(a). The process shown consists of applying metal on the wafer, spinning resist over the metal, exposing and developing the resist to define the pattern, etching away all exposed metal, and finally removing the remaining resist. The metal on the wafer now has the desired pattern. This is the etching process used in GaAs fabrication for patterning dielectric films, but it is rarely used to pattern metals, since many metal etchants will also attack the GaAs substrate. Silicon processing, however, often uses aluminum metalization, which can be etched with relatively innocuous etchants, whereas GaAs processes generally include metals

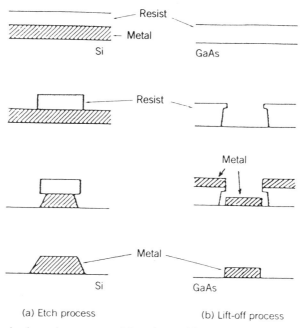

Fig. 5.1 The two basic resist process: (*a*) etch masking process typically used in silicon processing; (*b*) lift-off process typically used in GaAs processing. [Williams, R., *Modern GaAs Processing Methods*, Artech House (1990)]

such as gold that require for stronger etchants. There is also a tendency to use composite metalizations in GaAs processing, such as AuGeNi for ohmics and TiPtAu for gate or overlay metals, and these materials are difficult to etch.

Metal patterning in GaAs processing usually involves a "lift-off" procedure, as shown in Fig. 5.1(*b*). First, resist is applied to the wafer, exposed, and developed to define the desired pattern. Then metal is applied to the wafer, usually by evaporation, and a solvent is used to dissolve the resist. Now the metal that was on top of the resist has been lifted off. Unlike etching, the lift-off process is highly sensitive to the edge profile of the patterned resist. Most of the lithography techniques used in GaAs processing were developed to provide such an undercut resist edge profile, which is shown in Fig. 5.1(*b*). A variety of techniques have been reported for refining the lift-off process to meet linewidth/spacing demands of modern GaAs LSIs.

5.2 PHOTOLITHOGRAPHY

5.2.1 Photolithographic Techniques

A number of photolithography techniques are shown in Fig. 5.2, namely contact printing, proximity printing, projection printing, and optical stepping [Wil90]. We will discuss each technique in turn.

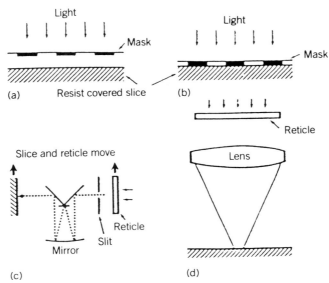

Fig. 5.2 The major categories of photolithography: (a) proximity; (b) contact; (c) projection; (d) optical stepping. [Sze, S. M., *Physics of Semiconductor Devices*, Wiley-Interscience (1985)]

Contact Printing

Contact printing is popular in GaAs device fabrication. In this technique a mask is aligned on the wafer and then vacuum clamped directly against the wafer for exposure. The resist is exposed by a nearly collimated beam of ultraviolet light through the back of the mask for a fixed time. The contact between the resist and mask provides very high resolution (~ 1 μm). However, the contact between wafer and mask tends to damage the mask, both from abrasion during the printing operation and from the frequent cleaning required to remove resist that gets attached to the mask. These problems make it necessary to replace masks after approximately 100 uses. Another disadvantage of contact printing is *runout*. Runout occurs when the mask and wafer are not perfectly flat. Any curvature results in imperfect registration between the pattern on the mask and the pattern on the wafer and thus in misalignment in some areas.

Even with these difficulties contact lithography is still a mainstay for GaAs processing for several reasons: First, the GaAs microwave devices, or even MMICs, generally employ fewer gates or critical active areas than Si VLSI circuits and can therefore tolerate a higher defect density. Critical geometries, such as FET gates, can be exposed by electron-beam lithography. This means that mask defects have minimal consequences. Second, the overlay requirements for analog GaAs microwave devices are usually not severe. The GaAs wafers are only 4 in. in diameter, and hence are not as big as silicon wafers. The runout problem is not as serious as for the larger silicon wafers [Wil90].

Higher resolution can be achieved by using *deep ultraviolet* (DUV) light. Xenon-mercury lamps are suitable for DUV light sources and emit in the 220–240-nm range. Quartz masks are employed, since glass is highly absorptive at these wavelengths.

Resists developed for electron-beam lithography, such as PMMA, tend to be sensitive to DUV light and are used for this purpose. The conventional near-UV (400-nm) optical lithography can achieve a resolution of FET's gate length around 1 µm.

Direct contact between mask and wafer is essential in obtaining good resolution and pattern definition in contact printing. Optical interference fringes should be visible (if using see-through masks) when the mask is clamped to the wafer. Because the entire wafer is exposed at one time, it is fast in contrast to the field-by-field exposure typical of optical steppers or e-beam machines. The cost of contact printers is also much less than the alternative choices. Contact lithography is suitable for fabricating GaAs analog devices and MMICs; GaAs digital circuits require a higher-yield technique [Wil90].

Proximity Printing

In proximity printing the mask is placed in close proximity to the wafer but not in contact with it. Light passes through the mask and exposes the resist on the wafer. The diffraction that occurs at the pattern's edges causes the light to blur. The amount of divergence is dependent on the spacing between the mask and the wafer. This technique is incapable of achieving the resolution required in modern lithography.

Optical Stepper

The optical stepper is becoming the main machine used for high-yield photolithography. This technique is shown in Fig. 5.2(d). The mask, called a *recticle*, contains the image of one or two fields. This pattern is imaged onto the wafer, and then the wafer is moved and the exposure repeated. The "step-and-repeat" procedure continues until the entire wafer is exposed. Optical steppers are available in essentially three configurations: The pattern on the recticle can be the same size as the pattern on the wafer, five times as large, or ten times as large. Steppers are very popular for low-defect photolithography, which usually is required for fabrication of digital circuits. Both projection printers and steppers share the advantage that the mask, or recticle, never touches the wafer. Correct handling procedure can result in virtual damage-free usage [Wil90].

The optical imaging can be performed by using either reflective optics (mirrors) or refractive optics (lenses). The accuracy demands that the optical components be *diffraction limited*. This requirement generally means that surfaces must be accurate to at least one-quarter wavelength, or about 0.1 µm.

The necessary accuracy for steppers requires extremely precise construction, low thermal expansion of critical elements, temperature compensation, vibration isolation, highly accurate stage stepping, and complex alignment and focusing systems. These considerations make steppers expensive.

5.2.2 Photoresists

The photoresist is a radiation-sensitive compound. It consists of photo-active material, sensitizer, and organic solvent. To match the semiconductor processing developments, there are photoresists, mid-UV resists, DUV resists, electron-beam resists, X-ray resists, and ion-beam resists. Resists can be either positive or negative depending whether the exposed portion is removed or remains during development. As a general rule, positive resists have more advantageous properties. They offer higher resolution and allow more appropriate edge profiles for lift-off processes. Positive resists are normally the preferred choice for most aspects of GaAs processing. A good resist should include the following requirements: (1) a high resolution, (2) a high sensitivity, (3) a broad process latitude, and (4) excellent process latitude. Figure 5.3 shows the difference between the positive and negative resists [Sze85]. The resist should also have the following characteristics:

1. Excellent dry etch resistency,
2. High purity (narrow molecular distribution),
3. Good stability,
4. Good coating uniformity,
5. Good step coverage,
6. Good adherence.

Positive Resists

A typical positive resist (e.g., Hoechst Celanese's AZ1350J) has three major components: a photosensitive compound (inhibitor), a base resin, and a suitable organic solvent. It will also contain other proprietary ingredients. Most of the solvent is lost after spin and bake. The base resin alone is moderately soluble in the aqueous alkaline

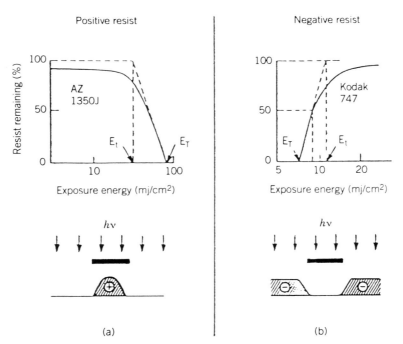

Fig. 5.3 Exposure response curve and cross section of the resist image after development: (a) positive photoresist; (b) negative photoresist. [Sze, S. M., *Physics of Semiconductor Devices*, Wiley-Interscience (1985)]

developer, the removal rate being approximately 15 nm/s [Dil75]. When the photosensitive compound is present (usually to about 25 to 30 wt%), the removal rate of unexposed resist is reduced to approximately 0.1 nm/s. This is why the photosensitive component is sometimes referred to as an inhibitor. Radiant energy in the wavelength range of 300 to 450 nm destroys the photosensitive component and results is an increased removal rate, up to 100 to 200 nm/s.

Resist is usually applied to the wafer by spinning. An appropriate amount of resist is placed on the wafer and the wafer is spun at a specified speed (2000 to 8000 rpm) for a specified time (20 to 40 s) to provide a uniformly thick coating. Final thickness is also affected by subsequent heating steps. Resist manufacturers supply curves showing thickness as a function of spin speed. The resist will require a postspin bake to harden it by removing more of the remaining solvents. Cassette-to-cassette, microprocessor controlled commercial equipment exists to prebake, coat, dry, and postbake wafers. Resist films are usually chosen to be between 0.3 to 2.5 µm thick depending on the application. Substantial local variations in the thickness of the resist film can occur if the wafer has topographical features [Wid75], such as etched mesas or metal patterns. Resist thickness variations translate directly into linewidth variations during exposure. The variations are worse for high-reflectivity substrates. Multilevel resist techniques can be used to alleviate the nonuniformity of topographical features [Wil90].

Most positive resists are sensitive to light between 300 and 450 nm. Absorption is low below 200 or above 500 nm. Mercury lamps are a common light source. The major lines in the Hg spectrum are designed as follows:

e line—546 nm,
g line—436 nm,
h line—405 nm,
i line—365 nm.

The g or the h line are most commonly employed in optical steppers, although i-line steppers are becoming more common and are of interest for GaAs work [Wil90].

A certain amount of contrast must be present to expose one area of resist while leaving another unexposed. Diffraction effects make 100% contrast impossible. Diffraction occurs at pattern edges in contact lithography. Diffraction effects lead to blurring in optical steppers, which means that some light will be diverted from areas intended to be exposed. Nevertheless, the pattern can be successfully exposed in the resist if there is sufficient contrast between the intensity of light in both areas.

Baking in resist operations serves a number of purposes. Oven bakes are generally between 60° and 140°C for periods of 5 to 60 min. Two different types of bakes are used. One immediately follows resist application and is sometimes referred to as a "softbake." The other is a bake following exposure, but before developing—it is called a "postbake." The softbake serves to remove water and solvents that remained in the film after spin-on. This hardens the resist film and improves adhesion. The removal of solvent also decreases the film's thickness. Note that resist thickness decreases rapidly over the lower temperature range, making the final thickness especially sensitive to temperature variations. Typical data showing the dissolution rate of unexposed resist as a function of bake temperature is shown in Fig. 5.4.

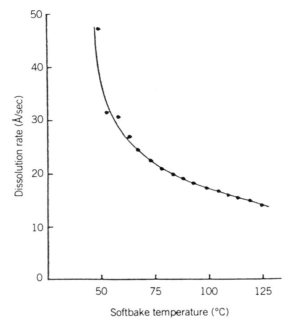

Fig. 5.4 Average dissolution rate of unexposed photoresist as a function of softbake temperature. [Williams, R., *Modern GaAs Processing Methods*, Artech House (1985)]

The second possible bake is after exposure but before development. The postbake can result in several advantages [Bat83; Wal75]. These include improved linewidth control, elimination of standing wave effects, increased contrast, increased edgewall angle, reduction of scumming, and improved adhesion.

The postbake can also reduce the phenomenon known as scumming [Wil90]. Even unexposed resist will slowly dissolve in the developer. The unexposed resist can be redeposited in developed-out spaces. Such material must then be removed by a "descumming" operation such as etching in oxygen plasmas. The hardening of the resist surface that occurs during postbake results in decreased solubility, and hence decreased scumming.

Resist patterns are developed either by immersing the wafer in the developer or by spraying the developer on the wafer. Some form of agitation is usually used in immersion developing. The wafer must be rinsed to remove the developer, usually using DI water, and then dried. Spin drying is done in commercial machines.

Acetone is the major solvent used to remove positive resist either in lift-off or stripping operations. It can be sprayed on the wafer, or the wafer can be immersed in hot acetone. Methanol and DI water are used to remove the acetone.

Negative Resists

Negative resists may also be used in some applications in GaAs processing, but the results are not as good. Generally the resolution, edge profiles, and linewidth control are poorer than those of the positive resist. This difference in quality is due partly to the tendency of the negative resist to swell in the presence of solvents. The poor edge profiles make negative resists unsuitable for lift-off operations. Negative resists are also often more difficult to remove from the wafer. J-100 is commonly used to remove negative resists [Wil90].

Multilevel Resist Techniques

Single layer resist (SLR) materials and processes are currently used the most in the manufacture of semiconductor devices but not in laboratory research. Research workers have chosen to use multilevel resist (MLR) techniques to achieve their advanced experimental results. Probably the most widespread MLR technique is the tri-layer approach which has a long history and many areas of application. The original patent was issued in 1973 to IBM [Fra73]. Subsequently the approach has been studied at Hewlett-Packard [Kru84; O'To83; O'To81], AT&T Bell Laboratories [Mor79], IBM [Fra75; Car76; Kit79; Und85], NTT [Kak77; Shi84; Suz84], and Hitachi [Hom81].

Tri-layer MLR Systems   Tri-layer MLR systems involve an imaging layer, a barrier layer, and a planarizing or base layer [McD86], as shown in Fig. 5.5 [Wil90]. The planarizing layer must be sufficiently thick to cover the topography. This highly planar resist layer can eliminate standing wave effects. Dyes can be added to this layer to eliminate effects due to reflectivity from the substrate in optical lithography. The barrier layer has a dual function: It physically separates the top and bottom layers which, due to their composition, have a tendency to intermix, and it provides an etch barrier for RIE patterning of the base layer and the underlying substrate.

Additionally, since the surface now is relatively planar, a thin layer of resist can be used to improve resolution without being optimized for etch resistance and thermal stability. The improvement in resolution is achieved without increasing defect levels.

Tri-layer systems offer the advantages of improvements in resolution, contrast, and etch resistance, as well as an ability to tailor the resist-wall and electron-beam exposure system [Hat84]. Additional advantages and disadvantages may arise as tri-layer MLR systems are considered for certain applications. As in electron-beam lithography, the presence of additional layers under a top imaging layer changes the effects of electron scattering. An MLR system can minimize the proximity effect and improve resolution and linewidth tolerance [McD86]. Ralph [Ral82] reported that the use of a poorly conducting material leads to image degradation and shifting due

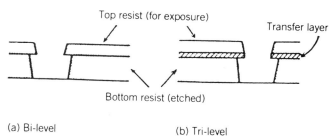

Fig. 5.5 Multilevel resist techniques: (*a*) bilevel; (*b*) trilevel. [Williams, R. H., *Modern GaAs Processing Methods*, Artech House (1985)]

to heating during exposure. However, if one or more of the layers is a good conductor, charging effects that are observed in electron-beam lithography may be avoided [Suz84]. The use of a tri-layer resist process is undesirable when the electron or optical lithography of the films is thick enough to interfere with the registration of patterns through detection of alignment masks in the underlying substrate.

Bi-layer MLR Systems In bi-layer methods there are only two resist layers. Intermixing is a major problem in this system. The solvent in the top resist tends to dissolve the lower resist. If the lower resist is baked sufficiently to prevent intermixing, it will be difficult to remove. An exception is the use of an electron-beam resist such as PMMA as the lower resist. This resist can be baked at a high temperature, around 180°C and can still be chemically removed [Wil90]. The resist is not sensitive to ordinary light but usually is sensitive to DUV light. Therefore one bi-layer process consists of spinning PMMA onto the wafer, baking it, spinning a positive resist over it, exposing and developing the upper resist in the usual manner, exposing the entire wafer to DUV, and finally using a developer that dissolves the exposed PMMA but leaves the upper resist.

When using ordinary positive resists, the surface of the first layer (top layer) may be hardened by exposure to a plasma. Then the second layer (bottom layer) may be applied. After the top layer is exposed and developed, dry etching techniques can be used to open up the bottom layer. Then comes the developing action which, after opening the exposed pattern in the top layer, yields an undercut profile. The amount of overhang obtained is very sensitive to the exact process chosen. The practical application of thin technique may have difficulties in uniformity and reproducibility.

Quasi-MLR Systems Occasionally a lithography process is described as a single-layer method when in fact at least one other layer of material is required for its satisfactory performance. For example, in using single-layer electron-beam resists, alignment errors due to charging are resolved only with an additional conductive layer of evaporated metal [Mat85].

The term "quasi-MLR" is used to describe MLR systems that have been developed to extend the utility of single-layer resists to include additional layers of materials but without requiring an image transfer step for image formation.

Antireflection Coating (ARC), combined with a single-layer, optical resist, has been shown to give improved resolution, tolerance, and linewidth control over topography. In 1970 Ilten et al. [Ilt70] first suggested controlling reflectivity effects by optimizing the combined thickness of resist and insulator films and by using an underlying film. Inorganic insulator and metal films have been used to reduce the effects of reflection from substrates [Kou70]. These films required additional deposition, etching, and removal steps and were only moderately effective in reducing such reflections in the near UV region.

Recent developments using commercially available dyed polymer films have been more effective in eliminating reflection [Tin83; Bre81; Lin84], and the thermal and dissolution characteristics of these materials have introduced the effective process window. Lin et al. [Lin88] has observed that film thicknesses, baking conditions, exposure doses, and development processes interact in complicated ways to affect the patterning accuracy of the resist/ARC layers. Listvan et al. [Lis84] used an ARC film for the production of 1.2 µm images over a maximum of 0.8-µm topography and

5.2.3 Advanced Image Reversal Techniques

Resolution, sidewall angles, and process latitude may be improved by image reversal of a positive photoresist. The reversible image system has the added advantage of enabling a negative image to be produced by aqueous development [Tay89]. In 1978 Moritz and Paul [Mor78] reported that the addition of monazoline to a positive working photoresist formulation allowed image reversal to occur. MacDonald and colleagues [Mac82] modified this process by using a similar additive, imidazole. Other basic chemicals such as trithanol amine and gaseous amines [All88, Jon88] are also reported to cause image reversals. Some commercial processes treat photoresist films with gaseous amines after exposure to cause image reversal. The general scheme is illustrated in Fig. 5.6(a). For a positive image the film is developed after exposure. For a negative image the film is exposed and then baked in the presence of an amine. The process results in the decarboxylation of indene carboxylic acid groups, which insolubilizes the exposed area. Flood exposure followed by development gives a negative image. The g-line and i-line dual-tone photoresists of positive- and negative-tone images with a 1.2-µm line are shown in Figs. 5.6(b) and (c).

The interest in image reversal stems from several possible applications including lift-off for metallization, tape-automated bonding, and complementary masking [Jon88]. Additional advantages include resolution improvements, the elimination of standing waves, and the formation of crosslinked negative images having good thermal stability [Mor88].

Experimental Results

New chemistry has been developed for an image reversal process that allows stable dual-tone photoresists to be formulated. Positive and negative images have been demonstrated on g-line and i-line exposure tools. Work is in progress to develop high-resolution dual-tone photoresists that are optimized for exposure at 248 nm.

The g-line photoresist was processed as follows:

	POSITIVE	NEGATIVE
Coat	1.2 µm	1.2 µm
Softbake	100°C, 60 s	100°C, 60 s
	Hot plate	Hot plate
Expose	13-s contact	11-s contact
	Cobilt aligner	Cobilt aligner
	100 mJ/cm^2	85 mJ/cm^2
Post-exposure bake	None	120°C, 60 s
Flood expose	None	1000 mJ
Develop	Puddle develop	Puddle develop
	60% KTI 934 (0.20 N)	70% KTI 934 (0.23 N)
	180 s	75 s

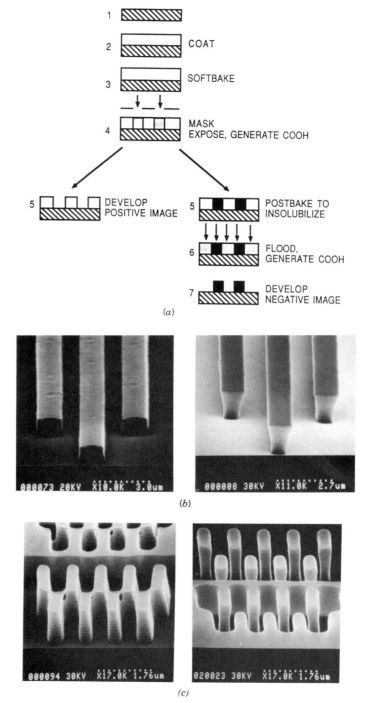

Fig. 5.6 Process routes for producing image reversal. [Taylor, J. W., T. L. Brown, and D. R. Bassett, Advanced image reversal techniques, vol. 1086: *Advances in Resist Technology and Processing VI*, p. 585, SPIE (1989)]

The i-line resist was processed as foloows:

	POSITIVE	NEGATIVE
Coat	1.23 μm	1.23 μm
Softbake	100°C, 60 s Hot plate	100°C, 60 s Hot plate
Expose	200 mJ 0.4 N.A. Stepper ASM	170 mJ 0.4 N.A. Stepper ASM
PEB	None	120°C, 60 s
Flood expose	None	400 mJ
Develop	Spray/puddle develop 47.5% KTI 934 (0.16 N) Two 25-s puddles	Puddle develop 70% KTI 934 (0.23 N) Two 40-s puddles

Phase-Shifting Masks

In recent years work on optical phase-shifting masks has shown promise in improving resolution and depth of focus [Flo91; Lin91, Lin92]. The shifting masks are two-level mask structures that can improve resolution in optical lithography systems 25% to 100% by an interference effect to cancel diffraction. The phase-shift idea originated in the United States by Dale Flanders and Hank Smith of MIT [Bur92]. However, most of today's work refers to a paper by Marc Levenson and coworkers at IBM in 1982 [Lev 82].

Phase shifting uses the interference effects of coherent light to form a higher-contrast image on the wafer plane. This produces higher resolution, larger exposure latitude, and greater depth of field. Figure 5.7 shows the difference between the phase-shifting mask and the conventional mask. The Levenson and the rim shift techniques are leading in the field. The attenuated phase-shifted marsk system, invented by H.I. Smith mainly for X-ray proximity printing, is getting more attention. It use a slightly transmissive absorber with a π phase shift. The attenuated phase-shifting system can be used for any arbitrary mask pattern. Greater variation at the phase-shifted wafer provides a considerably wider range of both intensity and amplitude. The intensity, which is proportional to the square of the amplitude, restores the spatial frequency of the mask openings, but with greater contrast and resolution. Phase-shifting masks can extend i-line technology (see below) to the 0.35-μm limit. Oki Electric applied phase-shift technology for sub-halfmicron lithography, using an i-line stepper and a high-resolution negative-working, low-molecular-weight UV resist (LMR-UV) to fabricate n^+ self-aligned MESFETs [Kim92].

Short wavelength KrF excimer laser light allows greater resolution without restrictions in pattern layout [Pol86]. Excimer laser source and phase-shifting mask technologies are now being intensively investigated for the purpose of delineating such fine line patterns ($\sim 0.2\,\mu m$) [Oka91].

Results and Discussion

There are two approaches being explored to develop dual-tone photoresists. The first approach uses thermal deblocking chemistry. Exposure of a dual-tone photoresist

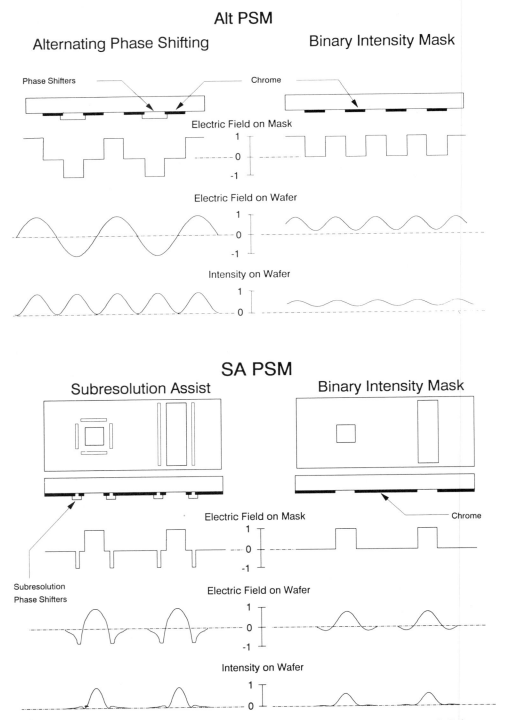

Fig. 5.7 Phase-shift masking techniques. [Lin, B. J., *Semiconductor International*, February (1992)]

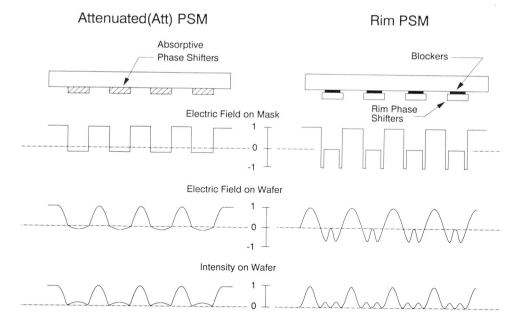

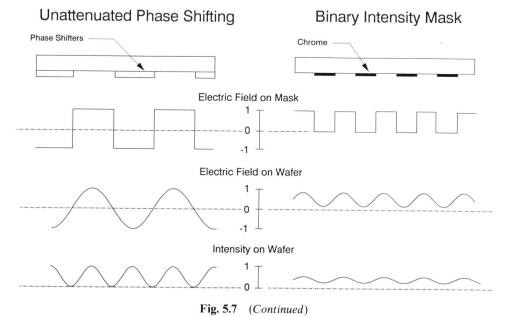

Fig. 5.7 (*Continued*)

178 LITHOGRAPHY

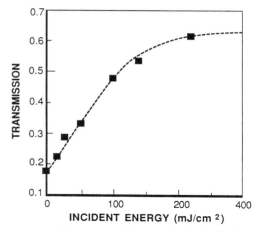

Fig. 5.8 I-line photoresist bleaching curve. [Taylor, J. W., T. L. Brown, and D. R. Bassett, Advanced image reversal techniques, vol. 1086: *Advances in Resist Technology and Processing VI*, p. 585, SPIE (1989)]

film converts the diazonaphthoquinone to indene carboxylic acid groups, which make the film base soluble. Development with an aqueous base solution gives a positive image. However, exposure of the film followed by a postbake, flood exposure, and development gives a negative image. A negative image forms because a reactive agent is liberated during the postbake step. A reactive agent, which is not amine based, reacts with indene carboxylic acid groups deactivating their acidity. The bleaching characteristics of the i-line photoresist at 365 nm is shown in Fig. 5.8.

A second approach, referred to as Dual-PAC imaging, is under active investigation. In this approach both the reactive agent and indene carboxylic acids are photochemically generated during imaging. The chemistry is schematically shown in Fig. 5.9.

Model photoactive compounds, which decomposed at various wavelengths and are reactive with carboxylic acid groups, have been prepared. The UV-spectra of representative compounds are shown in Fig. 5.10. A representative bleaching curve of a photoactive compound reactive at 254 nm is shown in Fig. 5.11.

5.3 NONOPTICAL LITHOGRAPHY

Various types of advanced lithographies for IC fabrication are shown in Fig. 5.12 [Sze85]. Optical exposure tools are capable of approximately 0.35 µm resolution and 0.1 µm registration. Advanced techniques using multilevel resist or procedures such as angle evaporation can achieve submicron resolution over a small area. But processes that press the very limits of optical lithography tend to be difficult to use and have a low yield. Optical lithography is not sufficient to pattern the smallest geometries used in many modern GaAs devices. For microwave applications, features such as the gates of GaAs FETs may require lithographic resolution of 0.5 µm or less. Registration requirements may be as stringent as 0.1 µm.

Such severe requirements demand other forms of lithography be used. They are electron beam (*e*-beam) lithography, X-ray lithography, or ion-beam lithography.

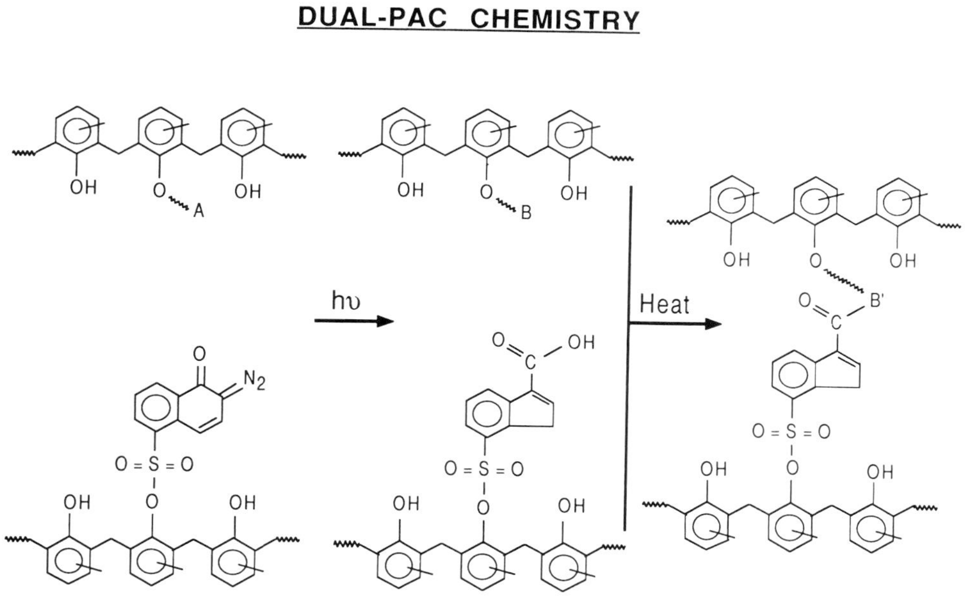

DUAL-PAC CHEMISTRY

Fig. 5.9 Dual-PAC chemistry. [Taylor, J. W., T. L. Brown and D. R. Bassett, Advanced image reversal techniques, vol. 1086: *Advances in Resist Technology and Processing VI*, p. 585, SPIE (1989)]

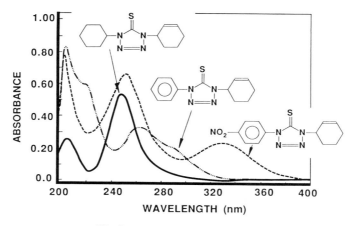

UV Spectra of tetrazole-5-thiones

Fig. 5.10 UV spectra of photogenerated inhibitors. [Taylor, J. W., T. L. Brown, and D. R. Bassett, Advanced image reversal techniques, vol. 1086: *Advances in Resist Technology and Processing VI*, p. 585, SPIE (1989)]

E-beam lithography is the most popular one. X-ray lithography has 0.25 μm or better resolution and 0.1-μm registration. However, it requires a complicated mask and is not yet used to produce ICs in volume. Ion-beam lithography offers patterned-doping capability and very high resolution (~ 100 Å); this method is still in its initial development stage.

180 LITHOGRAPHY

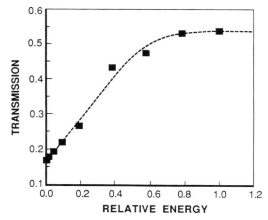

Fig. 5.11 Deep-UV bleaching curve for new photoactive compound in PMMA. [Taylor, J. W., T. L. Brown, and D. R. Bassett, Advanced image reversal techniques, vol. 1086: *Advances in Resist Technology and Processing VI*, p. 585, SPIE (1989)]

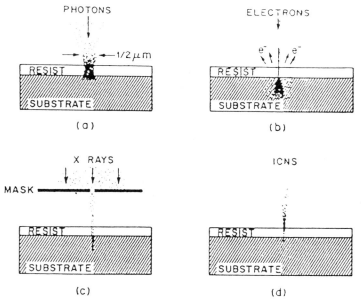

Fig. 5.12 Advanced lithographic methods: (*a*) optical lithography; (*b*) electron-beam lithography; (*c*) X-ray lithography; (*d*) ion-beam lithography. [Sze, S. M., *Physics of Semiconductor Devices*, Wiley-Interscience (1985)]

5.3.1 Electron-Beam Lithography

Electron-beam lithography is used to form patterns on photomasks used in optical lithography, and for exposing patterns directly on wafers. The latter is known as direct wafer writing (DSW) and is often used in GaAs device processing to define small features such as gate patterns on FETs. Figure 5.13 shows a schematic of an

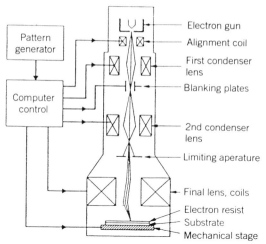

Fig. 5.13 Schematic of an electron-beam machine. [Sze, S. M., *Physics of Semiconductor Devices*, Wiley-Interscience (1985)]

electron-beam lithography system [Ker84]. The electron gun is a device that can generate a beam of electrons with a suitable current density; a tungsten thermionic-emission cathode or single-crystal lanthanum boride (LaB_6) can be used for the electron gun. Condenser lenses are used to focus the electron beam to a spot size of 0.01 to 0.1 μm in diameter. Beam-blanking plates (to turn the electron beam on and off) and beam deflection coils are computer-controlled and operated at high rates (in MHz or higher rates) to direct the focused beam to any location in the scan field on the substrate. Because the scan field (typically 1 cm) is much smaller than the substrate diameter, a precision mechanical stage is used to position the substrate to be patterned.

Electron-beam systems have two major components: The beam-forming and beam-deflection system, and the pattern-generation and pattern-control systems [Cha88]. The more stringent requirements of nanolithography applications occur in the beam-forming and beam-deflection system.

Beam Formation The beam-forming systems for electron-beam lithography use either a Gaussian round-beam approach or shape-beam approach. The Gaussian approach which uses the conventional probe-forming concept of the scanning electron microscope is the most widely used for nanolithography. Two or more lenses focus the electron beam onto the surface of the workpiece by demagnifying the electron source. High flexibility can be achieved. To ensure good line definition, the beam size is generally adjusted to a fraction, usually a quarter, of the minimum pattern linewidth.

The optical properties of the Gaussian beam system in the nanometer regime are determined by a combination of lens and deflection aberrations, diffraction, and source brightness [Bro72; Cha76]. In the axial position the principal aberrations are spherical, chromatic, and diffraction. When the beam is deflected off-axis, additional aberrations (transverse chromatic, coma, astigmatism, and field curvature) are introduced. The final beam diameter is given by a combination of these two main groups of aberrations. To improve resolution, it is necessary to reduce the axial aberrations, which necessitates the use of a short working distance in the final lens. A short working distance leads, however, to an increase in the deflection aberrations. The design of the system therefore requires careful trade-off and optimization of a multitude of parameters. A typical system, with a working distance of around 1 cm using a LaB_6 gun [Bro67; Hoh82] at 25 kV, can form a minimum axial beam diameter of approximately 5 nm. This fundamental limit is set by chromatic and diffraction effects. As the beam size approaches this limit, both the beam current and the current density decrease drastically due to the effect of the aberrations. Therefore a practical minimum beam-diameter limit is set by the need to achieve some minimum level of beam current density.

A field-emission source [Cre66; But66; Ven78] allows the e-beam system to achieve a limit in minimum beam size of approximately 1 nm, with a current density in excess of $1000 \, A/cm^2$ at a beam size of 8 nm. The significant advantage of a field-emission source over the LaB_6 source has been known for some time. It is not widely practiced mainly because of concerns over the stability and noises issues associated with such a source. Recent advances in thermal field-emission (TFE) sources [Swa69; Tug79; Wol79; Sam85; Sai86; Tho87] based on either zirconiated tungsten or titanium tungsten emitters have increased the stability to practical levels.

Round-beam systems based on critical [Var74] and Koehler [Bro79] illumination principles have been developed. They offer some improvement in sharpness of beam profile over the Gaussian beams, but the basic design considerations are not significantly different.

The shaped-rectilinear-beam approach [Pfe78; Got78; Tho78] offers the advantage of higher throughput than the round-beam approach. In this system a rectangular beam is formed by overlapping the images of two square apertures in the electron-optical column. By varying the amount of this overlap, a variable shaped rectangular beam can be formed. Such systems have been successfully developed and used for lithography applications in the submicron and deep-submicron regimes.

Pattern-Generation and Proximity Effects The two basic pattern-generation techniques are raster and vector. For nanolithography, vector scan is the method of choice primarily by reason of proximity-effect correction [Cha75; Par79]. Proximity effects are created by forward-scattered electrons and backscattered electrons in the resist and backscattered electrons from the substrate, which partially expose the resist up to several microns from the point of impact. As a result serious variations of exposure over the pattern area occur when pattern geometries fall below the 1-μm range. Correction for these effects can be readily achieved in vector scan by adjusting the beam-stepping rate and hence the exposure intensity for each pattern element.

Rishton and Kern [Ris87] developed a technique, using a very high-contrast resist, to measure experimentally the normalized point exposure distribution, both on solid substrates which cause backscattering and on thin substrates where backscattering is negligible. The data sets obtained can be applied directly to proximity correction, and they meet the practical conditions of pattern writing.

Electron exposure is usually modeled by assuming that the total exposure distribution is given by the sum of two Gaussian distributions [Cha75]. The first Gaussian results from the beam diameter and forward scattering. The second Gaussian results from backscattering. For some materials an additional exponential term must be added to the Gaussian terms to obtain a better fit with the measured exposure profiles. In the exposure distributions obtained on silicon, gallium arsenide, and thin silicon nitride substrates, significant deviations from the commonly assumed double Gaussian distributions may occur. For GaAs substrates the backscatter distribution may not be adequately described by the Gaussian function, as shown in Fig. 5.14(a). Even on silicon a significant amount of exposure is found in the transition region between the two Gaussian terms [Fig. 5.14(b)]. This exposure, which may be due to non-Gaussian tails primarily in the beam and to forward scattering in the resist, must certainly be accounted for in the accurate proximity correction for lithography on a sub-100-nm scale. The modified Gaussian proximity equation used to fit the experimental data is

$$f(r) = \frac{1}{\pi(1+\eta+\nu)} \left[\frac{1}{\alpha^2} \exp\left(\frac{-r^2}{\alpha^2}\right) + \frac{\eta}{\beta^2} \exp\left(\frac{-r^2}{\beta^2}\right) + \frac{\nu}{2\gamma^2} \exp\left(\frac{-r}{\gamma}\right) \right], \qquad (5.1)$$

where α is the half-width of the forward-scattering Gaussian profile, β the backscattered Gaussian profile half-width, and γ the decay constant for the exponential term, η the ratio of the backscattered to the forward-scattered intensity, and ν similarly the ratio of the exponential term intensity to the forward-scattered intensity. Table 5.1 [Ris87] summarizes the parameters that best fit the exposure profiles obtained from these measurements.

Electron Optics

A simple basic probe-forming electron-optical system is shown in Fig. 5.15. Two or more magnetic lenses form a demagnified image of the source on the wafer image plane. Provisions for scanning the image and blanking the beam are not included in the figure. The cathode is generally a thermionic emitter—either a tungsten hairpin or a pointed LaB_6 rod. Field emitters are also being used. In a field emitter a strong electric field pulls the electrons out. The rod may be a sintered material or a crystal. Emission current density from the cathode J_c is given by

$$J_c = AT^2 \exp(-E_w/kT). \qquad (5.2)$$

LaB_6 has a lower work function E_w than tungsten with a Richardson constant A of 14. The electrons are accelerated through the voltage $V = 10$ to $50\,kV$ and focused by the gun to a spot called the "crossover" whose diameter ranges from 10 to 100 µm. The crossover is seen near the anode in Fig. 5.14.

The thermal field-emission (TFE) source consists of a tungsten tip, of 0.5- to 1-µm radius, that is heated $1100°$ to $1400°C$ to provide annealing of the sputtering damage. With the TFE source vacuum requirements can be relaxed, and there is better stability than with a cold field emitter. The source is small (100 ro 1000 Å) and located inside the tip.

In the electron-optical column the lenses are magnetic. The field of such a lens, $\mathbf{B} = \mathbf{e}_r B_r(r,z) + \mathbf{e}_z B_z(r,z)$, has cylindrical symmetry about the optical axis z. If a parallel beam of radius r_0 enters the field of the lens, the electrons experience a force that causes those not on axis to rotate about the axis and turn toward it. For a thin lens

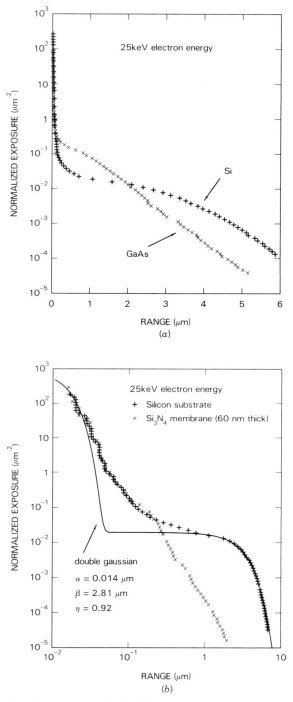

Fig. 5.14 Measured point exposure distributions: (a) on silicon and gallium arsenide; (b) on solid silicon and thin silicon nitride, compared to a double Gaussian distribution. [Rishton, S. A., and D. P. Kern, *J. Vac. Sci. Tech.* B **5** (1987): 135.]

TABLE 5.1 Parameters for the Approximation of Exposure Distributions on Silicon and Gallium Arsenide by Double Gaussian and Exponential Terms

Substrate	Energy (keV)	Approximation	α μm	β μm	η	γ μm	ν
Silicon	25	Double gaussian	0.014	2.81	0.92	—	—
Silicon	50	Double gaussian	0.033	8.80	0.75	—	—
10 μm AZ on SI	25	Double gaussian	0.020	5.22	0.35	—	—
GaAs	25	Double gaussian	0.017	1.21	3.24	—	—
GaAs	25	Gaussian + exponential	0.014	—	—	0.57	2.52
GaAs	50	Double gaussian	0.013	3.28	1.07	—	—
GaAs	50	Double Gaussian + exponential	0.013	3.78	0.75	1.28	0.54

Source: Rishton, S. A., and D. P. Kern, *J. Vac. Sci. Tech.* B **5** (1987): 135.

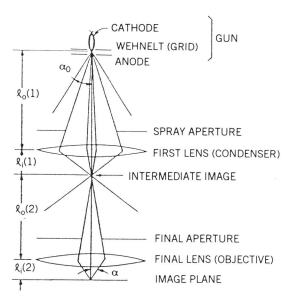

Fig. 5.15 A simple two-lens probe-forming electron optical system. [Watts, R. K., in *VLSI Technology*, ed. by S. M. Sze, 2d ed., McGraw-Hill (1988)]

the electron path beyond the lens is given by

$$\frac{dr}{dz} = -\left(\frac{r_0 e}{8\,mV}\right)\int_{-\infty}^{\infty} B_z^2\, dz. \tag{5.3}$$

The constant $dr/dz = -r_0/f$, where f is the focal length of the lens and e is the electronic charge. The thin lens law relating object distance, image distance, and focal length gives

$$\frac{1}{l_o} + \frac{1}{l_i} = \frac{1}{f}. \qquad (5.4)$$

The magnification of a lens is $M = l_i/l_o$. An intermediate image is formed by the first lens with magnification M_1, and this is the object that is further demagnified by the second lens to form a spot in the image plane of diameter

$$d_i = M d_0, \qquad M = M_1 M_2, \qquad (5.5)$$

where d_0 is the object diameter and M_2 is the magnification of the second lens. Typically $M = 10^{-3} - 10^{-1}$ [Wat88]. The lengths $l_o^{(1)}$ and $l_i^{(2)}$ are fixed, but $l_i^{(1)}$ and $l_o^{(2)}$ are variable. If the current through the windings of the first lens of Fig. 5.14 is increased, $l_i^{(1)}$ decreases, M_1 is reduced, and the beam current passing through the final aperture is reduced because of the increased divergence of the beam at the intermediate image. The current desnity and current in the image plane are

$$J \leqslant \pi \beta \alpha^2 \quad \text{and} \quad I = J\left(\frac{\pi d_i^2}{4}\right). \qquad (5.6)$$

The spot sizes of interest are in the range 0.1 to 2 µm. This is far from the diffraction limit. For 15-keV electrons the wavelength $\lambda = 0.1$ Å. Taking $\alpha = 10^{-2}$ radians, we have a diffraction spot width

$$d_{\text{diff}} = \frac{1.2\lambda}{\alpha} = 10^{-3} \, \mu\text{m}. \qquad (5.7)$$

Although the diffraction may be ignored, aberrations of the final lens and of the deflection system may increase the size of the spot and change its shape. Figure 5.16 shows a typical double-deflection system above the final lens; the system is arranged to that the deflected beam always passes through the center of the lens.

Another source of spot broadening is the mutual Coulomb repulsion of the electrons as they traverse the column. For total column length L this contribution is approximately given by

$$d_{\text{ee}} = \left[\frac{LI}{(\alpha V^{3/2})}\right] \times 10^8 \, \mu\text{m}. \qquad (5.8)$$

The spreading becomes a concern if at large currents, small spot sizes or sharp edges of large spots are to be distinguished.

Advantages and Disadvantages of E-Beam Lithography

The advantages to using *e*-beam lithography include improved resolutions, highly automated and precisely controlled operations, great depth of focus, excellent level-to-level overlay tolerance (*registration*), easy pattern modification, and absence of mask defects. Because of the nature of electron-beam physics, the practical depth of

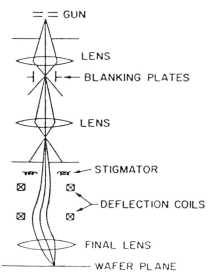

Fig. 5.16 Electron optical system with double deflection. [Watts, R. K., in *VLSI Technology*, ed. by S. M. Sze, 2d ed., McGraw-Hill (1988)]

focus is enormous compared to optical standards, approximately 25 μm [Cha82; Wil90]. This means that wafer flatness is not a limiting factor.

Since the *e*-beam process does not use a mask, there are no mask defects to degrade the image. The pattern is stored as computer instructions. This feature means that *e*-beam lithography can be an attractive technique even for defining geometries within the ability of photolithography.

The major problems associated with *e*-beam lithography are complexity, expense, and throughput. Although *e*-beam machines can be complex, technological advances have improved them so that there is not significant downtime in modern machines. Low throughput is adequate for the production of photomasks in situations that require small numbers of custom circuits or for design verification. Many analog GaAs devices and MMICs have only a few, low-area patterns that require *e*-beam exposure. But, for maskless direct writing, the machine must have the highest possible throughput, and hence as large a beam diameter as possible that would be consistent with the minimum device dimensions.

Charge buildup can arise as patterns are exposed on GaAs wafers with semi-insulating substrates. If electrons from the electron beam are not drained away, enough charge will build up to deflect the beam. This charge buildup will cause problems in establishing accurate alignment marks and in exposing patterns, which will be misaligned. For GaAs MMICs a thin coating of a metal such as Al or Au has been suggested in the literature [Lin83] to drain the charge away. The metals must be removed at an appropriate point in the process. If multilevel resist processes are used, the middle layer may be chosen to be a conductive material [Lin83].

E-Beam Resist and Exposure Characteristics

A focused beam of high-energy electrons is used to expose *e*-beam resists. The electron energy is generally between 10 to 25 keV, the beam current is less than 50 nA, and the spot size can be well under 1 µm. The exposure is controlled by adjusting the size of the spot, the beam current, and the speed with which the beam is moved across the wafer. When a high-energy electron enters the resist, it undergoes electromagnetic interactions with the electrons of the resist atoms and rapidly loses energy. The energy is transferred to molecules in the resist, resulting in either bond breaking or crosslinking. In positive resists bond breaking occurs, increasing the solubility of the resist polymer in the developer. In negative resists crosslinking occurs, and this decreases solubility. The resist will become exposed over an area wider than that of the incoming electron beam. This is caused by three effects: electron scattering within the resist, secondary electrons generated in the resist, and backscattered electrons from the substrate. These effects are shown in Fig. 5.17. Because of the backscattering, electrons effectively can irradiate several micrometers away from the center of the exposure beam. Since the dose of a resist is given by the sum of the irradiations from all surrounding areas, the electron-beam irradiation at one location will affect the irradiation in neighboring locations. This lateral exposure gives rise to what is called the *proximity effect*, as illustrated in Fig. 5.18. The proximity effect places a limit on the minimum spacings between pattern features. To correct for the proximity effect, patterns are divided into small segments because of the additional computer time required to expose the subdivided resist patterns.

The common *e*-beam resists are PMMA (polymethyl methacrylate) and PBS (polybutene-1 sulfone). Properties of some modern *e*-beam resists are compared in

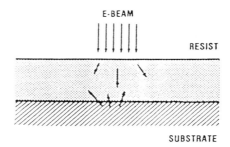

Fig. 5.17 Lateral exposure of *e*-beam resist resulting from forward scattering in the resist, secondary electron emission, and backscattering from the substrate. [Williams, R., *Modern GaAs Processing Methods*, Artech House (1985)]

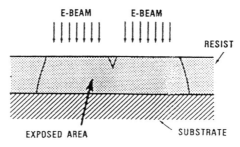

Fig. 5.18 Proximity effect resulting from electron scattering. [Williams, R., *Modern GaAs Processing Methods*, Artech House (1985)]

Martel and Thompson [Mar79]. PMMA is a positive resist having good resolution. It can be developed using MIBK methylisobutylketone). Development time will affect the final profile. At a constant exposure longer development times will tend to produce the undercut profile; shorter development times will not dissolve all of the lower material.

Lin et al. [Lin88] have investigated the lithographic aspects (contrast, resolution, exposure latitude, throughput, and statistics) of two-layer resist systems consisting of a sensitive positive electron beam resist poly(3-butenyltrimethylsilane sulfone) (PBTMSS) as the top layer, and a variety of materials such as diamondlike carbon film or polyimide as the bottom layer. In these systems exposed and developed PBTMSS is first oxidized in SiO_2, which masks the etching of the bottom layer. The pattern is in turn transferred to the substrate by either wet or dry etching with the bottom layer serving as a mask. Grating patterns with 0.2- and 0.1-μm pitch can be routinely fabricated on semiconductors, and 0.05-μm pitch patterns have been demonstrated. To achieve such high resolution, a low-strength developer is used, but a throughput about 30 to 50 times higher than that of PMMA can still be obtained. Gratings can be patterned for PMMA over square-centimeter areas in a few hours rather than days.

Multilayer resist techniques are also used in *e*-beam lithography. These techniques are relevant for very small linewidths sub-halfmicron) or linewidth control over topography. Because the lower resist layer acts as the substrate, the backscatter and proximity effect is greatly reduced. Several bi-layer processes have been reported [Wil90]. For example, one uses conventional PMMA as the lower layer resist. After baking, a copolymer (methylmethacrylate and methacrylic acid) is applied as the upper resist layer [Hat81; Hat84]. After *e*-beam exposure the two layers are developed with different developers, such as ethoxyethanol for the top layer and chlorobenzene for the bottom layer [Hat81]. The tri-layer techniques are more flexible and reliable than bi-layer techniques as described in optical lithography. In this case the top resist can be very thin, making it easier to define sub-micron or sub-halfmicron lines [Oro82]. Patterns in thin resist are also easier to assess.

5.3.2 X-Ray Lithography

X-ray lithography has been a theoretically attractive technique for years [Spe72]. Although the practical difficulties have delayed progress, X-ray lithography has been used for limited production. The basic process can be very effective. Soft X ray (2 to 10 Å) are used to expose certain resists. The short wavelength makes diffraction essentially nonexistent, allowing higher resolutions to be obtained. There are some other advantages that can be mentioned [Wat88]. First, simple thick single-layer resists can be used to image down to at least 0.25 μm lines. Second, resist profiles are essentially vertical, and excellent linewidth control is achieved. Third, some contaminants such as light organic materials on marks or wafers do not print as defects. Fourth, chip size is limited by mask size. Fifth, throughputs from one to two orders of magnitude larger than those in *e*-beam technology are possible, and throughput is independent of device design rules. Sixth, X-ray technology can be extended to submicrometer resolution.

In large-volume production X rays are in fact economical to use than electron beams. But there are a number of problems to consider: (a) Can we build a bright

Sources

X rays are produced by the interaction of incident electrons and a target material. The maximum X-ray energy is the energy E of the incident electrons. If E is greater than the excitation energy E_C of the characteristic line radiation of the atoms of the material, the X-ray spectrum will contain these lines.

In early X-ray lithography experiments, the X-ray source was often an electron-beam evaporator with the chamber modified to accept a mask and wafer [Wat88]. The target metal could be changed easily to modify the X-ray spectrum. With X-ray generation by electron bombardment, which is a very inefficient method, most of the input power is converted into heat in the target. The X-ray flux is generally limited by the heat dissipation in the target.

Much higher X-ray fluxes are available from generators that have high-speed rotating targets [Yos77]. The heat dissipation is spread over a larger area; the electron focal spot is on the rim of a spinning, water-cooled wheel, The wheel may be tilted or the rim beveled so that an elliptical focal spot appears round. Because more power can be dissipated in the larger spot, source brightness can be increased in this way, at the expense of a relative enhancement of the higher-energy portion of the spectrum and some increase in the flux spatial nonuniformity.

The synchrotron radiation X-ray source can produce flux densities in excess of $100 \, mW/cm^2$ [McG81; Spi76] at the wafer plane. Electromagnetic radiation is emitted by electrons in response to radial acceleration which keeps them in orbit in storage rings or synchrotrons. The bunches of electrons in a synchrotron are continuously injected into a ring, raised in energy, and then removed at a 50- to 60-Hz rate. In the storage ring a single bunch of electrons is injected, raised in energy, and kept stable for several hours. Synchrotron radiation is rich in long X-ray wavelengths ranging between 10 and 50 Å.

Another type of source, which is capable of an order of magnitude greater flux, is the plasma discharge source. X radiation is produced by heating the plasma to a sufficiently high temperature. The radiation consists of strong lines superimposed on a weak continuum [Wat88]. The source is pulsed at a low repetitious rate. Special problems with such a source are reliability and contamination produced in the plasma chamber.

Masks

The fabrication of X-ray masks is quite similar to that of masks used in conventional lithographic techniques. An X-ray mask consists of an absorber on a transmissive membrane substrate. The thickness of the absorbing material is determined by the X-ray wavelength of interest, the absorption coefficient of the material, and the contrast required by the resist to form an image [McG81]. Of the heavy metals with larger ρZ^4 absorbers, gold has been widely used. The thicknesses of gold necessary for absorption of 90% of the incident X-ray flux are $0.7\,\mu m$, $0.5\,\mu m$, $0.2\,\mu m$, and $0.08\,\mu m$ for the X-ray wavelengths $4.4\text{Å}(Pd_L)$, $8.3\text{Å}(Al_K)$, $13.3\text{Å}(Cu_L)$, and $44.8\,\text{Å}(C_K)$, respectively [Wat88]. Methods for patterning the gold with high resolution include electroplating and ion milling. Electroplating produces excellent definition with vertical walls but requires a vertical-wall primary pattern in a resist that has a thickness equal to that of the metal to be plated. The refractive metal serves as a mask for ion milling the underlying gold. With this method it has been possible to form walls that depart from the vertical by 20° or less. The minimum linewidth attainable by ion milling in 0.5-μm-thick gold is ~0.4 μm. For higher resolutions longer wavelengths such as the 13 Å Cu_L radiation may be used where the gold thickness can be reduced.

The membrane forming the mask substrate should be highly transparent to the X rays so that exposure times are minimized. It should be smooth, flat, dimensionally stable, reasonably rugged, and transparent to visible light if an optical registration scheme is used. Materials that have been used include polymers such as polyimide, polyethylene terephthalate, silicon, SiC, Si_3N_4, Al_2O_3, and a $Si_3N_4/SiO_2/Si_3N_4$ sandwich layer structure.

There are some research groups that have worked out specific steps to produce patterned mask membranes that show little distortion, have reasonable defect densities, and can be handled without fracture. Shimkunas [Shi84] in a review article concludes that "substantial gains" have been made in X-ray mask technology. For example, distortions are reported to be less than 0.1 μm for a 20-cm mask and defects are on the order of 1 or less per cm^2 of finished patterning. Improvements are still needed in the electron-beam tooling used to write the X-ray mask. Refinements have been made in both the additive and subtractive methods of X-ray absorber deposition so that the absorber does not adversely affect membrane positional accuracy.

Alignment

Alignment in lithography systems is a topic of considerable interest because it strongly influences the design rules by which a device can be made [Wil86]. Automatic alignment in optical lithography systems has a dismal record of achievement. This can be attributed to the limited flexibility available in optical lithography because of massive lens systems, wavelength restrictions and so forth. The final overlay is also a function of the lens or optical system distortions.

Several attempts have been made with X-ray lithography to find the position of the mask mark relative to the wafer mark. In one incoherent light optical approach, which has been developed at IBM, the relative mark positions have been determined to within 0.06 μm, 3 [Fay85]. Fay and Hasan [Fay86a] indicate that the overlay of a mask to itself can be as good as 0.1 μm, 3 for the x and y axes [Fay86b]. A characterizaion of a full-field lithography system with an X-ray aligner, using commercially available, fully automated electrical measurement and analysis equipment, has been developed [Sie88]. Figure 5.19 shows the X-ray lithography system [Sie88]. A 25-keV, 5-kW electron beam generated by an electron gun is incident upon a palladium target that emits X rays of a 4.4-Å wavelength. The X rays pass through a beryllium window into a helium-filled chamber to the mask and wafer. Helium is used in the chamber because air is a strong absorber of X rays. As shown at the left-hand side of Fig. 5.19, the X-ray mask and semiconductor wafer are first aligned with each other and then moved to the exposure position, as shown at the right-hand side of the figure.

Figure 5.20 [Sze85] shows the geometric effects on X-ray lithography. Because of the finite size of the X-ray source with diameter a and the finite mask-to-wafer gap

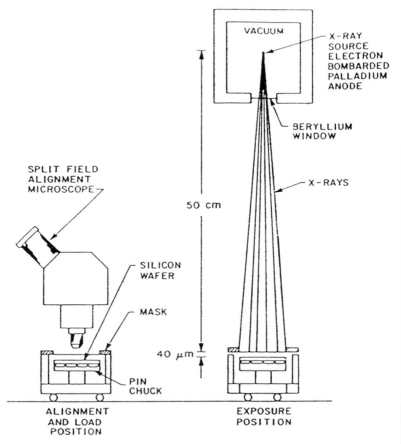

Fig. 5.19 Schematic of an X-ray lithographic system. [Sze, S. M., *Physics of Semiconductor Devices*, Wiley-Interscience (1985)]

g, a penumbral effect occurs. The penumbral blur δ at the edge of the resist image is

$$\delta = a \frac{g}{L}, \tag{5.9}$$

where L is the distance from the source to the X-ray mask. If $a = 3$ mm, $g = 40\,\mu$m, and $L = 50$ cm, the penumbral blur is about $0.2\,\mu$m. Another geometric effect is the lateral magnification error, due to th finite gap g and the nonvertical incidence of the X-ray flux. The projected images of the mask are shifted laterally by an amount d, called *runout* [Sze85]:

$$d = r \frac{g}{L}, \tag{5.10}$$

where r is the radial distance from the center of the wafer. The runout is zero in the

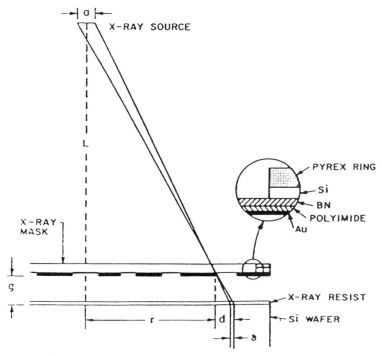

Fig. 5.20 Geometric effects in X-ray lithography. Insert shows the X-ray mask structure. [Sze, S. M., *Physics of Semiconductor Devices*, Wiley-Interscience (1985)]

center but increases linearly toward the wafer's edge. This runout error must be compensated for during the mask-making process.

E-beam resists can be used as X-ray resists. When an X ray is absorbed by an atom, the atom goes to an excited state with the emission of an electron. The excited atom returns to its ground state by emitting an X ray having a different wavelength than the incident X ray. This X ray is absorbed by another atom, and the process repeats. Once the resist film is irradiated, chain crosslinking or chain scission will occur, depending on the type of resist. One of the most attractive X-ray resist is DCOPA (dichloropropyl acrylate and glycidyl methacrylate-co-ethyl acrylate) because it has a very low threshold ($\sim 10\,\text{mJ/cm}^2$).

5.3.3 Ion-Beam Lithography

Ion-beam lithography (IBL) can achieve higher resolution than optical, X-ray, or electron-beam lithographic techniques because ions have a higher mass and therefore scatter fewer electrons. The ions produce very few, low-energy secondary electrons. There is no backscattering and minimal proximity effects, allowing high-density architecture to be patterned in VLSI device. In exposures with the lightest ion, such as H^+, the resolution is limited by the beam diameter rather than by the scattering effects characteristic of *e*-beam lithography. The energetic ions either break bonds or cause crosslinking. Since an ion can supply much greater energy to the resist than an

194 LITHOGRAPHY

electron, there is significantly greater resist sensitivity, and writing time is greatly decreased.

Processing with Ion Beams

The lithographic process generally involves patterning a layer of radiation-sensitive resist material that has been deposited on a substrate [Sie88]. Ion-beam exposure of resists can be either parallel or serial with broad ion beams focused to high resolution probes [Sie88]. In parallel processing the resist is exposed through a mask by projection or by proximity printing. There are limitations on resolution and flexibility. Serial exposure methods, which use a focused probe of ions scanned vectorially or in a raster, offer the possibility of critical alignment, correction for wafer distortion, and easy modification of the exposure pattern features. Focused ion-beam writing has the potential of producing very high resolution patterning with high-density structures and complex architectures because linewidths and spacings are minimally limited by proximity effects. At this stage these capabilities are much more important than very high throughput rates.

Methods of patterning using the focused ion beam are shown in Fig. 5.21. Ions such as Si^+ can be directly implanted in a resist such as PMMA to produce areas that are much more resistant to ion plasma etching [Ade82], or the focused ion beams can be used to write patterns in substrates by ion-beam-assisted etching [Gam82; Och83].

Ion-Beam Resists

The nature of the ion-resist interactions provides possibilities of new approaches to resist characteristics and types of resists that can be used in ion-beam lithography.

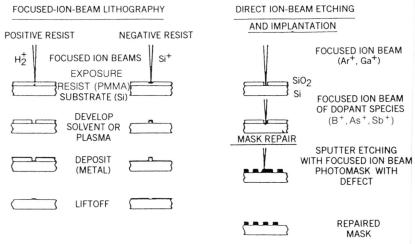

Fig. 5.21 Methods of patterning using focused ion beams. Exposed with a positive resist (e.g., PMMA) to produce a negative resist effect. Direct implantation of dopant through a stencil mask produced by focused ion-beam sputter etching. Mask repair by sputter etching out unwanted metal areas. Masks can also be repaired by ion deposition which will fill the holes in the masks. [Siegel, B. M., in *VLSI Electronics, Microstructure Science*, vol. 16, p. 147, Academic Press (1988)]

Typical organic resists are some two orders of magnitude more sensitive to ions than to electrons. The energy of the incident ion is deposited in a very narrow range both laterally and longitudinally. Secondary electrons have low energy (50 to 100 eV), so they will not broaden the energy deposited by the primary ions. Nor will the secondary electrons that may be backscattered from the substrate interface. These characteristics suggest the possibility of tailoring resists to optimize the resist material for processing considerations other than sensitivity.

Table 5.2 lists the proton and electron sensitivities of several resists compiled by Jensen [Jen84]. Tables 5.3 and 5.4 give the effects of ion mass on sensitivity and penetration, respectively. Considering the range and sensitive of ions to mass and

TABLE 5.2 Proton and Electron Sensitivities of Selected Resists[a]

Resists	$\bar{M}_w (\times 10^{-5})$[b]	E(keV)	Sensivitity H^+/cm^2	e^-/cm^2
Negative				
Polystyrene	1.6	1500	3.7×10^{13}	3.3×10^{14}
	3.2	100	2.4×10^{12}	2.9×10^{14}
Poly(glycidyl methacrylate cochlorostyrene)	2.0	1500	2.6×10^{12}	1.4×10^{13}
Novolac	0.07	1500	9.0×10^{14}	1.0×10^{16}
AZ1350J	—	180	1.9×10^{13}	6.2×10^{13}
Poly(glycidyl methacrylate coethylacrylate)	—	100	3.4×10^{11}	2.6×10^{12}
Poly(4-chlorostyrene)	2.9	100	1.3×10^{12}	2.9×10^{13}
Poly(4-bromostyrene)	5.3	100	6.6×10^{11}	1.1×10^{13}
Iodinated polystyrene	5.5	180	1.9×10^{12}	7.9×10^{12}
Poly(chloromethyl-styrene)	1.2	100	9.0×10^{11}	1.3×10^{13}
Poly(methylmetha-crylate)	8.0	1500	1.3×10^{13}	1.2×10^{14}
	9.5	1500	3.9×10^{13}	—
	6.0	180	1.0×10^{13}	—
	—	150	2.0×10^{13}	1.2×10^{15}
	4.0	100	2.6×10^{13}	—
	—	50	1.3×10^{13}	4.7×10^{12}
Poly(butenesulfone)	20	1500	3.1×10^{12}	—
Poly(tetrafluoropropyl methacrylate) (FPM)	—	50	1.9×10^{12}	2.4×10^{14}
Poly(methyl α-chloro acrylate)	15.5	100	1.8×10^{12}	6.0×10^{12}
Poly(trifluoroethyl α-chloroacrylate)	2.7	100	3.2×10^{12}	

From Jensen, J. E., *Solid State Technol.* **27**, 147 (1984).
[a] Electron exposures were at 20 keV; resist thicknesses 0.5–1.0 μm.
[b] Molecular weight data are the same for both H^+ and electron exposures and is the weight-average molecular weight.

TABLE 5.3 Effect of Ion Mass on Sensitivity of Resists

Resist	Ion	E(keV)	Sensitivity (ions/cm^2)
Polystyrene	O^+	1500	4.3×10^{11}
	He^+	1500	1.4×10^{12}
	H^+	1500	3.7×10^{13}
	e^-	20	3.3×10^{14}
PMMA	Ar^+	120	2.0×10^{11}
	He^+	120	5.2×10^{11}
	H^+	120	3.6×10^{12}
	H^+	20	3.0×10^{14}
	e^-		

From Jensen, J. E., *Solid State Technol.* **27**, 147 (1984).

TABLE 5.4 Penetration Range of Ions in Organic Resists[a]

Ion	E(keV)	Range (μm)
Ga^+	40	0.046
	55	0.06
	120	0.12
Ar^+	120	0.2
Si^{2+}	140	0.35
O^+	1500	3[b]
B^+	100	0.58
Be^+	100	0.75
Li^+	100	0.85
He^+	40	0.44
	100	0.92
	120	0.96
	1500	8[c]
H^+	40	0.52
	100	1.08
	120	1.12
	240	1.85
	1500	40[c]

From Jensen, J. E., *Solid State Technol.* **27**, 147 (1984).
[a] Range values are estimated from experimental results unless otherwise indicated.
[b] Calculated range.

energy, it may be possible to utilize different ions at different energies for specific applications and processes.

Ion-Beam Lithography Systems

In method and instrumentation ion-beam lithography depends primarily on the characteristics of the ion sources available. Until very high brightness field-ion sources were developed, IBL was limited to exposure of resists and substrates by proximity printing through stencil masks using duoplasmatron or similar ion sources [Bro82; Ren79]. The important advance that has made it possible to do IBL by direct writing with focused ion beams has been the development of field ionization sources [Sie88]. These sources have high brightness (10^6 to 10^8 A/cm^2 sr) and produce ion beams with high angular current densities and low-energy broadening. The beams can be focused to high resolution, and high-current-density probes can be scanned to expose substrates in the desired patterns for lithographic processing.

Figure 5.22 gives a schematic of typical optical configuration for an instrument using a field ionization source in a focused-ion-beam lithography system [Pai85]. The characteristics of the source determine the design of the ion optics and the limits of the IBL system's capabilities. The system column consists of the ion lenses, the alignment deflectors and stigmator, the beam blanking, the scanning deflection section, and the laser interferometer controlled stage on which the substrate is mounted.

In designing an ion optical system that will produce the desired ion probe at the substrate plane, some general considerations must be taken into account. The first lens focuses the source to a crossover at a blanking aperture that is placed midway between the blanking deflector plates. Since the effective source size is very small in field-ion sources and the wavelength of the ions is so small that diffraction effects

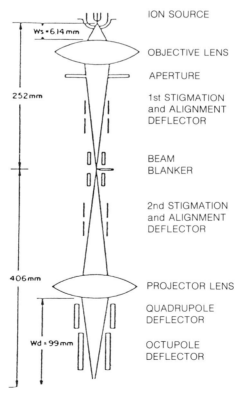

Fig. 5.22 Schematic of an electrostatic optical system for a direct-write IBL instrument using an H_2^+ field-ion source. [Paik, H., G. N. Lewis, E. J. Kirkland, and B. M. Siegel, *J. Vac. Sci. Tech.* **B 3** (1985) 75]

are negligible, the axial image figure is determined by the chromatic and spherical aberrations, usually of the first lens. To obtain the highest-resolution probe with the largest current density, this lens is carefully designed to have low chromatic and spherical aberration coefficients and a short focal length. This lens is placed as close to the source as possible to obtain a large acceptance angle of the ion beam from the source. With the first lens adjusted to focus the source to an image at the blanking aperture, the crossover at the blanking aperture is then focused by the projector lens on the substrate. A lens of relatively long focal length is needed for projector lenses to give the long working distances desired in IBL optical systems. The deflection aberrations will determine the image figure as the beam is scanned to the peripheral areas of the field and set the size of the field that can be exposed without stepping the stage. The stage position, monitored and under the control of laser interferometers, will have the highest accuracy if the full potential of IBL is to be realized.

REFERENCES

[Wil84] Williams, R., *Modern GaAs Processing Methods*, Artech House (1984).

[Wil90] Williams, R., *Modern GaAs Processing Methods*, 2d ed., Artech House (1990).

[Sze85] Sze, S., *Semiconductor Devices—Physics and Technology*, Wiley-Interscience (1985).

[Dil75] Dill, F. H., W. P. Hornberger, P. S. Hauge, and J. M. Shaw, *IEEE Trans. Electron Devices* **22** (1975): 445.

[Wid75] Widmann, D. W., and H. Binder, *IEEE Trans. Electron Devices* **22** (1975): 467.

[Bat83] Batchelder, T., and J. Piatt, *Solid State Tech.* **26**, 8 (1983): 211.

[Wal75] Walker, E. J., *IEEE Trans. Electron Devices* **22** (1975): 464.

[Fra73] Franco, J. R., J. R. Havas, and H. A. Levine, U.S. Patent 3,873,361 (1973).

[Kru84] Kruger, J. B., M. M. O'Toole, and P. Rissman, *VLSI Electronics Microstructure Science*, vol. 8, p. 91, Academic Press (1984).

[O'To83] O'Toole, M., E. Liu, and M. Chang, U.S. Patent 4,370,405 (1983).

[O'To81] O'Toole, M., E. Liu, and M. Chang, *Proc. SPIE*, **275** (1981): 128.

[Mor79] Moran, J. M., and D. M. Maydan, *J. Vac. Sci. Tech.* **16**, 6 (Nov./Dec. 1979): 1620.

[Fra75] Franco, J. R., J. R. Havas, and L. J. Rompala, U.S. Patent 4,003,044 (1975).

[Car76] Carruthers, R., T. Nagasaki, M. R. Poponiak, and T. Zielinski, *IBM Tech. Discl. Bull.* **19** (1976): 1214.

[Kit79] Kitcher, J. R., *J. Vac. Sci. Tech.* **16**, 6 (Nov./Dec. 1979): 2030.

[Und85] Underhill, J., V. Nguyen, M. Kerbaugh, and D. Sundling, *Proc. SPIE*, **539** (1985): 83.

[Kak77] Kakachi, M., S. Sugaware, K. Murasi, and K. Matsuyama, *J. Electrom. Soc.* **124** (1977): 1648.

[Shi84] Shimaya, M., O. Nakajima, C. Hashimoto, and Y. Sakakibara, *J. Electrochem. Soc.* **131** (1984): 1391.

[Suz84] Suzuki, M., H. Namamatsu, and A. Yashikawa, *J. Vac. Sci. Tech.* B **2**, 6 (Nov./Dec. 1984): 665.

[Hom81] Homma, Y., A. Yajuma, and S. Herada, *Proc. IEDM* (1981): 570.

[McD86] McDonnell, L. P., L. V. Gregor, and C. F. Lyons, *Solid State Tech.* (1986): 133.

[Hat84] Hatzakis, M., D. Hofer, and T. H. Chang, *J. Vac. Sci. Tech.* **16**, 6 (Nov./Dec. 1984): 1631.

[Ral82] Ralph, H. I., G. Duggan, and R. J. Elliott, *Electrochem. Soc. Ext. Abs.* **82-1**, Abs. 314, Montreal (May 1982).

[Suz84] Suzuki, M., H. Hamamatsu, and A. Yashikawa, *J. Vac. Sci. Tech.* B **2**, 6 (Nov./Dec. 1984): 665.

[Mat85] Matsuda, T., K. Miyosimi, R. Yamaguchi, S. Moryis, T. Hosoya, and K. Narada, *IEEE Trans. Electr. Dev.* **ED-32**, 2 (1985): 168.

[Ilt70] Ilten, D., and K. Patel, *Proc. SPSE Sem. on Appl. of Photopolymers*, p. 79, SPSE (1970).

[Ilt71] Ilten, D., and K. Patel, *Image Tech.* **2**, 9 (1971).

[Kou70] Koury, H. A., and K. V. Patel, *IBM Tech. Discl. Bull.* **13**, 1 (1970): 38.

[Tin83] Ting, C. H., *Proc. SPIE*, vol. 394 (1983).

[Bre81] Brewer, T., R. Carlson, and J. Arnold, *Appl. Photo. Engrg.* **7**, 6 (1981): 184.

[Lin84] Lin, Y. C., V. Marriott, K. Orvek, and G. Fuller, *Proc. SPIE*, vol. 394, p. 33 (1984).

[Lin88] Lin, P. S. D., and A. S. Gozdz, *J. Vac. Sci. Tech.* B **6**, 6 (1988): 2290.

[Lis84] Listvan, M. A., M. Swanson, A. Wall, and S. A. Campbell, *Proc. SPIE*, vol. 470, p. 85 (1984).

[Tay89] Taylor, J. W., T. L. Brown, and D. R. Bassett, *Advanced Image Reversal Techniques*, SPIE vol. 1086: *Advances in Resist Technology and Processing VI*, p. 585 (1989).

[Mor78] Moritz, H., and G. Paul, U.S. Patent 4,104,070 (1978).

[Mac82] MacDonald, S. A., R. D. Miller, C. G. Wilson, G. M. Feinberg, R. T. Gleason, R. M. Halveson, M. W. MacIntyre, and W. T. Mostsiff, *Proc. Kodak Microelectron. Seminar*, INTERFACE '82, p. 114 (1982).

[All88] Alling, E., and C. Stauffer, *Solid State Tech.* 37 (June 1988).

[Jon88] Jones, S. K., R. C. Chapman, and G. Dishon, *Regional Tech. Conf. of Photopolymers*, p. 279, Ellenville, NY (1988).

[Mor88] Moreau, W. M., *Semiconductor Lithography*, ch. 2, Plenum Press (1988).

[Ker84] Kern, D. P., P. J. Coane, P. J. Houzego, and T. H. P. Chang, *Solid State Tech.* **27**, 2 (1984): 127.

[Cha88] Chang, T. H. P., D. P. Kern, E. Kratschmer, K. Y. Lee, H. E. Luhn, M. A. McCord, S. A. Rishton, and H. Vladimirsky, *IBM J. Res. Dev.* **32**, 4 (July 1988).

[Bro72] Broers, A. N., in *Proc. 5th Int. Conf. Electron and Ion Beam Science and Tech.*, p. 3, ed. by R. Bakish, Electrochem. Soc., Pennington, NJ, (1972).

[Cha76] Chang, T. H. P., A. J. Speth, C. H. Ting, R. Viswanathan, M. Parikh, and E. Munro, in *Proc. 7th Int. Conf. Electron and Ion Beam Sci. and Tech.*, p. 377, ed. by R. Bakish, Electrochem Soc., Pennington, NJ (1976).

[Bro67] Broers, A. N., Long life LaB_6 cathodes, *J. Appl. Phys.* **38** (1967): 1991.

[Hoh82] Hohn, F. J., T. H. P. Chang, A. N. Broers, G. S. Frankel, E. T. Peters, and D. W. Lee, *J. Appl. Phys.* **53** (1981): 1283.

[Cre66] Crewe, A. V., D. N. Eggenberger, J. Wall, and L. M. Welter, *Science* **154** (1966): 729.

[But66] Butler, J. W., *Proc. 6th Int. Conf. on Electron Microscopes*, p. 119, Tokyo (1966).

[Ven78] Veneklasen, L., N. Yew, and J. Wiesner, *Proc. 8th Int. Conf. on Electron and Ion Beam Sci. and Tech.*, p. 880, Electrochemical Society, Pennington, NJ (1978).

[Swa69] Swanson, L. W., and L. C. Crouser, *J. Appl. Phys.* **40** (1969): 4741.

[Tug79] Tuggle, D., L. W. Swanson, and J. Orloff, *J. Vac. Sci. Tech.* **16** (1979): 1699.

[Wol79] Wolfe, J. E., *J. Vac. Sci. Tech.* **16** (1979): 1704.

[Sam85] samoto, N., R Shimizu, H. Hashimoto, N. Tamura, K. Gamo, and S. Namba, *Jap. J. Appl. Phys.* **24** (1985): 766.

[Sai86] Saitou, N., S. Hosoki, M. Okumura, T. Matsuzaka, G. Matsuoka, and M. Okyama, *Microelectron. Eng.* **5** (1986): 123.

[Tho87] Thompson, M. G. R., R. Liu, R. J. Collier, H. T. Carroll, E. T. Doherty, and R. G. Murray, *J. Vac. Sci. Tech.* B **5** (1987): 53.

[Var74] Varnell, G. L., D. F. Spicer, A. C. Rodger, and R. D. Holland, *Proc. 6th Int. Conf. on Electron and Ion Beam Sci. and Tech.*, p. 97, ed. by R. Bakish, Electrochemical Soc., Pennington, NJ (1974).

[Bro79] Broers, A. N., *J. Vac. Sci. Tech.* **16** (1979): 1692.

[Pfe78] Pfeiffer, H. C., *J. Vac. Sci. Tech.* **15** (1978): 883.

[Got78] Goto, E., T. Soma, and M. Idesawa, *J. Vac. Sci. Tech.* **15** (1978): 883.

[Tho78] Thompson, M. G. R., R. J. Collier, and D. R. Harriot, *J. Vac. Sci. Tech.* **15** (1978): 891.

[Cha75] Chang, T. H. P., *J. Vac. Sci. Tech.* **12** (1975): 1271.

[Par79] Parikh, M., and D. F. Kyser, *J. Appl. Phys.* **50** (1979): 1104.

[Ris87] Rishton, S. A., and D. P. Kern, *J. Vac. Sci. Tech.* B **5** (1987): 135.

[Wat88] Watts, R. K., in *VLSI Technology*, 2d ed., ed. by S. M. Sze, McGraw-Hill (1988).

[Cha82] Chang, T. S., D. F. Kyer, and C. H. Ting, *Solid State Tech.*, **15**, 5 (1982): 60.

[Lin83] Lin, B. J., *Solid State Tech.* **26**, 5 (1983): 105.

[Mar79] Martel, G. W., and W. B. Thompson, *Semiconductor Int.* (Jan./Feb. 1979): 69.

[Lin88] Lin, P. S. D., and A. S. Gozdz, *J. Vac. Sci. Tech.* B **6**, 6 (1988): 2290.

[Hat81] Hatzakis, M., *Solid State Tech.* **14** (Aug. 1981): 79–80.

[Hat84] Hatzakis, M., *J. Vac. Sci. Tech.*, **16** (1984): 1979.

[Oro82] Oro, J. A., and J. C. Wolfe, *J. Appl. Phys.* **53** (1982): 7379.

[Spe72] Spears, D. L., and H. I. Smith, *Electronics Lett.* **8** (1972): 102.

[Wat88] Watts, R. K., in *VLSI Technology*, 2d ed., ed. by S. M. Sze, McGraw-Hill (1988).

[Yos77] Yoshimatsu, M., and S. Kozaki, in *X-ray Optics*, ed. by H. J. Queisser, Springer-Verlag (1977).

[McG81] McGillis, D. A., in *VLSI Technology*, ed. by S. M. Sze, McGraw-Hill (1981).

[Spi76] Spiller, E., D. E. Eastman, R. Feder, W. D. Grobman, W. Gudat, and J. Topalian, *J. Appl. Phys.* **47** (1976): 5450.

[Pol86] Pol, V., *Proc. SPIE*, **633** (1986): 6.

[Oka91] Okazaki, S., *Solid State Tech.* (Nov. 1991): 77.

[Shi84] Shimkunas, A. R., *Solid State Tech.* **27**, 9 (Sept. 1984): 192–199.

[Kim92] Kimura, T., in *Semiconductor Int.* p. 13 (Feb. 1992).

[Flo91] Flores, G. E., and B. Kirkpatrick, *IEEE Spectrum* (Oct. 1991): 24–27.

[Lin91] Lin B. J., Phas-shifting and other challenges in optical mask technology, in *Bacus News* (Feb. 1991).

[Lin92] Lin, B. J., *Solid State Tech.* (Jan. 1992): 43.

[Bur92] Burgraaf, P., *Semicond. Int.* (Feb./Mar. 1992).

[Lev82] Levenson, M.D., *IEEE Trans. Electron Dev.* **29**, 12 (1982): 1828.

[Wil86] Wilson, A. D., *Solid State Tech.* (May 1986): 249.

[Fay85] Fay, B., L. Tai, and D. Alexander, in *SPIE*, vol. 537, pp. 57–68 (1985).

[Fay86a] Fay, B., and T. Hasan, *Solid State Tech.* (May 1986): 239–243.

[Fay86b] Fay, B. S., and W. T. Novak, *Proc. SPIE*, **632** (1986): 146.

[Sie88] Siegel, B. M., in *VLSI Electronics, Microstructure Science*, vol. 16, p. 147, Academic Press (1988).

[Ade82] Adesida, I., J. D. Chin, L. Rathburn, and E. D. Wolf, *J. Vac. Sci. Tech.* **21** (1982): 666.

[Gam82] Gamo, K., Y. Ochiai, and S. Namba, *Jap. J. Appl. Phys.* **21** (1982): 1.

[Och83] Ochiai, Y., K. Gamo, and S. Namba, *J. Vac. Sci. Tech.*, B **1** (1983): 1047.

[Jen84] Jensen, J. E., *Solid State Tech.* **27** (1984): 145.

[Bro82] Brodie, I., and J. J. Muray, *Physics of Microfabrication*, Plenum Press (1982).

[Ren79] Rensch, D. B., R. L. Seliger, G. Csanky, R. D. Olney, and H. L. Stover, *J. Vac. Sci. Tech.* **16** (1979): 1897.

[Pai85] Paik, H., G. N. Lewis, E. J. Kirkland, and B. M. Siegel, *J. Vac. Sci. Tech.* B **3** (1985): 75.

6

DEVICE-RELATED PHYSICS AND PRINCIPLES

6.1 SOME BASIC ELECTRONIC CONCEPTS IN LOW-DIMENSIONAL PHYSICS

In the last decade the technology drive toward smaller microelectronic devices has given rise to the concept of low-dimensional physics [Kel87]. Low-dimensional structures [Sak80; Pet82], such as the quantum-well wire (QWW) or the quantum-well box (QWB), have quantum confinement (QC) of two or three dimensions and have in the last few years attracted much attention not only for their potential in uncovering new phenomena in solid state physics but also for their potential application in semiconductor devices. Theoretically extremely high electron mobility in QWWs and high performance of QWW or QWB lasers [Ara82] and modulators [Sue88] can be expected. Recent experiments in QWB resonant-tunneling devices [Ree88] have further claimed to demonstrate new structures within a zero-dimensional system. This chapter will discuss the physics of low-dimensional devices.

In this section we consider the range of electronic states that are encountered in various microstructure systems. The ideal systems come straight from the quantum mechanics—potential wells of different shapes bounded by infinitely high barriers [Kel87]:

1. A three dimensional system without bounding (3D).
2. A purely two-dimensional system bounded by $Z = 0$ and L (2D), whenever L is small.
3. A purely one-dimensional system bounded by $X, Y = 0$ and L (1D), whenever L is small.
4. A quasi two-dimensional system bounded by $Z = 0$ and L (Q2D).
5. A quasi one-dimensional system bounded by $X, Y = 0$ and L (Q1D).
6. A zero-dimensional system bounded by $X, Y, Z = 0$ and L (0D).

SOME BASIC ELECTRONIC CONCEPTS IN LOW-DIMENSIONAL PHYSICS

The one-electron Schrödinger equation, relating the kinetic and potential energy to the total energy of that electron, is given by

$$\frac{\hbar^2}{2m^*}(\nabla^2 + V)\psi = E\psi,$$

where ∇ and V are 3, 2, 1, 2, 1, 0 dimensional in the six cases above, with V being of zero inside and infinity outside the well. The eigenfunctions and eigenvalues are given, respectively, by

$$\psi_n = \left(\frac{2}{L}\right)^{1/2} \sin\left(\frac{n\pi z}{L}\right), \quad E_n = \frac{\hbar^2}{2m^*}\left(\frac{n\pi}{L}\right)^2, \quad \text{(2D)} \quad (6.1)$$

$$\psi_{lm} = \left(\frac{2}{L}\right) \sin\left(\frac{l\pi x}{L}\right)\sin\left(\frac{m\pi y}{L}\right), \quad E_{l,m} = \frac{\hbar^2}{2m^*}\left[\left(\frac{l\pi}{L}\right)^2 + \left(\frac{m\pi}{L}\right)^2\right] \quad \text{(1D)} \quad (6.2)$$

$$\psi_{lmn} = \left(\frac{2}{L}\right)^{3/2} \sin\left(\frac{l\pi x}{L}\right)\sin\left(\frac{m\pi y}{L}\right)\sin\left(\frac{n\pi z}{L}\right),$$

$$E_{lmn} = \frac{\hbar^2}{2m^*}\left[\left(\frac{l\pi}{L}\right)^2 + \left(\frac{m\pi}{L}\right)^2 + \left(\frac{n\pi}{L}\right)^2\right]. \quad \text{(0D)} \quad (6.3)$$

where $n, l, m = 1, 2, 3, \ldots$.

Similar results are obtained for shapes 4 and 5 where appropriate alterations from L to λ are made. When $\lambda \ll L$, the kinetic energy of confinement in the λ dimension is very high, and in systems containing many electrons the usual quantum number to vary is in the L dimension. In **k**-space the density of modes is uniform at L/π in each spatial dimension. If we place many electrons into the wells satisfying Fermi-Dirac statistics, the distribution of energy states yields an important quantity, which is obtained by equating the number of states between E and $E + dE$ to the length/area/volume of **k**-space bounded by the two energies and the constant density of states in **k**-space. The results are [Kel87]

$$N(E) = \frac{2L^3}{\pi^2}\frac{(2m^*)^{3/2}}{\hbar^2}\sqrt{E}, \quad \text{whenever } L \text{ is very small} \quad \text{(3D)} \quad (6.4)$$

$$N(E) = \frac{L^2}{\pi}\frac{2m^*}{\hbar^2}, \quad \text{whenever } L \text{ is very small} \quad \text{(2D)} \quad (6.5)$$

$$N(E) = \frac{L}{2\pi}\frac{(2m^*)^{1/2}}{\hbar^2}\frac{1}{\sqrt{E}}, \quad \text{whenever } L \text{ is very small} \quad \text{(1D)} \quad (6.6)$$

$$N(E) = \sum_n \frac{L^2}{\pi}\frac{2m^*}{\hbar^2}H(E - E_n), \quad \text{whenever } L \text{ is very small} \quad \text{(Q2D)} \quad (6.7)$$

$$N(E) = \sum_{lm} \frac{L}{2\pi}\frac{(2m^*)^{1/2}}{\hbar^2}(E - E_{lm})^{-1/2}H(E - E_{lm}), \quad \text{(Q1D)} \quad (6.8)$$

$$N(E) = \delta(E - E_{lmn}) \quad \text{(0D)}$$

where $H(\sigma)$ is the Heaviside function $[H(\sigma) = 1(\sigma > 0); = 0(\sigma < 0)]$, E_n and E_{lm} are as defined in Eqs. (6.1) and (6.2). These functions are shown in Fig. 6.1 [Kel87]. In the limit $L \gg \lambda$, the density of energy states is high for the L dimensions, and the continuum limit is used.

The density of states is very useful in calculating many response functions, and changes to individual electron states or to their occupation. Specific heat, magnetization, thermoelectric power, and conductivity, all depend on the density of states at the Fermi energy or its energy derivative there. A number of points can be made about the density of states in Fig. 6.1:

1. In dimensionality analysis the differences in energy dependence are used to explain several of the optical and electrical properties. In the 1-D and quasi-1-D cases, there are square-root divergences that help to destabilize simple structures. The energy independence of the density of states in 2-D and quasi-2-D systems is exploited in several optical devices.
2. In the quasi-1-D and 2-D systems, the quantum energy of confinement associated with the thickness λ is large, though in practical semiconductor or metal systems several of the replicas (subbands or minibands) associated with the higher-energy states in the narrow direction are filled. The scattering of electrons between minibands contributes to the resistance. In systems where the Fermi energy can be moved, strong fluctuations are associated with the occupation of further minibands, and these fluctuations are measured as magnetoresistance and magnetization, and so forth.
3. In semiconductors the offset in the valence band edges results in hole levels that mirror the electron levels. The optical properties reflect the size quantization effects [E_n and E_{lm} in Eqs. (6.7) and (6.8)] in that optical transitions couple the electron and hole states across the gap, as well as where the narrow gap material forms the wells.
4. Perturbations to the simple picture of the electron states come from several sources: (a) Fluctuations in the dimensions (particularly in the λ direction) can

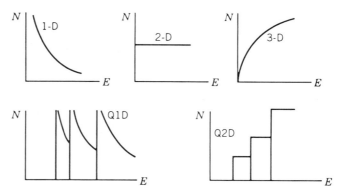

Fig. 6.1 The density of states for the 1-D, 2-D, 3-D, quasi-1-D and quasi-2-D systems. [Kelly, M. J., in *Physics and Technology of Submicron Structures*, pp. 174–196, ed. by G. Bauer, F. Kuchar, and H. Heinrich, *Springer Series in Solid State Sciences*, Springer-Verlag (1987)]

cause relatively large fluctuations in E_{lm} and E_n, and the Schrödinger equation no longer decouples easily. (b) Fields applied in the lateral dimensions alter the Schrödinger equation in the λ direction, and this technique is used to induce large relative changes in both the energy of the optical transitions and their associated optical nonlinearities. In special 3-D systems it can induce one-dimensional transition from L to a λ-length scale. (c) Fields applied in the L direction when one or both of the other dimensions are of size λ with dimensionality dependent hot-electron transport or quenching of optical nonlinearities.

5. Many thermodynamic quantities are related to the electronic density of states. The self-consistent analysis of

$$U = \int_0^{E_f} f(\varepsilon)n(\varepsilon)d\varepsilon,$$

gives a specific heat $U \propto N(E_f)T$. The low-temperature phonon contribution varies as T^3 for three-dimensional systems or low-dimensional electronic systems embedded in most 3-D material, but for free-standing membranes (wires) the variation is T^2 and T, since the lateral modes of vibration are quenched.

Magnetoresistance is a function of dimensionality. It depends on the density of states and the nature of the available final states for scattering. An extreme example is the quantum hall regime where magnetoresistance and magnetoconductivity of the two-dimensional system vanish. In systems where the dimensionality can be varied, any magnetoresistance is a sign of dimensionality.

Dimensionality is also suggested by the shape of the optical absorption edge. The total absorbing power of a system expressed as energy $\hbar\omega$ contains a factor

$$\int |M|^2 J(E, \hbar\omega) f(E)(1 - f(E + \hbar\omega))dE,$$

where f is the occupation factor, $J(E, \hbar\omega)$ is the joint density of states, a convolution of initial and final states, and M is a dipole matrix element coupling the initial and final states. The difference between the $E^{1/2}$ and E^0 band edges is apparent in the joint density of states and the shape of the optical absorption edge. The sharper features of the $E^{-1/2}$ edge in 1-D have been seen in some molecular solids, where very weak intermolecular interactions simulate the 1-D condition, now contribute to excitons.

6.2 BAND STRUCTURE, IMPURITIES, AND EXCITONS IN SUPERLATTICES

Advances in the area of semiconductor superlattices have proceeded at a remarkable pace during the past few years. Many theoretical methods [Car77; Gel86; Sch77; Sch81; Kri85; Sai77; Bas84; Mai84; Smi86; Fas84; Alt85; San85; San87] have been developed for calculating the electronic structures of superlattices. These theoretical methods can be classified into three categories: pseudopotentials [Car77; Gel86], tight bindings [Sch77; Sch81; Kri85], and envelope-functions (k.p) [Sai77; Bas82; Mai84; Smi86; Fas84; Alt85; San85; San87].

The GaAs–AlAs monolayer structure has a tetragonal unit cell [Car77]. It contains four basis atoms and has D_{2d} space group symmetry [Esa70]. The Brillouin zone (BZ) of this structure is one-half the size of the zincblende BZ. Caruthes and Lin-Chung gave the GaAs–AlGa$_{1-x}$Al$_x$As crystal pseudopotential in terms of the atomic form factors [Car77]. However, among these methods the envelope-function approach is most widely adopted because of its simplicity. With various refinements this approach can become quite sophisticated and can be used to study many problems such as band mixing, the effects of external fields, impurities, and exciton states. The major drawback of this method is that the boundary conditions become extremely complicated when many bands are involved [Smi86; Alt85]. In this section we will describe the envelope-function approximation developed by Altarelli [Alt86] for the heterostructure devices and a new bond-orbital model developed by Chang [Cha88].

6.2.1 The Tight-Binding Approximation

The electronic structure of a solid begins with an arbitrary atom and grows by including more and more neighbors in an appropriate configuration. The eigenfunctions of such a cluster are determined by solving its Schrödinger equation [Böe90]. In this approximation the electrons are not nearly free electrons in a periodic lattice. The electrons with relatively low energy are strongly bound to the atomic core. We combine the atomic orbitals to represent a state running throughout the crystal, with each orbital on a particular atom [Wan89]. This scheme is called the *tight-binding approximation*.

We use Bloch's representation, to find the wave function of the electron:

$$\psi(x, k) = \sum_n u_a(x - nc)\exp(iknc), \qquad (6.9)$$

where $u(x - nc)$ is the atomic orbital of the nth atom and $\phi = knc$. This wave function represents a series of localized atomic orbitals multiplied by a phase factor. As x varies from cell to cell, a sinusoidal variation in the amplitudes of Re(ψ) and Im(ψ) is superimposed on the atomic structure of each cell [Ash76]. We can also write

$$\psi(x + c, k) = \exp(ikc)\sum u_a(x - n'c)\exp(ikn'c) = \psi(x, k)\exp(ikc), \qquad (6.10)$$

where $n' = n - 1$. $\psi(x, k)$ is the Bloch expression.

Consider the one-dimensional case. The Schrödinger equation for the nuclear potential $V_0(x - nc)$ of the nth atom is

$$(H_{0n} - H_{1n})\psi = E\psi, \qquad (6.11)$$

where

$$H_{0n} = \frac{-\hbar^2 \partial^2}{2m\,\partial x^2} + V_0(x - nc), \qquad (6.12)$$

and
$$H_{1n} = V_0(x - nc) - V(x), \tag{6.13}$$

are the unperturbed and perturbation Hamiltonian, respectively, for the nth atom. For the atomic orbitals u_a we expect it to satisfy the local Schrödinger equation:

$$H_{0n} u_a(x - nc) = E_a u_a(x - nc), \tag{6.14}$$

where E_a is the eigenenergy of the atomic state. Substituting Eq. (6.9) in Eq. (6.11) and using Eq. (6.14), we find that

$$E\psi(x, k) = E_a \psi(x, k) - H_{1n}\psi(x, k). \tag{6.15}$$

The expectation value of E is equal to

$$\langle E \rangle = E_a - \frac{\int \psi^* H_{1n} \psi \, dx\, dy\, dz}{\int \psi^* \psi \, dx\, dy\, dz}. \tag{6.16}$$

Equation (6.16) is obtained by multiplying Eq. (6.15) by $\psi^*(x, k)$ and then integrating the product over space coordinates.

A typical form in the numerator of Eq. (6.16) is

$$I_m = \exp[(ikc)(n - m)] \int u_a^*(x - mc) H_{1n} u_a(x - nc) d\tau, \tag{6.17}$$

where $d\tau$ is the volume element and the integral is over the volume of the crystal. Therefore the purturbation potential H_{1n} changes with $u_a(x - nc)$ in Eq. (6.17). Since the atomic orbital $u_a(x - nc)$ falls very rapidly as x moves away from the nuclear position $x = nc$, the integral I_m has an appreciable value only if there is a sufficient overlap between $u_a(x - nc)$ and $u_a(x - mc)$. Only three terms are important with $m = n$ and $m = n \pm 1$ in Eq. (6.17). We let

$$\int u_a^*(x - nc) H_{1n} u_a(x - nc) d\tau = \alpha \int |u_a(x - nc)|^2 d\tau, \tag{6.18}$$

$$\int u_a^*(x - nc \pm c) H_{1n} u_a(x - nc) d\tau = \beta \int |u_a(x - nc)|^2 d\tau. \tag{6.19}$$

Since any atom has circumstances identical with those of other atoms, there are N such terms in the numerator and N terms of $\int |u_a| d\tau$ in the denominator of Eq. (6.16) for a linear chain of N atoms. If we neglect the contributions from the cross products $u_a(x - nc)u_a(x - mc)$ in the denominator, we obtain [Wan89]

$$\langle E \rangle = E_a - \alpha - 2\beta \cos(kc). \tag{6.20}$$

This analysis can be extended to a three-dimensional lattice. For an atomic s state with $E_a = E_1$,

$$\langle E \rangle = E_1 - \alpha_1 - 2\beta_1[\cos(k_x a) + \cos(k_y a) + \cos(k_z a)]. \tag{6.21}$$

For an atomic p state,

$$\langle E \rangle = E_2 - \alpha_2 + 2\gamma_2 \cos(k_x a) - 2\beta_2[\cos(k_y a) + \cos(k_z a)]. \tag{6.22}$$

Because the atomic orbitals p_x, p_y, p_z are directed along the x, y, and z axes, the value of α_2 is greater than α_1 in Eq. (6.21).

In the superlattice system there are many interfaces between different layers of semiconductors. Schulman and Chang [Sch83] have developed a reduced Hamiltonian method for solving the tight-binding model of interfaces. With the correct boundary conditions the tight-binding equations in this model are solved exactly. The reduced Hamiltonian has three major components: transfer matrices, an expansion in bulk states with complex wave-vector values, and a quickly convergent iterative method for finding energy eigenvalues.

6.2.2 The Envelope-Function Approximation

The envelope function has proved to be simple, accurate, and versatile approach for calculating electron levels [Alt86]. Electronic states in three types of heterostructures are considered: superlattices, isolated quantum wells, and heterojunctions, as shown in Fig. 6.2. The latter two are regarded as limiting cases of the first. When the thickness of each layer becomes much larger than some characteristic wave function decay

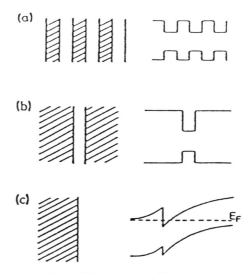

Fig. 6.2 Schematic representation of three types of heterostructures and (*on the right*) the corresponding band-edge profile: (*a*) superlattice; (*b*) isolated quantum well; (*c*) single heterojunction between an *n*-type material (*to the left of the junction*) and a *p*-type one (*on the right*). [Altarelli, M., *Heterojunctions and Semiconductor Superlattices*, ed. by G. Allan, G. Bastard, N. Boccara, M. Lannoo, and M. Voos, Springer-Verlag (1986)]

The Effective-Mass Equation

The envelope-function approximation is based on effective-mass theory (see Section 2.4.4). Consider the motion of an electron in a semiconductor in presence of some additional potential $U(\mathbf{r})$. The Schrödinger equation can be written

$$\left\{\frac{p^2}{2m_0} + V_{\text{per}}(\mathbf{r}) + U(\mathbf{r}) - E\right\}\psi(\mathbf{r}) = 0. \tag{6.23}$$

Here m_0 is the free electron mass, V_{per} is the periodic potential of the perfect bulk semiconductor, and $U(\mathbf{r})$ is the additional potential, which we assume to be slowly varying and weak. $U(z)$ represents the band-bending potential on the right-hand side of Fig. 6.2(c) (z being the coordinate normal to the interface); it arises from depleted acceptors. Equation (6.23) describes the motion of an electron on the right-hand side of the junction. If the potential $U(\mathbf{r})$ is vanishing, the solution of Eq. (6.23) becomes a Bloch function $\psi_{n\mathbf{k}}(\mathbf{r})$, with

$$\left\{\frac{p^2}{2m_0} + V_{\text{per}}(\mathbf{r})\right\}\psi_{n\mathbf{k}}(\mathbf{r}) = E_n(\mathbf{k})\psi_{n\mathbf{k}}(\mathbf{r}) \tag{6.24}$$

and

$$\psi_{n\mathbf{k}}(\mathbf{r}) = e^{i\mathbf{k}\cdot\mathbf{r}} u_{n\mathbf{k}}(\mathbf{r}), \quad (u_n \text{ periodic}). \tag{6.24'}$$

In Eq. (6.24) $\mathbf{k}$ is restricted to the first Brillouin zone. For $U(\mathbf{r}) \neq 0$ we expand Eq. (6.23) in the form

$$\psi(\mathbf{r}) = \sum_{n\mathbf{k}} \phi_n(\mathbf{k})\psi_{n\mathbf{k}}(\mathbf{r}). \tag{6.25}$$

Replacing $\psi(\mathbf{r})$ in Eq. (6.23) gives

$$[E_n(\mathbf{k}) - E]\phi_n(\mathbf{k}) + \sum_{n'\mathbf{k}'}\langle\psi_{n\mathbf{k}}|U(\mathbf{r})|\psi_{n'\mathbf{k}'}\rangle\phi_{n'}(\mathbf{k}') = 0, \tag{6.26}$$

where the potential matrix element is

$$\int \psi_{n\mathbf{k}}^*(\mathbf{r})U(\mathbf{r})\psi_{n'\mathbf{k}'}(\mathbf{r})d^3r = \int e^{-i(\mathbf{k}-\mathbf{k}')\cdot\mathbf{r}} u_{n\mathbf{k}}(\mathbf{r})u_{n'\mathbf{k}'}(\mathbf{r})U(\mathbf{r})d^3\mathbf{r}$$

$$= \sum_{\mathbf{G}} \tilde{U}(\mathbf{k}' - \mathbf{G})c(n\mathbf{k}, n'\mathbf{k}'; \mathbf{G}). \tag{6.27}$$

The last step is derived by expanding the function $u_{n\mathbf{k}}^* u_{n'\mathbf{k}'}$, which is periodic, in a Fourier series of reciprocal lattice vectors,

$$u_{n\mathbf{k}}(\mathbf{r})u_{n'\mathbf{k}'}(\mathbf{r}) = \frac{1}{(2\pi)^3}\sum_{\mathbf{G}} c((n\mathbf{k}, n'\mathbf{k}'; \mathbf{G})e^{-i\mathbf{k}\cdot\mathbf{r}}), \tag{6.28}$$

so that the d^3r integration in Eq. (6.27) yields the Fourier transformation $\tilde{U}$ of U. Assume that the $U(\mathbf{r})$ is slowly varying on the length scale of the unit cell, the Fourier transform $\tilde{U}$ is very small unless its argument is much smaller than the Brillouin zone. Here only $\mathbf{k}$ vectors in a small region of the zone are mixed by U and contribute to the expansion (6.25) for a given state. We can ignore all terms with $\mathbf{G} \neq 0$ and write

$$\langle \psi_{n\mathbf{k}} | U(\mathbf{r}) | \psi_{n'\mathbf{k}'} \rangle = \tilde{U}(\mathbf{k} - \mathbf{k}') c(n\mathbf{k}, n'\mathbf{k}'; 0). \tag{6.29}$$

In the evaluation of $c(n\mathbf{k}, n'\mathbf{k}'; 0)$ we may use the fact that $\mathbf{k}$ and $\mathbf{k}'$ are near $\mathbf{k} = 0$ and express $u_{n\mathbf{k}}$ and $u_{n'\mathbf{k}'}$ in terms of the $\mathbf{k} = 0$ periodic functions u_{m0}, by means of the standard $\mathbf{k} \cdot \mathbf{p}$ perturbation theory:

$$u_{n\mathbf{k}}(\mathbf{r}) = u_{n0}(\mathbf{r}) + \sum_{m \neq n} \frac{\mathbf{k} \cdot \mathbf{p}_{mn}}{m_0(E_n(0) - E_m(0))} u_{m0}(\mathbf{r}) + \ldots. \tag{6.30}$$

The higher-order terms are omitted. If band n is well separated from the others, the denominator is always much larger than the numerator, and the approximation is very good for small $\mathbf{k}$. On the same order of accuracy, we find from Eq. (6.30) that

$$c(n\mathbf{k}; n'\mathbf{k}'; 0) = \delta_{nn'} + \frac{(\mathbf{k} - \mathbf{k}') \cdot \mathbf{p}_{nn'}}{m_0(E_n(0) - E_{n'}(0))} (1 - \delta_{nn'}). \tag{6.31}$$

Therefore from Eq. (6.29) we obtain

$$\langle \psi_{n\mathbf{k}} | U(\mathbf{r}) | \psi_{n'\mathbf{k}'} \rangle = \tilde{U}(\mathbf{k} - \mathbf{k}')(\delta_{nn'} + A_{nn'}(1 - \delta_{nn'})), \tag{6.32}$$

where $A_{nn'}$ is the off-diagonal contribution of the second part of Eq. (6.31). If the potential is weak, however, this off-diagonal term can be ignored. When treated as a perturbation, it gives a second-order correction to the energy, and a correction of order

$$\frac{(\mathbf{k} - \mathbf{k}') \cdot \mathbf{p}_{nn'}}{m_0(E_n(0) - E_{n'}(0))} \frac{\tilde{U}(\mathbf{k} - \mathbf{k}')}{(E_n(0) - E_{n'}(0))} u_{n'} \tag{6.33}$$

to the wave function [i.e., smaller by a factor $U/\Delta E$ than the first-order $\mathbf{k} \cdot \mathbf{p}$ correction of Eq. (6.30)]. Here ΔE is a typical energy gap of the system ($\sim 1\,\text{eV}$) and a matrix element of the potential; by "weak" potential we mean that the ratio $U/\Delta E$ is small.

Equation (6.26) reduces to

$$[E_n(\mathbf{k}) - E]\phi_n(\mathbf{k}) + \sum \tilde{U}(\mathbf{k} - \mathbf{k}')\phi_n(\mathbf{k}') = 0. \tag{6.34}$$

In the small region around $\mathbf{k} = 0$ of interest here, $E(\mathbf{K})$ is well approximated by the second-order $\mathbf{k} \cdot \mathbf{p}$ expansion, which generally reads as

$$E_n(\mathbf{k}) = E_n(0) + \sum_{\alpha,\beta=1}^{3} \frac{\hbar^2}{2m_0} \left(\delta\alpha\beta + \frac{1}{m_0} \sum_{m \neq n} \frac{p_{nm}^\alpha p_{mn}^\beta + p_{nm}^\beta p_{mn}^\alpha}{E_n(0) - E_m(0)} \right) k_\alpha k_\beta, \tag{6.35}$$

where α, β run over x, y, and z. For the simple, isotropic nondegenerate band extremum (e.g., the conduction band minimum of the GaAs-related direct semiconductor), Eq. (6.35) gives only the nonvanishing diagonals.

Equation (6.35) reduces to

$$E_n(\mathbf{k}) = E_n(0) + \frac{\hbar^2 k^2}{2m^*}$$

where

$$\frac{1}{m^*} = \frac{1}{m_0} + \frac{2}{m_0^2} \sum_{m \neq n} \frac{p_{nm}^\alpha p_{mn}^\alpha}{E_n(0) - E_m(0)} \qquad (\alpha = x, y, \text{ or } z) \qquad (6.35')$$

is the inverse effective mass of band n.

Taking a Fourier transform of Eq. (6.24) and using Eqs. (6.35) and (6.35') as if they were valid in the whole $\mathbf{k}$-space, we find

$$\left[\frac{-\hbar^2 \nabla^2}{2m^*} + U(\mathbf{r}) \right] F(\mathbf{r}) = (E - E_n(0)) F(\mathbf{r}), \qquad (6.36)$$

where $F(\mathbf{r})$ is the Fourier transform of $\phi_n(\mathbf{k})$. Equation (6.36) is the effective-mass equation. The total wave function, in this approximation, is derived from Eqs. (6.25) and (6.30):

$$\psi(\mathbf{r}) = \sum_{\mathbf{k}} \phi_n(\mathbf{k}) \psi_{n\mathbf{k}}(\mathbf{r})$$

$$= \sum_{\mathbf{k}} \psi_n(\mathbf{k}) e^{i\mathbf{k}\cdot\mathbf{r}} \left[u_{n0}(\mathbf{r}) + \sum_{m \neq n} \frac{\mathbf{k} \cdot \mathbf{p}_{mn}}{m_0(E_n(0) - E_m(0))} u_{m0}(\mathbf{r}) \right]$$

$$= F(\mathbf{r}) u_{n0}(\mathbf{r}) + \sum_{m \neq n} \frac{-i(\nabla F(\mathbf{r})) \cdot \mathbf{p}_{mn}}{m_0(E_n(0) - E_m(0))} u_{m0}(\mathbf{r}). \qquad (6.37)$$

Equation (6.37) shows that to the lowest order $F(\mathbf{r})$ is a slowly varying "envelope" that modulates the rapidly varying Bloch component $u_{n0}(\mathbf{r})$. The following correction term shows that there is a contribution from other $\mathbf{k} = 0$ Bloch functions that is proportional to the gradient of the envelope $F(\mathbf{r})$.

Suppose that the two parameters m^* and $E_n(0)$ in Eq. (6.36) assume different values in the two semiconductors, say, A and B, that make up an interface system. The modified equation (6.36) would read

$$\left[\frac{-\hbar^2 \nabla}{2} \cdot \left(\frac{1}{m^*(z)} \nabla \right) + U(\mathbf{r}) + E_n(0, z) \right] F(\mathbf{r}) = E F(\mathbf{r}). \qquad (6.38)$$

The kinetic energy term has been rewritten, for a z-dependent mass, in a way which restores the Hermitian character of the hamiltonian, following Harrison [Har61] and Ben Daniel and Duke [Ben66]. For z well to the left of the interface, $m^* = m_A^*$, $E_n(0) = E_n^A(0)$, and for z well on the right side $m^* = m_B^*$ and $E_n(0) = E_n^B(0)$.

The potential term in Eq. (6.38) varies much too rapidly for the effective-mass formalism to be valid. But we can learn something about the boundary conditions from this differential equation. Taking the limit in which the effective-mass and band-edge energy variations occur over an infinitesimal thickness 2ε and integrating Eq. (6.38) between $z = -\varepsilon$ and $z = +\varepsilon$, we obtain

$$F^A(-\varepsilon) = F^B(+\varepsilon)$$
$$\frac{1}{m_A^*} \frac{\partial F^A}{\partial z}\bigg|_{-\varepsilon} = \frac{1}{m_B^*} \frac{\partial F^B}{\partial z}\bigg|_{\varepsilon} \tag{6.39}$$

From Eq. (6.37), we notice that for both F and ψ to be continuous, we must assume that

$$u_{n0}^A = u_{n0}^B, \tag{6.40}$$

that the second term of the wave function Eq. (6.37) is small, and that the **k**-dependence of the Bloch function $u_{n\mathbf{k}}$ about $\mathbf{k} = 0$ is weak.

For the probability current operator, the existence of stationary states implies that the z component of the current is the same on all planes parallel to the interface. Therefore its average over a microscopic volume Ω, including one or few unit cells, will be the same on both sides of the interface. Let us calculate this average on the A side. We use the wave form given by Eq. (6.37) and recall that

$$\int_{\text{cell}} u_{n0}^* \frac{\partial u_{m0}}{\partial z} = \frac{i}{\hbar} p_{nm} \tag{6.41}$$

and that this matrix element vanishes for $m = n$ because of symmetry [Ata86]. The average current J can be written

$$\bar{J}_A = \frac{\hbar}{m_0} \text{Im} \int_\Omega d^3\mathbf{r} \psi_A^* \frac{\partial}{\partial z} \psi_A = \frac{\hbar}{m_A^*} \text{Im}\left(F_A^*(0) \frac{\partial}{\partial z} F_A(0) \right). \tag{6.42}$$

Therefore the continuity of J implies that

$$\frac{\hbar}{m_A^*} \text{Im}\left(F_A^*(0) \frac{\partial}{\partial z} F_A(0) \right) = \frac{\hbar}{m_B^*} \text{Im}\left(F_B^*(0) \frac{\partial}{\partial z} F_B(0) \right). \tag{6.43}$$

It is now apparent that the boundary conditions of Eq. (6.39) on the envelope functions imply that the average of the probability current is constant, as stated in Eq. (6.43).

Coupled Bands

There may be various reasons why the conditions for the validity of the simple-band approach are violated [Alt86]:

1. The band degeneracy is near an extremum, as in the case of the valence band maximum at Γ in all cubic semiconductors shown in Fig. 6.3.

BAND STRUCTURE, IMPURITIES, AND EXCITONS IN SUPERLATTICES 213

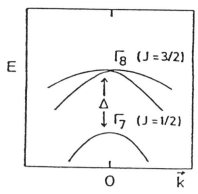

Fig. 6.3 Schematic representation of the top of the valence band of the cubic semiconductors. △ denotes spin-orbit splitting. [Altarelli, M., *Heterojunctions and Semiconductor Superlattices*, ed. by G. Allan, G. Bastard, N. Boccara, M. Lannoo, and M. Voos, Springer-Verlag (1986)]

2. The coupling between bands produces deviations from parabolicity, as in the conduction band of direct gap semiconductors. For narrow gap materials, like InAs or InSb the nonparabolicity of the conduction band, due to coupling with the valence bands, is quite large for energies very near the band minimum [Kan66], but even in GaAs it has a sizable effect on levels of energy > 0.1 eV above the band minimum.

3. The single-band approach for heterostructures fails. If two materials have a "staggered" energy gap configuration (as shown in Fig. 6.4) in a large and interesting energy range, then the wave function has conduction band character on one side of the junction and valence band character on the other. InAs–GaSb superlattices are one example of this case.

We treat bulk bands the same way as our analysis of the simple-band case, which was modeled on the theory of donor impurities:

$$H_{lm}(\mathbf{k}) = E_l(0)\delta_{lm} + \sum_{\alpha=1}^{3} p_{lm}^{\alpha} k_{\alpha} + \sum_{\alpha,\beta=1}^{3} D_{lm}^{\alpha,\beta} k_{\alpha} k_{\beta}, \quad (6.44)$$

where α, β run over the x, y, and z directions. Given matrix $H_{lm}(\mathbf{k})$, the n band energies $E_1(\mathbf{k})$ are given by the eigenvalues of the $n \times n$ matrix $H_{lm}(\mathbf{k})$. The direct $\mathbf{k} \cdot \mathbf{p}$ coupling between the n bands is thus retained in the terms $p_{lm}^{\alpha} \mathbf{k}_{\alpha}$, where the matrix p^{α} is given by

$$p_{lm}^{\alpha} = \frac{\hbar}{m_0} \langle u_l | p^{\alpha} | u_m \rangle \quad (6.45)$$

The k-quadratic terms proportional to the matrix $D^{\alpha,\beta}$, on the other hand, represent the indirect $\mathbf{k} \cdot \mathbf{p}$ coupling between two of the n bands via the other bands ($n + 1$ to ∞) not included in the set.

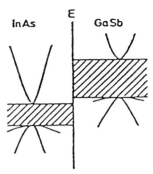

Fig. 6.4 Schematic representation of the energy gap at the InAs–GaSb interface. Hatched areas denote the energy gap. [Altarelli, M., *Heterojunctions and Semiconductor Superlattices*, ed. by G. Allan, G. Bastard, N. Boccara, M. Lannoo, and M. Voos, Springer-Verlag (1986)]

For the multiple band case we can obtain a system of n differential equations:

$$\sum_{m=1}^{n} [H_{lm}(-i\nabla) + U(\mathbf{r})\delta_{lm}] F_m(\mathbf{r}) = E F_l(\mathbf{r}) \tag{6.46}$$

for the n-component envelope function $F_l(\mathbf{r})$, $l = 1, 2, \ldots, n$. The total wave function $\psi(\mathbf{r})$ is expressed as

$$\psi(\mathbf{r}) = \sum_{l=1}^{n} \left\{ F_l(\mathbf{r}) u_{l0} + \sum_{m>n} -\frac{i(\nabla F_l(\mathbf{r})) \cdot \mathbf{p}_{lm}}{m_0 (E_l(0) - E_m(0))} u_{m0}(\mathbf{r}) \right\}. \tag{6.47}$$

In Eq. (6.46) the "kinetic energy" component is just obtained by replacing $\mathbf{k}$ with $-i\nabla$ in the $\mathbf{k} \cdot \mathbf{p}$ matrix [Eq. (6.44)] and the potential term is diagonal in the band index. It is also straightforward to generalize the boundary conditions. We find that

$$F_l^A(-\varepsilon) = F_l^B(+\varepsilon), \qquad l = 1, 2, \ldots, n,$$

and

$$\sum_{m=1}^{n} \left[p_{lm}^z - i \sum_{\alpha=1}^{3} (D_{lm}^{z\alpha} + D_{lm}^{\alpha z}) \nabla_\alpha \right] F_m \tag{6.48}$$

continues at $z' = 0$ for $l = 1, 2, \ldots, n$. Notice in Eq. (6.48) that terms without derivative come from P_{lm}, and those involving derivatives with respect to x and y come from the mixed terms in $k_x k_z$ and $k_y k_z$ and $k_y k_z$ in H_{lm}.

Discussion of the Envelope-Function Approximation

To derive the envelope-function approximation, we need to make certain critical assumptions. Consider a GaAs/AlGaAs superlattice. One could argue that a low-lying quantum well state in the conduction band has an energy close to the GaAs band minimum, but deep into the gap of AlGaAs. The corresponding wave function decays very rapidly into the AlGaAs barrier and does not satisfy the assumption of slow variation. On the other hand, the amplitude of the wave function in AlGaAs is quite small, so an inaccurate treatment of this region has little influence on the calculation.

Another possible source of uncertainty is in the boundary conditions. The discussion of boundary conditions is based on the idea of perfect periodicity of both media up to a geometrical plane defining the interface. This is certainly an idealization, since it is impossible to define a plane if one of the components is an alloy (e.g., GaAs–Al$_x$Ga$_{1-x}$As). It would be more realistic to talk about an interface layer of several atomic planes separating the two semiconductors [Alt86].

6.2.3 Bond-Orbital Models

The basic idea of the bond-orbital method is to use a minimum number of bond orbitals to describe, as accurately as possible, the most relevant portion of the band structure for the bulk materials that constitute the superlattice. For example, if we want to calculate the valence-subband structures of a GaAs–AlGaAs superlattice, the minimum information needed is the energy dispersion of the top four valence bands near the zone center for GaAs and AlGaAs. The four bond orbitals per unit cell are used to obtain the top four valence bands for each bulk material. We will assume that all of these bond orbitals are sufficiently localized so that the interaction between orbitals separated farther than the nearest neighbor distance can be ignored.

We first consider a valence-band model (VBM) in which only the top three valence bands (in the absence of spin-orbit interaction) are included. This model requires only three p-like bond orbitals (labeled x, y, z) per unit cell. Here a bond orbital is defined as the proper linear combination of two atomic orbitals within a unit cell of a diamond or zincblende crystal that best describes the valence-band states near the zone's center. We denote an α-like ($\alpha = x, y, z$) bond orbital localized at lattice site $\mathbf{R}$ as $|\mathbf{r}, \alpha\rangle$. The interaction between orbitals $|\mathbf{R}, \alpha\rangle$ and $|\mathbf{R}', \alpha'\rangle$ for a face-centered-cubic lattice is given by

$$\langle \mathbf{R}, \alpha | H | \mathbf{R}', \alpha' \rangle = E_p \delta_{\mathbf{R},\mathbf{R}} \delta_{\alpha,\alpha'} + \delta_{\mathbf{R}'-\mathbf{R},\tau} \{ E_{xy}\tau_\alpha \tau_{\alpha'}(1 - \delta_{\alpha,\alpha'}) + [E_{xx}\tau_\alpha^2 + E_{zz}(1 - \tau_\alpha^2)]\delta_{\alpha,\alpha'} \}, \tag{6.49}$$

where E_p denotes the on-site energy and E_{xx}, E_{xy}, and E_{zz} are three independent nearest-neighbor interaction parameters. Here $E_{\alpha,\alpha'}$ is the interaction between an α-like orbital at the origin and an α'-like orbital located at $(1, 1, 0)(a/2)$, a being the lattice constant. The remaining interaction parameters can be related to E_{xx}, E_{xy}, and E_{zz} by symmetry. τ_α denotes the α component of the twelve nearest-neighbor position vectors in units of $(a/2)$,

$$\tau = (\pm 1, \pm 1, 0), (\pm 1, 0, \pm 1), (0, \pm 1, \pm 1).$$

The top three valence bands (in the absence of spin-orbit interaction of a typical zincblende or diamond crystal can be obtained by diagonalizing the tight-binding hamiltonian:

$$H_{\alpha,\alpha'}(\mathbf{k}) = E_p \delta_{\alpha,\alpha'} + \sum_\tau e^{i\mathbf{k}\cdot\tau}\{E_{xy}\tau_\alpha t_{\alpha'} + [(E_{xx} - E_{xy})\tau_\alpha^2 + E_{xx}(1 - \tau_\alpha^2)]\delta_{\alpha,\alpha'}\}. \tag{6.50}$$

Taking the Taylor expansion over $\mathbf{k}$ and omitting terms higher than the second order, we obtain

$$H(\mathbf{k}) = \begin{bmatrix} E_v - \lambda_1 k_x^2 - \lambda_2 k^2 & -\lambda_3 k_x k_y & -\lambda_3 k_x k_z \\ -\lambda_3 k_x k_y & E_v - \lambda_1 k_y^2 - \lambda_2 k^2 & -\lambda_3 k_y k_z \\ -\lambda_3 k_x k_z & -\lambda_3 k_y k_z & E_v - \lambda_1 k_z^2 - \lambda_2 k^2 \end{bmatrix}, \quad (6.51)$$

where

$$E_v = E_p + 8E_{xx},$$

$$\lambda_1 = (E_{zz} - E_{xx})\frac{a^2}{2},$$

$$\lambda_2 = (3E_{xx} - E_{zz})\frac{a^2}{2},$$

$$\lambda_3 = E_{xy} a^2. \quad (6.51')$$

Comparing $H(\mathbf{k})$ with the $\mathbf{k} \cdot \mathbf{p}$ hamiltonian for the valence band, we have

$$(L - M) = \lambda_1, \quad M = \lambda_2, \quad N = \lambda_3, \quad (6.52)$$

where L, M, and N are the $\mathbf{k} \cdot \mathbf{p}$ band parameters defined by Kane [Kan56]. The above equations provide a one-to-one correspondence between the band parameters L, M, and N measured by the cyclotron-resonance experiment [Hen63] and the bond-orbital interaction parameters E_{xx}, E_{xy}, and E_{zz}. The on-site energy E_p plays the role of adjusting the energy position of the valence-band top (E_v).

For most semiconductors the spin-orbit interaction is important. We choose the proper linear combinations of the product of the electron spin and the three p-like orbitals as the new bond-orbital basis. The spin-orbit coupled bond orbitals (SOBOs) can be written

$$|\mathbf{R}, u_M^J\rangle = \sum_{\alpha, \sigma} C(\alpha\sigma; JM)|\mathbf{R}, \alpha\rangle \chi_\sigma, \quad (6.53)$$

where $J = 3/2, 1/2$, $M = -J, \ldots, J$, and χ_σ, $\sigma = 1/2, -1/2$, denote the electron spinors. The coupling coefficient $C(\alpha\sigma; JM)$ are given in Table 6.1. By including the spinor

TABLE 6.1 Coupling Coefficients for p-like States Coupled to the Electron Spin

(J, M)	(α, σ)	$(x, \frac{1}{2})$	$(y, \frac{1}{2})$	$(z, \frac{1}{2})$	$(x, -\frac{1}{2})$	$(y, -\frac{1}{2})$	$(z, -\frac{1}{2})$
$(\frac{3}{2}, \frac{3}{2})$		$-1/\sqrt{2}$	$-i/\sqrt{2}$	0	0	0	0
$(\frac{3}{2}, \frac{1}{2})$		0	0	$2/\sqrt{6}$	$-1/\sqrt{6}$	$i/\sqrt{6}$	0
$(\frac{3}{2}, -\frac{1}{2})$		$1/\sqrt{6}$	$i/\sqrt{6}$	0	0	0	$2/\sqrt{6}$
$(\frac{3}{2}, -\frac{3}{2})$		0	0	0	$1/\sqrt{2}$	$-i/\sqrt{2}$	0
$(\frac{1}{2}, \frac{1}{2})$		0	0	$-1/\sqrt{3}$	$-1/\sqrt{3}$	$-i/\sqrt{3}$	0
$(\frac{1}{2}, -\frac{1}{2})$		$-1/\sqrt{3}$	$i/\sqrt{3}$	0	0	0	$1/\sqrt{3}$

Source: Chang, Y. C., Phys. Rev. B **37**, 14 (1988): 8215–8222.

orbit interaction, we obtain the tight-binding hamiltonian in the new SOBO basis:

$$H_{JM,J'M'}(\mathbf{k}) = \left(J - \frac{3}{2}\right)\Delta\delta_{JM,J'M'} + \sum_{\alpha,\alpha',\sigma} C^*(\alpha\sigma; JM)C(\alpha'\sigma; J'M')H_{\alpha,\alpha'}(\mathbf{k}), \qquad (6.54)$$

where Δ is the spin-orbit splitting. Taking the Taylor expansion over $\mathbf{k}$ and omitting terms higher than the second order in $\mathbf{k}$, we immediately obtain the Kohn-Luttinger hamiltonian [Lut56]. There is a one-to-one correspondence between the Luttinger parameters, γ_1, γ_2, and γ_3 and the bond-orbital interaction parameters E_{xx}, E_{xy}, and E_{zz}. The relations are given in Table 6.2 in the column labeled VBM.

6.2.4 Applications of the Bond-Orbital Model to Superlattices

An appropriate bond-orbital model is chosen for each participating material in the superlattice. The interactions between any two bond orbitals located in the same material are taken to be the same as those in the bulk. The interaction between two bond orbitals located in two different materials may be taken to be the average of the corresponding matrix elements in the two participating bulk materials. Because of the simplicity of the interaction matrix in the bond-orbital model, the transfer-matrix method is considered to be a powerful method. This method can be applied to general situations in which the interaction parameters between bond orbitals vary from one atomic layer to another within the period of the superlattice. Thus superlattices under external fields can be treated in the same way.

Transfer-Matrix Method

For any kind of superlattice with period d, the superlattice state with wave vector $\mathbf{k} = (\mathbf{k}_\parallel, q)$ can be written

$$\phi_{\mathbf{k}_\parallel, q} = \sum_L \exp(iqLd)\sum C_\alpha(l)|\mathbf{k}_\parallel, l, L; \alpha\rangle, \qquad (6.55)$$

where L stands for the superlattice unit cells (SUCs), l the atomic layer within a given

TABLE 6.2 Relations between Bond-orbital Parameters and $\mathbf{k}\cdot\mathbf{p}$ Parameters. $R_0 = \hbar^2/2m_0 a^2$, $X_{h1} = 4\,\text{eV}$

VBM	CVBM
	$E_{zx}^2 = E_g(12\gamma_2 R_0 - X_{h1}/8)/32$
$E_{xy} = 6\gamma_3 R_0$	$E_{xy} = 6\gamma_3 R_0 - 16E_{xx}^2/E_g$
$E_{xx} = (\gamma_1 + 4\gamma_2)R_0$	$E_{xx} = \gamma_1 R_0 - 16E_{xx}^2/(3E_g) + X_{h1}/24$
$E_{zz} = (\gamma_1 - 8\gamma_2)R_0$	$E_{zz} = E_{xx} + X_{h1}/8$
$E_p = (E_v - 12\gamma_1 R_0)$	$E_p = E_v - 12E_{xx} + X_{h1}/2$
	$E_{zx} = -\dfrac{m_0}{m_c}R_0 + 64E_{zx}^2\left[\dfrac{2}{E_g} + \dfrac{1}{E_g+\Delta}\right]\bigg/3$

Source: Chang, Y. C., *Phys. Rev. B* **37**, 14 (1988): 8215–8222.

SUC, and N the number of atomic layers per SUC. Here the atomic layer contains all the bulk unit cells in the lattice plane perpendicular to the growth direction; $|\mathbf{k}_\parallel, l, L; \alpha\rangle$ denotes a Bloch sum of α-like bond orbitals in the lth atomic layer in the Lth period associated with the in-plane wave vector $\mathbf{k}_\parallel$. Substituting $\phi_{\mathbf{k}_\parallel, q}$ into Schrödinger equation yields

$$\sum_L e^{iqLd} \sum_{l=1}^{N} C_\alpha(l)(H - E)|\mathbf{k}_\parallel, l, L; \alpha\rangle = 0, \tag{6.56}$$

where H is the hamiltonian for the superlattice and E is the energy of the superlattice state. Projecting Eq. (6.56) in bond orbitals located at atomic layer l leads to an equation relating the coefficients at layer $l + 1$ to those at layers l and $l - 1$, we have

$$C(l+1) = -[H^{(+)}(l)]^{-1}[H^{(0)}(l)C(l) + H^{(-)}(l)C(l-1)], \tag{6.57}$$

where $C(l)$ is a column vector whose components are the coefficients $C_\alpha(l)$ and $H^{(+)}$, $H^{(0)}$, and $H^{(-)}$ are matrices with elements

$$H^{(+)}_{\alpha,\alpha'}(l) = \langle \mathbf{k}_\parallel, l, L; \alpha | H | \mathbf{k}_\parallel, l+1, L; \alpha' \rangle,$$
$$H^{(0)}_{\alpha,\alpha'}(l) = \langle \mathbf{k}_\parallel, l, L; \alpha | H | \mathbf{k}_\parallel, l, L; \alpha' \rangle - E\delta_{\alpha,\alpha'},$$
$$H^{(-)}_{\alpha,\alpha'}(l) = \langle \mathbf{k}_\parallel, l, L; \alpha | H | \mathbf{k}_\parallel, l-1, L; \alpha' \rangle.$$

We define a transfer matrix $T(l)$ as

$$T(l) = \begin{pmatrix} -[H^{(+)}(l)]^{-1} H^{(0)}(l) & -[H^{(+)}(l)]^{-1} H^{(-)}(l) \\ 1 & 0 \end{pmatrix}.$$

Then we have

$$\begin{pmatrix} C(l+1) \\ C(l) \end{pmatrix} = T(l) \begin{pmatrix} C(l) \\ C(l-1) \end{pmatrix}. \tag{6.58}$$

Repeatedly applying Eq. (6.58) from 1 to N, we obtain

$$\begin{pmatrix} C(N+1) \\ C(N) \end{pmatrix} = T_N \begin{pmatrix} C(1) \\ C(0) \end{pmatrix}, \tag{6.59}$$

where $T_N = T(N)T(N-1)\ldots T(2)T(1)$. From the definition of the coefficients $C(l)$ in Eq. (6.55), we see that $C(l + N) = \exp(iqLd)C(l)$. Thus

$$T_N \begin{pmatrix} C(1) \\ C(0) \end{pmatrix} = \exp(iqLd) \begin{pmatrix} C(1) \\ C(0) \end{pmatrix}. \tag{6.60}$$

Since T_N is a function of the energy E, the eigenvalue Eq. (6.60) can be solved to find the dispersion relation (E vs. q).

Subband Structure Calculated by Valence-Band Model

The valence-band model is used to calculate the subband structure of GaAs–$Al_xGa_{1-x}As$ (with infinite spin-orbit interaction). The conduction-subband structures are trivial, and they can obtained by a one-band model (e.g., one s-like bond orbital per unit cell).

Figure 6.5(a) shows the bulk valence-band structures of GaAs obtained by the VBM (solid curves) and by the **k·p** model (dashed curves). The band structures obtained from the bond-orbital model and the **k·p** model are nearly identical for small values of **k**. For large values of **k**, the bond-orbital model gives more satisfactorily results than the **k·p** model. Figure 6.6 shows the valence-subband structures of a (30, 20) GaAs–$Al_{0.1}Ga_{0.9}As$ superlattice for wave vectors (**k**) along $[00\chi]$ (the growth direction), $[\chi00]$ (the in-plane direction), and $[\chi]$ ($\chi = 0 - 1$). The units for wave vectors are π/d, where d is the width of the superlattice unit cell in the growth direction. Here the symbol (30, 20) indicates that the superlattice has 30 GaAs atomic layers and 20 $Al_{0.1}Ga_{0.9}As$ atomic layers per superlattice unit cell. The symbols HH_n and LH_n denote the nth heavy hole and light hole subbands, respectively.

6.2.5 Impurity and Exciton States in Heterostructures

Impurity States in Quantum Wells and Superlattices

Impurities are a central concern of semiconductor physics. Impurities influence the properties of semiconductors and hence the performance of devices. Many optical and transport measurements were developed to identify the impurity states in semiconductors.

Most theoretical work on impurity states is based on effective-mass approximation. This method is only valid for shallow impurities since one has to assume that the potential is weak and slowly varying. These conditions are satisfied by the Coulomb

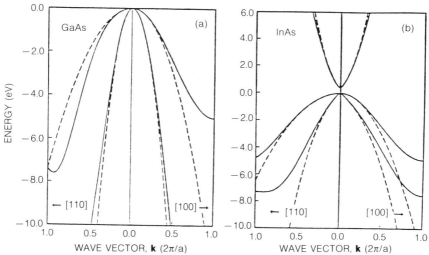

Fig. 6.5 (a) Valence-band structures of GaAs obtained in the VBM (*solid line*) and in the **k·p** theory (*dashed line*); (b) band structures of InAs obtained in the CVBM (*solid line*) and in the **k·p** theory (*dashed line*). [Chang, Y. C., *Phys. Rev. B*, **37**, 14 (1988): 8215–8222]

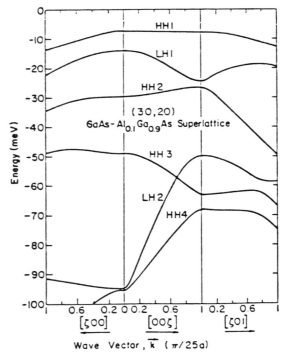

Fig. 6.6 Valence-subband structures of a (30, 20) GaAs–Al$_{0.1}$Ga$_{0.9}$As superlattice obtained in the VBM. [Chang, Y. C., *Phys. Rev. B*, 37, 14 (1988): 8215–8222]

potential, screened by the dielectric function of the superlattice, of simple donor and acceptor impurities.

Consider the problem of donor impurities, which does not suffer from the complications of a degenerate band edge. In bulk GaAs a purely Coulombic donor impurity has a hydrogenic spectrum

$$E_n = E_c - \frac{R^*}{n^2}, \quad (6.61)$$

with

$$R^* = \frac{m^* e^4}{2\hbar^2 \varepsilon^2} \quad (6.62)$$

and a Bohr radius, for the ground state effective-mass wave function,

$$a^* = \frac{\hbar^2 \varepsilon}{m^* e^2}.$$

Since for GaAs [Lan82] $m^* = 0.067$, $\varepsilon = 12.5$, one finds $R^* = 5.8$ meV, $a^* = 98.7$ Å. The energy scale of Coulomb binding is comparable to or smaller than typical quantization energies of quantum wells. When the doping densities are very high in a semiconductor ($>10^{18}$ cm^3), their eigenfunctions overlap significantly and permit the exchange of carriers directly. The impurity level is split and develops into a narrow impurity band [Böe90].

Excitons in Quantum Wells and Superlattices

Excitons are bound electron-hole pairs and are the lowest electronic excited states of nonmetallic crystals. They are easily detected in optical spectra, because they produce sharply defined lines. In III-V compound semiconductors, the orbit of an exciton extends over a great many lattice sites. For such excitons, the electron-hole interaction can be treated as Coulomb interaction between two point charges. Therefore the quantized exciton energy can be written as [Wan89]

$$E_{xn} = \frac{-m_r e^4}{2(4\pi\varepsilon\hbar)^2} \frac{1}{n_x^2} = \frac{-13.6 m_r}{n_x^2 m_0} \left(\frac{\varepsilon_0}{\varepsilon}\right)^2 \text{eV}, \qquad (6.63)$$

where $m_r = m_c^* m_v^*/(m_c^* + m_v^*)$ is the reduced mass and n_x an integer (1, 2, 3, ...) representing the quantum number. Since it takes an energy E_g to excite the electron-hole pair, the exciton absorption occurs at

$$(h\nu)_x = E_g - E_x. \qquad (6.64)$$

Using the effective masses and dielectric constant for GaAs, we find the lowest energy $E_{x1} = -4.4$ meV. Hence GaAs bulk excitons have a binding energy of about 4 meV. Clearly at low temperatures the exciton absorption peak will not be broadened by phonon scattering.

This situation, however, is changed in a quantum well. Figure 6.7 compares the

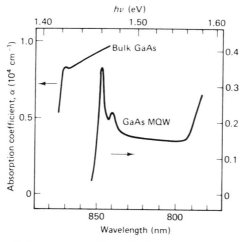

Fig. 6.7 Absorption coefficient measured in a GaAs sample (*right scale*) containing multiple quantum wells as compared to that measured in an ordinary GaAs sample (*left scale*). [Miller, D. A. B., D. S. Chemla, D. J. Eilenberger, P. W. Smith, A. C. Gossard, and W. T. Tsang, *Appl. Phys. Lett.* **41** (1982): 679]

absorption spectrum of GaAs multiple quantum wells (MQW) with that of a bulk GaAs at room temperature. Whereas the bulk spectrum shows only a slight hump, the MQW spectrum shows two distinct resonances, one at 1.463 eV from electron–heavy-hole exciton and the other at 1.474 eV from electron–light-hole exciton [Mil82]. One important parameter is the layer's thickness in comparison with the exciton's dimensions. In bulk GaAs the exciton's radius is given by

$$a_{xn} = \frac{4\pi\varepsilon\hbar^2}{m_r e^2} n_x^2 = \frac{0.53(m_0)\varepsilon n_x^2}{(m_r)\varepsilon_0} \text{ Å},\qquad(6.65)$$

which yields a value of $a_{x1} = 150$ Å for $n_x = 1$. In this experiment by Miller et al., a GaAs well thickness of 102 Å was used, which is much smaller than the exciton's diameter of 300 Å. For a purely two-dimensional exciton, the binding energy is equal to $4E_{xn}$ of Eq. (6.63). However, in a quantum well the penetration of the exciton's wave function into the large-gap AlGaAs is not negligible. For very thin well layers the large extension of the wave function makes the exciton again behave like a three-dimensional exciton. Therefore the binding energy of a QW exciton E_{xQW} peaks between $2E_{xn}$ and $3E_{xn}$ for well thicknesses in the range $0.5a_{x1}$ to $3a_{x1}$.

6.2.6 Miniband Structure of a Superlattice

There is much interest in the properties of carriers moving in the growth direction of epitaxially grown structures. When the wave functions of carriers in the neighboring wells of a multilayered heterostructure overlap significantly, the energy levels broaden into minibands, with extended Bloch states. These minibands are expected to allow the transport of carriers perpendicular to the layers (Bloch transport [Dev87]). The electronic transport includes tunneling, resonant tunneling, ballistic transport, and miniband transport [Has86; Sol84; Hay85; Dev87; Dav85].

Minibands

When a semiconductor with a larger band gap is interspaced with another one of a smaller gap, the former acts as a barrier for electrons in the conduction band of the latter [Böe90]. The periodic alternation of the deposition sequence on a substrate produces a potential of the same form as the Kronig-Penney potential. The eigenvalue spectrum is similar to the spectrum of free electrons exposed to the periodic potential of a crystal, except that now the periodic potential is imposed on *Bloch* electrons with an effective mass m_n. Within the conduction band, one observes a subband structure of *minibands* located within the valleys of this band; the higher minibands extend beyond the height of the potential barriers. The lower minibands are separated by minigaps (as shown in Fig. 6.8) in the direction of the superlattice periodicity z. Within the plane of the superlattice layers (x, y), however, the electron eigenfunction experiences only the regular lattice periodicity. Therefore the dispersion relations $E(k_x)$ and $E(k_y)$ are much like those for the unperturbed lattice except for the mixing with the states in the z direction. This results in lifting the lowest energy (at $k = 0$) of the $E(k)$ parabola above E_c of the bulk well material, as shown in Figs. 6.8(b) and (c). The second miniband results in a second, shifted parabola.

Figure 6.9 shows the computed widths of minibands and intermittent gaps as a

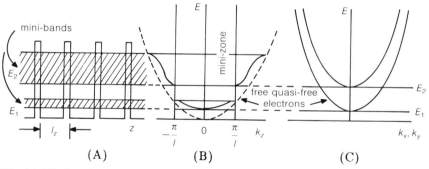

Fig. 6.8 (a) Minibands and (b) minizones of the conduction band in the k_z direction for a superlattice. Carriers are confined to the z-direction in the lower minibands. There is no confinement in the x- and y-directions. In the k_x- and k_y-directions the ordinary band picture applies, but with the band minimum lifted to the respective miniband minimum, shown in (c). [Böer, K. W., *Survey of Semiconductor Physics*, Van Nostrand Reinhold (1990)]

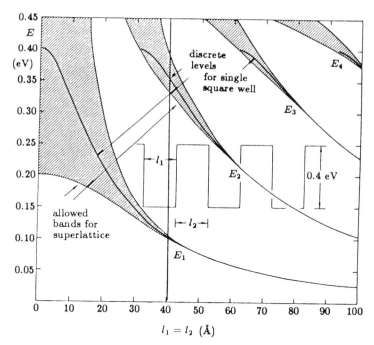

Fig. 6.9 Computed minibands for a symmetrical superlattice. [Esaki, L., in *The Technology and Physics of Molecular Beam Epitaxy*, p. 143, ed. by E. M. C. Parker, Plenum Press (1985); Böer, K. W., *Survey of Semiconductor Physics*, Van Nostrand Reinhold (1990)]

function of the period length (2l) for a symmetrical well/barrier structure with a barrier height of 0.4 eV [Esa85]. For $l_1 = l_2 = 40$ Å the lowest band is rather narrow and lies at 100 mV above the well's bottom. The second band extends from 320 to 380 mV. Higher bands overlap at the top of the barrier [Böe90].

224 DEVICE-RELATED PHYSICS AND PRINCIPLES

Superlattices that show such a beautiful illustration of quantum mechanical behavior have been fabricated from a number of compound semiconductor pairs using MBE or MOCVD methods. Here we discuss the electronic properties of strongly coupled GaAs/AlGaAs superlattices [Eng87]. The large unit cell of the artificially periodic structure means that the GaAs Brillouin zone is folded to produce minibands and band gaps. The minibands and band gaps with energies both below and above the conduction band offset in the AlGaAs superlattice barriers can be resolved. An envelope function approximation calculation of the position and width of the minibands is in a quantitative agreement with the features seen. Experimental data [Eng87] shows that the electron transport is coherent over several superlattice periods and is not consistent with sequential tunneling.

The samples under investigated were MBE-grown structures consisting of a strongly coupled superlattice together with a wide AlGaAs tunneling barrier and an electron injector, as shown in Fig. 6.10. The active region is sandwiched between two n^+ contacts. The doping is kept low in the injector to permit high resolution at low temperatures. The tunnel barrier is undoped to reduce DX centers in AlGaAs. "Strongly coupled" means that small variations in the quantum well's parameters throughout the superlattice are not sufficient to cause localization and destroy the band structures.

There are two factors that influence the choice of doping in the superlattice. For reduced scattering by ionized impurities, we expect the doping to be low. However, large electric fields in the sample can be sufficient to destroy the band's structure by Stark confinement. The depletion region close to the barrier can extend into the superlattice. For a miniband structure to be defined, we require that the voltage drop across a single quantum well be much less than the miniband's width.

Density of States in Minibands

For an isolated well the miniband has the character of a staircase. Each level can be occupied by the number of electrons given by its degeneracy multiplied by the number of atoms in the well. When significant tunneling becomes possible, each level splits into bands, and the staircase behavior (dashed steps) becomes somewhat abated, as shown in Fig. 6.11.

The effective density of states near the bottom of the first miniband is given by [Böe90]

$$N_{2c} = \frac{m^* kT}{\pi \hbar^2} \tag{6.66}$$

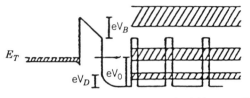

Fig. 6.10 Schematic of a tunneling structure under typical bias conditions. Also shown are the positions of the minibands. [England, P., J. R. Hayes, J. P. Harbison, and D. M. Hwang, in *Hot Carriers in Semiconductors*, ed. by J. Shah and G. J. Iafrate, Pergamon Press (1987): 735–737]

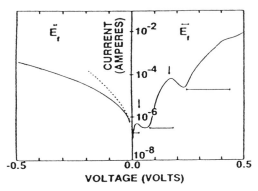

Fig. 6.11 The current voltage characteristic at 4.2 K of a tunneling structure with a bulk GaAs electron injector. Note the positions of the minibands (bars) and the resonance energies of the first well in the superlattice (*arrows*). [England, P., J. R. Hayes, J. P. Harbison, and D. M. Hwang, in *Hot Carriers in Semiconductors*, ed. by J. Shah and G. J. Iafrate, Pergamon Press (1987): 735–737]

measured in cm^{-2}. The subscript $2c$ refers to the two-dimensionality of the structure and to a miniconduction band.

Perpendicular Transport in a Superlattice: Miniband Conduction, Localization, and Hopping

Barrier thicknesses in a superlattice are comparable to the de Broglie carrier wavelength; thus the wave functions of the individual wells overlap due to tunneling, and an energy miniband of width 2Δ is formed [Esa70]. The width 2Δ is proportional to the tunneling probability through the barriers which, for rectangular barriers and weak coupling between wells, can be approximated by

$$T_e = \exp\{-\sqrt{(8m_e^*/\hbar)(\Delta E_c - E_{i,e})}L_B\} \qquad (6.67)$$

in the case of electrons and by

$$T_{hh} = \exp\{-\sqrt{(8m_{hh}^*/\hbar)(\Delta E_v - E_{i,hh})}L_B\} \qquad (6.68)$$

in the case of heavy holes where m_e^* and m_{hh}^* are the electron and heavy hole masses, ΔE_c and ΔE_v are the conduction and valence band discontinuities, $E_{i,e}$ and $E_{i,hh}$ are the lowest ground state electrons and heavy-hole minibands, and L_B is the barrier thickness [Cap86].

The picture above assumes a perfect superlattice, with no thickness or potential fluctuations and no scattering by impurities or phonons. In reality one must contend with such fluctuations and with the unavoidable presence of scattering. Such effects tend to perturb the wave function and the formation of extended Bloch states with the miniband result.

Consider the weak electric field limit in which the potential energy drop across the superlattice period is smaller than a miniband's width. Transport proceeds by miniband conduction if the low-field mean free path of the carriers appreciably exceeds the superlattice period [Esa74]. We can derive a phenomenological expression of the mobility $\mu_\parallel$ along the superlattice axis. Consider a one-dimensional model, and describe the band structure along the superlattice axis by the energy dispersion relationship [Esa70]

$$E(k) = \Delta[1 - \cos k_\parallel d] \tag{6.69}$$

in which $k_\parallel$ is the component of the wave front parallel to the superlattice axis and d is the superlattice period.

The average group velocity along the superlattice axis is obtained from Eq. (6.69):

$$v_d = \left(\frac{1}{\hbar}\frac{dE}{dk_\parallel}\right)_{k=\bar{k}_\parallel} = \Delta d \sin(\bar{k}_\parallel d), \tag{6.70}$$

where $\bar{k}_\parallel$ is the steady state average wave vector obtained from the momentum rate

$$\hbar \frac{d\bar{k}_\parallel}{dt} = eF - \frac{\hbar \bar{k}_\parallel}{\tau}, \tag{6.71}$$

where F is the electric field and τ is the relaxation time for the momentum $p_\parallel = \hbar k_\parallel$. Substituting Eqs. (6.69) and (6.71) into Eq. (6.70), we obtain

$$v_d = \frac{\Delta d}{\hbar} \sin\left(\frac{eF\tau d}{\hbar}\right). \tag{6.72}$$

For small electric fields, Eq. (6.72) reduces to

$$v_d = \frac{e\Delta}{\hbar^2} d^2 F. \tag{6.73}$$

The mobility $\mu_\parallel$ is then

$$\mu_\parallel = \frac{e\Delta d^2 \tau}{\hbar^2}. \tag{6.74}$$

The factor $\hbar^2/\Delta d^2$ represents the band-edge effective mass in the direction parallel to the superlattice axis, as is clear from Eq. (6.69) and the definition

$$m_\parallel^* = \left(\frac{\hbar^2}{d^2 E/dk^2}\right)_{k_\parallel = 0}. \tag{6.75}$$

For superlattices in which the barriers seen by holes are not much lower than those seen by electrons, the electron mobility $\mu_{e\parallel}$ can be made much greater $\mu_{hh\parallel}$, since the tunneling probability depends exponentially on the effective mass. This implies that the superlattice can act as a filter for effective masses [Cap85; Cap85b] by easily transmitting the light carriers (electrons) and effectively slowing down the heavy carriers (heavy holes).

As the barrier thickness increases, the miniband's width 2Δ decreases exponentially. The maximum group velocity in the band ($v_{max} = \Delta d/\hbar$) and the mobility decrease proportionally with the bandwidth. The relaxation time, however, is practically independent of d, since it is dominated by intralayer processes. Eventually $v_{max}\tau$, which is always greater than the mean free path λ, becomes smaller than d even for an ideal superlattice without layer thickness or compositional fluctuations [Doh75]. This can be written approximately as

$$\frac{\Delta d\tau}{\hbar} < d \tag{6.76}$$

from which it follows that

$$\frac{\hbar}{\tau} > \Delta. \tag{6.77}$$

In other words, the collisions broadening is greater than the miniband's width, and this implies that $\lambda < d$. If this condition or the one on the mean free path is satisfied, the states of the superlattice are no longer Bloch waves; they are localized in the wells along the direction perpendicular to the layer. The states, however, will be delocalized in the plane of the layers. If the nonhomogeneous broadening exceeds the intrinsic miniband's width, there are no Bloch states, and the wave function becomes localized in the wells. This type of localization occurring in superlattices is of the Anderson type [And58] and has profound effects on perpendicular transport. In this case conduction proceeds by phonon-assisted tunneling (or hopping) between adjacent layers. For superlattices of III–V materials such as $Al_{0.48}In_{0.52}As/Ga_{0.47}In_{0.53}As$, $AlGa_{1-x}As/GaAs$, $InP/Ga_{0.47}In_{0.53}As$, with equal barrier and well thicknesses, the localization of the electrons occurs when d is on the order of 100 Å.

Heavy holes, on the other hand, in these and most other superlattices become localized for much smaller δ (15 Å) due to their much larger effective mass. This implies that the miniband's width is much smaller than the size of an electron, so the localization criterion is easily satisfied. A superlattice tends to selectively localize carriers, acting as an effective mass filter.

The mobility perpendicular to the layers then is very small and cannot be described by Eq. (6.74). Given the limit of strong localization where one can neglect transitions other than those between adjacent wells, the mobility is written [Cal84]

$$\mu = \frac{ed^2}{kT} \langle W \rangle, \tag{6.78}$$

where $\langle W \rangle$ is the thermodynamically averaged phonon-assisted tunneling rate between adjacent wells, which is proportional to the tunneling probability.

At high electric fields, heavy holes tend to be much less localized than at low fields, due to hot carrier effects and barrier lowering (enhanced thermionic emission), which dramatically increase the tunneling probability and reduce the filtering effect of the superlattice [Cap84; Wei86].

6.3 SCATTERING THEORY

In a perfect crystal the motion of electrons is free. The application of an external field uniformly accelerates the electron, linearly increasing its drift velocity with time in the direction of the field. Such a linear increase in drift velocity with time is not observed in any real crystal. The average electron drift velocity reaches a limiting value, which at low fields is proportional to the magnitude of the field [Nag80]. The limit is set by the interaction of the electron with imperfections in the crystal through the so-called scattering processes. Scattering theory is fundamental for electron transport in solids. The interaction of carriers in a solid with imperfections of the crystal lattice such as impurities, lattice defects, lattice vibrations, strain [Hin88], and alloy [Sin93] contributed to scattering [See85]. In this section we will discuss the basic scattering mechanisms, Fermi's golden rules, and scattering rates and mobility.

6.3.1 Basic Scattering Mechanisms

We will focus on the important scattering mechanisms for carriers in silicon and gallium arsenide. The silicon conduction band has six equivalent ellipsoidal minima close to the Brillouin zone boundary at the connection line from H to X. Within each minima there are acoustic phonon scattering, ionized impurity scattering, and scattering by neutral impurities. At room temperature ionized impurity scattering starts to be important if the impurity density exceeds 10^{17} cm^{-3} [Hes88]. Acoustic phonons are always significant in silicon at room temperature. Optical deformation potential scattering is negligible for electron scattering within the particular minimum, since the matrix element vanishes for reasons of symmetry. The scattering between the minima is significant.

The scattering mechanisms in the conduction band of gallium arsenide are different from those of silicon [Hes88]. Because of the spherical symmetry of the wave function at Γ, optical deformation potential scattering is zero in the lowest GaAs minimum. The effective mass in this minimum is very small $m^* = 0.067m_0$. Therefore acoustic phonon scattering contributes little. These two facts are the very reason why GaAs exhibits high electron mobility compared to silicon. The most important scattering mechanisms for GaAs are polar optical phonon scattering, ionized impurity scattering if the density of impurities is significant ($\geqslant 10^{17}$ cm^{-3} at room temperature), and intervalley scattering in a very high electric field [Shu87]. A list of coupling constants that describes well a large set of experimental results [Shi81] is given in Table 6.3, and the total scattering rate is shown in Fig. 6.12 [Hes88].

The maxima of the valence bands of both Si and GaAs are at Γ, and the wave function does not have spherical symmetry. Therefore optical deformation potential scattering [Hes87] is important for holes in both GaAs and Si. Both acoustic phonon scattering and impurity scattering are important in the valence band [Rod75].

The scattering mechanisms are enlarged by the presence of interfaces. Interfaces usually necessitate dealing with interface roughness [Fer81] and with a host of "remote" scattering mechanisms [Hes79]. "Remote" here means that the scattering agent and the electrons are in different types of semiconductors, as at the heterojunction interface. Remote impurity scattering is an important mechanism in modulation-doped AlGaAs–GaAs layers. Typically the AlGaAs is doped with donors, and the electrons move to the GaAs, where, because of their high density compared to the

TABLE 6.3 Material Parameters for Gallium Arsenide

Bulk Material Parameters		
Lattice constant	5.642	Å
Electron affinity	4.07	eV
Piezoelectric constant	0.16	C/m^2
LO phonon energy	0.036	eV
Longitudinal sound velocity	5.24×10^4	cm/s
Optical dielectric constant	10.92	
Static dielectric constant	12.90	

Valley-Dependent Material Parameters			
	$\Gamma[000]$	L[111]	X[100]
Effective mass (m^*/m_0)	0.067	0.222	0.58
Nonparabolicity (eV^{-1})	0.610	0.461	0.204
Energy band gap relative to valence band (eV)	1.439 (0)	1.769 (0.33)	1.961 (0.522)
Acoustic deformation potential (eV)	7.0	9.2	9.7
Optical deformation potential (eV/cm)	0	3×10^8	0
Optical phonon energy (eV)	—	0.0343	—
Number of equivalent valleys	1	4	3
Intervalley deformation potential D (eV/cm)			
Γ	0	1×10^9	1×10^9
L	1×10^4	1×10^9	5×10^8
X	1×10^4	5×10^8	7×10^8
Intervalley phonon energy (eV)			
Γ	0	0.0278	0.0299
L	0.0278	0.0290	0.0293
X	0.0299	0.0293	0.0299

Source: Hess, K., *Advanced Theory of Semiconductor Devices*, Prentice Hall (1988).

unintentional GaAs "background doping," screening is effective and the remote impurities contribute only little to the screening rate [Hes79]. Therefore very low total scattering rates and high mobilities can be achieved in these materials.

In the following section we describe the Fermi golden rule (in the semiclassical picture) because it plays a fundamental part in many applications. The transition probabilities for the various kinds of scattering are evaluated using this theory. When quantum effects take place on a short-time scale, Fermi's golden rule breaks down. Quantum effects will be discussed in Section 6.4.

6.3.2 Scattering Theory from the Golden Rule

A central idea in scattering theory is the scattering probability per unit time $S(\mathbf{k}, \mathbf{k}')$, which is calculated using Fermi's golden rule [Fer50]. Suppose that we have a uniform homogeneous segment of semiconductor without any band bending. A stream of electrons is incident along the z-axis from the left at a localized time-independent potential fluctuation $U_s(\mathbf{r})$ that may be caused by an impurity or a defect, as shown in Fig. 6.13 [Hes88]. A detector is placed at the point D. If the potential fluctuation

230   DEVICE-RELATED PHYSICS AND PRINCIPLES

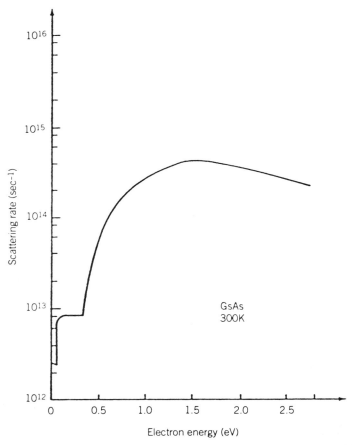

Fig. 6.12 Approximate phonon scattering rate of gallium arsenide using the material parameters of Table 6.5. [Hess, K., *Advanced Theory of Semiconductor Devices*, Prentice Hall (1988)]

were absent, we would not detect any electrons, since D lies outside the path of the incident beam. However, the potential scatters electrons in various directions so that there is a nonzero probability for an incident electron to be detected at D. We would like to know the probability $P(\theta, \phi)$ for an electron to be detected as a function of the location (θ, ϕ) of the detector D. The result is accurate if the total probability (P_S) of an electron to be scattered is small [Dat89a]:

$$P_S = \int_0^\pi d\theta \sin\theta \int_0^{2\pi} d\phi P(\theta, \phi) \ll 1. \tag{6.79}$$

To obtain scattering probability per unit time $S(\mathbf{k}, \mathbf{k}')$, we need to calculate the matrix element of the scattering potential $(U_S)_{\mathbf{k},\mathbf{k}'}$:

$$(U_S)_{\mathbf{k}',\mathbf{k}''} = \int \frac{d^3 r}{\Omega} U_S(\mathbf{r}) e^{i(\mathbf{k}'' - \mathbf{k}') \cdot \mathbf{r}}. \tag{6.80}$$

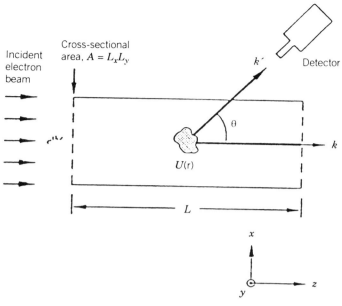

Fig. 6.13 Scattering of electrons by a potential fluctuation $U(\mathbf{r})$. [Hess, K., *Advanced Theory of Semiconductor Devices*, Prentice Hall (1988)]

The volume of the box Ω to which we normalize all the wave functions is completely arbitrary. The Schrödinger equation can be written as

$$\frac{d}{dt}\psi_{\mathbf{k}'} + \frac{iE_{\mathbf{k}'}}{\hbar}\psi_{\mathbf{k}'} = \sum_{\mathbf{k}''\neq\mathbf{k}'}\frac{(U_S)_{\mathbf{k}',\mathbf{k}''}}{i\hbar}\psi_{\mathbf{k}''}. \tag{6.81}$$

The diagonal matrix elements $(U_S)_{\mathbf{k},\mathbf{k}}$ are assumed zero; if they are not zero, we can include them in $E_{\mathbf{k}}$.

At $t=0$, $\psi_{\mathbf{k}}$ is one, and all other components $\psi_{\mathbf{k}}$ with $\mathbf{k}'\neq\mathbf{k}$, are zero:

$$\psi_{\mathbf{k}'}(t=0) = \begin{cases} 1, & \mathbf{k}'=\mathbf{k}, \\ 0, & \mathbf{k}'\neq\mathbf{k}. \end{cases} \tag{6.82a}$$
$$\tag{6.82b}$$

Now, if we consider the first-order solution of Eq. (6.81) subject to the initial condition in Eq. (6.82b) and neglect all higher-order solutions, we get

$$\psi_{\mathbf{k}'}(T) = e^{-iE_{\mathbf{k}}\cdot T/\hbar}\frac{(U_S)_{\mathbf{k}',\mathbf{k}}}{i\hbar}\int_0^T dt\, e^{-i(E_{\mathbf{k}}-E_{\mathbf{k}'})t/\hbar}. \tag{6.83}$$

The probability $P(\mathbf{k},\mathbf{k}')$ that an electron with initial momentum $\mathbf{k}$ has a final momentum $\mathbf{k}'$ at time T is given by

$$P(\mathbf{k},\mathbf{k}') = |\psi_{\mathbf{k}'}(T)|^2 = \frac{|(U_S)_{\mathbf{k}',\mathbf{k}}|^2}{\hbar^2}\left|\int_0^T dt\, e^{-i(E_{\mathbf{k}}-E_{\mathbf{k}'})t/\hbar}\right|^2. \tag{6.84}$$

The upper limit of the integration T is equal to the time it takes for the electron to cross the box of volume $\Omega = L_x L_y L_z$, to which we have normalized all the functions:

$$T = \frac{L_z}{v} = \frac{m^*\Omega}{\hbar k A}, \tag{6.85}$$

where $A = L_x L_y$ is the cross-sectional area of the electron beam.

We can rewrite Eq. (6.84) as

$$P(\mathbf{k}, \mathbf{k}') = T \frac{|(U_S)_{\mathbf{k}',\mathbf{k}}|^2}{\hbar^2} g(E_{\mathbf{k}'} - E_{\mathbf{k}}), \tag{6.86}$$

where

$$g(\varepsilon) = \frac{1}{T} \left| \int_0^T dt\, e^{i\varepsilon t/\hbar} \right|^2$$

$$= T \left(\frac{\sin(\varepsilon T/2\hbar)}{\varepsilon T/2\hbar} \right)^2 \tag{6.87}$$

and

$$\lim_{T \to \infty} g(\varepsilon) = 2\pi\hbar\delta(\varepsilon). \tag{6.88}$$

We can write Eq. (6.86) as

$$P(\mathbf{k}, \mathbf{k}') = TS(\mathbf{k}, \mathbf{k}'), \tag{6.89}$$

where

$$S(\mathbf{k}, \mathbf{k}') = \frac{2\pi}{\hbar} |(U_S)_{\mathbf{k}',\mathbf{k}}|^2 \delta(E_{\mathbf{k}'} - E_{\mathbf{k}}). \tag{6.90}$$

Equations (6.86) and (6.87) give us the probability that an electron originally having a momentum $\mathbf{k}$ is scattered into a momentum state $\mathbf{k}'$ by a time-independent potential $U_S(\mathbf{r})$. Equation (6.90) is a special case of Fermi's golden rule for stationary potentials. The result is easily extended to potentials that are sinusoidally varying in time with a frequency ω:

$$U_S(\mathbf{r}, t) = U^a(\mathbf{r}) e^{-i\omega t} + U^e(\mathbf{r}) e^{i\omega t}, \tag{6.91}$$

where the superscripts a and e represent absorption and emission of phonons (or photons), respectively. Fermi's golden rule for scattering potentials that vary sinusoidally with time is

$$S(\mathbf{k}, \mathbf{k}') = \frac{2\pi}{\hbar} [|U^a_{\mathbf{k}',\mathbf{k}}|^2 \delta(E_{\mathbf{k}'} - E_{\mathbf{k}} - \hbar\omega) + |U^e_{\mathbf{k}',\mathbf{k}}|^2 \delta(E_{\mathbf{k}'} - E_{\mathbf{k}} + \hbar\omega)]. \tag{6.92}$$

There is an important difference between the time-independent scattering potential and scattering potentials that vary sinusoidally with time. For a time-independent scattering potential, the energy E_k of the scattered electron is the same as the energy E_k of the incident electron. But, if the scattering potential varies with time, this is no longer true. The final energy E_k can be either less or more than the incident energy E_k by $\hbar\omega$. This is reminiscent of the side bands that arise when the amplitude of a carrier frequency is modulated with a single frequency.

$$E_{k'} = E_k + \hbar\omega \quad \text{(absorption)},$$

$$E_{k'} = E_k - \hbar\omega \quad \text{(emission)}.$$

When the scattered energy is greater than the incident energy, the electron is said to have absorbed a phonon or photon, depending on the origin of the scattering potential $U_s(\mathbf{r}, t)$; when the scattered energy is less than the incident energy, the electron is said to have emitted a phonon or photon. More complicated multiparticle processes require higher-order solutions which are not discussed in this book.

6.3.3 Scattering Rates and Mobility

The electron in a semiconductor is accelerated by an electric field E. The electron is scattered to a different point in **k**-space. It may or may not have lost energy during the scattering event, depending on the scattering agent. The new wave vector could point in any direction and the electron can now again be accelerated or decelerated by the electric field.

The governing equation for the description of transport is the Boltzmann equation, which will be discussed in the next section. The Boltzmann equation expresses the total time rate of change of the distribution $f(\mathbf{k}, \mathbf{r}, t)$, which describes the occupancy of allowed energy states involved in transport processes. At an elevated carrier temperature, $T_e > T$ (T = lattice temperature), the distribution function can be written

$$f(E) = f_0(E, T_e) \exp\left(\frac{-E}{kT_e}\right). \tag{6.93}$$

The increase in carrier temperature can be estimated from the incremental carrier energy in the field's direction. The field-dependent mobility can be expressed in terms of the increased electron temperature. For instance, for *warm electrons*

$$\mu_n = \mu_{n0}(1 + (T_e - T)\varphi(T) + \ldots), \tag{6.94}$$

where $\varphi(T)$ is a function of the lattice temperature alone and depends on the scattering mechanism. The concept of hot electrons will be discussed further in the next section.

A particular simple solution of the Boltzmann equation becomes possible in those cases where the effects of scattering can be described in terms of a relaxation time τ. The relaxation time for scattering is a useful parameter in describing transport. The relaxation time is defined as the average time that a carrier moves freely between collisions or scattering events [Bub74]. For most scattering processes the relaxation time is a function of energy of the carrier. The dependence of relaxation time on carrier energy is manifest as a characteristic temperature dependence of *mobility*, defined as the velocity of the carrier in the direction of the electric field per unit field.

Let $\tau_{tot}(\mathbf{k})$ be the total relaxation time for all important scattering mechanisms. If these mechanisms are independent, then the probabilities for an electron being scattered add:

$$\frac{1}{\tau_{tot}} = \frac{1}{\tau^I} + \frac{1}{\tau^{Ph}} + \ldots \tag{6.95}$$

where the superscript "I" denotes impurity and "Ph" phonon scattering.

The mobility μ can be defined as

$$\mu = \frac{e\langle \tau_{tot} \rangle}{m^*}, \tag{6.96}$$

where m^* is the effective mass of the carrier. The transport of the carriers will be described in the next section.

6.3.4 Intravalley Scattering at High Fields

When electrons are heated only slightly, the main scattering will be the low-energy acoustic phonons determining the momentum relaxation.

Scattering with the Acoustic Phonons

The field dependence of the mobility can be estimated from the energy balance equation. In steady state the collision term must equal the incremental electron energy between collisions:

$$\left\langle \frac{-\partial E}{\partial t} \right\rangle_{coll} = \frac{mv_D^2}{2} \langle \tau_m^{-1} \rangle = e\mu F^2, \tag{6.97}$$

where F is the electric field. When this energy is dissipated with acoustic phonons, the collision term can be estimated as [See73]

$$\left\langle \frac{-\partial E}{\partial t} \right\rangle_{coll} = \frac{mv_{s,l}^2}{2} \langle \tau_m^{-1} \rangle c_a \left(\frac{T_e - T}{T} \right) \tag{6.98}$$

where $v_{s,l}$ is the velocity of longitudinal acoustic phonons and c_a is a proportionality factor $[32/(3\pi)]$.

Combining Eqs. (6.97) and (6.98), we obtain

$$\frac{T_e - T}{T} = \frac{1}{c_a} \left(\frac{\mu F}{v_{s,l}} \right)^2. \tag{6.99}$$

Introducing the approximation

$$\mu = \mu_0 \sqrt{\frac{T}{T_e}} \tag{6.100}$$

for the field dependence of the mobility for warm electrons ($\mu_0 F \ll v_{s,l}$) yields

$$\mu = \frac{\mu_0}{\sqrt{1 + (1/c_a)(\mu_0 F/v_{s,l})^2}}. \tag{6.101}$$

At higher fields ($\mu_0 F \gg v_{s,l}$), the electron temperature becomes

$$\frac{T_e}{T} = \frac{1}{2}(1 + \sqrt{1 + (4/c_a)(\mu_0 F/v_{s,l})^2}), \tag{6.102}$$

which is show in Fig. 6.14 as a function of the low-field drift velocity $\mu_0 F$. Eliminating T_e/T from Eqs. (6.100) and (6.102), we have the Shockley approximation [Sho51]

$$\mu = \mu_0 (c_a)^{1/4} \sqrt{\frac{v_{s,l}}{\mu_0 F}}. \tag{6.103}$$

The field dependence of the mobility for scattering with acoustic phonons is given in Fig. 6.15.

Scattering with Optical Phonons

Scattering with Optical Deformation Potential At sufficiently high fields a large number of electrons have enough energy to dissipate this energy through the creation of longitudinal optical phonons. A more substantial reduction of the average relaxation time τ^- results, and $\mu = (e/m)\tau^-$ decreases more rapidly with the electron temperature [See73]:

$$\mu(T_e) = \frac{3(\pi)^{3/2} \hbar^2 \rho \sqrt{k\Theta}}{2 m_n^{3/2} D_0^2 \Theta^{3/2}} \varphi_e(T_e), \tag{6.104}$$

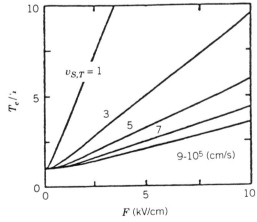

Fig. 6.14 Electron temperature for acoustic deformation potential scattering as a function of the electric field with the sound velocity as the family parameter. [Böer, K. W., *Survey of Semiconductor Physics*, Van Nostrand Reinhold (1990)]

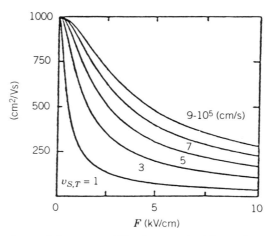

Fig. 6.15 Field dependence of electron mobility in acoustic deformation scattering. The sound velocity is the family parameter. [Böer, K. W., *Survey of Semiconductor Physics*, Van Nostrand Reinhold (1990)]

with

$$\varphi_e(T_e) = \frac{(T^{3/2}/2\Theta)\sinh(\Theta/2T)}{\cosh\{((T_e - T)/T_e)(\Theta/2T)\}K_2(\Theta/2T_e) + \sinh\{((T_e - T)/T_e)(\Theta/2T)\}K_1(\Theta/2T_e)}.$$

(6.105)

The K_1 and K_2 are Bessel functions. With increasing electron temperature, the electron mobility decreases. Introducing the field dependence of the electron temperature from Eq. (6.99), we obtain a drift velocity that first increases linearly with the field and then levels off at the saturation velocity:

$$v_{Ds} = \sqrt{\frac{3k\Theta}{4m_n}\tanh\left(\frac{\Theta}{2T}\right)}.$$

(6.106)

The saturation drift velocity can be easily obtained by eliminating F from the energy balance $eF = 2m^*v_D(\exp x + 1)$ with $x = \Theta/T$.

However, because of the spherical symmetry of the wave function at Γ, optical deformation potential scattering is zero in the lowest GaAs minimum. The most important scattering mechanisms for GaAs are polar optical phonon scattering, ionized impurity scattering if the density of impurities is significant ($\geqslant 10^{17}\,\text{cm}^{-3}$ at room temperature) and intervalley scattering in a very high electric field [Shu87].

Polar Optical Scattering at High Fields In polar semiconductors the interaction of carriers with the optical mode of lattice vibrations is known as *polar optical scattering* [See73]. The mobility as a function of the electron temperature is similar to the optical deformation potential scattering in the warm electron range. For the zero-field mobility μ_0, we have [Sea73]

$$\mu_0 = \frac{3\pi^{1/2}}{2^{5/2}}\frac{e}{m\alpha\omega_0}\frac{\sinh(\Theta/T)}{(\Theta/T)^{3/2}K_1(\Theta/T)}.$$

(6.107)

The drift velocity does not saturate at high fields but shows an increase above the threshold field before the hot-electron range is reached. Here the dielectric breakdown effects begin (see Fig. 6.16).

6.3.5 Intervalley Scattering

In intervalley scattering there are the equivalent and nonequivalent valleys. Equivalent valleys have their minima at the same energy, the conduction band edge. Nonequivalent valleys have a slightly higher energy, and hence need additional carrier energy to become populated.

Equivalent Intervalley Scattering

Intervalley scattering with optical phonons, using a deformation potential approximation, can be evaluated for warm electrons. The resulting mobility is given by [Str70]

$$\mu_i = \frac{4}{3} \frac{\alpha_c k \Theta_i}{\hbar} \sqrt{\frac{T_e}{\Theta_i}} \frac{\sinh(\Theta_i/2T)}{\cosh[(1-(T/T_e))(\Theta_i/2T)] K_1(\Theta_i/2T_e)}, \qquad (6.108)$$

where K_1 is the modified first-order Bessel function.

At high electric fields the anisotropy of the valleys must be taken into consideration. The direction of the field relative to the valley axis identifies the relevant effective mass, which changes with different alignment. It is smallest in the direction of the short axis, and largest in the direction of the long axis of the rotational ellipsoid. The electron heating is most effective when the field is aligned with the short axis of

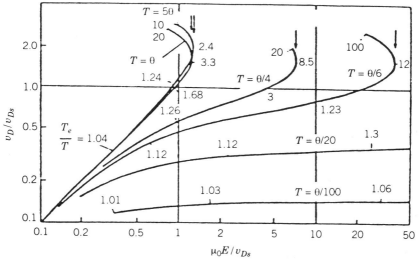

Fig. 6.16 Drift velocity as a function of the field for polar optical scattering. The lattice temperature is the family parameter, indicating the onset of dielectric breakdown effects (*arrows*). Values of the relative electron temperature are indicated along the curve. [Seeger, K., *Semiconductor Physics*, Springer-Verlag (1973)]

the ellipsoid. The ellipsoid is called the *hot ellipsoid*, and the corresponding valley, the *hot valley*; the other with the long axis in the field direction is called the *cold ellipsoid*, and the corresponding valley, the *cool valley*. [Böe90]. Scattering proceeds preferentially from the hot to the cool valley, since hot electrons have a higher average and consequently generate more intervalley phonons.

Intervalley Scattering into Nonequivalent Valleys

In an external field of sufficient magnitude, electrons in a valley with a small effective mass are heated very efficiently. Scattering into higher satellite valleys (see Fig. 6.17) requires a higher effective mass, so heating is reduced and backscattering lowered. Since a substantial fraction of the conduction electrons is pumped into the high valley, the average electron mobility is reduced. The current therefore increases less than ohmically with the increased field [Alb71].

If the ratio of the mobilities is high enough, *negative differential conductivity* occurs, as shown in curve 2 of Fig. 6.18(a). Corresponding experimental results are shown in Fig. 6.18(b) for several III–V compounds. Here a field range exists in which the current decreases with increasing field. In this range stationary solutions of the transport equation may not exist, and high-field domains may develop that move through the semiconductor and cause current oscillations. These oscillations are called the *Gunn effect*.

Such a repopulation, with an increasing electric field, competes with thermal excitation at higher temperatures, so only a change in the slope of the current-voltage characteristic is observed, without going through a maximum, as shown in curve 1 of Fig. 6.18(a). Increased doping in the high-doping range results in a decrease in the threshold field and, finally, in the disappearance of negative differential conductivity [See73].

In GaAs the L satellite minimum lies $0.36\,\text{eV}$ above the Γ-point $E(\mathbf{k})$ minimum at $\mathbf{k} = 0$, as shown in Fig. 6.19. The effective mass in the central minimum is $m_n = 0.07 m_0$, and $m_n \cong m_0$ in the satellite minima. This cause a reduction of the mobility in *n*-type GaAs from $\sim 8000\,\text{cm}^2/\text{Vs}$ at low fields to $\sim 200\,\text{cm}^2/\text{Vs}$ at high fields, when most

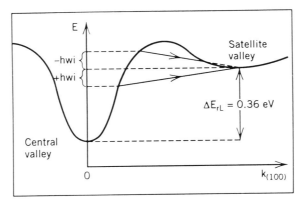

Fig. 6.17 Electron scattering into a higher satellite valley with higher effective mass of GaAs. [Böer, K. W., *Survey of Semiconductor Physics*, Van Nostrand Reinhold (1990)]

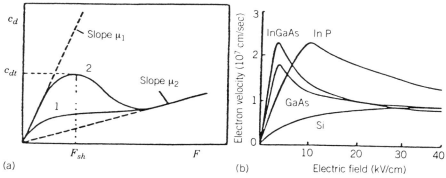

Fig. 6.18 (a) Drift velocity as a function of the field when major repopulation of the higher satellite valley with substantially lower effective mass become effective at the threshold field F_{th}; (b) measured field dependence of the drift velocity for several semiconductors [Böer, K. W., *Survey of Semiconductor Physics*, Van Nostrand Reinhold (1990)]

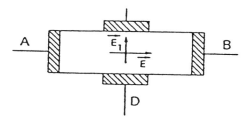

Fig. 6.19 Geometry of the samples for studying the transverse components of μ' and T_n. [Reggiani, L., in *Topics in Applied Physics*, vol. 58: *Hot-Electron Transport in Semiconductors*, pp. 7–83, ed. by L. Reggiani, Springer-Verlag (1985)]

of the electrons are pumped into the valley. Fields of 3 kV/cm are sufficient for achieving the necessary pumping.

6.4 BALLISTIC TRANSPORT IN GaAs SCATTERING

A system initially driven from the equilibrium state can be restored to equilibrium by scattering forces. For a semiconductor device the usual approach to the steady state transport of carriers across a device under a constant driving force is based on the drift-diffusion equation. As the device's size shrinks from micrometers to nanometers, semiclassical transport becomes important. The hot-electron and quantum effects invalidate the drift-diffusion equation. There are two ways of describing the motion of any collection of particles. One is the single-particle approach in which the coordinates of an individual electron evolve with time. The other is the collective approach where collective variables, like electron density, current density, and the probability distribution function, are calculated by averaging over the ensemble of electrons.

The space and time evolution of the distribution function is described by the Boltzmann equation (BE). In the semiclassical transport theory little work has been done on solving the BE for a particular device [Dat89], but Monte Carlo simulation has been used in a corresponding single-particle approach for submicron devices. Most of the analyses of quantum devices, like resonant tunneling diodes, also take the single-particle approach using the one-electron Schrödinger equation. As devices shrink to dimensions below 100 nm, the wave nature of electrons is affected and quantum transport analyses become critical.

6.4.1 Hot-Electron Concept

The term "hot electrons," which refers to the problem of transport at high electric fields in semiconductors (as pointed out in a fundamental work by Fröhlich [Frö47]), is associated with a rise of the electron mean energy above its thermal equilibrium value $(3/2)kT_0$, where T_0 is the thermodynamic temperature. As we noted in Section 6.3, physicists speak of an electron temperature T_e higher than the lattice temperature of the host crystal. It is now known that this idea does not correctly describe reality, since the electronic distribution function very often prevents an unambiguous definition of an electron temperature [Reg85]. The concept of an electron temperature has, however, greatly helped the understanding of high-field transport and still gives a useful terminology for a heuristic investigation of this problem.

When hot-electron conditions are determined by application of a static high electric field E, the three basic macroscopic quantities of interest are the drift velocity v_d (and the associated differential mobility $\mu' = dv_d/dE$), the diffusion coefficient D, and the white-noise temperature T_n (a measure of the noise associated with carrier velocity fluctuations). Depending on the direction of the applied electric field, there are one longitudinal and two transverse components for μ', D, and T_n. As illustrated in Fig. 6.19, by superimposing on the longitudinal field E between terminals $A-B$ the application of a transverse electric field E_1 between terminals $C-D$, one can measure the transverse components of μ' and T_n.

In the linear response regime (i.e., $E \to 0$) the drift velocity depends linearly on the electric field through the mobility μ as

$$v_d = \mu E. \qquad (6.109)$$

Einstein's relation is used to express diffusion in terms of mobility and the thermal equilibrium temperature:

$$D = \frac{\mu k T_0}{e}, \qquad (6.110)$$

where e is the electron charge. Nyquist's relation is used to express the available noise power P_{av} of a two-terminal network for unit bandwidth of frequency Δf in terms of the thermal equilibrium temperature:

$$\frac{P_{av}}{\Delta f} = kT_0. \qquad (6.111)$$

At thermal equilibrium, and thus for continuity at $E \to 0$, it is

$$\frac{P_{av}}{\Delta f} = kT_0 = \frac{eD}{\mu} \qquad (6.112)$$

Equation (6.112) gives a macroscopic version of the fluctuation-dissipation theorem.

Generally in a linear-response regime information about diffusion, the white-noise temperature, and drift velocity is not important in determining the transport properties of a material whose ohmic mobility is already known. Under hot-electron conditions, however, Einstein's and Nyquist's relations no longer hold, and drift, diffusion, and white noise are necessary components of the analysis. Deviations from linearity of drift velocity and departures from thermal equilibrium values of diffusion and white-noise temperature have been experimentally measured at increasing field strengths. Figure 6.20 gives the evidence of the effect of these components on experimental hot-electron conditions.

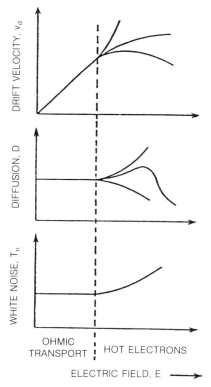

Fig. 6.20 Schematic illustration of the dependence on electric field strength of drift velocity, diffusion coefficient, and noise temperature as taken along the field direction (scale is assumed to be logarithmic). [Reggiani, L., in *Topics in Applied Physics*, vol. 58, *Hot-Electron Transport in Semiconductors*, pp. 7–83, ed. by L. Reggiani, Springer-Verlag (1985)]

6.4.2 Transport of a Single Particle

With the transport of the single particle, we have an individual carrier (electron) that behaves according to the Schrödinger function or Newton's law. Consider the problem of finding the current across a device shown in Fig. 6.21. The contacts are assumed to be in local thermodynamic equilibrium. The device thus connects two reservoirs (contacts), each in its equilibrium state but with a different chemical potential. The electrons in each contact are distributed according to Fermi-Dirac statistics. As the left contact

$$f_L(\mathbf{k}) = \frac{1}{\exp[E_{CL} - E_{FL} + \varepsilon(\mathbf{k})]/kT + 1}, \qquad (6.113)$$

where

$$\varepsilon(\mathbf{k}) = \frac{\hbar^2 k^2}{2m^*} = \frac{\hbar^2(k_x^2 + k_y^2 + k_z^2)}{2m^*}. \qquad (6.113')$$

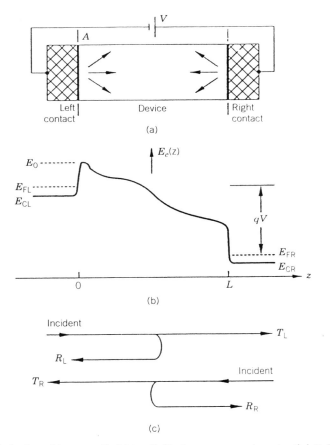

Fig. 6.21 (a) A device with an applied bias V; (b) the macroscopic potential $E_C(z)$; (c) incident, reflected, and transmitted currents from the contacts. [Hess, K., *Advanced Theory of Semiconductor Devices*, Prentice Hall (1988)]

At the right contact

$$f_R(\mathbf{k}) = \frac{1}{\exp[E_{CR} - E_{FR} + \varepsilon(\mathbf{k})]/kT + 1}. \quad (6.114)$$

If the contacts are identical, then regardless of the applied bias V, $E_{CL} - E_{VL} = E_{CR} - E_{FR}$. Hence

$$f_L(\mathbf{k}) = f_R(\mathbf{k}) = f(\mathbf{k}). \quad (6.115)$$

With an applied bias V we can see electrons incident from the contact A onto the device. The incoming current density J_i is calculated by summing the currents carried by electrons in different states $|\mathbf{k}\rangle$ and converting the summation into an integral:

$$J_i = -\frac{q}{4\pi^3} \int_{k_z > 0} d^3k \frac{\hbar k_z}{m^*} f(\mathbf{k}). \quad (6.116)$$

An electron incident from the left contact with a wave vector $\mathbf{k}$ has a certain probability $T_L(\mathbf{k})$ of being transmitted across the device into the right contact, and a probability $R_L(\mathbf{k}) = 1 - T_L(\mathbf{k})$ of returning to the same contact. The current J_L flowing from the left to the right contact is given by

$$J_L = -\frac{q}{4\pi^3} \int_{k_z > 0} d^3k \frac{\hbar k_z}{m^*} f(\mathbf{k}) T_L(\mathbf{k}). \quad (6.117)$$

Similarly we can calculate the current J_R flowing from the right to the left contact:

$$J_R = -\frac{q}{4\pi^3} \int_{k_z > 0} d^3k \frac{\hbar k_z}{m^*} f(\mathbf{k}) T_R(\mathbf{k}). \quad (6.118)$$

$T_R(\mathbf{k})$ is the probability that an electron incident from the right contact with wave vector $\mathbf{k}$ is transmitted to the left contact. The net current J through the device is given by

$$J = -\frac{q}{4\pi^3} \int_{k_z > 0} d^3k \frac{\hbar k_z}{m^*} f(\mathbf{k})[T_L(\mathbf{k}) - T_R(\mathbf{k})]. \quad (6.119)$$

The problem of calculating current is thus reduced to calculating the transmission coefficient $T_L(\mathbf{k})$ and $T_R(\mathbf{k})$ for a given applied voltage V. If $V = 0$, then $T_L = T_R$, and no net current flows. When $V > 0$, it makes it easier for electrons with a given wave vector $\mathbf{k}$ to go from left to right than to go from right to left; that is, $T_L(\mathbf{k}) > T_R(\mathbf{k})$. There is now a net electron flow from left to right, so the conventional current flows from right to left.

In the single-particle picture, device analysis is essentially a problem of calculating the transmission probabilities $T(\mathbf{k})$. In quantum transport theory we calculate $T(\mathbf{k})$ from the Schrödinger equation, whereas in semiclassical transport theory we calculate it from Newton's law [Dat89]. Any device can be viewed as a complex array of scatters. The transmission probability $T(\mathbf{k})$ can be obtained by solving the Schrödinger equation

$$i\hbar \frac{\partial \Psi}{\partial t} = -\frac{\hbar^2 \nabla^2 \Psi}{2m^*} + E_c \Psi + U_S(\mathbf{r}, t)\Psi. \tag{6.120}$$

The microscopic three-dimensional time-varying scattering potential $U_S(\mathbf{r}, t)$ represents the entire array of scatterers.

Most of our progress in understanding conduction processes comes from using semiclassical transport theory. The moving electron has an initial momentum $\hbar \mathbf{k}$. The effect of the macroscopic potential $E_C(z)$ is based on Newton's law:

$$\frac{d\mathbf{r}}{dt} = \mathbf{v}(\mathbf{k}), \tag{6.121}$$

$$\frac{d}{dt}(\hbar \mathbf{k}) = -\nabla E_C + \sum_i \hbar \beta \delta(t - t_i). \tag{6.122}$$

The velocity $\mathbf{v}(\mathbf{k})$ of an electron with momentum $\hbar \mathbf{k}$ is equal to the group velocity of a wave packet centered around the state $\mathbf{k}$:

$$\mathbf{v}(\mathbf{k}) = \frac{1}{\hbar}\nabla_{\mathbf{k}} E. \tag{6.123}$$

For parabolic bands with $E = E_{co} + (\hbar^2 k^2/2m^*)$ we have the usual relation

$$\mathbf{v}(\mathbf{k}) = \frac{\hbar \mathbf{k}}{m^*}. \tag{6.124}$$

The first term in Eq. (6.122) stands for Newton's law for a force $-\nabla E_c = -e\mathbf{E}$ acting on an electron given parabolic bands with $m^*\mathbf{v} = \hbar\mathbf{k}$. The second term at the right-hand side of Eq. (6.122) represents a series of impulsive forces whose strength, direction, and location in time are random variables whose distribution reproduces the correct scattering function $S(\mathbf{k}, \mathbf{k}')$ calculated from Fermi's golden rule. Consider the probability $P(t')dt'$ that the time difference between two impulses $(t_{i+1} - t_i)$ lies between t' and $t' + dt'$. If we integrate this probability from $t' = 0$ to $t' = t$, the result should equal the probability $P_S(t)$ that the electron is scattered within a time t:

$$P(t')dt' = P_S(t) = 1 - e^{-t/\tau}, \tag{6.125}$$

where τ is the lifetime of a state. Equation (6.125) gives us the probability $P(t)$ that one impulse follows the preceding one after time t. The strength (β) and direction (θ_p, ϕ) of the impulse are similarly distributed so as to match the phonon emission and absorption rates, $R^e(\mathbf{k}, \beta)$ and $R^a(\mathbf{k}, \beta)$.

If the random variables β and t_i characterizing the stochastic force in Eq. (6.122)

$[\sum_i \hbar\beta\delta(t - t_i)]$ have the correct probability distribution, the single-particle equation

$$\frac{d}{dt}(\hbar\mathbf{k}) = \sum_i \hbar\beta\delta(t - t_i) \quad (6.126)$$

is equivalent to the collective master equation

$$\frac{\partial f(\mathbf{k}, t)}{\partial t} = S_{\text{op}} f. \quad (6.127)$$

Equations (6.121) and (6.122) are simply the single-particle version of the (collective) Boltzmann equation [Dat89]. The individual particles of electron transport are at the heart of Monte Carlo simulations, which are gaining in popularity in the modeling of submicron devices. Monte Carlo methods will be discussed in Section 6.5.

6.4.3 The Boltzmann Equation

The Time-Dependent Boltzmann Equation

In the semiclassical approach the kinetic equation that used to describe hot-electron transport is the Boltzmann equation. The main assumptions that justify such an approach are the following [Gan79; Dud79]:

1. Carrier density should be sufficiently low so that only binary collisions occur.
2. The time between successive collisions τ should be long enough with relation to the duration of a collision τ_{coll} (i.e., $\tau_{\text{coll}}/\tau \ll 1$).
3. Density gradients should be small over the range of the interparticle potential.

We will discuss transport in the collective picture. The Boltzmann equation gives the total time rate of change of the distribution $f(\mathbf{r}, \mathbf{k}, t)$. The function f has six independent variables (three components of $\mathbf{r}$ plus three components of $\mathbf{k}$) in addition to the time t. The six-dimensional space is known as the *phase space*. Conceptually we divide up the semiconductor into little volumes Ω in each of which the electrons are distributed among the states $|\mathbf{k}\rangle$ with probability $f(\mathbf{k})$. Consider first the master equation from which the effects of the electric field and spatial diffusion are excluded:

$$\frac{\partial f(\mathbf{r}, \mathbf{k}, t)}{\partial t} = S_{\text{op}} f(\mathbf{r}, \mathbf{k}, t). \quad (6.128)$$

We need to extend Eq. (6.128) to include two additional physical process that affect the probability distribution within volume Ω. The first is the electric field, which continually acts to increase the wave vectors of the individual electrons according to Newton's law. The second is the diffusion term, which is the flow of electrons from adjacent volumes. The Boltzmann equation can be written as

$$\frac{df}{dt} = S_{\text{op}} f + \frac{\partial f}{\partial t}\bigg|_{\text{field}} + \frac{\partial f}{\partial t}\bigg|_{\text{diffusion}}, \quad (6.128a)$$

where the scattering operator S_{op} acts only on the variable $\mathbf{k}$ and not on $\mathbf{r}$ and t:

$$S_{op}f(\mathbf{r},\mathbf{k},t) = \sum_{\mathbf{k}'} S(\mathbf{k}',\mathbf{k})f_{\mathbf{k}'}(1-f_{\mathbf{k}}) - S(\mathbf{k},\mathbf{k}')f_{\mathbf{k}}(1-f_{\mathbf{k}'}). \quad (6.129)$$

In Eq. (6.129) we have written $f(\mathbf{r},\mathbf{k},t)$ as $f_{\mathbf{k}}$ on the right-hand side. Equation (6.128) is a collective version of the single-particle equation. There are two physical processes that also affect the probability distribution within the volume Ω. The first is the electric field, which continually acts to increase the wave vectors of the individual electrons according to Newton's law. The second is the flow of electrons from adjacent volumes. Equation (6.129) can be extended to

$$\frac{df}{dt} = S_{op}f + \left.\frac{\partial f}{\partial t}\right|_{\text{field}} + \left.\frac{\partial f}{\partial t}\right|_{\text{diffusion}}. \quad (6.130)$$

To evaluate two additional terms in Eq. (6.130), suppose that there are no scattering forces, and observe how the distribution function evolves within a volume due to the electric field and diffusion. Consider a one-dimensional distribution function $f(z,k_z,t)$. In time Δt the coordinate and momentum of an individual particle i changes by

$$\Delta z_i = v_{zi}\Delta t, \quad (6.131)$$

$$\Delta k_{zi} = -\frac{eE_z\Delta t}{\hbar}. \quad (6.132)$$

Consequently in time Δt the distribution function must be shifted along the z-axis by Δz_i and along the k_z axis by Δk_{zi}. This is illustrated in Fig. 6.22 [Dat89]:

$$f(z,k_z,t+\Delta t) = f\left(z - v_z\Delta t, k_z + \frac{q\mathscr{E}_z\Delta t}{\hbar}, t\right). \quad (6.133)$$

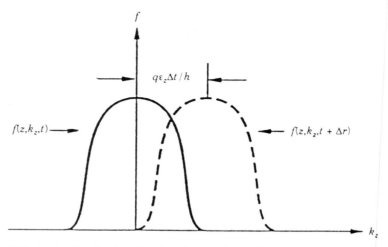

Fig. 6.22 Shift in the distribution function due to electric field. [Hess, K., *Advanced Theory of Semiconductor Devices*, Prentice Hall (1988)]

For small Δt we can keep only the first term in a Taylor series expansion. We rewrite the left-hand side of Eq. (6.133) as

$$f(z, k_z, t + \Delta t) = f(z, k_z, t) + \frac{\partial f}{\partial t} \Delta t \qquad (6.134\text{a})$$

The right-hand side can be rewritten as

$$f\left(z - v_z \Delta t, k_z + \frac{q\mathcal{E}_z \Delta t}{\hbar}, t\right) = f(z, k_z, t) - v_z \Delta t \frac{\partial f}{\partial z} + \frac{q\mathcal{E}_z \Delta t}{\hbar} \frac{\partial f}{\partial k_z}. \qquad (6.134\text{b})$$

Substituting Eqs. (6.134) into Eq. (6.133), we have

$$\frac{\partial f}{\partial t} = -v_z \frac{\partial f}{\partial z} + \frac{q\mathcal{E}_z}{\hbar} \frac{\partial f}{\partial k_z} \qquad (6.135)$$

Comparing with Eq. (6.130), we have

$$\left.\frac{\partial f}{\partial t}\right|_{\text{field}} = \frac{q\mathcal{E}_z}{\hbar} \frac{\partial f}{\partial k_z}, \qquad (6.136)$$

$$\left.\frac{\partial f}{\partial t}\right|_{\text{diffusion}} = -v_z \frac{\partial f}{\partial z}. \qquad (6.137)$$

After these results have been generalized to three dimensions, we get the Boltzmann transport equation:

$$\frac{\partial f}{\partial t} + \mathbf{v}(\mathbf{k}) \cdot \nabla f - \frac{q\mathcal{E}}{\hbar} \cdot \nabla_\mathbf{k} f = S_{\text{op}} f,$$

where S_{op} is given by Eq. (6.129) and $v(\mathbf{k})$ by

$$\mathbf{v}(\mathbf{k}) = \frac{1}{\hbar} \nabla_\mathbf{k} E = \frac{\hbar \mathbf{k}}{m^*} \qquad \text{(for parabolic bands)}.$$

This is the central equation of semiclassical transport theory. The only quantum mechanical inputs to this equation are the evaluation of the scattering operator from Fermi's golden rule and the calculation of the velocity from the band structure [Dat89].

The Current Equation

To study the electrical behavior of materials when high electric fields are applied, a phenomenological current equation is usually introduced and the current density is given by [Reg85]

$$j(\mathbf{r}, t) = en(\mathbf{r}, t)\mathbf{v}_{d0}(\mathbf{E}) - eD(\mathbf{E})\frac{\partial}{\partial \mathbf{r}} n(\mathbf{r}, t) \qquad (6.138)$$

where $\mathbf{E} = \mathbf{E}(\mathbf{r}, t)$. The drift velocity $\mathbf{v}_{d0}$ is that obtained in absence of concentration gradients, and the diffusion term accounts for concentration gradient effects through the diffusion coefficients $D(\mathbf{E})$. For the case of an electric field Eq. (6.138) is usually called *Fick's law*.

By substituting Eq. (6.138) into the continuity equation

$$\frac{\partial n}{\partial t}(\mathbf{r}, t) = \frac{1}{e} \operatorname{div} j(\mathbf{r}, t), \qquad (6.139)$$

the diffusion equation is obtained:

$$\frac{\partial n}{\partial t}(\mathbf{r}, t) = -v_{d0,i}\frac{\partial n}{\partial x_i}(\mathbf{r}, t) + D_{ij}\frac{\partial^2 n(\mathbf{r}, t)}{\partial x_i \partial x_j}, \qquad (6.140)$$

where D_{ij} is the diffusion tensor. With the initial distribution

$$n(\mathbf{r}, 0) = \delta(x)\delta(y)\delta(z) \qquad (6.141)$$

the solution of Eq. (6.140) is straightforward. By assuming an x direction for the isotropic crystals, we obtain

$$n(x, y, z; t) = A_0 \exp\left[\frac{-(x - v_{d0}t)^2}{(4D_{lo}t)}\right] \exp\left[\frac{-(y^2 + z^2)}{(4D_{tr}t)}\right], \qquad (6.142)$$

where A_0 is the appropriate normalizing constant and D_{lo} and D_{tr} are, respectively, the longitudinal and the transverse components of the diffusion coefficient.

We define the mth moment of $n(x, t)$ as

$$M^{(m)} = \frac{1}{N}\int_{-\infty}^{\infty} x^m n(x, t) dx = \langle x^m \rangle, \qquad (6.143)$$

where N stands for the total number of particles. By using the diffusion equation (6.140), the following recursion relations are found:

$$M^0 = 1,$$

$$\frac{dM^{(1)}}{dt} = v_{d0},$$

$$\frac{dM^{(2)}}{dt} = 2(v_{d0}M^{(1)} - D),$$

$$\frac{dM^{(m \geq 2)}}{dt} = mv_{d0}M^{(m-1)} + m(m-1)DM^{(m-2)}.$$

For the second central moment $\langle (x - \langle x \rangle)^2 \rangle$, we have

$$D_{ij} = \lim \frac{d}{dt}[\langle x_i(t)x_j(t)\rangle - \langle x_i(t)\rangle\langle x_j(t)\rangle], \qquad (6.144)$$

where the long time limit ensures that Fick's law is satisfied.

By carrying out the time derivative of Eq. (6.144), and using the fact that an ensemble average commutes with the time derivative, an equivalent expression of diffusion is obtained:

$$D_{ij} = \lim[\langle x_i(t)v_j(t)\rangle - \langle x_i(t)\rangle\langle v_j(t)\rangle]. \tag{6.145}$$

In this way diffusion is correlated to the achievement of a stationary value for the space velocity covariance of the initial particle distribution.

By writing the trajectory in terms of velocity, a third equivalent expression of diffusion is obtained as

$$D_{ij} = \lim \int \langle \delta v_i(t)\delta v_j(t-t')\rangle dt' + \langle \delta x_i(0)\delta_i(t)\rangle. \tag{6.146}$$

Equation (6.146) relates diffusion to the autocorrelation function of velocity fluctuations. At very short time periods, comparable to or less than average scattering time, the time dependence of the quantities inside the limit in Eqs. (6.144) through (6.146) can be used to describe a diffusion coefficient that will depend upon time and initial conditions of motion. Figure 6.23 shows the dependence on time of the quantities

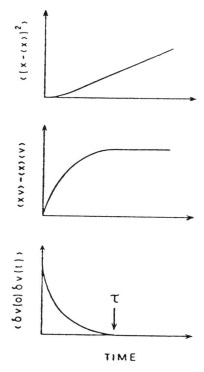

Fig. 6.23 Schematic representation of the time dependence of spatial spreading, space-velocity covariance, and velocity-fluctuation autocorrelation function. [Reggiani, L., in *Topics in Applied Physics*, vol. 58: *Hot-Electron Transport in Semiconductors*, pp. 7–83, ed. by L. Reggiani, Springer-Verlag (1985)]

defined by Eqs. (6.144) through (6.146) when there is no applied electric field and each particle is assumed to start at $t = 0$ at position $x = 0$ with an equilibrium Maxwell-Boltzmann momentum distribution.

Figure 6.23 indicates that τ is the time necessary to achieve the correct space-velocity correlation for Fick's law holds. Since the space-velocity covariance and the velocity-fluctuations autocorrelation function remain constant over the long time period, they are better suited to check the validity of Fick's law than the second central moment which exhibits a long-time trend.

Figure 6.24 shows the transient behavior of diffusion in space and the related diffusion coefficient, which depends on time, when at $t = 0$ an electric field is applied. Curve 1 results from an applied small electric field, and curves 2 and 3 from an applied high electric field by which hot-electron conditions are established. Under transient conditions hot-electron diffusion is shown to exhibit values greater than those under long-time limit conditions.

The Generalized Diffusion Equation

In real physical systems the particle density $n(x, t)$ never exactly takes the Gaussian form given by Eq. (6.142) [Van82]. If $n(x, t)$ satisfies the diffusion equation, it does so only in an asymptotic sense. A generalization of the diffusion equation with relation to arbitrary space time variations of the density gradient is presented below [Reg85]. This allows us to define a frequency and wave vector–dependent diffusion coefficient $D(q, \omega)$ [Zwa64] which, in general, is a complex number. The limit $D(0,0)$ is real, since it relates to ordinary diffusivity, as described in the previous sections.

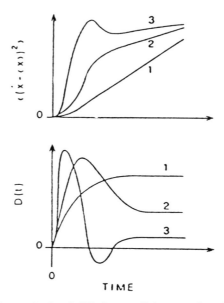

Fig. 6.24 Spatial spreading and related diffusion coefficient as a function of time when transient conditions are analyzed: (*1*) refers to an ohmic case; (*2*) and (*3*) refer to possible hot-electron conditions (see text). [Reggiani, L., in *Topics in Applied Physics*, vol. 58: *Hot-Electron Transport in Semiconductors*, pp. 7–83, ed. by L. Reggiani, Springer-Verlag (1985)]

Consider a system of particles subject to an external uniform static electric field **E** along x. A particle source of sinusoidal type,

$$S(x,t) = S_0[1 + \cos(qx - \omega t)], \qquad (6.147)$$

is then introduced together with a trapping mechanism at a constant rate τ_d^{-1}. The source amplitude S_0 and τ_d are connected through the condition of particle conservation over a wavelength λ:

$$\int_\lambda S_0[1 + \cos(qx - \omega t)]dx = \int_\lambda \tau_d^{-1} n(x,t)dx. \qquad (6.148)$$

The particles are thus trapped and redistributed in space. By making use of the more convenient complex formulism, the diffusion equation (6.140) becomes

$$\frac{\partial n}{\partial t} = D\frac{\partial^2 n}{\partial x^2} - v_{d0}\frac{\partial n}{\partial x} + S_0\{1 + \exp[i(qx - \omega t)]\} - \frac{n}{\tau_d}. \qquad (6.149)$$

We want to solve Eq. (6.149) of the type

$$n(x,t) = n_0 + n_1 \exp[i(qx - \omega t + \varphi)], \qquad (6.150)$$

where n_0 is a constant average term, n_1 and φ are the amplitude and the phase shift of the harmonic disturbance, respectively. Substituting Eq. (6.150) into Eq. (6.148), we obtain

$$S_0 = \frac{n_0}{\tau_d}, \qquad (6.151)$$

and Eq. (6.149) becomes

$$-i\omega + q^2 D + iqv_{d0} + \frac{1}{\tau_d} = \left(\frac{n_0}{n_1}\right)\frac{\exp(-i\varphi)}{\tau_d}. \qquad (6.152)$$

When n is linear, Eq. (6.152) can be used at frequencies and wavelengths so high that Fick's law does not hold. Then D would be taken as a function of q and ω. Thus from Eq. (6.152) we obtain an expression for $D(q,\omega,\tau_d^{-1})$ in terms of the relative amplitude n_1/n_0 and the phase shift ψ of the disturbance

$$D = \frac{1}{q^2}\left[\left(\frac{n_0}{n_1}\right)\tau_d^{-1}\exp(-i\psi) + i(\omega - qv_{d0}) - \tau_d^{-1}\right]. \qquad (6.153)$$

Now, using the path integral method [Cha52], we find the following integral expression for $n(x,t)$:

$$n(x,t) = n_0\int dt'\tau_d^{-1}\int dx'\{1 + \exp[i(qx't' - t')]\}\cdot P(x',t';x,t)\exp[-\tau_d^{-1}(t-t')]\}, \qquad (6.154)$$

where $P(x', t'; x, t)dx$ is the probability that without trapping, a particle in x' at t' will be in dx around x at t. The integral of the first term in brackets in Eq. (154) is equal to n_0, since from the time $-\infty$ to t all particles have been trapped and redistributed by the source. Here $P(x', t'; x, t)$ is a function of x', t' and x, t only through the differences $x'' = x - x'$ and $t'' = t - t'$ so that Eqs. (6.150) and (6.154) yield

$$n(x,t) - n_0 = n_0 \exp[i(qx - \omega t)] \int \tau_d^{-1} dt \int dx \exp[i(qx - \omega t) - t\tau_d^{-1}] P(x,t). \quad (6.155)$$

From Eqs. (6.150) and (6.154) we obtain

$$R = n_1 \exp(i\varphi) = \int_0^\infty \tau_d^{-1} dt \int_0^\infty dx \exp[-i(qx - \omega t) - t\tau_d^{-1}] P(x,t), \quad (6.156)$$

and Eq. (6.153) finally yields

$$D(q, \omega, \tau_d^{-1}) = \frac{1}{q^2 \tau_d R} [1 - R + i\tau_d R(\omega - qv_{d0})]. \quad (6.157)$$

Equations (6.156) and (6.157) provide an expression for $D(q, \omega, \tau_d^{-1})$ that is valid for any wave vector, frequency, and particle lifetime τ_d. When τ_d is much greater than both ω^{-1} and any microscopic time, $(q, \omega) = D(q, \omega, 0)$ is obtained for steady state conditions. $P(x, t)$ must be derived from knowledge of particle dynamics.

Figures 6.25 and 6.26 [Reg85] give the results for the real part of $D(q, \omega, \tau)$ as a function of q and ω, respectively. As a general trend $D(q, \omega)$ decreases at asymptotically high values of both q and ω because the average over several wavelengths, as well as over several periods of carrier concentrations in the past, is not affected by contributions from electrons coming from far away [Jac81; Jac74]. The peaks of $D(q, \omega)$ at the intermediate frequency (Fig. 6.26) are related to the presence of high-energy intervalley phonons whose emission at high fields produces the streaming character of carrier motion.

6.4.4 Density Matrix Formulation

The discussion of carrier transport in the previous sections has been predicated on a semiclassical picture. Quantum effects [Phi82] take place on a short-time scale where Fermi's golden rule breaks down and where spatial feature sizes are on the order of tens of angstroms. The quantum mechanical effects become prominent when the feature size is on the order of a thermal de Broglie wavelength or shorter, as shown in Fig. 6.27 [Iaf81; Gru89].

The quantum transport formulation for devices needs new approaches. Theoretical approaches developed by researchers include the use of density matrix, Wigner distribution function, Feynman path-integral, and moment equations. These formulations incorporate a variety of important phenomena, including many-body interactions, dissipation in quantum systems [Cal83; Cal84], Pauli exclusion, finite-temperature effects [Cal84], interactions between open systems [Fre85], and self-consistent time-dependent potential variations [Fre85; Str89].

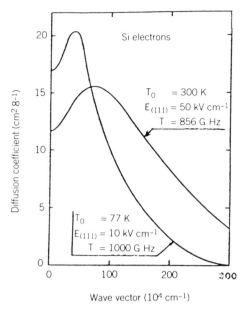

Fig. 6.25 High-field, low-frequency diffusion coefficient as a function of the wave vector at the indicated temperatures for electrons in Si. [Reggiani, L., in *Topics in Applied Physics*, vol. 58: *Hot-Electron Transport in Semiconductors*, pp. 7–83, ed. by L. Reggiani, Springer-Verlag (1985)]

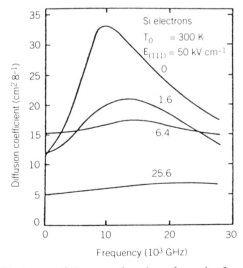

Fig. 6.26 High-field diffusion coefficient as a function of angular frequency ω at the different values of wave vector q indicated by the numbers on the curves in units of 10^5 cm^{-1} for electrons in Si. [Reggiani, L., in *Topics in Applied Physics*, vol. 58: *Hot-Electron Transport in Semiconductors*, pp. 7–83, ed. by L. Reggiani, Springer-Verlag (1985)]

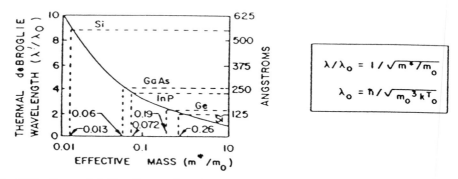

Fig. 6.27 Thermal de Broglie wavelength versus effective mass. [Grubin, H. L., in *Introduction to Semiconductor Technology, GaAs and Related Compounds*, ed. by C. T. Wang, Wiley Interscience (1989)]

The density matrix description of quantum mechanics is well-suited for describing transport in ultrasmall electronic devices. The density operator $\hat{\rho}$ for a quantum mechanical system described by m independently prepared state vectors $|\psi_m\rangle$ is defined by

$$\hat{\rho} = \sum_m W_m |\psi_m\rangle\langle\psi_m|, \qquad (6.158)$$

where W_m is the statistical probability for the system to be in state $|\psi_m\rangle$.

The density matrix ρ_{ij} is obtained from the density operator through the relationship

$$\rho_{ij} = \langle\phi_i|\hat{\rho}|\phi_j\rangle, \qquad (6.159)$$

where $|\phi_n\rangle$ is a complete orthonormal basis set for $|\psi_m\rangle$:

$$|\psi_m\rangle = \sum_n a_n^{(m)} |\phi_n\rangle \qquad (6.160a)$$

and

$$\langle\phi_m| = \sum_{n'} a_{n'}^{(m)*} \langle\phi_{n'}|. \qquad (6.160b)$$

From these equations it follows that

$$\begin{aligned}
\rho_{ij} &= \langle\phi_i|\hat{\rho}|\phi_j\rangle \\
&= \left\langle\phi_i \Big| \sum_m W_m \sum_{nn'} a_n^{(m)} a_{n'}^{(m)*} \Big| \phi_n \right\rangle \langle\phi_{n'}|\phi_j\rangle \\
&= \sum_m W_m a_i^{(m)} a_j^{(m)*}.
\end{aligned} \qquad (6.161)$$

The equation of motion for the density matrix may be derived by considering the time derivative of a pure state $|\psi\rangle$:

$$|\psi\rangle = \sum_n a_n(t)|\phi_n\rangle. \tag{6.162}$$

That is,

$$i\hbar \sum_n \frac{\partial}{\partial t} a_n|\phi_n\rangle = \sum_n a_n H|\phi_n\rangle. \tag{6.163}$$

Operating on this equation from the left with $\langle\phi_p|$,

$$i\hbar \frac{\partial a_p}{\partial t} = \sum_n H_{pn} a_n, \tag{6.164a}$$

$$-i\hbar \frac{\partial a_p^*}{\partial t} = \sum_n H_{pn}^* a_n^*, \tag{6.164b}$$

where

$$H_{pn} = \langle\phi_p|H|\phi_n\rangle.$$

For a pure state

$$\begin{aligned}
i\hbar \frac{\partial \rho_{np}}{\partial t} &= i\hbar \frac{\partial (a_p^* a_n)}{\partial t} \\
&= i\hbar \left(\frac{\partial a_p^*}{\partial t} a_n + a_p^* \frac{\partial a_n}{\partial t} \right) \\
&= \sum_j (H_{nj}\rho_{jp} - \rho_{nj}H_{jp}),
\end{aligned} \tag{6.165}$$

where $H_{jp} = H_{pj}^*$; this result may be expressed in terms of the density operator as

$$i\hbar \frac{\partial \hat{\rho}}{\partial t} = [\hat{H}, \hat{\rho}] \tag{6.166}$$

Equation (6.166) for the time evolution of the density operator is known as the *quantum Liouville equation*.

The expectation value of $\langle a(x,p)\rangle$ of an observable $a(x,p)$ may be constructed on the $|\phi_k\rangle$ basis as follows:

$$\begin{aligned}
\langle a(x,p)\rangle &= \langle\psi|a(x,p)|\psi\rangle \\
&= \sum_{l,k} \langle\psi|\phi_l\rangle\langle\phi_l|a(x,p)|\phi_k\rangle\langle\phi_k|\psi\rangle \\
&= \sum_{l,k} \langle\phi_k|\psi\rangle\langle\psi|\phi_l\rangle\langle\phi_l|a(x,p)|\phi_k\rangle \\
&= \sum_{l,k} \rho_{k,l} a(x,p)_{l,k} \\
&= \operatorname{Trace}[\hat{\rho}(x,p)].
\end{aligned} \tag{6.167}$$

The density matrix contains all of the physical information needed to calculate the expectation value of an observable. This result is of special significance in quantum transport theory where one of major aims is to predict measurable values of observables such as current and charge density.

6.4.5 Feynman Path-Integral Formulation

In the previous section we discussed the quantum transport formalisms based on the density matrix theory. While this theory is extremely useful in treating quantum mechanical systems, it is convenient to introduce path-integral techniques to formulate models for [Str89] quantum dissipation, quantum Monte Carlo transport, and strong electron-LO-phonon coupling in polar semiconductors.

These important phenomena are essential to the full description of transport in quantum-based electronic devices. In 1948 Richard Feynman published a paper [Fey48] on the path-integral formulation of quantum mechanics. In 1963 Feynman and Vernon published a key work that established the influence-functional approach to describing interaction in complex systems [Fey63]. This approach is well suited to describing interacting systems such as that of the small superlattice structure in contact with a large wafer.

Before the path-integral formulation of quantum mechanics is presented, it is advantageous to review the principle of least action in the context of classical mechanics as well as the propagator method of quantum mechanics [Str89]. The principle of least action may be used to describe the behavior of conservative classical systems (e.g., see [Fey64]). For such system the action S along a path $x(t)$ is defined by

$$S[x(t)] = \int_{t_1}^{t_2} [T(t) - V(t)] dt, \tag{6.168}$$

where $T(t)$ and $V(t)$ are the kinetic and potential energies, respectively. The action $S[x(t)]$ is a functional of the path $x(t)$. According to the principle of least action, the classical path is the path $\underline{x}(t)$ for which the action is a minimum.

In practice, the path $\underline{x}(t)$ is determined by exploiting the following two facts:

1. First-order variations in x about the minimum value of a function $f(x)$ cause only second-order variations in $f(x)$.
2. First-order variations in x about positions away from the minimum value of a function $f(x)$ cause first-order variations in $f(x)$.

To determine $\underline{x}(t)$, these facts may be applied formally through the calculus of variations [Fey65] or in a unique manner for a given action.

Consider the following example as an application of the principle of least action:

$$S[x(t)] = \int_{t_1}^{t_2} \left[\frac{m}{2}\left(\frac{dx(t)}{dt}\right)^2 - V(x)\right] dt. \tag{6.169}$$

Take

$$x(t) = \underline{x}(t) + \varepsilon(t). \tag{6.170}$$

Then

$$S = \int_{t_1}^{t_2} \left[\frac{m}{2} \left(\frac{d\underline{x}}{dt} + \frac{d\varepsilon}{dt} \right)^2 - V(\underline{x} + \varepsilon) \right] dt$$

$$= \int_{t_1}^{t_2} \left[\frac{m(d\underline{x})^2}{2\, dt} - V(\underline{x}) + m \frac{d\underline{x}}{dt} \frac{d\varepsilon}{dt} - \varepsilon V'(\underline{x}) + \text{higher-order terms} \right] dt, \quad (6.171)$$

where

$$V'(\underline{x}) = \frac{\partial V}{\partial \underline{x}}(\underline{x}). \quad (6.172)$$

Thus

$$S = \bar{S} + \delta S^{(1)}, \quad (6.173)$$

where

$$\bar{S} = \int_{t_1}^{t_2} \left[\frac{m}{2} \left(\frac{d\underline{x}}{dt} \right)^2 - V(\underline{x}) \right] dt, \quad (6.174)$$

$$\delta S^{(1)} = \int_{t_1}^{t_2} \left[m \frac{d\underline{x}}{dt} \frac{d\varepsilon}{dt} - \varepsilon V'(\underline{x}) \right] dt. \quad (6.175)$$

Here $\delta S^{(1)}$ denotes the first-order variation in the action. The integral $\delta S^{(1)}$ must be zero when evaluated along the classical path $\underline{x}(t)$, since the variation in δS along $\underline{x}(t)$ is second order or higher. Since $\delta S^{(1)}$ equals zero when evaluated along $\underline{x}(t)$, this must be true independently of the value of $\varepsilon(t)$. Upon integrating by parts and using the fact that all paths must begin at $x_1 = x(t_1) = \underline{x}(t_1)$ and end at $x_2 = x(t_2) = \underline{x}(t_2)$, we have

$$\delta S^{(1)} = m \frac{d\underline{x}(t)}{dt} \varepsilon(t) \Big|_{t_1}^{t_2} - \int_{t_1}^{t_2} \left[m \frac{d^2\underline{x}}{dt^2} + V'(\underline{x}) \right] \varepsilon(t)\, dt \quad (6.176)$$

$$= \int_{t_1}^{t_2} \left[-m \frac{d^2\underline{x}}{dt^2} - V'(\underline{x}) \right] \varepsilon(t)\, dt = 0. \quad (6.177)$$

Since this integral must vanish independent of $\varepsilon(t)$, it follows that

$$\frac{m d^2 \underline{x}}{dt^2} = -\frac{\partial V}{\partial \underline{x}}(\underline{x}) = F(\underline{x}). \quad (6.178)$$

Therefore, for a conservative system, the principle of least action selects the path $\underline{x}(t)$ for which Newton's law holds.

Feynman described the path-integral formulation using the quantum mechanical propagator $K(x_2, t_2; x_1, t_1)$:

$$K(x_2, t_2; x_1, t_1) = \sum_{\substack{\text{All paths} \\ \text{from } x_1, t_1 \\ \text{to } x_2, t_2}} \phi[x(t)], \quad (6.179)$$

where the functional $\phi[x(t)]$ is defined by

$$\phi[x(t)] = \text{constant} \exp\left\{\frac{iS}{\hbar}[x(t)]\right\}. \tag{6.180}$$

In the classical limit the action $S[x(t)]$ is large compared to $\hbar$, and $\phi[x(t)]$ varies rapidly as a function of the phase $S[x(t)]/\hbar$. In this classical limit the terms in the sum over $\phi[x(t)]$ are nearly all canceled by rapid phase variations in adjacent paths. For arbitrary values of $S[x(t)]$, contributing paths are generally not restricted to those near $\underline{x}(t)$. In the quantum mechanical limit many paths contribute to $K(x_2, t_2; x_1, t_1)$.

For the wave function of a particle in an external field, a complete formal solution from quantum mechanics is

$$\psi(x,t) = \psi(x',t')K(x',t';x,t)dx', \tag{6.181}$$

where

$$K(x',t';x,t) = \sum_n \psi_{E_n}^*(x')\psi_{E_n}(x)e^{-iE_n(t-t')/\hbar}. \tag{6.182}$$

The arbitrary wave function $\psi(x,t)$ is expressed as a superposition

$$\psi(x,t) = \sum_n a_{E_n}(t)\psi_{E_n}(x), \tag{6.183}$$

where $\psi_{E_n}(x)$ are eigenfunctions of E and $|a_{E_n}(t)|^2$ is the probability that the particle will be found in state n at time t. Equations (6.179) and (6.180) can be rewritten as [Fey65]

$$K(x_2,t_2;x_1,t_1) = \int_{t_1}^{t_2} \exp\left\{\frac{iS}{\hbar}[x(t)]\right\} Dx(t) \tag{6.184}$$

Here $Dx(t)$ indicates that the integral is taken over all paths $x(t)$ from t_1 to t_2.

Consider a charge carrier of coordinate x interacting with an ensemble of n interaction centers located at positions defined by the n components of the vector $\mathbf{R}$. The quantum mechanical propagator for such a system is defined by $K(x_2, \mathbf{R}_2, t_2; x_1, \mathbf{R}_1, t_1)$; at time t_1, the system coordinates are $(x_1, \mathbf{R}_1)$, and at time t_2 they are $(x_2, \mathbf{R}_2)$. The propagator $K(x_2, \mathbf{R}_2, t_2; x_1, \mathbf{R}_1, t_1)$ may be written in terms of a Feynman path integral as follows:

$$K(x_2, \mathbf{R}_2, t_2; x_1, \mathbf{R}_1, t_1) = \int_{t_1}^{t_2}\int_{t_1}^{t_2} Dx(t)D\mathbf{R}(t)\exp\left\{\frac{iS}{\hbar}[x,\mathbf{R}]\right\}. \tag{6.185}$$

The time-dependent density operator for such a system is

$$\rho(t) = \exp\left(\frac{-iHt}{\hbar}\right)\rho(0)\exp\left(\frac{iHt}{\hbar}\right), \tag{6.186}$$

and the density matrix in coordinate representation may be written [Cal83]

$$\langle x, \mathbf{R} | \hat{\rho}(t) | y, \mathbf{Q} \rangle = \int dx' dy' d\mathbf{R}' d\mathbf{Q}' \left\langle x, \mathbf{R} \left| \exp\left(\frac{-iHt}{\hbar}\right) \right| x', \mathbf{R}' \right\rangle$$

$$\times \langle x', \mathbf{R}' | \hat{\rho}(0) | y', \mathbf{Q}' \rangle \left\langle y', \mathbf{Q}' \left| \exp\left(\frac{iHt}{\hbar}\right) \right| y, \mathbf{Q} \right\rangle, \quad (6.187)$$

where

$$\langle x, \mathbf{R} | \exp(-iHt) | x', \mathbf{R}' \rangle = K(x, \mathbf{R}, t; x', \mathbf{R}', 0)$$

$$= \iint Dx D\mathbf{R} \exp\left\{\frac{iS}{\hbar}[x, \mathbf{R}]\right\}, \quad (6.188a)$$

$$\langle y', \mathbf{Q}' | \exp(iHt) | y, \mathbf{Q} \rangle = K^*(y, \mathbf{Q}, t; y', \mathbf{Q}', 0)$$

$$= \iint Dy D\mathbf{Q} \exp\left\{\frac{-i}{\hbar}S[y, \mathbf{Q}]\right\}. \quad (6.188b)$$

The endpoints of these path integrals are $x = x(t)$, $x' = x(0)$, $y = y(t)$, $y' = y(0)$, $\mathbf{R} = \mathbf{R}(t)$, $\mathbf{R}' = \mathbf{R}(0)$, $\mathbf{Q} = \mathbf{Q}(t)$, and $\mathbf{Q}' = \mathbf{Q}(0)$.

6.4.6 Wigner Distribution Function

One of the earliest formulations of quantum transport was given by Wigner in 1932 [Wig32]. The Wigner function have been applied recently [Plo85] to describe the dynamical evolution of many-body quantum mechanical systems in a closed self-consistent set of moment equations. This approach has been extended [Str86] to include dissipative quantum phenomena [Cal83] associated with the interaction of a charge carrier with a heat bath comprised of an ensemble of harmonic oscillators. Here we discuss the basic of the Wigner distribution function.

For an n-particle system with a pure-state wave function $\psi(x_1, \ldots, x_n)$, Wigner made the assertion that the probability distribution function of the simultaneous values of $x_1, \ldots, x_n$ for the coordinates, and $p_1, \ldots, p_n$ for the momenta, is

$$P(x_1, \ldots, x_n; p_1, \ldots, p_n) = \left(\frac{1}{2\pi\hbar}\right)^n \int_{-\infty}^{\infty} \cdots \int_{-\infty}^{\infty} dy_1, \ldots, dy_n \psi^*\left(x_1 + \frac{y_1}{2}, \ldots, x_n + \frac{y_n}{2}\right)$$

$$\times \psi\left(x_1 - \frac{y_1}{2}, \ldots, x_n - \frac{y_n}{2}\right) e^{i(p_1 y_1 + \cdots + p_n y_n)/\hbar}. \quad (6.189)$$

This distribution function was assumed despite the well-known fact that both the position and momentum variables overspecify the quantum mechanical system. In the equation the distribution function P may take on negative values. Although the Wigner distribution function $P(x, p)$ cannot be considered a rigorous probability distribution function, meaningful probabilities have been extracted by taking appropriate moments [Str89].

For a single coordinate and momentum, the Wigner distribution function reduces to

$$P(x, p) = \frac{1}{2\pi\hbar} \int_{-\infty}^{\infty} dy \psi^*\left(\frac{x+y}{2}\right) \psi\left(\frac{x-y}{2}\right) \exp\left(\frac{ipy}{\hbar}\right), \qquad (6.190)$$

where ψ is the coordinate representation of the wave function. Wigner determined the dynamical behavior of $P(x, p)$ through evaluation of the expression

$$E\psi(x, t) = \left[\frac{p^2}{2m} + V(x)\right]\psi(x, t), \qquad (6.191)$$

or

$$i\hbar \frac{\partial}{\partial t} \psi(x, t) = \left[\frac{-\hbar^2}{2m} \frac{\partial^2}{\partial x^2} + V(x)\right]\psi(x, t).$$

Straightforward mathematical manipulation yields

$$\frac{\partial P(x, p)}{\partial t} = -\frac{p}{m} \frac{\partial P(x, p)}{\partial x} + \frac{1}{(2\pi\hbar)} \int_{-\infty}^{\infty} dy\, e^{ipy/\hbar} \frac{i}{\hbar}\left[V\left(x + \frac{y}{2}\right) - V\left(x - \frac{y}{2}\right)\right]$$
$$\times \psi^*\left(x + \frac{y}{2}\right)\psi\left(x - \frac{y}{2}\right), \qquad (6.192)$$

or

$$\frac{\partial P(x, p)}{\partial t} + \frac{p}{m} \frac{\partial P(x, p)}{\partial x} = \frac{i}{\pi\hbar^2} \int_{-\infty}^{\infty} dy [V(x + y) - V(x - y)]$$
$$\times \psi^*(x + y)\psi(x - y) e^{2ipy/\hbar} \qquad (6.193)$$

Expanding $V(x + y)$ in a Taylor series about x, we have

$$V(x + y) = \sum_{\lambda=0}^{\infty} \frac{y^\lambda}{\lambda!} \frac{\partial^\lambda V(x)}{\partial x^\lambda} = \sum_{\lambda=0}^{\infty} \frac{y^\lambda}{\lambda!} V^{(\lambda)}(x). \qquad (6.194)$$

It follows that

$$V(x + y) - V(x - y) = 2 \sum_{\substack{\lambda=1 \\ (\text{odd})}}^{\infty} \frac{y^\lambda}{\lambda!} V^{(\lambda)}(x). \qquad (6.195)$$

The dynamic behavior of $P(x, p)$ is

$$\frac{\partial P(x, p)}{\partial t} + \frac{p}{m} \frac{\partial P(x, p)}{\partial x} = \frac{2i}{\pi\hbar^2} \int_{-\infty}^{\infty} dy \sum_{\lambda=0}^{\infty} \frac{y^\lambda}{\lambda!} V^{(\lambda)}(x) \psi^*(x + y)\psi(x - y) e^{2ipy/\hbar}$$
$$= \sum_{\substack{\lambda=1 \\ (\text{odd})}}^{\infty} \frac{1}{\lambda!} \left(\frac{\hbar}{2i}\right)^{\lambda-1} \frac{\partial^\lambda V(x)}{\partial x^\lambda} \frac{\partial^\lambda P(x, p)}{\partial p^\lambda}. \qquad (6.196)$$

This is a Boltzmann-like transport equation. Through order λ^3, the Wigner distribution function $P(x, p)$ satisfies

$$\frac{\partial P(x,p)}{\partial t} + \frac{p}{m}\frac{\partial P(x,p)}{\partial x} - \frac{\partial V(x)}{\partial x}\frac{\partial P(x,p)}{\partial p} = \frac{1}{3!}\left(\frac{\hbar}{2i}\right)^2 \frac{\partial^3 V}{\partial x^3}\frac{\partial^3 P(x,p)}{\partial p^3} + \mathcal{O}(\hbar^4) \quad (6.197)$$

The left-hand side of this equation is in the form of the collisionless Boltzmann equation. At the right-hand side are the lowest-order quantum mechanical corrections to the Wigner–Boltzmann-like transport equation.

For the multidimensional case of an n particle, in n-dimensional configuration space, Wigner demonstrated that

$$\frac{\partial P(x_1,\ldots,x_n;p_1,\ldots,p_n)}{\partial t} + \sum_{k=1}^{n}\frac{p_k}{m_k}\frac{\partial P(x_1,\ldots,x_n;p_1,\ldots,p_n)}{\partial x_k}$$
$$= \sum \frac{\partial^{\lambda_1+\cdots+\lambda_n} V}{\partial x_1^{\lambda_1},\ldots,\partial x_n^{\lambda_n}} \frac{(\hbar/2i)^{\lambda_1+\cdots+\lambda_n-1}}{\lambda_1!,\ldots,\lambda_n!} \times \frac{\partial^{\lambda_1+\cdots+\lambda_n} P(x_1,\ldots,x_n;p_1,\ldots,p_n)}{\partial p_1^{\lambda_1},\ldots,\partial p_n^{\lambda_n}}, \quad (6.198)$$

where summation on the right-hand side is extended over all positive integer values of $\lambda_1,\ldots,\lambda_n$ for which the sum $\lambda_1+\lambda_2+\cdots+\lambda_n$ is odd.

Consider an observable $a(x,p)$ that depends on the position operator alone, on the momentum operator alone, or on an additive combination thereof. From the expressions for the coordinate-space and momentum-space integrals over the Wigner distribution function, it is evident that the expectation value $\langle A \rangle$ of the observable, $a(x,p)$

$$\langle A \rangle = \int_{-\infty}^{\infty}\int_{-\infty}^{\infty} a(x,p)P(x,p)dx\,dp. \quad (6.199)$$

6.4.7 Moment Equation Approach

A straightforward way of illustrating the basic quantum mechanical corrections to classical transport phenomena is provided by the moment equation [Str90]. These corrections are evident in the moment equations obtained through the density matrix, path-integral, and Wigner function techniques.

The use of moment equations to describe quantum transport in ultrasubmicron devices has been discussed previously by Iafrate et al. [Iaf81]. This section will summarize their major findings. Consider the Wigner distribution function that obeys the equation

$$\frac{\partial P(x,p)}{\partial t} + \frac{p}{m}\frac{\partial P(x,p)}{\partial x} - \sum_{\substack{\lambda=1 \\ (\text{odd})}}^{\infty} \frac{1}{\lambda!}\left(\frac{\hbar}{2i}\right)^{\lambda-1} \frac{\partial^\lambda V(x)}{\partial x^\lambda}\frac{\partial^\lambda P(x,p)}{\partial p^\lambda} = 0 \quad (6.200)$$

or

$$\frac{\partial P(x,p)}{\partial t} + \frac{p}{m}\frac{\partial P(x,p)}{\partial x} - \frac{2}{\hbar}\sum_{n=0}^{\infty}(-1)^n\frac{(\hbar/2)^{2n+1}}{(2n+1)!}\frac{\partial^{2n+1} V(x)}{\partial x^{2n+1}}\frac{\partial^{2n+1} P(x,p)}{\partial p^{2n+1}} = 0. \quad (6.200')$$

This equation may be written

$$\frac{\partial P(x,p)}{\partial t} + \frac{p}{m}\frac{\partial P(x,p)}{\partial x} + \Theta \cdot P(x,p) = 0, \qquad (6.201)$$

where

$$\Theta \cdot P(x,p) = -\frac{2}{\hbar}\left[\sin\frac{\hbar}{2}\left\{\frac{\partial^{(V)}}{\partial x}\frac{\partial^{(P)}}{\partial p}\right\}\right]V(x)P(x,p). \qquad (6.202)$$

Here $\partial^{(V)}/\partial x$ and $\partial^{(P)}/\partial p$ operate only on the potential $V(x)$ and the Wigner distribution function $P(x,p)$, respectively. By including a term to describe the time rate of change due to collisions of $P(x,p)$, the Wigner-Boltzmann equation may be written as

$$\frac{\partial P(x,p)}{\partial t} + \frac{p}{m}\frac{\partial P(x,p)}{\partial x} + \Theta \cdot P(x,p) = \left(\frac{\partial P(x,p)}{\partial t}\right)\bigg|_{\text{collisions}} \qquad (6.203)$$

Iafrate et al. [Iaf81] have found that when Eq. (6.203) is multiplied by a function of momentum $\chi(p)$ and integrated over all values of momentum, the resulting expression is

$$\frac{\partial \langle \chi \rangle}{\partial t} + \frac{1}{m}\frac{\partial}{\partial x}\langle \chi p \rangle - \sum_{n=0}^{\infty}\left(\frac{\hbar}{2i}\right)^{2n}\frac{1}{(2n+1)!}\frac{\partial^{2n+1}}{\partial x^{2n+1}}V(x)\int \chi(p)\frac{\partial^{2n+1}}{\partial p^{2n+1}}P(x,p)dp$$

$$= \left\langle \chi\left(\frac{\partial p(x,p)}{\partial t}\right)_{\text{collisions}}\right\rangle, \qquad (6.204)$$

where the momentum integration denoted by $\langle \rangle$. After it is integrated by parts, Eq. (6.204) becomes

$$\frac{\partial \langle \chi \rangle}{\partial t} + \frac{1}{m}\frac{\partial}{\partial x}\langle \chi p \rangle + \sum_{n=0}^{\infty}\left(\frac{\hbar}{2i}\right)^{2n}\frac{1}{(2n+1)!}\left(\frac{\partial^{2n+1}}{\partial x^{2n+1}}V(x)\right)\left\langle \frac{\partial^{2n+1}\chi}{\partial p^{2n+1}}\right\rangle$$

$$= \left\langle \chi\left(\frac{\partial P}{\partial t}\right)_{\text{collisions}}\right\rangle. \qquad (6.205)$$

For $\chi = 1$, Eq. (6.205) yields the continuity equation

$$\frac{\partial \rho}{\partial t} + \frac{1}{m}\frac{\partial}{\partial x}\langle p \rangle = 0, \qquad (6.206)$$

where $\rho = \psi^*\psi$. Eq. (6.206) reduces to the momentum equation:

$$\frac{\partial \langle p \rangle}{\partial t} + \frac{1}{m}\frac{\partial}{\partial x}\langle p^2 \rangle + \rho\frac{\partial V}{\partial x} = \left\langle p\left(\frac{\partial P(x,p)}{\partial t}\right)_{\text{collisions}}\right\rangle. \qquad (6.207)$$

These results may be reduced to

$$\langle p^n \rangle = \left(\frac{\hbar}{2i}\right)^n \sum_{j=0}^{n} (-1)^j \frac{n!}{(n-j)!j!} \frac{\partial^j \psi^*(x)}{\partial x^j} \frac{\partial^{n-j} \psi(x)}{\partial x^{n-j}}, \qquad (6.208)$$

and a wave function of the form

$$\psi(x,t) = A(x,t) \exp\left(\frac{iS(x,t)}{\hbar}\right) \qquad (6.209)$$

may be invoked to demonstrate that

$$\langle p^0 \rangle = \rho(x,t) = \rho, \qquad (6.210\text{a})$$

$$\langle p^1 \rangle = mv\rho, \qquad (6.210\text{b})$$

$$\langle p^2 \rangle = (mv)^2 \rho - \frac{\hbar}{4} \rho \frac{\partial^2}{\partial x^2} \ln \rho, \qquad (6.210\text{c})$$

$$\langle p^3 \rangle = (mv)^3 \rho - \frac{\hbar^2}{4} \rho \left\{ 3mv \frac{\partial^2}{\partial x^2} \ln \rho + \frac{\partial^2}{\partial x^2}(mv) \right\} \qquad (6.210\text{d})$$

The expressions for both $\langle p^2 \rangle$ and $\langle p^3 \rangle$ contain quantum mechanical corrections to moments of the momentum. Since there is no first-order terms in $\hbar$ in the dynamical equations for the Wigner distribution, the lowest-order quantum corrections in $\langle p^n \rangle$ are of the order $\hbar^2$.

6.5 THE MONTE CARLO METHOD

The Monte Carlo method is at present widely used in hot-electron studies. In this section we will describe the basic principles of this method. It consists in a simulation of the motion of one or more electrons inside a crystal, subject to the action of an external applied electric field and of given scattering mechanisms. Underlying the simulation is the generation of a sequence of random numbers with given distribution probabilities.

When the charge transport is analyzed on submicrometer scales, and when very high electric fields are involved, the conventional semiclassical theoretical approach of transport processes in terms of the Boltzmann equation can be replaced by a full quantum description. The quantum Monte Carlo approach is described in Section 6.5.5.

Consider the case of a cubic semiconductor with an applied electric field E. The simulation begins with the given initial conditions for wave vector $\mathbf{k}_0$. The duration of the first free flight is computed stochastically from a probability distribution determined by the scattering probabilities. During the free flight the external force is made to act according to

$$\frac{d\mathbf{k}}{dt} = \frac{e}{\hbar} \mathbf{E}. \qquad (6.211)$$

264 DEVICE-RELATED PHYSICS AND PRINCIPLES

The simulation of all quantities of interest (velocity, energy, etc.) are recorded. A scattering mechanism is then chosen as responsible for the end of the free flight, according to the relative probabilities of all possible scattering mechanisms. From the transition rate of this scattering mechanism a new **k** state after scattering is stochastically determined as the initial state of the new free flight, and the entire process is iteratively repeated [Reg85]. The results of the calculation become more and more precise as the simulation ends when the quantities of interest are known with desired precision.

Figure 6.28 shows the basic Monte Carlo technique. Figure 6.28(a) shows the actual electron path in two dimensions, under the influence of a large external field. It is composed of eight segments of a parabola corresponding to the eight free flights. Figure 6.28(b) shows the same eight events as heavy line segments in momentum space. These heavy lines are interfaced by light lines representing the changes of momentum in each scattering event. Figure 6.28(c) gives the velocity of the carrier averaged at the nth point over all previous $(n-1)$ paths [Reg86]. This average velocity approaches the drift velocity (dash-dotted line) when enough paths are taken; 100 paths are shown in Fig. 6.28(c). The drift velocity is a direct measure of mobility.

The Monte Carlo method permits physical information to be extracted from simulated experiments, and for this reason it is a powerful tool for the analysis of stationary [Jac83] or transient [Leb71; Böe90] transport effects in semiconductor.

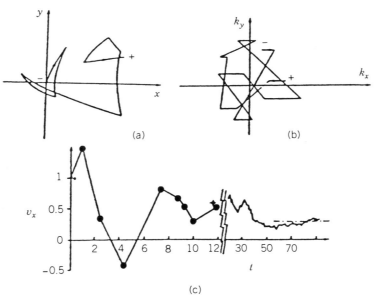

Fig. 6.28 Schematic for Monte Carlo method: (a) real space path of a carrier with a large field in the x-direction; (b) same path as in (a), in momentum space; (c) carrier velocity averaged over all steps starting from step 1 up to the running step number as a function of simulation time. [Brunetti, R., and C. Jacoboni, in *Hot Carriers in Semiconductors*, pp. 523–526, Pergamon Press (1987)]

6.5.1 The Initial Conditions of Motion

When successive scattering events of one electron are followed long enough, its behavior is equivalent to that of the average behavior of the entire ensemble, this is called *ergodicity*. In the steady state case under consideration, the time of simulation must be sufficiently long so that the initial conditions of the electron's motion do not influence the final result.

Choosing the length of time for the simulation is a compromise between the need for ergodicity ($t \to \infty$) and the need to save computer time. The longer the simulation time, the less the initial conditions will influence the average results. When the simulation is split into many subhistories, better convergence to steady state can be achieved by taking the initial state of each new subhistory as equal to the final state of the previous one. In this way only the initial condition of the first subhistory will influence the final results in a biased way.

On the other hand, when a simulation is used to study a transient phenomena or a transport process in a nonhomogeneous system, it is necessary to simulate many electrons separately. In this case the distribution of the initial electron states must be taken into account according to the particular physical situation under investigation.

6.5.2 Flight Duration and Self-scattering

The electron wave vector **k** changes continuously during a free flight because of the applied field of Eq. (6.211). We need to know the probability $P[\mathbf{k}(t)]dt$ that the electron in the state **k** suffers a collision during time dt. The probability that an electron is not scattered after a time t is

$$\exp\left\{-\sum_0^t P[\mathbf{k}(t')]dt'\right\}, \qquad (6.212)$$

where t' is duration of the free flight. The probability $\mathscr{P}(t)dt$ that the electron will suffer its next collision during dt around t is given by

$$\mathscr{P}(t)dt = P[\mathbf{k}(t)]\exp\left\{-\sum_0^t P[\mathbf{k}(t')]dt'\right\}dt \qquad (6.213)$$

Because of the complexity of the integral at the exponent in Eq. (6.213), it is not practical to start the generation of stochastic free flights from an evenly distributed random number r. Rees [Ree68; Ree69] has devised a very simple method to overcome this difficulty. If $\Gamma = 1/\tau_0$ is the maximum value of $P(\mathbf{k})$ in the region of interest in **k** space, a new fictitious "self-scattering" is introduced so that the total scattering probability, including the self-scattering, is constant and equal to Γ. If the carrier undergoes such a self-scattering, its state $\mathbf{k}'$ after the collision is taken to be equal to its state **k** before the collision so that the electron path continues unperturbed as if no scattering occurred at all. More generally, Γ is not less than the maximum value of $P(\mathbf{k})$.

With a constant $P(\mathbf{k}) = \tau_0^{-1}$, Eq. (6.212) reduces to

$$\mathscr{P}(t) = \frac{1}{\tau_0}\exp\left(\frac{-t}{\tau_0}\right), \qquad (6.214)$$

and random numbers r can be used very simply to generate stochastic free flights t_r through

$$t_r = -\tau_0 \ln(1-r) = -\tau_0 \ln(r), \qquad (6.215)$$

where we have made use of the fact that as r is uniformly distributed between 0 and 1, the same applies to $(1-r)$.

6.5.3 The Choice of the Scattering Mechanism

During free flight the electron's dynamics are governed by Eq. (6.211). At the end the electron's wave vector and energy are known, and each scattering rate λ_i (i refers to the ith mechanism) can be evaluated.

The probability of a self-scattering event is the complement of Γ of the sum of λ_i's. To choose a mechanism from all possible mechanisms given a random number r, one compares the product $r\Gamma$ with successive sums of λ_i's. The jth mechanism is chosen if j is such that the first of these successive sums that is greater than $r\Gamma$ is $\lambda_1 + \lambda_2 + \cdots + \lambda_j$.

If all real scattering events have been tried and none of them has been selected, which means that $r\Gamma > \lambda(E)$, then self-scattering occurs. In this procedure tracking the self-scattering event is the most time-consuming, since all λ_i's must be calculated. However, an expedient has been used to overcome this problem, and it is called *fast self-scattering* [Reg85].

Consider, for example, the hot-electron behavior in semiconductors of GaAs, whose energy momentum and structure ($E(k)$ versus $\pm k$) is shown in Fig. 6.29 [Boa78]. There are energy minima, located at the zone center and at the edges, which we will call *valleys*. There is one central valley with a small radius of curvature where electrons have a low effective mass equal to $0.067m_0$, and there is a shallow satellite valley at the equivalent (100) Brillouin zone edges. By symmetry, two other identical satellite valleys are located at the (010) and (001) edges as well. Electrons associated with these satellite valleys have an effective mass equal to $0.35m_0$. The conduction band has a valley structure, and at a zero field ($E = 0$) applied at room temperature or below, the free electrons are located in the bottom regions of the central valley. The valence band is well separated from the conduction band, so the transport of holes is not considered.

In a nondegenerate sample of GaAs, the electron density is quite small (10^{21} m^{-3}), so the Pauli exclusion principle does not have to be taken into account. Both types of valleys are parabolic. However, the central valley of GaAs is really not parabolic, and the equation becomes

$$E(\mathbf{k}) = \frac{\hbar^2 k^2}{2m^*} + \alpha k^4, \qquad (6.216)$$

where α is a constant.

If the applied electric field is zero, the electrons are in equilibrium with the vibrating solid atoms, whose mean energy is $(k_B T/1.6) \times 10^{19}$ eV where T is the absolute temperature of the semiconductor and k_B is Boltzmann's constant. For a material at 300 K the mean electron energy is therefore 0.03 eV. The application of an electric

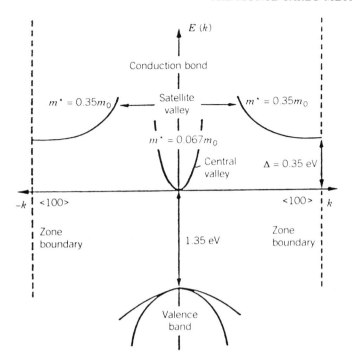

Fig. 6.29 Sketch of the band structure of GaAs. [Boardman, A. D., *Physics Programs*, ed. by A. D. Boardman, Wiley (1978)]

field causes the electrons to increase in energy until they acquire a temperature in excess of that solid. The electrons become hot and occupy the upper regions of the central valley. At sufficiently high electric fields the electrons may become hot enough to transfer from the central valley to the satellite valleys and switch from being light electrons to being heavy electrons. This transfer is achieved through the agency of phonons that are energy quanta of the lattice vibrations and is a scattering event in the history of the electron.

The intervalley and intravalley scattering events that leave the electron in the same valley, involve phonons that have an energy $\hbar\omega$, where ω is the frequency of a lattice vibration. In general, an electron that experiences a scattering event due to an interaction with a phonon field has an ante-scattering energy $E(\mathbf{k})$ and a post-scattering energy $E(\mathbf{k}')$, where $\mathbf{k}$ and $\mathbf{k}'$ are the initial and final $\mathbf{k}$ states such that $E(\mathbf{k}') - E(\mathbf{k}) \pm \hbar\omega = 0$. If the intravalley scattering occurs, via polar optical phonons, then to a good approximation the electron either emits or absorbes a phonon during the scattering event according to the following rules:

$$\text{Absorption:} \quad E(\mathbf{k}') = E(\mathbf{k}) + \hbar\omega_0,$$
$$\text{Emission:} \quad E(\mathbf{k}') = E(\mathbf{k}) - \hbar\omega_0, \quad (6.217)$$

where ω_0 is a constant frequency. If only acoustic phonons are involved, then to a good approximation the intravalley scattering can be assumed to be elastic with $E(\mathbf{k}') = E(\mathbf{k})$.

For GaAs the important intravalley scattering mechanism between states inside the central valley and between states inside a satellite valley is polar optical phonon scattering. Obviously acoustic phonon scattering also takes place within the central and satellite valleys, but it has a minor influence on the electron distribution.

If scattering between the central and satellite valley occurs, then the electrons must acquire at least an energy Δ before the transition becomes possible. However, in the satellite valley the electron energy is measured from its minimum, so it is necessary to add Δ to the electron energy if scattering from the satellite to the central valley occurs. Such nonequivalent intervalley transitions are also shown in Fig. 6.30 and are summarized as follows:

For central → satellite,

$$\text{Absorption:} \quad E_s(\mathbf{k}') = E_c(\mathbf{k}) - \Delta + \hbar\omega_n,$$
$$\text{Emission:} \quad E_s(\mathbf{k}') = E_c(\mathbf{k}) - \Delta - \hbar\omega_n.$$

(6.218a)

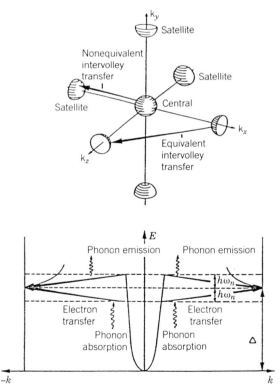

Fig. 6.30 Intervalley transitions on GaAs. For simplicity one-way transitions are shown. Obviously the reverse transitions also occur, with phonon emission or absorption changing to absorption or emission. [Boardman, A. D., *Physics Programs*, ed. by A. D. Boardman, Wiley (1978)]

For satellite → central,

$$\text{Absorption:} \quad E_c(\mathbf{k}') = E_s(\mathbf{k}) + \Delta + \hbar\omega_n,$$
$$\text{Emission:} \quad E_c(\mathbf{k}') = E_s(\mathbf{k}) + \Delta - \hbar\omega_n. \qquad (6.218b)$$

In the equations above ω_n is a constant frequency. The other transition that must be considered takes place between the satellite valleys and involves the emission or absorption of a phonon of energy $\hbar\omega_e$, where ω_e is a constant frequency. The energy relationship for satellite-to-satellite transitions has the same form as Eq. (6.217).

A complete list of electron–phonon scattering processes for plane wave states in parabolic valleys is given in Table 6.4. Each process is governed by an energy-conserving delta function condition $\delta(A)$ for which we write $E' = E(\mathbf{k}')$ and $E = E(\mathbf{k})$. Data for GaAs that are required to evaluate these formula are given in Table 6.5.

6.5.4 Time Average for the Collection of Results under Steady State Conditions

The time-average value of a quantity $A[\mathbf{k}(t)]$ (the drift velocity, the mean energy, etc.) during a single history of duration T is

$$\bar{A} = \frac{1}{T}\int_0^T A[\mathbf{k}(t)]dt = \frac{1}{T}\sum_i \int_0^{t_i} A[\mathbf{k}(t')]dt'. \qquad (6.219)$$

The bar over the A indicates the time average, and the integral over the whole simulation time T is separated into the sum of integrals over all flights of duration t_i. When a steady state is considered, T should be taken sufficiently long so that the estimator $\bar{A}$ in Eq. (6.219) gives an unbiased estimator of the average of the quantity A over the electron gas.

To obtain the electron distribution function, a mesh of $\mathbf{k}$ space (or of energy) is set up at the begining of the computer run [Reg86]. During the simulation the time spent by the sample electron in each cell of the mesh is recorded, and for large T this time conveniently normalized will represent the electron distribution function, which is the solution of the Boltzmann equation.

6.5.5 Quantum Monte Carlo Approach

New Monte Carlo techniques have been proposed for solving Boltzmann transport equations, including the ensemble Monte Carlo solution [Kiz87]. The original technique, which was developed for homogeneous systems in steady state condition, has been extended to study space- and time-dependent phenomena, to simulate high-speed devices, and to permit quantum corrections to the Monte Carlo solutions [Bru87].

When charge transport is analyzed on submicrometer scales, and when very high electric fields are involved, the conventional semiclassical theoretical approach to transport processes in the Boltzmann equation can be replaced by a full quantum description [Bru87]. The quantum Liouville equation for the density matrix of an electron gas in a semiconductor crystal offers an effective means of describing quantum transport features. The Liouville equation for the electronic density matrix can be

TABLE 6.4 Scattering Mechanisms in Gallium Arsenide

Scattering Mechanism	Transition Rate $(\hbar/2\pi)S(\mathbf{k}',\mathbf{k})$	Total Scattering Rate $\lambda(\mathbf{k}) \equiv \lambda(E)$	Definitions
Acoustic phonon for either absorption or emission (intravalley)	$\dfrac{\hbar D_a^2 \|\mathbf{k}-\mathbf{k}'\| N_a \delta(A)}{(2\rho s)(2\pi)^3}$ $N_a = \dfrac{k_A T}{\hbar s \|\mathbf{k}-\mathbf{k}'\|}$	$\dfrac{(2m_{c,s}^*)^{3/2} k_B T D_a^2 E^{1/2}}{4\pi \rho s^2 \hbar^4}$ absorption or emission $A = E' - E = 0$	Central or satellite valley effective mass $m_{c,s}^*$; density ρ; velocity of sound s; acoustic deformation potential, D_a; phonon occupation number N_a
Polar optical phonon (intravalley)	$\dfrac{2\pi e^2 \hbar \omega_0}{\|\mathbf{k}-\mathbf{k}'\|^2}\left(\dfrac{1}{\varepsilon_\infty}-\dfrac{1}{\varepsilon_0}\right)\dfrac{X}{4\pi\kappa_0 (2\pi)^3}$ $X = N_0 \delta(A_a)$ absorption $(N_0+1)\delta(A_c)$ emission $N_0 = [\exp(\hbar\omega_0/k_B T) - 1]^{-1}$	$\dfrac{Y e^2 m^{*1/2}\omega_0}{\sqrt{2}\hbar(4\pi\kappa_0)2 E^{1/2}}\left(\dfrac{1}{\varepsilon_\infty}-\dfrac{1}{\varepsilon_0}\right)\ln\left\|\dfrac{E^{1/2}+E'^{1/2}}{E^{1/2}-E'^{1/2}}\right\|$ $Y = N_0$ absorption (N_0+1) emission $A_{a,c} = E' - E \mp \hbar\omega_0 = 0$	High frequency and static dielectric constants ε_∞, ε_0; phonon occupation number N_0; permittivity of free space κ_0
Equivalent intervalley (satellite $\rightleftarrows$ satellite)	$\dfrac{(Z-1)\hbar D_c^2 X}{(2\pi)^3 (2\rho\omega_e)}$ $X = N_c \delta(A_a)$ absorption $(N_e+1)\delta(A_c)$ emission $N_c = [\exp(\hbar\omega_c/k_B T)-1]^{-1}$	$\dfrac{(Z-1)m_s^{*3/2} D_c^2 E^{1/2} Y}{\sqrt{2\pi}\rho\omega_e \hbar^3}$ $Y = N_c$ absorption (N_e+1) emission $A_{a,c} = E' - E \mp \hbar\omega_c$	Intervalley deformation potential, D_c; number of equivalent valleys, Z (for GaAs, $Z=3$); phonon occupation number N_c

Nonequivalent intervalley				Intervalley deformation potential D_n; phonon occupation number N_n
(a) central → satellite	(a)	$\dfrac{3\hbar D_n^2 X}{2\rho\omega_n(2\pi)^3}$	absorption emission	
		$X = N_n \delta(A_a)$ $(N_n+1)\delta(A_c)$		
		$\dfrac{3m_s^{*3/2} D_n^2 E^{1/2} Y}{\sqrt{2\pi\rho\omega_n \hbar^3}}$	absorption emission	
		$Y = N_n$ (N_n+1)		
		$A_{a,c} = E' - E + \Delta \mp \hbar\omega_n = 0$		
(b) satellite → central	(b)	$\dfrac{\hbar D_n^2 X}{2\rho\omega_n(2\pi)^3}$	absorption emission	
		$X = N_n \delta(A_a)$ $(N_n+1)\delta(A_c)$		
		$\dfrac{m_c^{*3/2} D_n^2 E^{1/2} Y}{\sqrt{2\pi\rho\omega_n \hbar^3}}$	absorption emission	
		$Y = N_n$ (N_n+1)		
		$A_{a,c} = E' - E - \Delta \mp \hbar\omega_n = 0$		
	$N_n = [\exp(\hbar\omega_n/k_B T) - 1]^{-1}$			

Source: Boardman, A. D., *Physics Programs*, ed. by A. D. Boardman, Wiley (1978).

TABLE 6.5 Data for Gallium Arsenide

Parameter	Value
Density	5.37 g/cm^3
Static dielectric constant (ε_0)	12.53
Polar optical phonon frequency (ω_0)	5.37×10^{13} rad·S^{-1}
Equivalent intervalley phonon frequency (ω_e)	4.54×10^{13} rad·S^{-1}
Nonequivalent intervalley phonon frequency (ω_n)	4.54×10^{13} rad·S^{-1}
Equivalent intervalley deformation potential (D_e)	10^9 eV/cm
Nonequivalent intervalley deformation potential (D_n)	10^9 eV/cm
Central valley effective mass (m_c^*)	0.067 m_0
Satellite valley effective mass (m_s^*)	0.35 m_0
Valley separation	0.36 eV

Note: If SI units are used to calculate $\lambda_n(E)$, then care must be exercised with quantities measured in eV/cm.
Source: Boardman, A. D., *Physics Programs*, ed. by A. D. Boardman, Wiley (1978).

solved using a numerical procedure developed by Brunetti and Jacobini [Bru86] that generates quantum interactions the way the classical Monte Carlo technique generates classical scattering events.

Consider an ensemble of electrons interacting with a phonon gas and subject to constant, uniform, arbitrarily high electric fields. We include in the hamiltonian the free-electron and free-phonon dynamics, electron interactions which the external field, and the coupling hamiltonian between electrons and optical phonons in the deformation potential approximation. Assume that both the electric field and the electron–phonon interaction are activated at $t = 0$, which is the equilibrium condition of the two subsystems. The phonon population is always assumed to be at equilibrium.

With an appropriate choice of base functions [Bru87], the Liouville equation is transformed into an integral equation, which, by iteration, takes the form of a perturbative series:

$$\rho(x, x', t) = \rho(x, x', 0) + \sum_n \left(\frac{1}{i\hbar}\right)^n \int_0^t dt_1 \int_0^{t_1} dt_2$$
$$\cdots \int_0^{t_n} dt_n [H_{ep}(t_1), [H_{ep}(t_2), \ldots, [H_{ep}(t_n), \rho(0)]\ldots](x, x'), \quad (6.220)$$

where x and x' are the base vectors.

With the zero-order approximation we obtain the effect of ballistic translation due to an electric field in time interval $(0, t)$. If we are interested in the evaluation of the diagonal elements of the density matrix, only even-order perturbative corrections contribute to the expansion in Eq. (6.220). To reduce the problem to the electronic part of the density matrix, a trace operation is performed over the phonon coordinates. The new equation retains the full many-body dynamics, since the trace does not commute with the interactive hamiltonian.

Through random selections of the different terms associated with each perturbative correction, we have a numerical procedure that permits an estimate of the solution to Eq. (6.220). These random terms can be represented by Feynman diagrams. For any order each term can be described through four basic processes: real and virtual absorptions, and real and virtual emissions. Each process includes two vertices, which indicate the beginning and the end of a transition. A real process is such that at its end a phonon is emitted or absorbed by an electron, while in a virtual process, which involves only one index of the density matrix, the final state is equal to the initial state. During each quantum interaction we keep track of the phonon involved, and the excess phonons are assumed to dissipate via interaction with the thermal bath at the end of each quantum process. In this section we consider first two perturbative corrections: second order and fourth order.

Second-Order Corrections (One Scattering) The second-order perturbative correction in the interactive hamiltonian involves only one of the processes discussed above. The explicit form of the corresponding integrals is

$$\int_0^t dt_1 \int_0^{t_1} dt_2 e^{\{ib(t_1 - t_2) + a(t_1^2 - t_2^2)\}},$$

$$a = \frac{e}{2m^*}(k_f - k_i) \cdot E,$$

$$b = \tfrac{1}{2}(\omega_f - \omega_i - \omega_q). \tag{6.221}$$

In Eq. (6.221), i and f refer to the interacting initial and final electron states and ω_q is the phonon frequency. This simple expression of the second-order correction after integration is

$$\frac{\pi}{4a}\left\{\left[C\left(\frac{at+b}{a}\right) - C\left(\frac{b}{a}\right)\right]^2 + \left[S\left(\frac{at+b}{a}\right) - S\left(\frac{b}{a}\right)\right]^2\right\}(4). \tag{6.222}$$

The coefficient a accounts for the effect of the field during the finite duration of the collision, while b is related to the energy of the quantum states involved in the transition.

Figure 6.31 compares the absolute value of the second-order correction as given in Eq. (6.222) with the same contribution without intracollisional field effect (ICFE). Also shown in the figure is the corresponding classical contribution to the perturbative expansion of the classical integral equation for the distribution function obtained from the Boltzmann equation. The classical correction concerns "one-scattering" trajectories.

If the ICFE is neglected, a higher effect of the "quantum one-collision" trajectories is obtained. By changing the energy of the carrier during the collision, the ICFE reduces the efficiency of the scattering because it reduces the time of positive interference that occurs when the energy difference between initial and final states is equal to the phonon energy [Bar78]. The classical contribution is higher because the field is so high that even at a short time duration the electrons can reach enough energy to activate classical instantaneous phonon emission.

Fourth-Order Corrections (Two Scatterings) An improvement in efficiency can be obtained also at higher orders by integrating over time every other vertex between the two adjacent vertices. The result is expressed in terms of Fresnel integrals. We

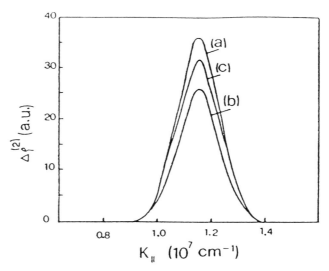

Fig. 6.31 Absolute value of quantum corrections at the second perturbative order compared with the absolute value of the classical one-scattering correction for the model semiconductor: (a) classical one-scattering contribution; (b) quantum correction of the second perturbative order; (c) the same as in (b) but without ICFE. [Brunetti, R., and C. Jacoboni, in *Hot Carriers in Semiconductors*, pp. 523–526, Pergamon Press (1987)]

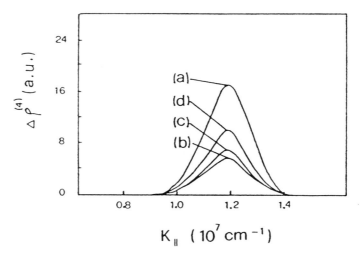

Fig. 6.32 Quantum corrections at the fourth perturbative order compared with the classical two-scattering correction for the model semiconductor: (a) classical two-scattering contribution; (b) quantum correction of the fourth perturbative order; (c) the same as in (b) with separate collisions; (d) the same as in (c) but without ICFE. [Brunetti, R., and C. Jacoboni, in *Hot Carriers in Semiconductors*, pp. 523–526, Pergamon Press (1987)]

can neglect the field effect in the interference exponentials and include the ICFE in the two-scattering trajectories. The integration of one vertex would lead to a function such as $\sin(bt)/b$ with which is obtained the delta of energy conservation for larger times.

Figure 6.32 shows the quantum corrections of the fourth perturbative order compared with the classical two-scattering contribution for the same model in Fig. 6.31. The quantum result is lower than the classical term for the second-order contribution. For separate collisions the correction term would be somewhat higher. Two separate collisions allow for a higher contribution of transitions that do not conserve energy. If the ICFE is turned off, the whole curve is higher and closer to the classical case.

From the preceding analysis we observe that multiple collisions can make an appreciable contribution to the transport picture when the classical collisions occur at time intervals on the order of 10^{-14} s and that the ICFE becomes effective at fields on the order of $100 \, \text{kV/cm}$.

REFERENCES

[Ree88] Reed, M. A., J. N. Randall, R. J. Aggarwal, R. J. Matyi, T. M. Moore, and A. E. Wetsel, *Phys. Rev. Lett.* **60** (1988): 535.
[Sak80] Sakaki, H., *Jap. J. Appl. Phys.* **19** (1980): 94.
[Pet82] Petroff, P. M., A. G. Gossard, R. A. Logan, and W. Wiegman, *Appl. Phys. Lett.* **41** (1982): 635.
[Ara82] Arakawa, Y., and H. Sakaki, *Appl. Phys. Lett.* **40** (1982): 939.
[Kel87] Kelly, M. J., in *Physics and Technology of Submicron Structures*, pp. 174–196, ed. by G. Bauer, F. Kuchar, and H. Heinrich, Springer-Verlag (1987).
[Car77] Caruthes, E., and P. J. Lin-Chung, *Phys. Rev. Lett.* **39** (1977): 1543.
[Gel86] Gell, M. A., D. Ninno, M. Jaros, and D. C. Herbert, *Phys. Rev.* B **34** (1986): 2416.
[Sch77] Schulman, J. N., and T. C. McGill, *Phys. Rev. Lett.* **39** (1977): 1680.
[Sch81] Schulman, J. N., and Y. C. Chang, *Phys. Rev.* B **24** (1981): 4445.
[Kri85] Krishnamurthy, S., and J. A. Moriarty, *Phys. Rev.* B **32** (1985): 1027.
[Sai77] Sai Halasz, G. A., R. Tsu, and L. Esaki, *Appl. Phys. Lett.* **30** (1977): 651.
[Bas84] Bastard, G., *Phys. Rev.* B **25** (1982): 7584.
[Mai84] Mailhiot, C., D. L. Smith, and T. C. McGill, *J. Vac. Sci. Tech.* B **2** (1984): 371.
[Smi86] Smith, D. L., and C. Maihiot, *Phys. Rev.* B **33** (1986): 8345.
[Fas84] Fasolino, A., and M. Altarelli, in *Two-Dimensional Systems, Heterojunctions, and Superlattices*, ed. by G. Bauer, F. Kucher, and H. Heinrich, Springer-Verlag (1984).
[Alt85] Altarelli, M., *Phys. Rev.* B **32** (1985): 5138.
[San85] Sanders, G. D., and Y. C. Chang, *Phys. Rev.* B **32** (1985): 5517.
[San87] Sanders, G. D., and Y. C. Chang, *Phys. Rev.* B **35** (1987): 1300.
[Car77] Caruthes, E., and P. J. C. Lin-Chung, *Phys. Rev. Lett.* **39** (1977): 1543.
[Esa70] Esaki, L., and R. Tsu, *IBM J. Res. Dev.* **14** (1970): 61.
[Smi86] Smith, D. L., and C. Maihiot, *Phys. Rev.* B **33** (1986): 8345.
[Alt85] Altarelli, M., *Phys. Rev.* B **32** (1985): 5138.
[Alt86] Altarelli, M., *Heterojunctions and Semiconductor Superlattices*, ed. by G. Allan, G. Bastard, N. Boccara, M. Lannoo, and M. Voos, Springer-Verlag (1986).

[Cha88] Chang, Y. C., *Phys. Rev.* B **37**, 14 (1988): 8215–8222.
[Böe90] Böer, K. W., *Survey of Semiconductor Physics*, Van Nostrand Reinhold (1990).
[Wan89] Wang, S., *Fundamentals of Semiconductor Theory and Device Physics*, Prentice Hall (1989).
[Sch83] Schulman, J. N., and Y. C. Chang, *Phys. Rev.* B **27**, 4 (1983): 2346–2354.
[Har61] Harrison, W. A., *Phys. Rev.* **123** (1961): 85.
[Ben66] Ben Daniel, D. J., and C. B. Duke, *Phys. Rev.* **152** (1966): 683.
[Kan66] Kane, E. O., in *Semiconductors and Semimetals* vol. 1, p. 75, ed. by R. K. Willardson and A. C. Beer, Academic Press (1966).
[Kan56] Kane, E. O., *J. Phys. Chem. Solids*, **1** (1956): 83.
[Hen63] Hensel, J. C., and G. Feher, *Phys. Rev.* **129** (1963): 1041.
[Lut56] Luttinger, J. M., and W. Kohn, *Phys. Rev.* **97** (1956): 896.
[Car77] Caruthes, E., and P. J. Lin-Chung, *Phys. Rev. Lett.* **39** (1977): 1543.
[Gel86] Gell, M. A., D. Ninno, M. Jaros, and D. C. Herbert, *Phys. Rev.* B **34** (1986): 2416.
[Sho51] Shockley, W., *Bell Sys. Tech. J.* **30** (1951): 990.
[Lan82] Landolt-Bornstein, New Series, III.17a and b, ed. by O. Madlung, M. Schulz, and H. Weiss, Springer-Verlag (1982).
[Mil82] Miller, D. A. B., D. S. Chemla, D. J. Eilenberger, P. W. Smith, A. C. Gossard, and W. T. Tsang, *Appl. Phys. Lett.* **41** (1982): 679.
[Dev87] Devaud, B., J. Shah, and T. C. Damen, *Phys. Rev. Lett.* **58**, 24 (1987): 2582.
[Has86] Hase, I., H. Kawai, K. Kaneko, and N. Watanabe, *J. Appl. Phys.* **59** (1986): 3792.
[Sol84] Sollner, T. C. L. G., P. E. Tannenwald, D. D. Peck, and W. D. Goodhue, *Appl. Phys. Lett.* **45** (1984): 1319.
[Hay85] Hayes, J. R., A. F. J. Levi, and W. Wiegmann, *Phys. Rev. Lett.* **54** (1985): 1570.
[Dav85] Davies, R. A., M. J. Kelly, and T. M. Kerr, *Phys. Rev. Lett.* **55** (1985): 1114.
[Esa85] Esaki, L., in *The Technology and Physics of Molecular Beam Epitaxy*, p. 143, ed. by E. M. C. Parker, Plenum Press (1985).
[Eng87] England, P., J. R. Hayes, J. P. Harbison, and D. M. Hwang, in *Hot Carriers in Semiconductors*, ed. by J. Shah, and G. J. Iafrate, Pergamon Press (1987).
[Esa70] Esaki, L., and R. Tsu, *IBM J. Res. Dev.* **14** (1970): 61–65.
[Cap86] Capasso, F., K. Mohammed, and A. Y. Cho, *J. Quantum Electronics*, vol. QE-22, No. 9 (1986).
[Esa74] Esaki, L., and L. L. Chang, *Phys. Rev. Lett.* **33** (1974): 495–498.
[Cap85] Capasso, F., K. Mohammed, A. Y. Cho, R. Hull, and A. L. Hutchinson, *Phys. Rev. Lett.* **55** (1985): 1152–1155.
[Cap85b] Capasso, F., K. Mohammed, A. Y. Cho, R. Hull, and A. L. Hutchinson, *Appl. Phys. Lett.* **47** (1985): 420–422.
[Doh75] Dohler, G. H., R. Tsu, and L. Esaki, *Solid State Commun.* **17** (1975): 317–320.
[And58] Anderson, P. W., *Phys. Rev.* **109** (1958): 1492–1505.
[Cal84] Calecki, D., J. F. Palmier, and A. Chomette, *J. Phys.* C **17** (1984): 5027–5030.
[Cap84] Capasso, F., H. M. Cox, A. L. Hutchingson, N. A. Olsson, and D. A. Hummel, *Appl. Phys. Lett.* **45** (1984): 1193.
[Wei86] Weil, T., and B. Vinter, *J. Appl. Phys.* **60** (1986): 3227.
[Nag80] Nag, B. R., *Electron Transport in Compound Semiconductors*, Springer-Verlag (1980).
[See85] Seeger, K., *Semiconductor Physics*, 3d ed., Springer-Verlag (1985).
[Hin88] Hinckley, J. M., and J. Singh, *Appl. Phys. Letts.*, **53** (9) (1988): 75.

[Sin93] Singh, J., *Physics of Semiconductors and Heterostructures*, p. 453, McGraw Hill, p. 508, 511 (1983).

[Hes88] Hess, K., *Advanced Theory of Semiconductor Devices*, Prentice Hall (1988).

[Dat89a] Datta, S., *Quantum Phenomena*, Addison-Wesley (1989).

[Shu87] Shur, M., *GaAs Devices and Circuits*, Plenum Press (1987).

[Shi81] Shichijo, H., and K. Hess, *Phys. Rev.* B **23** (1981): 4197–4207.

[Rod75] Rode, D. L., in *Semiconductors and Semimetals*, vol. 10, pp. 1–89, ed. by R. K. Willardson and A. C. Beer, Academic Press (1975).

[Fer81] Ferry, D. K., K. Hess, and P. Vogl, in *VLSI Electronics*, vol. 2, p. 72, ed. by N. Einspruch, Academic Press (1981).

[Hes79] Hess, K., *Appl. Phys. Lett.* **35** (1979): 484–486.

[Fer50] Fermi, E., *Nuclear Physics*, p. 142, University of Chicago Press (1950).

[Bub74] Bube, R. H., *Electronic Properties of Crystalline Solids*, Academic Press (1974).

[See73] Seeger, K., *Semiconductor Physics*, Springer-Verlag (1973).

[Str70] Streitwolf, H. W., *Phys. Stat. Sol.* **37** (1970): K47.

[Böe90] Böer, K. W., *Survey of Semiconductor Physics*, Van Nostrand Reinhold (1990).

[Alb71] Alberigi-Quaranta, A., C. Jacoboni, and G. Ottaviani, *Nuovo Cimento* **1** (1971): 445.

[Lun90] Lundstrom, M., and S. Datta, *IEEE Circuits and Devices* **6**, 4 (1990).

[Frö47] Fröhlich, H., *Proc. Roy. Soc.* A **188** (1947): 521, 532.

[Reg85] Reggiani, L., in *Topics in Applied Physics*, vol. 58: *Hot-Electron Transport in Semiconductors*, pp. 7–83, ed. by L. Reggiani, Springer-Verlag (1985).

[Gan79] Gantsevich, S. V., V. L. Gurevich, and R. Katilius, *Rivista Nuovo Cimento* **2**, 1 (1979).

[Dud79] Duderstadt, J. J., and W. R. Martin, *Transport Theory*, Wiley (1979).

[Van82] Van Beijern, H., *Rev. Mod. Phys.* **54** (1982): 195.

[Zwa64] Zwanzig, R., *Phys. Rev.* **133** (1964): A50.

[Cha52] Chambers, R. G., *Proc. Phys. Soc. London* A **65** (1952): 458.

[Jac81] Jacoboni, C., L. Reggiani, and R. Brunetti, *Proc. 3d Int. Conf. in Hot Carriers in Semiconductors* Pergamon Press (1981).

[Jac74] Jacoboni, C., *Phys. Stat. Sol.* **65**(b) (1974): 61.

[Hes88] Hess, K. *Advanced Theory of Semiconductor Devices*, Prentice Hall (1988).

[Phi82] Philippidis, C., D. Bohm, and R. D. Kaye, *Nuovo Cimento Soc. Ital. Fis.* B **71**, 11 (1982): 75.

[Iaf81] Iafrate, G. L., H. L. Grubin, and D. K. Ferry, *J. Phys.* **42** (1981): C7-307.

[Gru89] Grubin, H. L., in *Introduction to Semiconductor Technology, GaAs and Related Compounds*, ed. by C. T. Wang, Wiley-Interscience (1989).

[Cal83] Caldeira, A. O., and A. J. Leggett, *Physica* A **121** (1983): 587.

[Cal84] Caldeira, A. O., *Ann. Phys.* I **53** (1984): 445(E).

[Fre85] Frensley, W. R., *J. Vac. Sci. Tech.* B **2**, 3 (1985): 1261.

[Str89] Stroscio, M. A., in *Introduction to Semiconductor Technology, GaAs and Related Compounds*, ed. by C. T. Wang, Wiley-Interscience (1989).

[Str89] Stroscio, M. A., in *Introduction to Semiconductor Technology*, ed. by C. T. Wang, Wiley-Interscience (1989).

[Fey48] Feynman, R. P. *Rev. Mod. Phys.* **20** (1948): 367.

[Fey63] Feynman, R. P., and F. L. Vernon, Jr., *Ann. Phys.* **24** (1963): 118.

[Fey64] Feynman, R. P., *The Feynman's Lectures on Physics*, vol. 1, ch. 19, Addison-Wesley (1964).

[Fey65] Feynman, R. P., and A. R. Hibbs, *Quantum Mechanics and Path Integrals*, McGraw-Hill (1965).
[Wig32] Wigner, E. P., *Phys. Rev.* **40** (1932): 749.
[Plo85] Ploszajczak, M., and M. J. Rhoades-Brown, *Phys. Rev. Lett.* **55** (1985): 147.
[Str86] Stroscio, M. A., *Superlattices Microstrut.* **2** (1986): 45.
[Cal83] Caldeira, A. O., and A. J. Leggett, *Physica* A **121** (1983): 587.
[Iaf81] Iafrate, G. J., H. L. Grubin, and D. K. Ferry, *J. Phys. C.* **7** (1981).
[Reg85] Reggiani, L., General theory, in *Hot Electron Transport in Semiconductors*, p. 11, ed. by L. Reggiani, Springer-Verlag (1985).
[Jac83] Jacoboni, C., and L. Reggiani, *Rev. Mod. Phys.* **55** (1983): 645.
[Leb71] Lebwohl, P. A., and P. J. Price, *Solid State Commun.* **9** (1971): 1221.
[Böe90] Böer, K. W., *Survey of Semiconductor Physics*, Van Nostrand Reinhold (1990).
[Ree68] Rees, H. D., *Phys. Lett.* A **26** (1968): 416.
[Ree69] Rees, H. D., *J. Phys. Chem. Solids* **30** (1969): 643.
[Boa78] Boardman, A. D., *Physics Programs*, ed. by A. D. Boardman, Wiley (1978).
[Kiz87] Kizilyalli, I., K. Hess, T. Higman, M. Emanuel, and J. Coleman, *Hot Carriers in Semiconductors* Pergamon Press (1987).
[Bru87] Brunetti, R., and C. Jacoboni, in *Hot Carriers in Semiconductors*, pp. 523–526 Pergamon Press (1987).
[Bru86] Brunetti, R., C. Jacobini, P. Lugli, and L. Reggiani, *Proc. Int. Conf. on the Physics of Semiconductors*, p. 157, Stockholm, Sweden, World Publishing (1986).
[Bar78] Barker, J. R., *Solid State Electron.* **21** (1978): 267.

7

METAL-TO-GaAs CONTACTS

7.1 ELECTRICAL PROPERTIES OF METAL–SEMICONDUCTOR CONTACTS

Metallization systems are a fundamental component of all semiconductor devices and integrated circuits. They provide the electrical connection between the active regions of the semiconductor and other circuit elements of an IC and also serve as the link to the external circuit, by interconnecting with functional electrodes and bonding pads [Wel88]. Two types of metal–semiconductor contacts have been identified by researchers. These are ohmic (low resistance) and Schottky (rectifying) type contacts [Hen84].

The series resistance at the metal–semiconductor interface of the ohmic contact is very small and can be neglected during device operation. The Schottky barrier contact has a high-resistance region devoid of mobile carriers. This is the depletion region, and the contact exhibits rectification properties that are in some respects similar to those of a p–n junction diode. The choice of metal for a second-level interconnection is dictated by the desire for high conductivity, excellent bondability, adherence to oxides and nitrides, and compatibility with the ohmic and Schottky metal systems. This chapter will discuss the general electrical properties of the metal–GaAs systems, Schottky barrier metals, and refractory silicide metallization; then for ohmic contacts to n^+-GaAs and p-GaAs alloyed and nonalloyed ohmic contacts, interconnect design considerations, measurement of contact resistance r_c, interconnect metal systems, and surface passivation of compound semiconductors.

Classification of Metal–Semiconductor Contacts

Based on the current-voltage characteristics, contacts on semiconductors can be classified into three generally different groups [Hen84]:

280 METAL-TO-GaAs CONTACTS

1. *Blocking contacts.* As interpreted here, these contacts do not permit any current to pass.
2. *Semipermeable contacts.* The current in this case is allowed to pass, but not without localized high resistance. Contacts with rectifying characteristics are called *Schottky barriers.*
3. *Low resistance contacts.* Present is an "ohmic" impediment to carrier flow which is negligible in comparison with other impediments.

The chapter will concern itself with contacts (2) and (3); type (1) will not be discussed.

7.2 THE PHYSICS OF METAL–GaAs SYSTEMS

7.2.1 Classical Models of the Interface

Schottky Model

A Schottky barrier can be described as two infinite half-planes of material, one a metal and the other a semiconductor, brought into contact. Figure 7.1(a) gives the energy band diagrams for a metal and an *n*-type semiconductor separated by a large distance. The work function of the metal ϕ_m is the energy needed to remove an electron from the Fermi level of the metal (E_F) to the vacuum level. The vacuum level is the energy level of an electron just outside the metal with zero kinetic energy. The work function ϕ_m has a volume contribution due to the periodic potential of the crystal lattice and a surface contribution due to the possible existence of a dipole layer at the interface [Tya84]. The work function ϕ_s of the semiconductor is defined similarly and is a variable quantity because of the doping concentration. An important surface parameter that does not depend on doping is the electron affinity χ_s defined

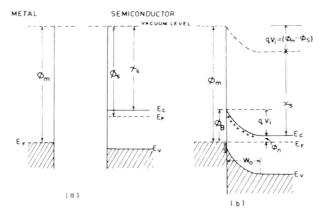

Fig. 7.1 Electron energy band diagrams of metal contact to *n*-type semiconductor with $\phi_m > \phi_s$: (a) neutral materials separated from each other; (b) thermal equilibrium situation after the contact has been made. [Tyagi, M. S., in *Metal-Semiconductor Schottky Barrier Junctions and Their Applications*, ed. by B. L. Sharma, Plenum Press (1984); with permission from the publisher]

as the energy difference of an electron between the vacuum level and the lower edge of the conduction band. The quantity ϕ_n is the energy difference between E_n and the Fermi level. We have the following relation:

$$\phi_s = \chi_s + \phi_n. \qquad (7.1)$$

Figure 7.1(b) shows the energy-band diagram after contact is made and equilibrium has been reached. Electrons in the semiconductor have higher energy than the metal electrons, and thus they flow into the metal until the Fermi level on both sides coincides. The electrons move out of the semiconductor and accumulate on the surface of the metal. The positive donor ions form near the boundary region. This results in a dipolar electric field which opposes further electron flow. As shown in Fig. 7.1(b), if the semiconductor is uniformly doped, the charge density is uniform to a depth w_0, called the *depletion width*. The energy bands in the semiconductor will bend upward. Since the band gap of the semiconductor is not changed by making contact with the metal, the valence band edge E_v will move up parallel to the conduction band edge E_c. The electron affinity of the semiconductor is unchanged, and hence the vacuum level in the semiconductor will follow the same variations as E_c. The amount of band bending is equal to the difference between the vacuum levels, which is equal to the difference between the two work functions. The difference is given by $qV_i = (\phi_m - \phi_i)$, where V_i is expressed in volts and is known as the built-in potential of the junction. The barrier height ϕ_B is given by

$$\phi_B = (\phi_m - \chi_s). \qquad (7.2)$$

Since

$$\phi_s = \chi_s + \phi_n,$$

we have

$$\phi_B = (qV_i + \phi_n), \qquad (7.3)$$

where $\phi_n = (E_c - E_F)$ represents the penetration of the Fermi level in the band gap of the semiconductor and q is the electronic charge. The description of a metal–semiconductor contact was first presented by Schottky [Sch38], so Eq. (7.2) is referred to as the *Schottky limit*. Equation (7.2) states that the barrier ϕ_B is directly proportional to the metal work function ϕ_m. However, Schottky diodes formed on many of the III–V compound semiconductors do not show this behavior. In particular, Eq. (7.2) shows that the barrier height ϕ_B increases linearly with the metal work function ϕ_m. Strong dependence of barrier height on ϕ_m is observed only in predominantly ionic semiconductors. In many covalent semiconductors the barrier height is a less sensitive function of ϕ_m than given by Eq. (7.1), or it is almost independent of ϕ_m.

Bardeen Model

In 1947 Bardeen [Bar47] proposed that if the localized surface states existed at the metal–semiconductor interface in sufficient numbers, the barrier height (ϕ_B) would be insensitive to the metal work function (ϕ_m). Surface states are electronic states localized at the surfaces of semiconductor crystals. They are produced by an interruption in the periodicity of a crystals lattice. In a covalent crystal the surface atoms have neighbors only on the semiconductor's side; on the vacuum's side there are no neighbors with whom the surface atoms can make covalent bonds. Each of the surface atoms has one broken covalent bond. The broken covalent bonds are known as *dangling bonds*.

Dangling bonds give rise to localized energy states at the surface of the semiconductor, with energy levels lying in the forbidden gap. These surface states are usually continuously distributed in the band gap and are characterized by a neutral level ϕ_0. There is no band bending in the semiconductor when the states are occupied by electrons up to ϕ_0. The states below ϕ_0 are donorlike because they are neutral when occupied and are positive when empty. The states above ϕ_0 behave as acceptors. The band bending in the Schottky model was due to entirely to the difference between ϕ_s and ϕ_m. However, if an interface layer is present, the potential difference ($\phi_m - \phi_s$) will appear entirely across the interfacial layer, for the charge in the surface states will fully accommodate the necessary potential difference. There is no change in the charge within the depletion region of the semiconductor when a metal is brought into contact with the semiconductor. Hence ϕ_B is independent of ϕ_m and is given by

$$\phi_B = E_g - \phi_0. \tag{7.4}$$

In this case the barrier height is said to be "pinned" by surface states. Equation (7.4) will be referred to as the *Bardeen limit*.

General Case

A general analysis of metal–semiconductor contact involving surface states and an interfacial layer was first made by Cowley and Sze [Cow65]. Consider the metal–semiconductor interface shown in Fig. 7.2. The semiconductor is *n*-type and has

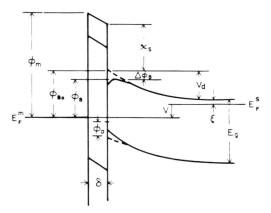

Fig. 7.2 Band diagram for a nonideal metal–semiconductor interface under bias V. An insulating layer of thickness δ exists between the metal and semiconductor and surface states are filled to the level ϕ_0. [Robinson, G. Y., in *Physics and Chemistry of III–V Compound Semiconductor Interfaces*, ed. by C. W. Wilmsen, Plenum Press (1985); with permission from the publisher]

permittivity ε_s, the interfacial layer has thickness δ and permittivity ε_i, and the surface states are characterized by the density N_{ss} (per unit area per eV) and the neutral level ϕ_0. Cowley and Sze [Cow65] have shown that the barrier energy with no electric field inside the semiconductor (i.e., the "flat band condition") is given by

$$\phi_B^0 = \gamma(\phi_m - \chi_{so}) + (1 - \gamma)(E_g - \phi_0), \qquad (7.5)$$

where

$$\gamma = \frac{\varepsilon_i}{\varepsilon_i + qN_{ss}\delta}. \qquad (7.6)$$

The dimensionless parameter γ varies between zero and unity. If $qN_{ss}\delta \ll \varepsilon_i$, then $\gamma = 1$, and Eq. (7.5) reduces to the Schottky limit of (Eq. (7.3). If $qN_{ss}\delta \gg \varepsilon_i$, then $\gamma \ll 1$, and Eq. (7.5) reduces to the Bardeen limit [Bar47]. Assuming that $\delta = 10$ Å (a value for ε_i typical of common insulating films), N_{ss} would have to exceed about 10^{13} states/cm^3-eV to pin the Fermi level at ϕ_0. Such high surface state concentrations have been measured directly on many III–V semiconductor prepared by chemical etching, and thus Schottky diodes formed on such surfaces do not depend on the metal work function.

The barrier height ϕ_B will not equal ϕ_B^0 at zero bias because part of the potential difference $\phi_m - \phi_s$ appears across the interfacial layer and the space-charge region of the semiconductor. The effect of the electric field at the metal–semiconductor interface can be written [Cow65]

$$\phi_B = \phi_B^0 - \alpha E_m, \qquad (7.7)$$

where E_m is the maximum value of the electric field in the semiconductor, and the parameter α has dimensions of length and is given by

$$\alpha = \frac{\delta \varepsilon_s}{\varepsilon_i + qN_{ss}\delta}, \qquad (7.8)$$

where $qN_{ss}\delta \gg \varepsilon_i$, $\alpha = 0$, and $\phi_B = \phi_B^0$. In this case the barrier energy is independent of the electric field and the band bending inside the semiconductor. Thus a large concentration of surface states screens the interior of the semiconductor from the metal. When $N_{ss} = 0$, $\alpha = \delta\varepsilon_s/\varepsilon_i$, and ϕ_B will depend on E_m. In the latter case, under reverse bias E_m is bias depend and is the largest in reverse bias. The dependence of E_m on the applied voltage V can be found from Poisson's equation and is

$$E_m = \left[\left(\frac{2qN_D}{\varepsilon_s}\right)\left(\frac{V_d - kT}{q}\right)\right]^{1/2}, \qquad (7.9)$$

where $V_d = \phi_B^0 - \xi - V$, for V positive with respect to the semiconductor, as shown in Fig. 7.2.

Another important mechanism that can lead to a reduction in ϕ_B^0 is the image-force lowering of a Schottky barrier. An electron just inside the semiconductor induces a sheet of charge on the surface of the metal such that the electrostatic force exerted on the electron is equivalent to that of a positive image charge located inside the metal. It takes less energy to remove the electron from the semiconductor. The image-force lowering $\Delta\phi_B$ is found by equating the force due to electric field in the space-charge region to that of the image force. The result is [Cow65]

$$\Delta\phi_B = \frac{(qE_m)^{1/2}}{4\pi\varepsilon'_s}, \tag{7.10}$$

where ε'_s is the dynamic permittivity of the semiconductor and is usually smaller than the static permittivity ε_s. The image-force lowering is shown in Fig. 7.2. Equation (7.10) was delivered assuming no interfacial layer present. If we include the interfacial layer effects, the barrier height becomes

$$\phi_B = \phi_B^0 - \alpha E_m - \Delta\phi_B. \tag{7.11}$$

Here, if we assume that

$$\phi_{B0} = \phi_B^0 - \alpha E_m, \tag{7.12}$$

we have

$$\phi_B = \phi_{B0} - \alpha\phi_B. \tag{7.13}$$

ϕ_{B0} is usually referred as "the barrier height at zero bias without the image-force lowering." However, the term αE_m does not depend weakly on the bias voltage. It has been pointed out by Rhoderick [Rho78] that the image-force lowering $\Delta\phi_B$ arises from an electron in the conduction band near the top of the barrier. If one determines the barrier energy by techniques that are not dependent on the presence of electrons at the barrier, the measured energy will be ϕ_{B0}.

7.2.2 Schottky Contacts

The first practical semiconductor device was the metal–semiconductor contact in the form of a point contact rectifier [Sze85]. In 1938 Schottky suggested that the rectifying behavior could arise from a potential barrier as a result of stable space charges in the semiconductor. For this reason we call the electrostatic potential barrier that exists at the boundary between a semiconductor and a metal the *Schottky barrier*. We will consider the energy band diagram and the current-voltage characteristics of the Schottky barrier.

Current-Voltage Characteristics

Thermionic Emission The current transport in Schottky barrier is due mainly to majority carriers [Cha70]. The current may be limited either by the drift and diffusion of carriers in the space charge region (in low mobility samples) or by thermionic emission over the barrier (in high-mobility semiconductors). The thermionic model is valid when the mean free path of electrons exceeds the distance over which the barrier decreases by $k_B T/q$:

$$\lambda > \frac{k_B T}{qE_{max}}, \qquad (7.14)$$

where E_{max} is the maximum electric field at the metal–semiconductor interface

$$E_{max} = \sqrt{\frac{2qN_0 V_{bi}}{\varepsilon_s}} \qquad (7.15)$$

and the mean free path

$$\lambda = \frac{\mu}{q}(1.5k_B T m)^{1/2}. \qquad (7.16)$$

The inequality (7.14) is fulfilled in GaAs at room temperature for $N_D > 10^{14}\,\text{cm}^{-3}$, assuming that $V_{bi} = 0.7\,\text{V}$ and $\mu = 0.5\,\text{m}^2/\text{V-s}$.

The band diagrams of the Schottky barrier under forward and reverse biases are shown in Fig. 7.3. The quasi-Fermi levels for electrons are practically flat throughout the depletion regions except at a narrow region near the interface.

The current-voltage characteristics predicted by the thermionic model are given by [Cro66]

$$j = j_0 \exp\left(\frac{qV}{nk_B T}\right)\left[1 - \exp\left(-\frac{qV}{k_B T}\right)\right], \qquad (7.17)$$

where

$$j_0 = A^* T^2 \exp\left(\frac{-q\phi_{Bn}}{k_B T}\right) \qquad (7.18)$$

and A^* is the Richardson constant:

$$A^* = f_p f_Q \frac{4\pi m q k_B^2}{h^3}. \qquad (7.19)$$

Here f_p is the probability of an electron reaching the metal without being scattered by an optical phonon after having passed the top of the barrier. f_Q is the transmission coefficient ($f_p f_Q = 0.5$), and m is the effective mass. For a nonparabolic semiconductor with N conduction minima,

$$m = \frac{N(m_{\text{eff}})^{3/2}}{(m_C)^{1/2}}, \qquad (7.20)$$

where m_{eff} is the density of states effective mass $m_{\text{eff}} = (m_t^2 m_l)^{1/3}$ and m_C is the conduction effective mass. For (111) surfaces of GaAs, A^* is equal to $4.4 \times 10^4\,\text{A/m}^2\text{K}^2$, respectively.

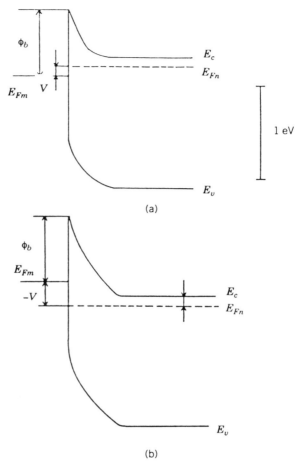

Fig. 7.3 Band diagram of a Schottky barrier under (a) forward and (b) reverse bias. [Shur, M., *GaAs Devices and Circuits*, Plenum Press (1987); with permission from the publisher]

The factor n in Eq. (7.17) is called the *ideality factor*. The factor is related to the voltage dependence of the barrier height. If we compare the value of the saturation current j_0 for the Schottky barrier with the saturation current in p–n junction J_{opn}, we get [Rho78]

$$\frac{j_0}{j_{opn}} = \frac{(\tau_{nl})^{1/2}}{(\tau_n)} \exp\left(\frac{V_{bi} - \phi_b}{k_B T} - \phi_b\right), \tag{7.21}$$

where τ_{nl} is the electron lifetime in the p-type region, τ_n is the average time between collisions, and V_{bi} is the build-in voltage. The ratio τ_{nl}/τ_n is on the order of 10^4 in GaAs.

Field Emission and Thermionic-Field Emission In highly doped semiconductors and Schottky barrier becomes so thin that electrons near the top of the barrier can tunnel through the barrier [Pad66, Cha70]. This process is called *thermionic-field emission*. In degenerate semiconductors, especially if there is present a small electron effective mass such as GaAs, electrons can tunnel through the barrier near the Fermi level (field emission). The current-voltage characteristic in case of the thermionic-field emission or field emission is determined by the competition between the thermal activation and tunneling and is given by [Pad66]

$$J = J_{\text{stf}} \exp\left(\frac{qV}{E_0}\right), \qquad (7.22)$$

where

$$E_0 = E_{00} \coth\left(\frac{E_{00}}{k_B T}\right), \qquad (7.23)$$

$$E_{00}(eV) = \frac{h}{4\pi}\left(\frac{N_D}{m\varepsilon_s}\right)^{1/2} = 1.85 \times 10^{-14} \left[\frac{N_D}{(m/m_e)(\varepsilon_s/\varepsilon_0)}\right]^{1/2}, \qquad (7.24)$$

where SI units are used. The pre-exponential term J_{stf} was calculated by Crowell and Rideout [Cro69]:

$$J_{\text{stf}} = \frac{A^*T[\pi E_{00}q(\phi_{Bn} - V - \xi)]^{1/2}}{k_B \cosh(E_{00}/k_B T)} \exp\left[\frac{-q\xi}{k_B T} - \frac{q}{E_0}(\phi_{Bn} - \xi)\right]. \qquad (7.25)$$

Here $\xi = (E_c - E_{fn})/q$ and is negative for a degenerate semiconductor. In GaAs the thermionic field emission occurs roughly for $N_D > 10^{17} \text{cm}^{-3}$ at 300 K and for $N_D > 10^{16} \text{cm}^{-3}$ at 77 K.

At very high doping levels the width of the depletion region becomes so narrow that direct tunneling from the semiconductor to the metal may take place, as shown in Fig. 7.4. This happens when E_{00} becomes much greater than $k_B T$. The current-voltage characteristics in this region are given by

$$J = J_{\text{sf}} \exp\left(\frac{qV}{E_{00}}\right), \qquad (7.26)$$

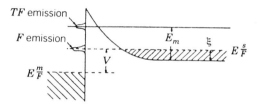

Fig. 7.4 Field and thermionic emission under forward bias. [Shur, M., *GaAs Devices and Circuits*, Plenum Press (1987); with permission from the publisher]

where

$$J_{sf} = \frac{\pi A^* T}{k_B C_1 \sin(\pi k_B T C_1)} \exp\left(\frac{-q\phi_{Bn}}{E_{00}}\right),$$

$$C_1 = (2E_{00})^{-1} \ln\left[\frac{-4(\phi_{Bn} - V)}{\xi}\right].$$

The effective resistance of the Schottky barrier in the field emission regime is quite low. Therefore the metal n^+-type Schottky barrier is used to make ohmic contacts.

7.2.3 Schottky Barrier Metals

Classification of Contact Metals

There are no well-established rules for reaction between a metal and a compound semiconductor [Kwo86]. However, based on experimental data and empirical approach, Sinha and Poate [Sin73] have classified reactions of commonly used contact metals and compound semiconductors into three groups, largely according to their electronegativities. Figure 7.5 reproduces a section of the periodic table for groups III and V of compound semiconductors, group II as p-type dopants, group VI as n-type dopants and group IV as amphoteric dopants. The corresponding electronegativities are also shown in the figure. Sinha and Poate [Sin73] classified common contact metals into three groups: M1, M2, and M3.

Group M1 consists of noble metals such as Cu, Ag, and Au that have large electronegativities. For these metals the dominant reaction is the out diffusion of electropositive group III elements into the metal. For example, in a Au/GaAs interface, due to the large electronegative difference, Au causes dissociation of GaAs and Ga migrates into Au under thermal annealing as low as 250°C. Figure 7.6 gives the RBS spectra of the Au/GaAs interface, where the Ga peak has moved toward Au (high-energy channel).

Group M2 consists of near-noble metals, near-transition metals, with large electronegativities such as Ni, Pd, Pt, Rh, Ir, and Os. Again, because of a large

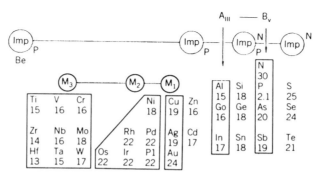

Fig. 7.5 Portion of the periodic table and electronegativity of important III–V compounds, dopants, and metals. (M_1 = noble metals; M_2 = near-noble metals; M_3 = early transition metals.) [Kwok, S. P., *J. Vac. Sci. Tech.* B **4**, 6 (1986): 1383–1391]

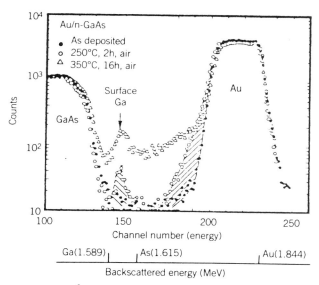

Fig. 7.6 RBS spectra of 700-Å Au/n-GaAs. [Kwok, S. P., *J. Vac. Sci. Tech.* B **4**, 6 (1986): 1383–1391]

electronegative difference, the electropositive group-III element diffuses to the metal. These metals form stable compounds with group-V metalloids and group-III metals. Figure 7.7 gives an example of the RBS spectra of a 950-Å Pt/n-GaAs system, where Ga diffuses toward Pt, and Pt forms stable $PtAs_2$ and PtGa intermetallic compounds. This reaction takes place at a low temperature of 350 °C.

Group M3 consists of early transition metals having small electronegativities such as Cr, Mo, W, V, Nb, Ta, Ti, Zr, and Hf. Because of small electronegative differences, the interfaces of these metals with III–V compounds are relatively inert. Ti/GaAs is stable at 300°C. Interface reactions take place at 500°C.

Interaction of Refractory Metal Silicides and GaAs

In GaAs MESFETs fabrication, multilayer contacts have been used for the gate material. Figure 7.8 shows the RBS spectra of Pt/Ti/GaAs [Kwo86]. At 350°C, no reaction at Ti/GaAs and Ti/Pt is evident. At 500°C, Ti tends to form TiAs which leads to a reduction in the Ti peak. The TiAs layer is apparently a poor diffusion barrier for a Ga/Pt reaction to take place at this temperature. The Pt/Ti/GaAs is more stable than the Pt/GaAs system. The Au/Pt/Ti system has been studied for gate material. However, at high temperatures the interdiffusion between GaAs and Ti degrades the device performance.

The Au/Ti–W system maintains its stability up to 600°C for up to 15 hours. An advantage of gold/refractory metal contacts is that they can withstand adverse sodium and chloride ion contamination. The limitation for the Au/Ti–W structure is the thermal expansion difference between the Ti–W and GaAs. W/GaAs was found to be metallurgically stable at 500°C [Sin73]. The Schottky barrier height and the diode ideality factor were found to be stable at 550°C. At temperatures above 600°C, the

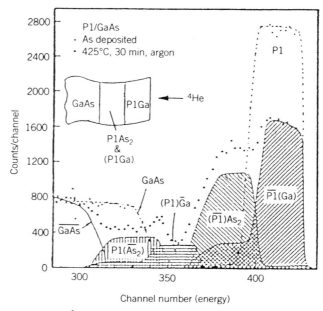

Fig. 7.7 RBS spectra of 950-Å Pt/n-GaAs reacted to form $PtAs_2$ and PtGa. [Kwok, S. P., *J. Vac. Sci. Tech.* **B 4**, 6 (1986): 1383–1391]

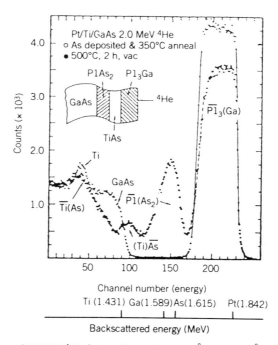

Fig. 7.8 RBS spectra of 2 MeV $^4He^+$ ions from Pt (1000 Å)–Ti (500 Å)–GaAs. [Kwok, S. P., *J. Vac. Sci. Tech.* **B 4**, 6 (1986): 1383–1391]

W film peels off due to poor adhesion and high stress in the interface [Ohn83]. Mixed TiW/GaAs is evident, as shown in Fig .7.9.

As packing density increases, transition metal silicides, whose main advantages are their stability at high temperatures during device processing, have attracted attention for use in Si integrated circuit technology. In GaAs IC technology metal silicides play an important role on low contact resistance under high annealing temperature.

Figure 7.10 shows the tungsten silicide deposition on GaAs by rf cosputtering technique [Lin86]. Annealing was carried out in an $N_2 + H_2$ gas mixture with a 100-nm-thick SiO_2 encapsulation film for 15 min. Figure 7.11 shows the RBS spectrum of a $WSi_{0.64}$/GaAs contact before and after annealing at 850°C for one hour. There is no change in these spectra, indicating that there is no metallurgical reaction between

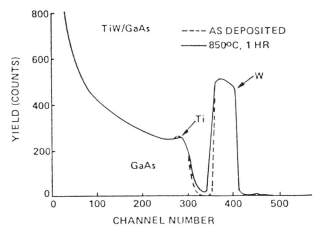

Fig. 7.9 RBS spectra of TiW/GaAs. [Kwok, S. P., J. Vac. Sci. Tech. B 4, 6 (1986): 1383–1391]

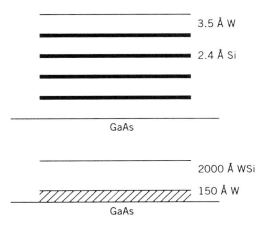

Fig. 7.10 Tungsten silicide deposition. [Lin, M. S., Lecture on Tungsten silicide deposition, Tainan (1986)]

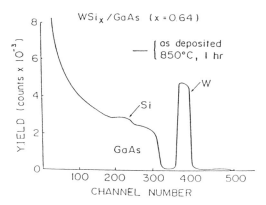

Fig. 7.11 RBS spectrum of a WSi$_{0.64}$/GaAs contact before and after annealing at 850°C for one hour. [Lin, M. S., Lecture on tungsten silicide deposition, Tainan (1986)]

WSi$_{0.64}$ and GaAs. This demonstrates that WSi$_{0.64}$–GaAs contacts are stable at high temperature. The Schottky barrier height and ideality factor of WSi$_x$ contacts on GaAs fabricated by both cosputtering and LPCVD methods are found at around $x = 0.6$ to be optimum [Cho86].

Molybdenum silicide (MoSi$_2$) [Kao86] is stable up to annealing temperature 800°C. Another metallization that has been shown to be stable is tantalum silicide [Pal85]. Figure 7.12 shows the backscattering spectra of a layered Si/Ta/Si/GaAs structure

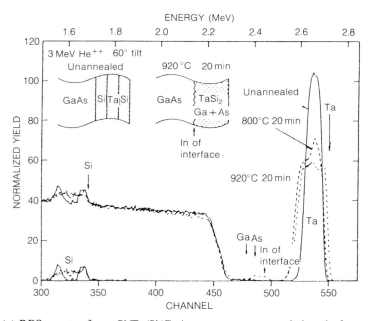

Fig. 7.12 (a) RBS spectra for a Si/Ta/Si/GaAs structure unannealed and after annealing at 800° and 920°C for 20 min in an arsenic overpressure furnace. [Palmstrøm, C. J., and D. V. Morgan, in *Gallium Arsenide, Materials, Devices, and Circuits*, p. 218, ed. by M. J. Howes and D. V. Morgan, Wiley (1985)]

before and after annealing at 800°C and 920°C for 20 min in an arsenic overpressure ambient. There is no change in the front edge of the GaAs in the 800°C annealed spectrum. The drop in the height of the Ta peak and the change in the Si peaks arise from the formation of $TaSi_2$ over the GaAs. Complete $TaSi_2$ formation is observed after the 920°C 20-min anneal.

TABLE 7.1 WAl and WN Compounds for the Self-aligned Gate Field-effect Transistor

Computer	Gate Metal	Temperature (°C)	Fet Gm (ma/mm)	Lg (μm)	Circuit	Self-aligned
Fujitsu	TiW	800	87	1.5	RO	Gate
Fujitsu	TiWSi	800	—	2.0	—	Gate
Toshiba	Pt	380	85	1.2	—	Buried gate
Hughes	TiW	800	140	0.8	RO	Gate
Fujitsu	TiWSi	800	—	2.0	6 × 6 mult.	Gate
NTT	TiPtAu	RT				N^+
AT&T	Al	RT	160	0.9	RO	Recessed gate
NTT	TiPtAu	RT	—	1.1	1K SRAM	N^+
Toshiba	Pt	380	85	1.2	500 GA	buried gate
Fujitsu	WSi	800	127	2.0	1 K SRAM 16B mult.	Gate
NTT	TiPtAu	RT	—	0.95	1K SRAM	N^+
Textrnx	TiPdAu	RT	200	1.0	1K GA	Recessed gate
OKI	WAl	800	112	1.3	RO	Gate
Toshiba	Pt	380	200	1.1	—	Gate
Toshiba	Pt	380	110	1.0	1K GA	Buried gate
NTT	TiPtAu	RT	132	1.0	4K SRAM	N^+
Fujitsu	Wsi	800	160	1.5	4K SRAM	Gate
NEC	Al	RT	123	1.0	4K SRAM	Ohmic
OKI	WAl	800	75	1.5	8B mult.	Gate
Hitachi	WSi	800	110	1.0	1K SRAM	Gate
Toshiba	Pt	380	200	1.0	4K SRAM	Buried gate
NTT	TiPtAu	RT	147	1.0	16K SRAM	N^+
Honeywell	TiWSi	800	170	1.0	8:1 MUX	Gate
Rockwell	TiPtAu	RT	280	1.0	Prescalar	N^+
OKI	WAl	800	150	1.0	1K GA	Gate
Toshiba	WN	800	200	1.5	2K GA	Gate
NTT	a-SiGeB	450	260	0.8	—	N^+
NEC	WSi	800	300	1.0	—	Gate
FMI	TiPtAu	RT	225	1.4	1 K GA (Eqv) 8 × 8 mult. ECL comp	Gate

Source: Kwok, S. P., J. Vac. Sci. Tech. B 4, 6 (Nov./Dec. 1986).

Others have reported successful applications of WAl and WN compounds for the self-aligned gate field-effect transistor (SAGFET) [Kwo86, Table 7.1]. Tungsten nitride (WN_x), for example, which is attracting much interest as a diffusion barrier for interconnecting metallization in Si VLSI technology [Sze88], is another promising refractory metal for the gate electrode because of its higher Schottky barrier height (0.84V) to n-type GaAs and lower film resistivity (70 Ω-cm) than those of WSi_x.

7.2.4 Techniques for Barrier Height Measurement

Capacitance-Voltage Measurement

In the capacitance-voltage method the diode's capacitance is measured as a function of applied reverse bias. When a small ac voltage of a few millivolts is applied to a reverse-biased diode, the depletion region capacitance C is given by the relation

$$C = A \left[\frac{\varepsilon_s q N_d}{2(V_i + V_R - kT/q)} \right]^{1/2}, \qquad (7.27)$$

where A is the diode cross-sectional area, ε_s is the permittivity of the semiconductor, V_R is the applied voltage, and all other symbols have their usual meanings. In this equation it is assumed that the diode does not have an appreciable interfacial oxide layer and that the n-type semiconductor has a uniform donor concentration N_d. A plot of $1/C^2$ versus V_R gives a straight line with slope $2/A^2 \varepsilon_s q N_d$ and an intercept on the voltage axis $V_0 = (V_i - kT/q)$. The slope of the straight line can be used to determine the dopant concentration N_d. Since $qV_i = (\phi_B - \phi_n)$, the barrier height ϕ_B is obtained as

$$\phi_B = (qV_0 + \phi_n + kT). \qquad (7.28)$$

The kT factor comes from the contribution of majority carriers to the space charge, and Eq. (7.28) does not include image-force lowering.

Current-Voltage Measurement

For the Schottky barrier on high-mobility semiconductors such as Si and GaAs, the current is due to thermionic emission of electrons over the barrier. The current is a function of applied bias V given by the relation

$$I = I_0 \left[\exp\left(\frac{qV}{nkT}\right) - 1 \right], \qquad (7.29)$$

where $I_0 = AA^*T^2 \exp(-\phi_B/kT)$. Here n is the diode ideality factor and A^* is the modified Richardson constant for the semiconductor. For forward biases of V in excess of $3kT/q$, a plot of $\ln I$ against V gives a straight line. The value of I_0 can be obtained by extrapolating the straight line to $V = 0$. Knowing I_0, A^*, the diode cross-sectional area A, and the temperature T, the barrier height ϕ_B can be determined. This is the zero-bias barrier height that includes the image-force barrier lowering $\Delta\phi_B$.

Photoresponse Measurement

One of the earliest and most direct techniques for measurement of ϕ_B consists of photoexcitation of electrons from the Fermi level in the metal to the conduction band of the semiconductor by illuminating the Schottky diode with monochromatic light, as shown in Fig. 7.13. When a monochromatic light is incident upon a metal surface, a photocurrent is generated. As the photon energy $h\nu$ is varied, the diode photocurrent will increase sharply as long as $h\nu > q\phi_B$ (process a in Fig. 7.13), and when $h\nu > E_g$ the photocurrent will increase even more rapidly as a result of band-to-band excitation (process b in Fig. 7.13).

The photocurrent per absorbed photon I_p for photon energies more than $3kT$ larger than qB but less than E_g is given by Fowler's theory [Fow31] for classical photoemission as

$$I_p = B(h\nu - q\phi_B)^2, \tag{7.30a}$$

or

$$\sqrt{I_p} = B^{1/2}(h\nu - q\phi_B), \tag{7.30b}$$

where B is a constant of proportionality. Thus by plotting the square root of I_p normalized to the photon flux, a straight line results with an extrapolated intercept which is $q\phi_B$. An example of the photoresponse data for Pd Schottky diodes [Rob85] on GaAs and InP is shown in Fig. 7.14.

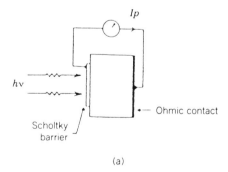

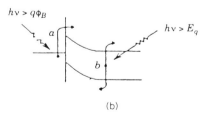

Fig. 7.13 Measurement of ϕ_B using internal photoemission: (*a*) experimental arrangement for determination of the short-circuit photocurrent I_p; (*b*) two processes for carrier excitation. [Robinson, G. Y., in *Physics and Chemistry of III–V Compound Semiconductor Interfaces*, ed. by C. W. Wilmsen, Plenum Press (1985); with permission from the publisher]

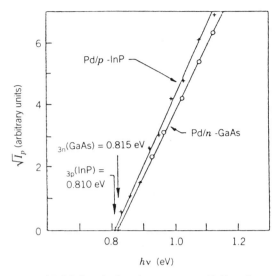

Fig. 7.14 Photo response of Pd Schottky barriers on *p*-type InP and *n*-type GaAs. [Robinson, G. Y., in *Physics and Chemistry of III–V Compound Semiconductor Interfaces*, ed. by C. W. Wilmsen, Plenum Press (1985); with permission from the publisher]

7.3 UNPINNED SCHOTTKY BARRIER FORMATION

In the previous section we discussed the formation of the Schottky barrier to GaAs. It is a common knowledge that metal contacts to GaAs produce only a narrow band of the Fermi level's stabilization energies, regardless of the metal work function, and that this behavior is representative of many metal–(III–V) semiconductor interfaces [Sze81; Mea66; Kur69; Mea70; Spi80]. The surface science measurements of metals on the $In_xGa_{1-x}As$ (100) ($0 \leqslant 1$) pseudobinary alloys [Bri86; Bri87; Bri86a], InAs (110) [Bri86a], and GaP (110) [Bri87] surfaces have revealed unpinned and even near-ideal, Schottky-like behavior. These results comprise a major part of the limited data base for clean metal–III–V compound semiconductor interfaces.

Brillison et al. [Bri88] discovered that the metal/GaAs interface can exhibit relatively unpinned E_f movements. Metals on clean MBE-grown GaAs obtained by thermal decapping of a protective As overlayer exhibit a 0.6–0.7 eV (or larger) range of E_f stabilization. E_f stabilization occurs at the same energies for *n*-type and *p*-type GaAs, as expected from a self-consistent analysis of junction electrostatics [Duk85; Mai86]. Such an analysis reveals that the presence of an acceptor state with mid-10^{13} cm^{-2} density at 0.2 eV above the valence band edge is compatible with the dependence of the measured Schottky barrier height on the metal work function. Furthermore, E_f movements are metal dependent and can be evolve over multimonolayer coverages. Varying surface Ga-to-As ratios indicate that initial surface composition, and reconstruction do not dominate the final metal/GaAs band bending. The cleaved, melt-grown GaAs has orders-of-magnitude higher densities of deep-trap levels than MBE-grown GaAs.

7.3.1 Experiment

Soft X-ray photoemission spectroscopy (SXPS) measurements were performed for Ga and As core levels with metal deposition in order to monitor the GaAs E_f movements as Schottky barriers formed [Bri88]. The bulk-sensitive SXPS spectra of As $3d$ and Ga $3d$ core levels were measured using $hv = 60$ and $40\,\text{eV}$ and surface-sensitive As $3d$ and Ga $3d$ core level spectra using 100 and $80\,\text{eV}$, respectively [Sea79]. These energy sets produce photoelectrons with identical escape depth and thereby identical depth resolution for both elements. Comparison of bulk- versus surface-sensitive spectra allow us to monitor E_f changes and band bending with metallization from rigid core level shifts (bulk sensitive) versus any chemical shifts (surface sensitive). Comparison of bulk- versus surface-sensitive spectra provide a means to identity and separate effects due to chemical bonding changes.

The initial Fermi level position relative to the valence band edge is determined by the valence band difference between the extrapolated leading edge of clean GaAs versus the midpoint of the leading edge of a thick Au film grounded to the photoelectron analyzer. The resolution of the core level spectra analyzed is 0.23–$0.25\,\text{eV}$ for 40–$60\,\text{eV}$ and 0.26–$0.37\,\text{eV}$ for $hv = 80$–$100\,\text{eV}$. A valence spectrum of a thick ($100\,\text{Å}$) Au film deposited on a Ta substrate in contact with the GaAs established the initial E_f position of the clean semiconductor. Metal evaporation took place in an UHV chamber from tungsten filament with a pressure rise no higher than mid-$10^{-9}\,\text{torr}$.

To obtain clean, ordered GaAs (100) surfaces, MBE-grown GaAs films were used and "capped" immediately after deposition with several hundred monolayers of "cracked" As as protection against ambient contamination. These caps were thermally desorbed in UHV to provide clean ordered surfaces, as determined by valence band (VB) photoemission and LEED measurements. This desorption procedure involved several stages at temperatures up to $600\,°\text{C}$ [Bri88]. The 40-eV valence band spectra exhibits a characteristic set of peak features that are particularly sensitive to ambient contamination and lattice disruption.

7.3.2 Results

The band bending produced by deposition of Au, Al, Cu, and In on clean, ordered GaAs (100) surfaces was measured. The rigid Ga $3d$ and As $3d$ core level shifts observed by SXPS provided a measure of the band bending that occurred during the initial stages of Schottky barrier formation. Figure 7.15 shows Ga $3d$ and As $3d$ core level spectra for p-type GaAs under both bulk-sensitive [Fig. 7.15(a)] and surface-sensitive [Fig. 7.15(b)] conditions as a function of increasing metal deposition. With initial metal deposition both substrate Ga $3d$ peaks shift to lower kinetic energy, corresponding to increasing p-type band bending. A deposition of only $2\,\text{Å}$ is sufficient to produce a second Ga $3d$ peak feature due to dissociated Al in both surface- and bulk-sensitive spectra. This feature continues to grow with increasing Al coverage, and the splitting between dissociated and undissociated peaks increases. At coverages of 10–$20\,\text{Å}$ Al, the bulk-sensitive Ga $3d$ spectra still provides distinct energies for the substrate Ga, while the surface-sensitive Ga $3d$ spectra reflect almost entirely the dissociated component. By considering only the undissociated Ga $3d$ component in the bulk-sensitive spectra, one obtains a $0.27\,\text{eV}$ total band bending. Since the starting E_f position was $E_v + 0.53\,\text{eV}$, E_f moves to $0.80\,\text{eV}$ above E_{VBM}.

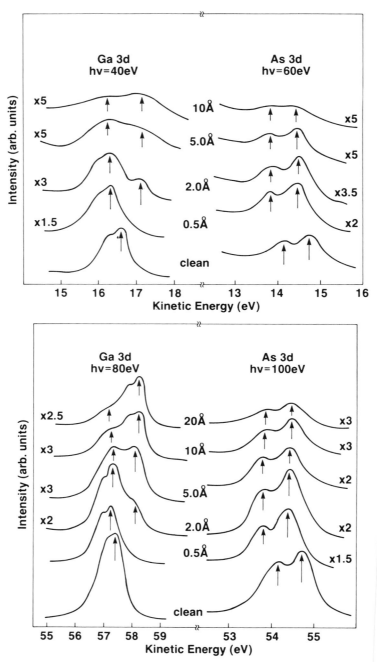

Fig. 7.15 Al/MBE-GaAs (100) *p*-type: (*a*) bulk and (*b*) surface-sensitive SXPS features for Ga 3*d* and As 3*d* core levels as a function of Al deposition on clean GaAs (100). Bulk-sensitive (40 eV) and surface-sensitive (80 eV) Ga 3*d* spectra both exhibit features due to substrate band bending and Ga dissociation. Bulk-sensitive Ga 3*d* and As 3*d* specta in (*a*) exhibit rigid shifts due to band bending. The dissociated component dominates the features above 2-Å coverage in the surface-sensitive but not the bulk-sensitive spectra. [Brillson, L. J., R. E Viturro, C. Mailhiot, J. L. Shaw, T. Tache, J. McKinley, G. Margaritondo, J. M. Woodall, P. D. Kirchner, G. D. Pettit, and S. L. Wright, *J. Vac. Sci. Tech.* B **5**, 4 (1988): 1263–1269]

Figure 7.16 illustrates the E_f movements induced by deposition of Au, Al, Cu, and In on both n-type and p-type GaAs surfaces. The E_f behavior is different from that reported consistently for UHV-cleaved GaAs (110) surfaces. First, this set of common metals produce a range of E_f stabilization energies extending over 0.7 eV from $E_v + 0.18$ eV to $E_v + 0.92$ eV. This is in contrast to the narrow 0.2 to 0.25 eV range reported for GaAs (110) [Spi80; New86] and GaAs (100) [Wal84]. Second, the E_f stabilization energies are the same for both n-type and p-type GaAs with the same metal. This is in contrast to the 0.2-eV separation seen for many adsorbates on UHV-cleaved GaAs (110) [Spi80; New86]. Third, the E_f stabilization occurs over 5 to 20 Å in all cases, and not within the submonolayer coverages reported for GaAs (110).

The GaAs specimens exhibit a range of 0.35 eV in E_f energies for their clean surfaces. The variation in starting energies may be related to differences in Ga:As stoichiometry, which is known to produce changes in the E_f position [Sve84; Bac81]. These variations correspond to reconstruction from (4 × 2) through (2 × 4). However, significant differences in starting energies produce little or no differences in final stabilization energies, as evidenced by the Au and Al curves. For Au deposition on three n-type and one p-type surfaces, Fig. 7.16 shows an E_f stabilization range of 0.3 eV but a final E_f spread of only 0.05 eV after 20-Å deposition. Likewise, for Al deposition on one n-type and two p-type surfaces, Fig. 7.16 displays a 0.2 eV initial E_f range but a

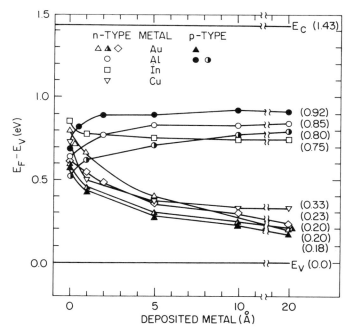

Fig. 7.16 Fermi level shifts within the GaAs band gap for deposition of Au, Al, In, and Cu on both n- and p-type GaAs (100) surfaces. The E_F shifts extend over 0.7 eV. The initial E_F positions for clean, ordered GaAs are located in a 0.35-eV window near midgap. [Brillson, L. J., R. E. Viturro, C. Mailhiot, J. L. Shaw, T. Tache, J. McKinley, G. Margaritondo, J. M. Woodall, P. D. Kirchner, G. D. Pettit, and S. L. Wright, *J. Vac. Sci. Tech.* B **6**, 4 (1988): 1263–1269]

final E_f spread of only 0.12 eV. The metal interaction rather than the starting surfaces appears to be dominant in determining final stabilization energies. This does not preclude compositional effects caused by variations in Ga:As out diffusion and resultant interfacial stoichiometry. Most important, Fig. 7.16 demonstrates that the metals on GaAs produce a wide range of E_f movements that is not constrained to a narrow window of midgap energies.

7.3.3 Discussion

Figure 7.17 illustrates the contrast between the band bending induced by metal deposition on clean MBE-grown GaAs (100) versus GaAs (110) surfaces. Plotted with relation to the metal work function [Bri88] Φ_M are barrier heights Φ_B (right ordinate) and E_f positions below the vacuum level (left ordinate). A comparison of the absolute valence band energies for UHV-cleaved (110) and thermally decapped (100) surfaces under identical conditions reveals that the same binding exists for both orientations to within the precision of the SXPS energy measurements (± 0.05 eV). The inset in the upper left-hand corner represents ideal Schottky behavior and corresponds to

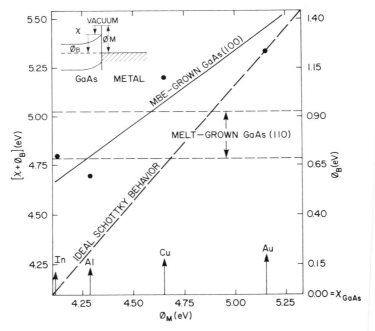

Fig. 7.17 GaAs barrier heights Φ_B and E_F positions below the vacuum level (Φ_B + electron affinity X_{SC}) plotted with respect to the metal work function. Melt-grown GaAs (110) surfaces exhibit a 0.2–0.25-eV range near midgap for a wide variety of metals. Only four metals on MBE-grown GaAs (100) surfaces exhibit a 0.7-eV range that overlaps the metal-grown band and extends to within 0.2 eV of the valence band. Ideal Schottky behavior appears in the upper left-hand inset and corresponds to the diagonal line. [Brillson, L. J., R. E. Viturro, C. Mailhiot, J. L. Shaw, T. Tache, J. McKinley, G. Margaritondo, J. M. Woodall, P. D. Kirchner, G. D. Pettit, and S. L. Wright, *J. Vac. Sci. Tech.* B **6**, 4 (1988): 1263–1269]

the diagonal line shown. The melt-grown GaAs (110) surface exhibits a 0.2- to 0.25-eV range near the midgap for a wide variety of metals. On the other hand, only four metals on MBE-grown GaAs (100) exhibit the 0.7-eV range found in Fig. 7.16. This 0.7-eV range overlaps the melt-grown band and extends to within 0.2 eV of the valence band edge. The line drawn through the four data points has a slope $S = 0.8$ eV, compared to $S \leqslant 0.25$ eV for the melt-grown data. Only the Au data point lies on the $S = 1$ ideal Schottky line. This behavior indicates that the GaAs (100) surfaces, while permitting a wide range Schottky barrier heights, still exhibit localized interface charge states.

Nearly identical E_f positions for the same metal on both n-type and p-type GaAs are expected to occur in a self-consistent electrostatic model of the interface E_f stabilization. Figure 7.18 illustrates the results of a generalized analysis by Duke and Mailhiot [Duk85]. The energy positions, surface densities, and donor/acceptor characteristics of the interface states are varied to provide an optimal fit to the data points. The curves in the figure correspond to the dependence of the barrier on the work function for a single acceptor level located 0.2 eV above E_{vbm}. The data points from Fig. 7.17 lie between the ideal Schottky limit $N < 10^{12}$ cm^{-2} and the strongly "pinned" limit $N \geqslant 10^{14}$ cm^{-2}. A single acceptor with density $N_A = 5 \times 10^{13}$ cm^{-2} appears to fit the data more closely.

Several factors can attribute to the pronounced difference in Schottky barrier formation for melt-grown versus MBE-grown GaAs. These include crystal surface

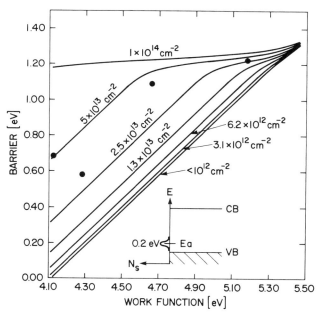

Fig. 7.18 Self-consistent electrostatic analysis of the metal/GaAs (100) data points. The generalized model of Duke and Mailhiot provides an optimal fit of interface state energies, densities of states, and donor/acceptor characteristics. [Brillson, L. J., R. E. Viturro, C. Mailhiot, J. L. Shaw, T. Tache, J. McKinley, G. Margaritondo, J. M. Woodall, P. D. Kirchner, G. D. Pettit, and S. L. Wright, *J. Vac. Sci. Tech.* B **6**, 4 (1988): 1263–1269]

orientation, excess As, thermal pretreatment, and bulk traps. Crystal orientation produces no major differences in deep levels for the relaxed GaAs (110) versus (100) [Cha78] clean surfaces.

A difference in the densities and energies of bulk traps can also account for Schottky barrier differences between melt versus MBE-grown GaAs. Melt-grown GaAs contains high concentrations of deep levels. For example, EL2 concentrations alone range up to 2×10^{16} cm^{-3} for LEC-grown [Hol82] and to 5×10^{16} cm^{-3} for HB-grown GaAs [Lag82] orders of magnitude higher than 10^{13} cm^{-3} trap densities for MBE-grown GaAs [Mir79]. Furthermore LEC-grown crystals frequently contain native defects with the potential for electrical activity exceeding 10^{18} cm^{-3} [Fuj84]. All of these electrically active sites have the potential to segregate at the GaAs surface [Yak84], thereby increasing the local deep-level concentration. When such concentrations exceed the bulk doping, these interface traps can restrict E_f movement [Zur83].

7.3.4 Implications for Schottky Barrier Models

The E_f measurements described in this section provide evidence for unpinned metal-GaAs interfaces. The large range of Schottky barrier heights for MBE-grown GaAs (100) surfaces indicates that any states induced by chemisorption or chemical interdiffusion play only a secondary role in the E_f stabilization. For melt-grown GaAs the existence of high densities of states near midgap provides a direct explanation for the rapid E_f movements to a correspondingly narrow energy range [Bri88]. Outside of the melt-grown material, a wide range of E_f movement is present.

The E_f results in Figs. 7.15 and 7.16 are consistent with the larger role of interfacial As in stabilizing the melt-grown versus MBE-grown GaAs/metal interface. The electrical activity of interfacial As is evident in cathodoluminescence (CLS) spectra of Viturro et al. [Duk81], and a comparison of melt versus MBE-grown GaAs, CLS spectra suggests that discrete, As-related states [Lil86] play a larger role in the former. Finally, when taken with results for InAs, InGaAs, GaP, and the MBE-grown GaAs, we discover that E_f pinning of metal–semiconductor interfaces in a narrow energy range, irrespective of the metal, is not characteristic of III–V compound semiconductors.

For many years the need to understand and control Schottky barrier formation has been a driving force for basic research into mechanisms of E_f pinning. While E_f pinning on melt-grown GaAs will likely continue to be a controversial issue, it appears not to have major consequences for MBE-grown GaAs (100) surfaces. The latter are prime candidates for future high-speed device structures. With the large range of barrier heights now available for III–V compound semiconductors, the main focus of interface barrier research may shift from a search for mechanisms to account for pinning to methods of semiconductor growth and processing that afford even greater control [Bri88].

7.4 OHMIC CONTACT

Low-resistance ohmic contacts are necessary in most III–V semiconductor devices. For light-emitting diodes, lasers, solar cells and Gunn diodes, values of the specific contact resistance r_c in the range of 10^{-2} to 10^{-5} cm^2 have been found to be adequate, since the ohmic contacts employed in these devices are relatively large in area and the resulting total contact resistance can easily be less than a few ohms. In the high-speed devices r_c should be even less ($\leqslant 10^{-6}$ Ω-cm^2). In GaAs ohmic contact can be formed using n^+ or p^+ layers directly in contact with the metal, with dopant concentrations of about 5×10^{18} cm^{-3} to 2×10^{19} cm^{-3} [Rob85, Cha71]. The metal may be a multilevel metal-alloyed system. New methods of forming ohmic contacts, such as epitaxial heterojunction structures, are now being explored. Metal silicide systems are also being studied, since metal silicides are stable after high-temperature annealing. In this section the technology of forming ohmic contacts to the III–V semiconductors is reviewed.

7.4.1 Methods of Forming Ohmic Contacts

In low-resistance ohmic contacts current flow across the metal–semiconductor interface is by field emission. According to the theoretical expression for field emission, the specific contact resistance r_c should be proportional to the factor $\exp(\phi_B/N)$, where N is the net concentration of the dopant at the semiconductor surface [Rob85]. In GaAs, ϕ_B depends only weakly on the choice of metal and on the method of surface cleanliness. Figure 7.19 shows the band diagrams for ohmic contact at zero bias. The thickness t of the heavily doped layer is chosen to be larger than the depletion width w of the metal–semiconductor barrier. Current flow for the structure is governed by field emission through the metal–semiconductor barrier if the contact potential of the n^+/n (or p^+/p) homojunction is approximately equal to or less than kT/q.

At the metal–semiconductor interface the low-resistance ohmic contacts to the wide-band-gap semiconductor in the form of a thin heavily doped layer [Yu70]. The

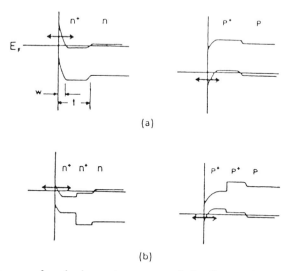

Fig. 7.19 Band diagrams for ohmic contacts at zero bais. Current flow by majority carriers is shown by the arrow: (a) homojunction n^+/n or p^+/p contacts; (b) heterojunction $n^+/n^+/n$ or $p^+/p^+/p$ contacts. (Robinson, G. Y., in *Physics and Chemistry of III–V Compound Semiconductor Interfaces*, ed. by C. W. Wilmsen, Plenum Press (1985); with permission from the publisher]

band diagrams for $n^+/n^+/n$ and $p^+/p^+/p$ heterojunction contacts are shown in Fig. 7.19(b). The heterojunction ohmic contact combines a low ϕ_B with a large N to ensure small values of r_c.

Thin highly doped layers of the III–V semiconductors can be formed under well-controlled conditions using MOCVD or MBE. The incorporation of the dopant is governed by surface kinetics during film growth in the MBE system. In situ metallization in MBE system can eliminate oxide contamination in the ohmic contact formation.

The most widely used method of forming ohmic contacts to the III–V semiconductors is the alloy regrowth technique. A thin metallic film containing a suitable dopant is deposited by thermal evaporation, sputtering, electroplating, or other means onto the surface to be contacted. The contact structure is then heated to a temperature above the melting point of the metal film, and a thin layer of the semiconductor is partially dissolved by the molten metal, forming an alloy containing the dopant in high concentration. Upon cooling, the dopant is incorporated in the semiconductor during epitaxial regrowth. The melting point of the metal film is chosen to be well below the melting point of the semiconductor. Metallic alloys of eutectic composition are often used to achieve ohmic-contact formation at relatively low alloying temperatures.

Another way of obtaining low-resistance ohmic contacts is to lower the barrier at the metal–semiconductor interface. The barrier energy is lowered by introducing a thin layer of a dissimilar semiconductor, with a characteristically low ϕ_B, between the semiconductor to be contacted and the metal electrode. The heterojunction contact was first proposed as an explanation for the ohmic behavior of In contacts alloyed to n-GaAs [Yu70], since InAs has a small band gap and is likely to form by epitaxial regrowth during low-temperature alloying. Woodall et al. [Woo81] have proposed a nonalloyed graded-gap ($Ga_xIn_{1-x}As$) structure grown between the GaAs substrate and the InAs adjacent to the metal contact. Epitaxial heterojunction nonalloyed ohmic contacts are important for heterojunction bipolar transistors and other high-speed devices. The alloyed and nonalloyed ohmic contacts will be discussed in later sections.

7.4.2 Alloyed Ohmic Contacts

The alloyed ohmic contact method is widely used in forming ohmic contacts to the III–V semiconductors. Rideout [Rid75] has reviewed methods of forming ohmic contacts to the III–V semiconductors and the associated state-of-the-art technology throughout 1974. Alloyed ohmic contacts to GaAs have been reported by many workers. More often alloys based on Au or Ag are used to form alloyed contacts in GaAs devices, and these alloys usually contain Zn as the dopant for contacting p-type material and Ge or Sn for contacting n-type GaAs. In this section alloyed ohmic contacts on GaAs and AlGaAs will be discussed.

The AuGeNi Contact

Tables 7.2 and 7.3 list several multilayered, alloyed thin film systems that have been developed as ohmic contacts to GaAs [Rod85]. The contact exhibiting the lowest r_c and the highest reliability is the Au–Ge–Ni system, which was introduced by Braslau

OHMIC CONTACT

TABLE 7.2 Alloyed Ohmic Contacts to the III–V Binary Compounds

Semi-conductor	Type	Contact Material	Minimum r_c ($\Omega\text{-cm}^2$)	Majority Carrier Concentration, n on p (cm^{-3})
AlP	n	Ga–Ag		
AlAs	n, p	In–Te		
	n, p	Au		
	n, p	Au–Ge		
	n	Au–Sn		
GaP	p	Au–Be–Ni	7.5×10^{-5}	2×10^{18}
	p	Au–Zn–Sb	1.5×10^{-2}	6×10^{17}
	p	Au–Be		5×10^{17}
	p	Au–Zn	6.5×10^{-4}	2×10^{19}
	n	Au–Ge–Ni	8×10^{-5}	3×10^{18}
	n	Au–Ge–Ni–Sb	3×10^{-3}	5×10^{16}
GaAs	p	Au–Zn	$r_c \approx (1.8 \times 10^{18})/p^{1.3}$	10^{17}–10^{19}
	p	Ag–Zn	2×10^{-5}	2×10^{17}
	p	Ag–In–Zn	$< 10^{-4}$	10^{18}
	n	Au–Ge–Ni	$r_c \approx (1.8 \times 10^{12})/n$	10^{15}–10^{19}
	n	Ag–In–Ge	6×10^{-4}	5×10^{15}
	n	Ni–Ge^a	3×10^{-5}	1×10^{17}
	n	Pd–Ge^a	3.5×10^{-4}	1×10^{16}
GaSb	p	Au–Zn		
	p, n	In		
	n	Ag–In		
	n	Au–Ge–Ni		
InP	p	Au–Be	$r_c \approx (1 \times 10^{14})/p$	10^{16}–10^{19}
	p	Au–Mg	10^{-4}	6×10^{17}
	p	Au–Zn	5×10^{-3}–1.1×10^{-4}	10^{16}–10^{18}
	p	In–Zn	1×10^{-2}	5×10^{16}
	n	Au–Ag–Sn	8×10^{-5}	5×10^{17}–2×10^{18}
	n	Ag–Sn–In	$< 10^{-4}$	3×10^{15}
	n	Au–Ge–Ni	3×10^{-5}–8×10^{-7}	3×10^{16}–8×10^{17}
	n	Au–Sn	1.8×10^{-6}	3×10^{18}
InAs	n	In		
	n	Sn–Te		
InSb	n	In		
	n	Sn–Te		

Source: Robinson, G. Y., in *Physics and Chemistry of III–V Compound Semiconductor Interfaces*, ed. by C. W. Wilmsen, Plenum Press (1985).

et al. [Bra67] for ohmic contacts to Gunn diodes. It was found that when the contact was heat treated above the Au–Ge eutectic temperature (360°C), an ohmic contact was formed to n-type GaAs for a wide range of GaAs resistivity. During alloying, the Ge apparently formed an n^+ layer [And72] sufficiently heavily doped to produce a linear current-voltage characteristic as a result of field emission at the contact inter-

TABLE 7.3 Alloyed Ohmic Contacts to III–V Alloy Semiconductors

Alloy	Type	Contact Material	Minimum r_c (Ω-cm^2)	Majority Carrier Concentration, n or p (cm^{-3})
$Al_xGa_{1-x}As$				
$x = 0.4$	p	Al	2×10^{-5}	2×10^{19}
	p	Au–Zn	8×10^{-6}	2×10^{19}
	n	Au–Ge–Ni	2×10^{-4}	1×10^{18}
$Ga_xIn_{1-x}As$				
$x = 0.47$	p	Au–Zn	2×10^{-5}	5×10^{18}
	n	Au–Ge–Ni	5×10^{-7}	1×10^{17}
	n	Au–Sn		
$GaAs_{1-x}P_x$				
$x = 0.4$	p	Au–Zn	6.5×10^{-4}	2×10^{19}
	n	Au–Ge–Ni	2×10^{-4}	4.5×10^{17}
	n	Au–Ge–NiCr	10^{-4}	10^{16}–10^{17}
$Ga_yIn_{1-y}As_{1-x}P_x$				
$x = 0.4, y = 0.3$	p	Au–Mg		2×10^{18}–5×10^{18}
vary x, y	p	Au–Zn	10^{-4}–10^{-5}	1×10^{17}
$x = 0.4, y = 0.3$	n	Au–Ge–Ni	5.8×10^{-6}	
$Al_yGa_{1-y}As_{1-x}Sb_x$				
$x = 0.9, y = 0.5$	p	Au–Zn		
	n	Au–Ge–Ni		

Source: Robinson, G. Y., in *Physics and Chemistry of III–V Compound Semiconductor Interfaces*, ed. by C. W. Wilmsen, Plenum Press (1985).

face [Rob85]. The Au provided a low eutectic temperature and was compatible with microelectronic processing and packaging techniques.

The Au–Ge alone did not wet the GaAs surface well during alloying, but the presence of a small amount of Ni (2–11 wt%) greatly improved surface uniformity of the alloyed Au–Ge contact. Ni improves the wetting of the molten Au–Ge film and increases the incorporation of the Ge in GaAs [Rob75]. The barrier energy was found to be about 0.68 eV before alloying [Rob75] and about 0.3 to 0.4 eV after alloying [Rob75, Par76]. Murakami et al. [Mur86] reported that in the Au–Ni–Ge system, a bottom Ni layer was found to improve adhesion between the metal and the GaAs and to lead to more uniform contact resistance.

Three different kinds of samples were prepared, and their cross sections are shown in Fig. 7.20. Sample A was prepared by sequentially depositing a 100-nm Au–Ge layer with a eutectic composition (27 at % Ge), followed by 35-nm Ni and 50-nm Au layers. Samples B and C were prepared by depositing 5 nm and 10 nm of Ni, respectively, as the first layer, followed by Au–Ge, Ni, and Au layers. The total thickness of Au–Ge, Ni and Au layers were kept at 100, 35, and 50 nm, respectively, corresponding to an average composition of 59 at .% Au, 28 at .% Ni, and 13 at .% Ge. The samples were alloyed in a furnace with a continuous flow of Ar/H$_2$ (10%) to form contacts. The contact resistance were measured by an automatic tester using

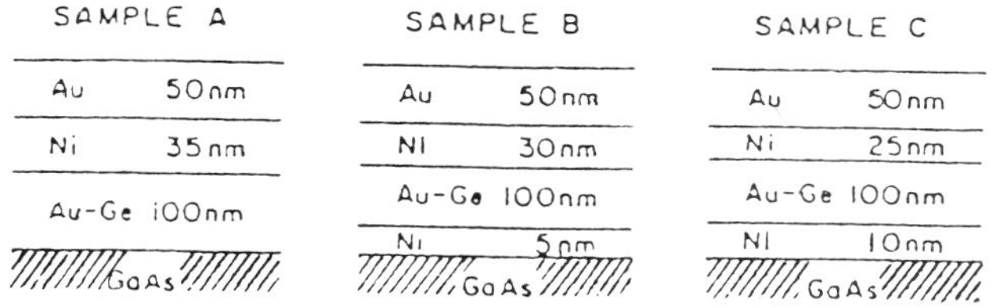

Fig. 7.20 Cross sections of samples *A*, *B*, and *C*. [Murakami, M., K. D. Childs, J. M. Baker, and A. Callegari, *J. Vac. Sci. Tech.* **B 4**, 4 (1986): 903–911]

the transmission line model (TLM) [Ber72]. The low r_c values were observed at temperatures at which the β-AuGa and NiAs(Ge) compounds were formed in the region of the metal–semiconductor interface. In addition the lowest r_c values were obtained when most NiAs(Ge) phases are in contact with the GaAs substrate and the β-AuGa phases are concentrated near the top of the contact. For sample *B* the low r_c values are obtained when the β-AuGa and NiAs(Ge) compounds coexist. Sample *B* has the smallest spread of the resistance values, as seen in Fig. 7.21.

The characteristics of the Au–Ge–Ni contact to GaAs have been reviewed by Braslau [Bra81] and by Heiblum et al. [Hei82]. Braslau [Bra81] noted that the r_c

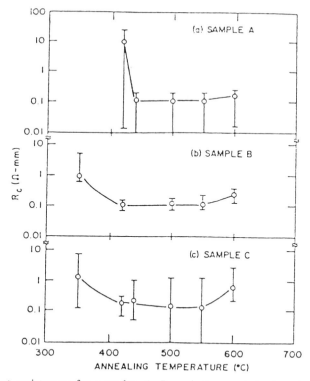

Fig. 7.21 Contact resistances for samples *A*, *B*, and *C* annealed at various temperatures. [Murakami, M., K. D. Childs, J. M. Baker, and A. Callegari, *J. Vac. Sci. Tech.* **B 4**, 4 (1986): 903–911]

data for the Au–Ge–Ni contact varied inversely with the substrate doping N_D, as shown in Fig. 7.22.

The $1/N_D$ dependence for r_c had been previously noted by Goldberg and Isarenkov [Gol70] and Edwards et al. [Edw72]. Such dependence is unexpected if during alloying it is assumed that the Ge is incorporated as a donor in concentration N^+ to form a n^+ layer of thickness t as shown in Fig. 7.19 [Rob85]. If $t > w$, where w is width of the depletion region at the metal–semiconductor interface, r_c will be controlled by field emission independent of N_D in the form of $\exp(\phi_B/N)$. This case is shown in Fig. 7.22 as horizontal lines, where it is assumed that $N^+ = 6 \times 10^{19}$ cm^{-3} and $uo/\phi_B = 0.3$ and 0.8 eV. If $t < w$, then r_c will depend on N_D, and in the limit of $t \to 0$, $r_c \propto \exp(uo/\phi_B/N_D^{1/2})$. Neither case explains the experimental measurements for the entire range of N_D.

An important consideration is finding an optimum alloy cycle to form the ohmic contact. The resulting metallurgical Au–Ge/GaAs interface has been analyzed using Auger depth profiling, X-ray photoelectron spectroscopy (XPS), scanning electron microscope (SEM), and other techniques by various researchers [Rob75; Ili83; Oga80].

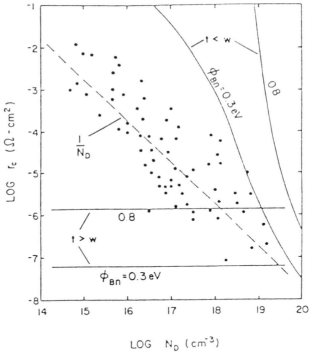

Fig. 7.22 Specific ohmic resistance r_c versus substrate doping N_D for Au–Ge–Ni alloyed contacts to n-type GaAs. the data points (from Braslau [Bra81] and references therein) follow a N_D^{-1} dependence (*dashed line*). Assuming uniform contact area, two theoretical cases (*solid lines*) are shown: for $t < w$, r_c is controlled by the substrate doping and varies with N_D and for $t > w$, r_c is controlled by the concentration of the Ge ($\sim 6 \times 10^{19}$ cm^{-3}) and independent of N_D. (Robinson, G. Y., in *Physics and Chemistry of III–V Compound Semiconductor Interfaces*, ed. by C. W. Wilmsen, Plenum Press (1985); with permission from the publisher]

In particular, Ogawa's microprobe AES and X-ray diffraction analyses on Ni/Au-Ge/GeAs ohmic contacts alloyed at various temperatures are important [Oga80]. Ogawa found the following properties:

1. At 300°C Ge diffuses rapidly toward the contact surface and is trapped or reacts with a part of the Ni. The remaining Ni diffuses inward. At the interface between the contact and the substrate, GaAs is decomposed partly through the reaction between Ni and GaAs but mainly through the Au and GaAs reaction catalyzed by Ni.
2. High reactivity of Ni with GaAs in a solid–solid phase explains the uniform alloying behavior of this contact.
3. Hexagonal reactants of β-AuGa, α-AuGa, and NiAs are formed during alloying at 300°C.
4. At 400°C the Ge that has been trapped in the surface layer diffuses inward and dopes the GaAs.
5. At 500°C the contact has a microscopic grain structure consisting of Ni–As–Ge grains and Au–Ga grains.
6. Contacts alloyed around 500°C are highly reliable due to the metallurgical stability of the grain structure.
7. Rapid heating is necessary to reduce the irregular penetration of Ni into the GaAs substrate.
8. The atmosphere in which the alloying is conducted can also affect the contact's quality. Arsenic overpressure has sometimes been used to suppress As out diffusion. Hydrogen and forming gas are more commonly used.

Besides these guidelines there are the wide variety of alloy temperatures and times used by different manufactures due to differences in equipment and methods of temperature measurement. At each process site the alloy parameters are optimized by running a series of experiments with different alloy parameters and times for their specific metal system. In each case the contact resistance is usually measured with a TLM test structure. Stability of the contact is checked by subjecting the sample to high-temperature (250°C) storage bakes for 500 to 1000 h. In the commercial market TriQuint Semiconductor uses Au/Ni/Ge structures for their MESFETs.

Other Alloyed Contacts

The search for reproducible, reliable, thermally stable, low-contact-resistance ohmic contact materials to GaAs is an important research area for material scientists. Various metals have been used, and different annealing schemes attempted [Mar90].

Thermal, flash-lamp, laser, and electron-beam annealing have been used successfully to obtain ohmic contacts. Heating of only the metal–semiconductor interface can be achieved by sub-band-gap laser illumination from the backside of the wafer. Since the GaAs is transparent to $\lambda = 1.06 - \mu$m radiation, all of the energy is deposited at the contact [Ora81]. Laser energies between 0.3 and 0.5 J/cm^2 produce reasonable contact resistances.

To circumvent the problems encountered with low-melting-point metals such as AuGe, other material systems have been investigated. One such system is AlNiGe, described by Zuleeg [Zul86]. Al has a higher melting point than Au, requiring an anneal at 500°C. These contacts exhibited high reliability because of the high activation energy required for phase migration. Ni is used to prevent AlGe from forming and balling up to degrade the surface morphology.

Refractory metals are described for use in ohmic contacts for several reasons. Contact formation occurs at higher temperatures and by sintering, as opposed to alloying. This type of contact does not penetrate into the GaAs deeply, which is better suited to thin epitaxial structures. These contacts are thermally stable.

Mo/Ge contacts were successfully fabricated by Tiwari [Tiw83]. Very little (~ 40 nm) intermixing of the MoGe-GaAs was observed at the relatively high anneal temperatures used ($\sim 745°C$). The Ge diffused into the GaAs to form the usual n^+ region, and $\rho_c = 10^{-6}$ was obtained without the thermal stability problem in the AuGeNi case.

7.4.3 Nonalloyed Ohmic Contacts

Alloyed contacts suffer from several drawbacks. Contact properties are not easily reproduced, the contact system poses reliability concerns, and the maximum achievable surface-doping density is limited. The Au–Ge–Ni alloy system makes it difficult to understand the underlying physical mechanisms that govern the formation of the ohmic contact. Other fabrication techniques have been studied as possible replacements to the contact alloying procedure. Contacts formed by these methods are commonly referred to as *nonalloyed contacts*. Such contacts are shallow, interfacially uniform, and have good surface morphology.

Methods of Forming Nonalloyed Contacts

There are three methods for forming nonalloyed contacts [Mar90]. They are doping to extremely high levels, sintering, and using low-barrier heterojunctions. These methods involve processes like ion implantation; epitaxial growth of heavily doped regions; dopant diffusion from a solid, a gas source, or an epitaxially grown suitable film prior to metal deposition; and the formation of several epitaxially grown compound semiconductor layers of varying composition. The epitaxial growth of a very heavily doped region can be achieved by MBE, ICB (ion cluster beam) [Ish76], or MOCVD techniques.

The dopants are activated by annealing at an elevated temperature. This can be done by rapid thermal annealing (RTA). The main advantages that nonalloyed contacts have to offer are good process controllability, the possibility of achieving high doping density levels (carrier concentration in excess of 5×10^{19} cm^{-3}), and the possibility in certain cases of avoiding the need to anneal the doped regions [Coh86]. Some of the disadvantages inherent to these techniques are the excessive damage caused to the substrate, the need to apply high-temperature cycles in most cases, and the difficulty to ensure that a given type of dopant would reside preferentially on the proper lattice site.

A common method is to epitaxially grow a semiconductor with a small band gap on top of the GaAs layer that requires a contact. One example is to insert a thin layer of degenerately doped (n^+) Ge between the metal and the GaAs reported by Stall et al. [Sta79; Sta81]. n^+-Ge has a negligible electronic barrier height of 60 mV at the interface to n-GaAs. By using MBE to deposit the Ge at low temperature with As doping, the donor concentration exceeds 10^{20} cm^{-3} in the Ge layer as compared to a maximum donor concentration of about $5 \times 10^{19}\,\text{cm}^{-3}$ obtainable by MBE in GaAs [Bar78].

The low-resistance ohmic contacts to n-GaAs are difficult to obtain due to a 0.8 eV Schottky barrier associated with the metal–GaAs interface. Woodall et al. [Woo81] first proposed and demonstrated a nonalloyed graded-gap scheme for obtaining ohmic contacts n-GaAs. Fermi-level pinning occurs at or in the conduction band on InAs surfaces, and Schottky barrier heights for metal contacts to $Ga_{1-x}In_xAs$, $0.8 < x < 1$, are less than or equal to zero [Woo81]. Thus, an epitaxial layer of n-type $Ga_{1-x}In_xAs$ grown by MBE on n-GaAs that is graded in composition from $x = 0$ at the GaAs interface to $0.8 < x < 1$ at the surface is expected to produce an "ohmic" structure. If the GaAs-to-InAs transition were not graded, it would act as a quasi-Schottky barrier with a barrier height close to the conduction-band offset E_c of the GaAs/InAs heterojunction, about 0.9 eV [Woo81]. The grading flattens out the heterojunction barrier and loads to an excellent ohmic contact. This method is an attractive alternative to the widely used Au/Ge/Ni/Au alloyed system.

The metal/n-GaAs contact can be represented by the energy band diagram of Fig. 7.23(a). The situation for a metal/n-InAs contact shown in Fig. 7.23(b) produces

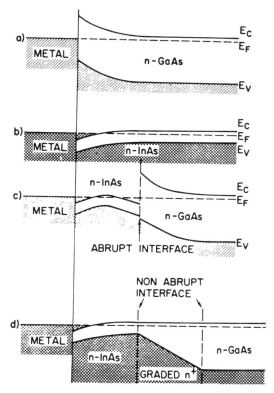

Fig. 7.23 Band diagram for InAs–gaAs heterojunction ohmic contacts: (a) n-GaAs only; (b) n-InAs only, (c) inAs-GaAs abrupt interface; (d) InAs–$Ga_xIn_{1-x}As$–GaAs graded interface. [Woodall, J. M., J. L. Freeoout, G. D. Pettit, T. Jackson, and P. Kirchner, *J. Vac. Sci. Tech.* **19** (1981): 626–627]

an ideal ohmic contact where ϕ_b is 0. In this case tunneling is not required, and low-resistance contacts can be made for a wide range of n-type doping without need of alloying to form n^+ surface layer. Figure 7.23(c) shows that the structure M/n-InAs/n-GaAs may not give a good ohmic contact for GaAs. There is a positive gap between the n-InAs and n-GaAs which, depending on the doping level, results in either rectifying or tunneling ohmic contacts. The barrier results from the large lattice's constant discontinuity and a "dirty" GaAs surface prior to epitaxial growth. This will produce midgap interface states and hence midgap Fermi-level pinning. Figure 7.23(d) shows that the abrupt n-InAs/n-GaAs junction is replaced by a layer of $Ga_{1-x}In_xAs$ graded in composition from $x = 0$ at the GaAs interface to $x = 1$ at the InAs interface. There is no abrupt discontinuity in the conduction band, and ϕ_b is 0 for the M/n-InAs contact.

Sintered contacts are formed by solid-phase migration and formation of metal constituents. Ge/Pd(Sb) contacts were successfully formed at 450°C [Mar85]. The Sb-doped Pd layer is deposited first. PdGe is formed, and the remaining Ge is transported through the PdGe to grow epitaxially onto the GaAs. The Sb effectively dopes the Ge, forming a highly doped Ge/GaAs heterojunction [Mar90].

For advanced electronic, photonic, and quantum effect devices, the development of ohmic contacts with greater thermal stability and shallower penetration depth is of critical need. Although the Ge/Pd and related contacts are shallow, contacts based on amphoteric dopants (e.g., Ge and Si in GaAs) may be inappropriate for applications that require long-term reliability or thermal stability at elevated temperatures [Wan87]. A stable, shallow, low-resistance contact that with low sheet resistance (2 to 3 Ω/sq. for a 90-nm-thick metallization) can be fabricated by electron-beam evaporation, photoresist lift-off patterning, and moderate temperature annealing without a refractory capping layer has been developed by Wang et al. [Wan90]. The annealed metallization consists of a single-phase PdIn with a high melting point of 1285°C. Figure 7.24 shows the thermal stability of Pd–In/Pd/n-GaAs contacts at 400° and 500°C. For the Ge-doped contacts the specific contact resistivity is relatively insensitive to anneal temperature.

7.4.4 Ohmic Contact to p-GaAs

In the past Ohmic contacts to p-GaAs did not receive as much attention as ohmic contacts to n-GaAs. However, in recent years with the tremendous growth of GaAs technology, reproducible and reliable ohmic contacts to p-GaAs in devices such as complementary FETs, heterojunction lasers, integrated optical devices, and solar cells have become extremely important. The performances of p-channel FETs in complementary logic are governed by the large gate-source resistance, particularly by the large contact resistance [Cas86; Hir86]. Thus a reduction of the contact resistance is essential for improving the performance of p-channel devices.

An AuZn alloy has been widely used to form ohmic contacts to p-GaAs in such optical devices as lasers and solar cells [Gop71; Hen77]. However, the AuZn-alloy contact at times exhibits poor adhesion, since Zn (which has poor adherence to GaAs) evaporates faster than Au due to its higher vapor pressure at the same evaporation temperature. An AnZnNi alloy, similar to AuZn, can also be applied to form ohmic contacts in p-type HFET devices [Oe85]. The contact resistance of such an alloy is

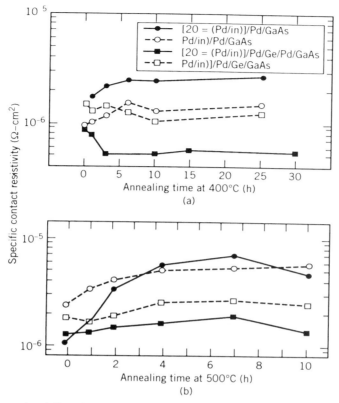

Fig. 7.24 Thermal stability of Pd–In/Pd/n–GaAs contacts at 400° and 500°C. All samples were annealed initially for 5 s at 650°–700°C. [Wang, L. C., X. Z. Wang, S. S. Lau, T. Sands, W. K. Chan, and T. F. Kuech, *Appl. Phys. Lett.* **56**, 21 (1990): 2129]

about ten times that of the usual *n*-channel devices and not sufficiently low for FET applications. New contact systems such as Au/Cr on Zn-doped (100) oriented GaAs [Cas86] and AuZnNi/Ti/Au [Hir86] are capable of yielding fine-pattern contacts with lower resistances.

Hirano et al. [Hir86] have developed a metal structure consisting of AuZnNi/Ti/Au that can reduce the "ball-up" effect in metallization by introducing over 100-nm thick Ti. The substrate material used in this experiment was a 450-um thick semi-insulating GaAs wafer with a (100) surface orientation. The wafer was implanted with 50-keV Be-ions mainly at a dose of 6×10^{13} cm^{-2}. The sample was than annealed at 850°C for 4 s with a lamp annealing system for dopant activation. A metal structure consisting of $Au_{0.85}Zn_{0.10}Ni_{0.5}$ (160 nm thick), Ti (0 to 200 nm thick) and Au (200 nm thick) was successively evaporated while the wafers were maintained at room temperature. Metal sintering was performed at 520°C for 30 s in a flowing N_2 atmosphere. The contact resistance was measured by a transmission-line model (TLM).

The measured contact resistance for *p*-GaAs as a function of the thickness of Ti metal is shown in Fig. 7.25 [Hir86]. The contact resistance R_c can be reduced by introducing Ti to about one order of magnitude smaller in value than that without

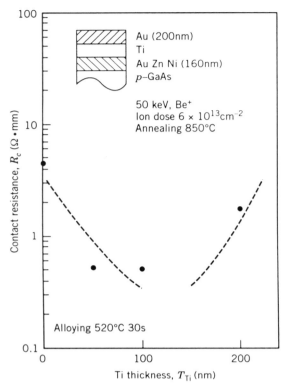

Fig. 7.25 Contact resistance versus Ti thickness. [Hirano, M., and F. Yanagawa, *Jap. J. Appl. Phys.* (August 1986): 1268–1270]

Ti, up to a Ti thickness of 150 nm. This reduction of the ohmic contact resistance is analogous to a resistivity reduction of AuGe/Ni ohmic contacts to n-GaAs by covering with Ti as observed by Ito et al. [Ito84]. The resistivity decrement in an ohmic contact to n-GaAs achieved by adding Ti originates mainly from preventing excess Au diffusion. The origin of the reduction of the p-type ohmic contact resistance was confirmed to be basically the same as that in the n-type case [Hir86].

Figure 7.25 gives the plot of the contact resistance (R_c) against Ti thickness (T_{Ti}). R_c again increased when the Ti thickness was greater than 150 nm. This is caused by an excess diffusion of Ti (itself a Schottky-metal on GaAs) that exists at the GaAs interface. Moreover, since Zn is localized (compared to the 150-nm Ti), not only the Au but also the Zn might be suppressed from diffusion into GaAs. Thus optimizing the tickness of Ti metal is essential in order to reduce R_c. With a Ti thickness of 150 nm, the best ohmic contact was formed with R_c value 0.3 Ω-mm for a GaAs wafer that was Be implanted to a dose of 6×10^{13} cm^{-2}. The R_c reduction with 150-nm thick Ti was cased by the suppression of excess diffusion of the Au metal to the GaAs interface.

The use of carbon as an intentional p-type dopant species for films of GaAs and $Al_xGa_{1-x}As$ has recently become a topic of great interest because of the extremely high substitutional concentration that can be achieved, with doping in excess of

1×10^{19} cm^{-3} having been demonstrated by MOCVD [Kue88, Cun89, Che93] and beyond 1×10^{21} cm^{-3} in the case of MOMBE [Kon89]. Such high levels of carbon doping facilitate the formation of ohmic contacts to *p*-type GaAs and AlGaAs layers and minimize the parasitic sheet and bulk resistances associated with *p*-type layers in device structures such as heterojunction bipolar transistors [deLy91]. We will discuss the carbon-doping effect in Chapter 10.

7.4.5 Measurement of the Specific Contact Resistance r_c

There are several methods that can be used to measure the specific contact resistance of the ohmic contact to semi-insulating GaAs substrate. For the III–V semiconductors accurate determination of r_c is most often carried out by the following methods: the technique of Cox and Strack or the transmission-line model. We will discuss these methods of determining r_c in this section.

Cox-Strack Method

Consider the structure of Fig. 7.26. The resistance of a circular contact of radius a on an *n*-type film of resistivity and thickness t is to be detrmined. The current is flowing through the film to the heavily doped n^+ substrate. The structure requires metallization of both the back and the front surfaces of the semiconductor wafer. The spreading resistance for the layer is

$$R_b = \frac{\rho}{a} F, \qquad (7.31)$$

where F is a function of the ratio a/t and was found experimentally by Cox and Strack to have the approximate form

$$F\left(\frac{a}{t}\right) = \frac{1}{\pi} \arctan\left(\frac{2t}{a}\right). \qquad (7.32)$$

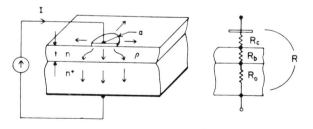

$$r_c = \pi a^2 \left[R - \frac{\rho}{a} F\left(\frac{a}{t}\right) - R_o \right]$$

Fig. 7.26 Measurement of the specific contact resistance r_c using the Cox-Strack method. The circular ohmic contact has radius a. [Robinson, G. Y., in *Physics and Chemistry of III–V Compound Semiconductor Interfaces*, ed. by C. W. Wilmsen, Plenum Press (1985); with permission from the publisher]

More accurate values of F can be found by evaluating numerically an integral expression for $F(a/t)$ in the form of a universal curve. With $F(a/t)$ known, we have

$$R = \frac{r_c}{\pi a^2} + \frac{\rho F}{a} + R_0. \qquad (7.33)$$

In practice, the resistance of an array of contacts with different areas are measured, the spreading resistance is calculated for each contact using Eq. (7.33), and a plot of $R - R_b$ versus $1/a^2$ is made. As shown in Fig. 7.27, a straight line fitted to the data points yields the values of r_c and R_0. For n-type GaAs epitaxial layers with a $5a$, it is possible to measure r_c values down to approximately 1×10^{-6} Ω-cm^2 with an error of about 25%; at higher values of r_c, the error can be much smaller.

Transmission-Line Model (TLM)

The theory of the transmission-line model, or transfer length model (TLM), was developed independently by Murrmann and Widmann [Mur69] and by Berger

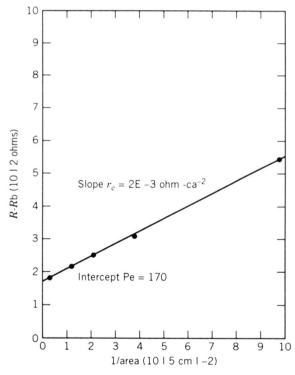

Fig. 7.27 Determination of r_c for an Au–Be alloyed ohmic contact to p-type InP using the Cox-Strack method. [Robinson, G. Y., in *Physics and Chemistry of III–V Compound Semiconductor Interfaces*, ed. by C. W. Wilmsen, Plenum Press (1985); with permission from the publisher]

[Ber72]. Figure 7.28 shows the measurement of a specific contact resistance r_c by the TLM technique. The planar contact is treated as a resistive transmission line with uniform sheet resistance R_s and specific contact resistance r_c.

The voltage and current distributions along the contact are given by the well-known transmission line equations

$$V(x) = V_1 \cosh(\gamma x) - I_1 Z \sinh(\gamma x),$$
$$I(x) = I_1 \cosh(\gamma x) - V_1 Z \sinh(\gamma x), \tag{7.34}$$

where Z is the characteristic impedance of the transmission line and γ is the propagation constant. These are defined by

$$Z^* = \frac{R}{\sqrt{G}} = \frac{\sqrt{R_{sc} r_c}}{W} \frac{1}{\sqrt{1 + j\omega C r_c}}, \tag{7.35}$$

$$\gamma = \sqrt{RG} = \sqrt{\frac{R_{sc}}{r_c}} \cdot \sqrt{1 + j\omega C r_c}.$$

If the analysis is restricted to $\omega = 0$ and the contact is assumed infinite in length

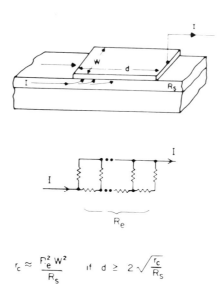

$$r_c \approx \frac{R_e^2 W^2}{R_s} \quad \text{if} \quad d \geq 2\sqrt{\frac{r_c}{R_s}}$$

Fig. 7.28 Measurement of the specific contact resistance r_c by the transmission-line model (TLM) technique. R_e is the total resistance of the metal–semiconductor interface and the epitaxial layer under the contact. [Robinson, G. Y., in *Physics and Chemistry of III–V Compound Semiconductor Interfaces*, ed. by C. W. Wilmsen, Plenum Press (1985); with permission from the publisher]

$(Z = R_c)$, then

$$Z = R_c = \frac{\sqrt{r_c R_{sc}}}{W},$$

$$\gamma = \frac{1}{L_T} = \sqrt{\frac{R_{sc}}{r_c}},$$

(7.36)

where γ is real, for $\omega = 0$ and obviously is the inverse of the transfer length L_T.

7.5 INTERCONNECT METAL SYSTEMS

The interconnect metal system comprises the first-level metal, the vias, and the second-level metal.

7.5.1 First-Level Metal

The metal to be chosen as the first-level metal must have good electrical conductivity, good adhesion, and compatibility with the ohmic contact, and Schottky barrier contacts. The first-level metal plays a major role as FET gate metallization and overlay for the ohmic contact. Some ohmic processes (e.g., Au–Ge/Ni/Au) include a thick overcoat of gold, but the sheet resistance of the composite film after alloying is significantly lower than that of the top layer of gold, prior to the alloy process. Since the total thickness of the ohmic metallization is typically limited to 1000 to 2000 Å, ohmic metal is not generally used as a first-level interconnect. In a similar manner the gate metal may also be limited in thickness. Therefore the first-level metal is often deposited over the ohmic and Schottky barrier contact annd gate metals to increase electrical conductivity. In MMICs the first-level metal may also be used to form capacitor plates and transmission-line elements.

Gold, aluminum, silver, and copper are the only metals with high enough electrical conductivity to be considered for interconnects. Silver is highly reactive and is rarely used in GaAs, except as plated heat sinks [Wel88]. Copper diffuses rapidly in GaAs and getters onto crystalline defects. Hence copper and its alloys are not used on GaAs devices. Aluminum forms intermetallics with gold at higher current densities. Such components start off as Au_2Al, called *white plague* [Wel88], which on further interaction results in a high-resistance intermetallic substance. Hence aluminum is not used as a first metal, leaving gold as the best choice for the first-level interconnect metal system on GaAs. Gold has poor adhesion on GaAs and the dielectrics. The first-level metal is normally a trilayer such as Ti/Pt/Au (or Ti/Pd/Au). Titanium provides the adhesion, and Pt serves as the diffusion barrier for Au. First-level metal is usually 4000 to 5000 Å thick, with Au being the major component. Titanium, Pt, and Au are deposited by *e*-beam deposition and replicated by an enhanced lift-off process.

7.5.2 Vias

The vias or contact windows through the interlevel dielectric, unite the first- and second-metal systems. The choice of a via metal is similar to the second-level metal and is based on high electrical conductivity compatibility with the first-level metal and on resistance to corrosion. In addition the via metal should be free from contact electromigration at relatively high electric fields and current density. A Ti/Au layer of total thickness ranging from 3000 to 7000 Å is normally used for the via metal in a typical IC process. Two methods for via metal are currently in used [Wel88]. The more common method is to deposit vias as part of the second-level metal process, usually by magnetron sputtering. The more sophisticated method now gaining popularity is to fill the vias with evaporated TiAu, using enhanced lift-off techniques followed by additional metal deposited at the second-level metal interconnect process.

7.5.3 Second-Level Metal

In addition to its high electrical conductivity, gold is also resistant to electromigration at high-current densities ($> 1 \times 10^6$ amp/cm^2) [Bla72]. The chip is normally bonded to a gold-plated package with gold wire in high reliability applications. Hence the choice of gold as the second metal, provides an ideal all-gold metal system, free from intermetallics.

Gold can be plated by either an "electron-assisted" or "electroless" process from acqueous solution, usually a gold-cyanide complex. For IC applications, gold in the form of $Au(CN)_2^-$ is used in an acidic bath with pH = 3 [Wel88]. The electroplating baths employed in GaAs IC manufacture invariably use a platinum anode at a current density of 5 mA/cm^2. Electroless plating on GaAs is less common and is mostly used to plate via holes and the backside of GaAs in a MMIC process.

The most common application of plating is used in conjunction with air bridges. Air bridges are used to interconnect the FET electrodes, to cross over a low level of metallization, or to connect the top plate of a MIM capacitor to adjacent metallization in certain GaAs IC processes. Low parasitic capacitances, freedom from edge profile problems, and the ability to carry large currents have been claimed as advantages of an air-bridge interconnect. Since air has a much lower dielectric constant than other dielectrics, lower capacitances are achieved, resulting in high speed.

A typical air-bridge process is as follows: A layer of resist is patterned to open areas over metal pads. A thin film of gold (100 to 500 Å) is sputtered over the entire wafer. Another coat of photoresist is applied and patterned, such that only the pads to be interconnected will be exposed during the plating operation. The thin gold layer provides conduction for the plating current to all parts of the wafer. After the plating is completed, both resist layers and the thin gold layer are removed, leaving the plated thick gold layer to form bridges of interconnect between support posts, with only air below the metal bridges.

The second-level metal in a typical GaAs IC process is Ti/Au with a total thickness 0.7 to 1.0 μm. Magnetron sputtering is used to deposit this thick layer, and an ion mill process is used to replicate the pattern.

REFERENCES

[Wel88] Welch, B. M., D. A. Nelson, Y. D. Shen, and Venkataraman, in *VLSI Electronics: Microstructure Science*, Vol. 15, Academic Press (1988).

[Hen84] Henisch, Heinz K., *Semiconductor Contacts, An Approach to Ideas and Models*, (1984).

[Tya84] Tyagi, M. S., in *Metal-Semiconductor Schottky Barrier Junctions and Their Applications*, ed. by B. L. Sharma, Plenum Press, New York, (1984).

[Sch38] Schottky, W., *Naturwissenschaften* **26** (1938): 843.

[Bar47] Bardeen, J., *Phys. Rev.* **71** (1947): 717–727.

[Cow65] Cowley, A. M., and S. M. Sze, *J. Appl. Phys.* **36** (1965): 3212–3220.

[Rho78] Rhoderick, E. H., *Metal-Semiconductor Contacts*, Clarendon Press, Oxford (1978).

[Cha70] Chang, C. Y., and S. M. Sze, *Solid State Electron.* **13** (1970): 727.

[Sze85] Sze, S. M., *Semiconductor Devices, Physics and Technology*, Wiley-Interscience, New York (1985).

[Cro66] Crowell, C. R., and S. M. Sze, *Solid State Electron.* **9** (1966): 1035.

[Pad66] Padovani, A., and R. Stratton, *Solid State Electron.* **9** (1966): 695.

[Cro69] Crowell, C. R., and V. R. Rideout, *Solid State Electron.* **12** (1969): 89.

[Kwo86] Kwok, S. P., *J. Vac. Sci. Techn.* B **4**, 6 (Nov/Dec 1986): 1383–1391.

[Sin73] Sinha, A. K., and J. M. Poate, *Appl. Phys. Lett.* **23** (1973) 666.

[Ohn83] Ohnishi, T., N. Yokoyama, H. Onodera, S. Suzuki, and A. Shibatomi, *Appl. Phys. Lett.* **43** (1983): 600.

[Lin86] Lin, M. S., Lecture on Tungsten Silicide Deposition, Tainan (1986).

[Cho86] Chou, H. C., and M. S. Lin, in *1986 Electron Devices and Materials Symposium*, IEEE Tainan, p. 284 (1986).

[Kao86] Kao, H.-L., and M. S. Lin, in *1986 Electron Devices and Materials Symposium*, Taiwan, p. 287 (1986).

[Pal85] Palmstrøm, C. J., and D. V. Morgan, in *Gallium Arsenide, Materials, Devices, and Circuits*, ed. by M. J. Howes and D. V. Morgan, Wiley, p. 218 (1985).

[Fow31] Fowler, R. H., *Phys. Rev.* **38**, (1931): 45–56.

[Rob85] Robinson, G. Y., in *Physics and Chemistry of III-V Compound Semiconductor Interfaces*, ed. by C. W. Wilmsen, Plenum Press, New York (1985).

[Sze81] Sze, S. M., *Physics of Semiconductor Devices*, 2nd ed., Wiley-Interscience, New York (1981).

[Mea66] Mead, C. A., *Solid State Electron.* **9** (1966): 1023.

[Kur69] Kurtin, S., T. C. McGill, and C. A. Mead, *Phys. Rev. Lett.* **22** (1969): 1433.

[Mea70] Mead, C. A., *Phys. Rev. Lett.* **22** (1970): 1433.

[Spi80] Spicer, W. E., I. Landau, P. Skeath, C. Y. Su, and P. Chye, *Phys. Rev. Lett.* **44** (1980): 420.

[Bri86] Brillson, L. J., M. L. Slade, R. E. Viturro, M. Kelly, N. Tache, G. Margaritondo, J. M. Woodall, G. D. Pettit, P. D. Kirchner, and S. L. Wright, *Appl. Phys. Lett.* **48** (1986): 1458.

[Bri86a] Brillson, L. J., M. L. Slade, R. E. Viturro, M. Kelly, N. Tache, G. Margaritondo, J. M. Woodall, G. D. Pettit, P. D. Kirchner, and S. L. Wright, *J. Vac. Sci. Tech.* B **4** (1986): 919.

[Bri87] Brillson, L. J., R. E. Viturro, M. L. Slade, P. Chiaradia, D. Kilday, M. Kelly, N. Tache, and G. Margaritondo, *J. Vac. Sci. Tech.* B **5** (1987): 1075.

[Bri88] Brillson, L. J., R. E. Viturro, C. Mailhiot, J. L. Shaw, T. Tache, J. McKinley, G. Margaritondo, J. M. Woodall, P. D. Kirchner, G. D. Pettit, and S. L. Wright, *J. Vac. Sci. Tech.* B **6**, 4 (July/Aug. 1988): 1263–1269.

[Duk85] Duke, C. B., and C. Mailhiot, *J. Vac. Sci. Tech.* B **3** (1985): 1970.

[Mai86] Mailhiot, C., and C. B. Duke, *Phys. Rev.* B **33** (1986): 1118.

[Sea79] Seah, S. P., and W. A. Dench, *Surf. Inter Anal.* **1** (1979): 2.

[Spi80] Spicer, W. E., I. Lindau, P. Skeath, C. Y. Su, and P. Chye, *Phys. Rev. Lett.* **44** (1980): 420.

[New86] Newman, N., Spicer, W. E., Kendelwicz, T., and Lindau, I., *J. Vac. Sci. Tech.* B **4** (1986): 931.

[Wal84] Waldrop, J. R., *Appl. Phys. Lett.* **44** (1984): 1002.
[Spi80] Spicer, W. E., I. Lindau, P. Skeath, C. Y. Su, and P. Chye, *Phys. Rev. Lett.* **44** (1980): 420.
[New86] Newman, N., W. E. Spicer, T. Kendelwicz, and I. Lindau, *J. Vac. Sci. Tech.* **B 4** (1986): 931.
[Sve84] Svensson, P., J. Kanski, T. G. Andersson, and P. O. Nilsson, *J. Vac. Sci. Tech.*, **B 2** (1984): 235.
[Bac81] Bachrach, R. Z., R. S. Bauer, P. Chiaradia, and G. V. Hansson, *J. Vac. Sci. Tech.* **18** (1981): 797
[Duk85] Duke, C. B., and C. Mailhiot, *J. Vac. Sci. Tech.* **B 3** (1985): 1970.
[Cha78] Chadi, D. J., *J. Vac. Sci. Tech.* **15** (1978): 1244.
[Hol82] Holmes, D. E., R. Y. Chen, K. R. Elliot, and C. G. Kirkpatrick, *Appl. Phys. Lett.* **40** (1982): 46.
[Lag82] Lagowski, J., H. C. Gatos, J. M. Parsey, K. Wada, M. Kaminska, and W. Walukiewicz, *Appl. Phys. Lett.* **40** (1982): 342.
[Mir79] Mirceau, A., and D. Bois, *Inst. Phys. Conf. Ser.* **46** (1979): 82.
[Fuj84] Fujimoto, I., *Jpn. J. Appl. Phys.* **23** (1984): L287.
[Yak84] Yakata, A., and M. Nakajima, *Jpn. J. Appl. Phys.* **23** (1984): L313.
[Zur83] Zur, A., T. C. McGill, and D. L. Smith, *Phys. Rev. B* **28** (1983): 2060.
[Duk81] Duke, C. B., A. Paton, R. J. Mey, L. J. Brillson, A. Kahn, D. Kanani, J. Carelli, J. L. Yeh, G. Margaritondo, and A. D. Katanani, *Phys. Rev. Lett.* **46** (1981): 440.
[Lil86] Liliental-Weber, Z., R. Gronsky, J. Washburn, N. Newman, W. E. Spicer, and E. R. Weber, *J. Vac. Sci. Tech.* **B 4** (1986): 912.
[Rob85] Robinson, G. Y., in *Physics and Chemistry of III–V Compound Semiconductor Interfaces*, pp. 73–164, ed. by C. W. Wilmsen, (1985).
[Yu70] Yu, A.C.Y., H. J. Gopen, and R. K. Waits, *Contacting Technology for GaAs*, Final Technical Report, No. AFAL-TR-70-196, AFAL, W-AFB (1970).
[Woo81] Woodall, J. M., J. L. Freeouf, G. M. Pettit, T. Jackson, and P. Kirchner, *J. Vac. Sci. Tech.*, **19** (1981): 626–627.
[Cha71] Chang, C. Y., Y. K. Fang, and S. M. Sze, *Solid State Electron.* **14** (1970): 541.
[Rid75] Rideout, V. L., *Solid State Electron.* **18**, (1975): 541–550.
[Bra67] Braslau, N., J. B. Gunn, and J. L. Staples, *Solid State Electron.* **10** (1967): 381–383.
[And72] Andrews, A. M., N. Holonyak, Jr., *Solid State Electron.* **15**, (1972): 601–604.
[Rob85] Robinson, G. Y., in *Physics and Chemistry of III–V Compound Semiconductor Interfaces*, Chap. 2, pp. 73–163, ed. by Carl W. Wilmsen, (1985).
[Rob75] Robinson, G. Y., *Solid State Electron.* **18** (1975): 331–342.
[Par76] Paria, H., and H. Hartnagel, *Appl. Phys.* **10** (1976): 97–99.
[Ber72] Berger, H. H., *Solid State Electron.* **15** (1972): 145.
[Mur86] Murakami, M., K. D. Childs, J. M. Baker, and A. Callegari, *J. Vac. Sci. Tech.* **B4(4)** (July/Aug. 1986): 903–911.
[Bra81] Braslau, N., *J. Vac. Sci. Tech.* **19** (1981): 803–807.
[Hei82] Heiblum, M., M. I. Nathan, and C. A. Chang, *Solid-State Electron.* **25** (1982): 185.
[Gol70] Goldberg, Yu., and B. V. Isarenkov, *Sov. Phys.—Semicond.* **3** (1970): 1448.
[Edw72] Edwards, W. D., W. A. Hartman, and A. B. Torrens, *Solid State Electron.* **15** (1972): 387–392.
[Ili83] Iliadis, A., and K. E. Singer, *Solid State Electron.* **26(4)** (1983): 7–14.
[Oga80] Ogawa, M., *J. Appl. Phys.* **51** (1) (1980): 406–412.

[Ora81] Oraby, H., K. Murakami, Y. Yuba, K. Gamo, S. Namda, and Y. Masuda, *Appl. Phys. Lett.*, **38** (7) (1981): 562–564.

[Zul86] Zuleeg, R., P. E. Freibertshauser, J. M. Stephens, and S. H. Watanabe, *IEEE Electron Dev. Lett.*, **EDL-7**(11) (1986): 603–604.

[Tiw83] Tiwari, S., T. S. Kuan, and E. Tierney, *Proc. Intl. Electron Dev. Mtg.* (1983): 115.

[Mar90] Maracas, G. N., in *Gallium Arsenide Technology*, Chapter 9, ed. by D. K. Ferry, H. W. Sams Company, (1990).

[Ish76]: Ishida, T., S. Waco, and S. Ushio, *Thin Solid Films* **39** (1976): 227.

[Coh86] Cohen, S. S., and Sh. Gildenblatt, G., *Metal-Semiconductor Contacts and Devices*, Academic Press, Orlando (1986).

[Sta79] Stall, R., C. E. C. Wood, and L. F. Eastman, *Electron. Lett.* **15** (1979): 800–801.

[Sta81] Stall, R., C. E. C. Wood, K. Board, N. Dandekar, L. F. Eastman, and J. Devlin, *J. Appl. Phys.* **52** (1981): 4062–4069.

[Bar78] Barnes, P. A., and A. Y. Cho, *Appl. Phys. Lett.* **33** (1978): 651–653.

[Woo81] Woodall, J. M., J. L. Freeouf, G. M. Petti, T. Jackson, and P. Kirchner, *J. Vac. Sci. Tech.* **19** (1981): 626–627.

[Mar85] Marshal, E. D., W. X. Chen, C. S. Wu, S. S. Lan, and T. F. Kuech, *Appl. Phys. Lett.* **47**(3) (1985): 298–300.

[Wan87] Wang, L. C., F. Fang, E. D. Marshall, and S. S. Lau, *Diffusion Processes in High Technology Materials*, Proceedings of the ASM Symposium, Ohio (1987).

[Wan90] Wang, L. C., X. Z. Wang, S. S. Lau, T. Sands, W. K. Chan, and T. F. Kuech, *Appl. Phys. Lett.* **56** (21) (1990): 2129.

[Cas86] Castanedo, R., R. Asomoza, G. Jimenez, S. Romero, and J. L. Pena, *J. Vac. Sci Tech. A* **4**, 3 (May/June 1986): 814–817.

[Hir86] Hirano, M., and F. Yanagawa, *Jpn. J. Appl. Phys.* (Aug. 1986): 1268–1270.

[Gop71] Gopen, H. J., and A. Y. C. Yu, *Solid State Electron.* **14** (1971): 515.

[Hen77] Henshall, G. D., *Solid State Electron.* **20** (1977): 595.

[Oe85] Oe, K., M. Hirano, K. Arai, and F. Yanagawa, *Jpn. J. Appl. Phys.* **24** (1985): L335.

[Ito84] Ito, H., T. Ishibashi, and T. Sugeta, *Jpn. J. Appl. Phys.* **23** (1984): L635.

[Mur69] Murrmann, H., and D. Widman, *IEEE Trans. Electron. Devices* **ED-16** (1969): 1022–1024.

[Ber72] Berger, H. H., *J. Electrochem. Soc.* **4** 507 (1972): 507.

[Kue88] Kuech, T. F., M. A. Tischler, P. J. Wang, G. Scilla, R. Potemski, and F. Cardone, *Appl. Phys. Lett.* **53** (1988): 1317.

[Cun89] Cunningham, B. T., M. A. Haase, M. J. McCollum, J. E. Baker, and G. E. Stillman, *Appl. Phys. Lett.* **54** (1989): 1905.

[Che93] Chen, H. D., C. Y. Chang, K. C. Lin, and S. H. Chen, *J. Appl. Phys.* **73**, 11 (1993): 7851.

[Kon89] Konagai, M., T. Yamada, T. Akatsuka, K. Saito, E. Tokumitsu, and K. Takahashi, *J. Crystal Growth* **98** (1989): 167.

[deLy91] de Lyon, T. J., N. I. Buchan, P. D. Kirchner, J. M. Woodall, D. T. McInturff, G. J. Scilla, and F. Cardone, *J. Crystal Growth* **111** (1991): 564–569.

[Wel88] Welch, B. M., D. A. Nelson, Y. D. Shen, and R. Venkataraman, in *VLSI Electronics, Microstructure Science*, Vol. 15, p. 392, Academic Press (1988).

[Bla72] Blair, J. G., and C. R. Fuller, *J. Appl. Phys.* **43** (1972): 307–315.

8

GaAs METAL–SEMICONDUCTOR FIELD-EFFECT TRANSISTOR

8.1 INTRODUCTION TO MESFET

The gallium arsenide transistor with a diffusion-gate structure, which was first reported by Turner [Tur67], yields useful gains in the lower megahertz frequency bands. In 1969 Middelhoek obtained, by projection masking, a silicon metal–semiconductor field-effect transistor (MESFET) with a 1-μm gate length and a high maximum frequency of oscillation, F_{max}. In 1971 a significant step was made by Turner et al. [Tur71], when 1-μm gate length FETs on GaAs were made with f_{max} equal to 50 GHz and useful gains up to 18 GHz.

The substantial improvement in FET performance over silicon bipolar transistors is due mainly to the material's properties [Pen86]: First, in gallium arsenide the conduction electrons have a six times larger mobility and twice the peak drift velocity of those in silicon [Ruc70]. Second, the active layer is grown on a semi-insulating GaAs substrate with resistivity larger than 10^7 Ω-cm. This compares with a typical value of 30 Ω-cm for intrinsic silicon. The first property results in lower parasitic resistances, larger transconductances, and shorter electron transit time. The second property results in lower parasitic capacitance when the gate pad is on the semi-insulating (SI) substrate.

Oxide growth has been tried on GaAs surfaces for more than 20 years. The quality of the oxide grown on GaAs has been poor, and a high density of surface states results at the GaAs-insulator interface. These effects make it difficult to fabricate GaAs MOSFETs. Schottky barrier MESFETs and junction field effect transistors (JFETs) are examples of practically used GaAs FETs. In many cases these devices are fabricated by direct ion implantation into a GaAs semi-insulating substrate. Since the enhancement/depletion mode MESFETs can give better speed and complexity for digital applications, the depletion and enhancement channel can be implanted separately.

In this chapter we will discuss the fabrication technology of the MESFET, models of the MESFET, parameter extraction, parasitic effects, and noise theory.

8.2 FABRICATION TECHNOLOGY

The fabrication technology of GaAs MESFET is different from that of Si MOSFET. However, for digital IC applications, the Vitesse Corporation has adopted the nMOS technology to fabricate their gate arrays. This approach is good for DCFL, since the NOR gate is the fundamental logic unit. GigaBit Logic (now merged with TriQuint) has developed their standard cell based on SCFL structures, which will be discussed in Chapter 14. It uses high-temperature refractory metals (Ti–W, W-silicide) for self-aligned gate process compatibility and Ti/Au for the interconnect. The enhancement/depletion MESFET devices give both speed and complexity advantages. For the MMIC application the technology will be different. In this section we will discuss the self-aligned ion implantation technique, D-FET and E-FET channel recess gate formation, and submicrometer T-gate formation.

8.2.1 Self-aligned Ion Implantation

Early MESFETs were fabricated using *n*-type epitaxial layers grown on semi-insulating GaAs with mesa etching employed to isolate the devices. Later developments involved the use of implantation of *n*-type dopants into epitaxial buffer layers or directly into semi-insulating GaAs to achieve continuous *n*-type regions with mesa etching employed for device isolation [Hig78]. With the development of depletion-mode FET-based integrated circuits, localized implantation of *n*-type dopants has been employed to form the active regions of FETs and diodes [Wel77].

Figure 8.1 shows the basic processing steps of the self-aligned gate technology. The first step is that the isolation mesa is defined by etching away the active *n*-layer until the semi-insulating substrate is reached. The gate metal is then deposited over the active area, as shown in Fig. 8.1(*a*) [Pen86].

Source and drain areas are defined in photoresist, and the exposed gate metal is removed by etching. Other etching is used to undercut the resist, as shown in Fig. 8.1(*b*), to allow the necessary space between the gate and drain and the gate and source. Gold

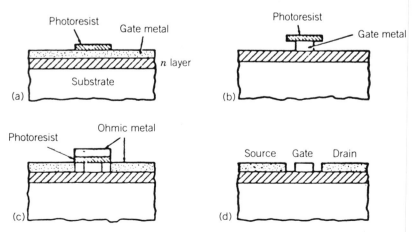

Fig. 8.1 Processing steps of self-aligned gate (SAG) technology. [Pengelly, R. S., *Microwave Field-Effect Transistors—Theory, Design and Applications*, Research Studies Press (1986)]

ohmic contact metallization, usually Au–Ge–Ni or In–Ge–Au is then evaporated, as shown in Fig. 8.1(c). The resist which was protecting the gate stripe is now covered with this ohmic metallization, but this is conveniently removed by "floating off" the gold in dissolving away the resist. The remaining thin gate is left situated between source and drain contacts, as shown in Fig. 8.1(d).

There are two different methods for producing a self-aligned structure. These methods are normally referred to as the *gate-priority* and *ohmics-priority* approaches. The first approach uses a temperature stable gate technique developed by Yokoyama et al. [Yok81] in which the gate metal is first deposited and patterned and then acts as an implantation mask for the self-aligned n^+ contact layers, as shown in Fig. 8.2. Since the gate metal is present during the high-temperature annealing of the n^+ implant, its composition must be carefully chosen to retain its Schottky barrier characteristics. Sputter-deposited Ti–W alloy film was the metal first reported [Yok81]. Thereafter sputter Ti–W silicide [Yok82] and W-silicide [Yok83] was reported. Ta-silicide has been reported as another stable refractory gate metal [Tse82] and electron-beam-evaporated W has also been reported to have good thermal stability up to 950°C [Mat82].

The second approach involves more complex processing, relying on dielectric lift-off using a tri-level photoresist technique to define the placing of the gate metal at a controlled distance from the selectively implanted n^+ regions. Enhancement mode (normally off) and depletion mode (normally on) FETs have been fabricated using this self-aligned FET technology known as *SAINT* (self-aligned implantation for n^+-layer

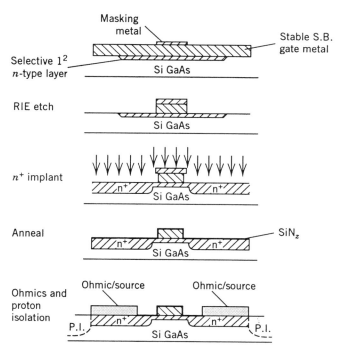

Fig. 8.2 Stable gate self-alignment for a GaAs MESFET. [Pengelly, R. S., *Microwave Field-Effect Transistors—Theory, Design and Applications*, Research Studies Press (1986)]

technology). A typical process sequence for SAINT is illustrated in Fig. 8.3. The main feature of the SAINT FET is that its n^+-layer is embedded between the source and drain electrodes beside the gate channel region. The distance between the gate metal and the n^+-layer can be controlled by undercutting the bottom layer of the tri-level resist. Gate-to-source capacitance can be reduced for high-frequency operation by reducing the distance.

As mentioned in Chapter 4, planar isolation techniques can be used with ion-implanted GaAs layers. A typical processing procedure for an ion-implanted planar FET is listed as follows:

1. The wafer is coated with a layer of silicon nitride that remains on the substrate during subsequent processing. The nitride is removed from the channel region by a plasma etching technique [Rod79].
2. A Si^+ ion implantation creates the active channel. The Si_3N_4 acts as an implantation mask.
3. Activation of the ions is accomplished by annealing using a Si_3N_4 cap.

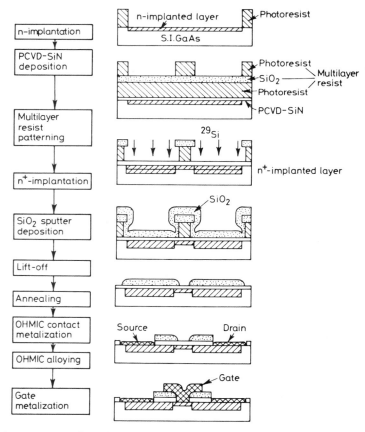

Fig. 8.3 SAINT process flow. [Pengelly, R. S., *Microwave Field-Effect Transistors—Theory, Design and Applications*, Research Studies Press (1986)]

4. The cap is removed only in the source and drain regions, which are defined by applying photoresist. The photoresist is used to lift off the ohmic metal. The source and drain ohmic contacts are alloyed to produce low specific contact resistance.
5. The gate metallization is put down by using a photolithographic technique to open up areas in the remaining Si_3N_4, followed by a resist stage and a float-off process to define the gate stripes resulting in the structure shown in Fig. 8.3.

8.2.2 Recessed Channel Technology

The active channel thickness in MESFET can be defined by etching under the gate region. This removes the high tolerance in thickness required for the epitaxial layer when the channel region is not etched. Figure 8.4 shows the possible gate recess shapes. A recess produces a fairly dramatic change to the field distribution in the channel, as shown in Fig. 8.5 [Lad86]. Once the depletion region reaches the drain-side edge of the recess as the gate-drain voltage is increased, further increases in V_{GD} give rise to only a very slow extension of the depletion edge toward the drain; in other words, the high field region becomes "trapped" by the recess edge.

Figure 8.6 shows the processing steps of etched-channel technology. Source and drain contacts are deposited first as in Fig. 8.6(a), and the gate is defined in the photoresist. A channel is etched in the GaAs until a specific current is measured between the source and drain contacts. The gate metal is then evaporated, and the excess metal is removed by the lift-off technique. Murai et al. [Mur77; Pen86] have reported a technique of intentional side etching of the gate metal to produce a cross-sectional shape much as shown in Fig. 8.7, where the effective gate length is 0.25 μm. The gate metal resistance is reduced because of the mushroom shape (T-gate). The T-gate structure fabricated by e-beam lithography will be discussed in the next section.

In power FETs the breakdown voltage between gate and drain is of importance in determining the RF power handling capability. The recessed channel structure improves this breakdown voltage. Figure 8.8 shows a power FET device structure that improves this breakdown voltage [Pen86]. Three regions are identified where high

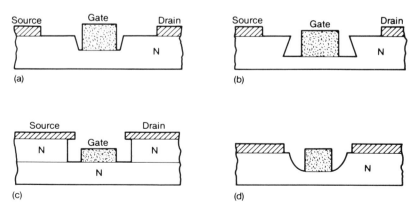

Fig. 8.4 Cross sections of various FET recessed channel structures. [Pengelly, R. S., *Microwave Field-Effect Transistors—Theory, Design and Applications*, Research Studies Press (1986)]

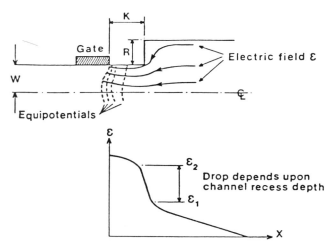

Fig. 8.5 Effect of gate recess on field. [Ladbrook, P. H., in *Gallium Arsenide for Devices and Integrated Circuits*, p. 197, ed. by H. Thomas, IEE Press (1986)]

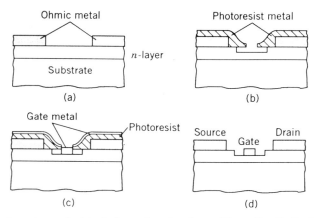

Fig. 8.6 Processing steps of etched-channel technology. [Pengelly, R. S., *Microwave Field-Effect Transistors—Theory, Design and Applications*, Research Studies Press (1986)]

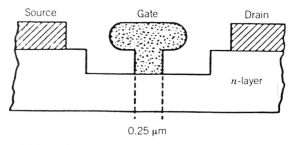

Fig. 8.7 GaAs FET with intentionally side etched gate. [Pengelly, R. S., *Microwave Field-Effect Transistors—Theory, Design and Applications*, Research Studies Press (1986)]

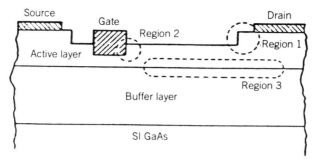

Fig. 8.8 Power GET geometry showing recessed gate and recessed channels. [Pengelly, R. S., *Microwave Field-Effect Transistors—Theory, Design and Applications*, Research Studies Press (1986)]

electric fields can occur. The field in region 1 is reduced by smoothing the channel near the drain, thus reducing the possibility of breakdown due to avalanching and electron-hole pair generation [Tiw79; Pen86]. The field in region 2 is reduced by recessing the gate. The electric field in region 3 occurs because of rising fringing fields toward the active-buffer depletion region, through which substantial space charge limited current may be flowing. This leads to current crowding of carriers as they approach the drain region of the channel. By recessing the active layer 0.5 to 1 µm away from the drain contact edge and making the thickness under the drain contact edge equal to the breakdown depletion width, these effects can be avoided.

8.2.3 Submicrometer Gate MESFET Fabrication

Optical Lithography Process

GaAs MESFETs with $\frac{1}{4}$-µm gate lengths can be developed by a reliable optical lithography process [Can88]. Fabrication involves a bilayer technique [Ban83] incorporating positive photoresist image reversal [Kim85; Can83] and subsequent creation of an Al lift-off mask. The critical process step involves the controlled erosion of a reverse polarity 0.65-µm positive photoresist line to 0.25 µm. The final line length is reproducible to within $\pm 10\%$.

Electron-Beam Lithography Technique

In GaAs MESFETs small gate resistance and short gate length are essential for high-gain and low-noise performance. In the adoption of a T-shaped cross-sectional gate (T-gate) structure, the small foot defines the small length and the wide top provides a low resistance. T-shaped metal lines have been fabricated through the use of a high-sensitivity resist on low-sensitivity resist (HI/LO) double-layer electron-beam (*e*-beam) resist system [Cha83; Cha85]. The resist profiles of a gate test pattern are formed using P(MMA-MAA)/PMMA double-layer resist films. Due to the proximity effect, unless proper resist development and dose compensation are adopted, it is very difficult to achieve sufficiently good undercut profiles for a lift-off through the whole pattern. The T-gate structure has been demonstrated using a resist sandwich structure

[Cha85], where the process's complexity is increased in the preparation of the resist structure.

In this section a LO/HI/LO trilayer system is described for the fabrication of very short T-gates in GaAs FETs [Smi88]. Unlike conventional HI/LO double-layer resist films, the undercut property of the new resist system is less sensitive to the electron dose due to the use of the top LO/HI structure. Therefore undercut T-shaped resist profiles in both small gate lines and large gate pads can be insured for good lift-off capability. The T-shaped resist profiles demonstrate that improved undercut is possible with the trilayer resist system [Cha85]. No barrier layers are needed between resist films in this system. As a result the fabrication process is simpler, more reliable, and more easily reproduced.

The idea behind the fabrication of T-gates in FETs is the use of a thin LO layer on the conventional HI/LO double-layer films. Figure 8.9 gives a schematic of the trilayer resist system. In this system the lower two resist layers provide the T-shaped profile. Forward- and backscattering is minimal in the LO upper layer due to its greater distance from the GaAs substrate. This effect, together with the enhanced development in the middle HI layer, makes the final profile ideal for use in a lift-off process for a wide range of electron doses.

In the process 496K poly methyl methacrylate (PMMA) and copolymers of methyl methacrylate and methacrylate acid P(MMA-MAA) were used as LO and HI resist layers. Since the copolymer is not soluble in the PMMA solvent (i.e., chlorobenzene), there is no mixing of the layers during the preparation of the resist structure, and therefore no barriers are needed between films [How81]. The gate lines were exposed using a Cambridge EBMF-6.5 system at an acceleration voltage of 20 kV, a field size of 3.2768 mm, and a single-pass line dose of 2.5 nC/cm. The composite layer was then developed in ethyl cellosolve acetate (ECA) and methyl isobuthyl ketone (MIBK).

GaAs MESFETs with 0.25-μm T-gates have been fabricated on wafers grown by MBE. New structures of 0.1 μm have been reported [Cha87b]. The material was grown in a Varian GEN-II system on (100)-oriented lightly Cr-doped LEC substrates. The FET material, grown at 600°C with a growth rate of 1 μm/h, consisted of a 1-μm-thick high-quality undoped buffer layer, and a $4 \times 10^{17} \, \text{cm}^{-3}$ Si-doped n-layer with a thickness of 0.3 μm. The high channel doping concentration reduces the parasitic channel resistances and is therefore suitable for low-noise applications.

E-beam lithography was employed for mesa, source-drain, and gate definitions. In the lithography, a "split-level" e-beam exposure technique was adopted: The source-drain ohmic level, for example, was divided into a fine-level A and a coarse-level B,

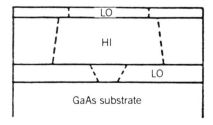

Fig. 8.9 Schematic configuration of the trilayer resist system. [Chao, P. C., P. M. Smith, S. C. Palmateer, and J. C. M. Huang, *IEEE Trans. Electron Devices* **ED-22** (1985): 1042–1046]

as illustrated in Fig. 8.10. The fine level, which consists of 2-µm-wide border of original ohmic levels, is exposed with a low *e*-beam current. A very high beam current and writing speed is then used to expose level *B*, consisting of the inner shaded areas. This allows a vast improvement in the exposure throughput while maintaining very fine edge definition of exposure patterns. In the exposure level *B* was slightly overdosed to ensure a complete exposure coverage over patterns.

In the process AZ resist was used as negative *e*-beam resist for mesa exposures. A chemical etch was adopted to isolate the mesas. PMMA positive resist was then used for source-drain definition. AuGeNi alloyed ohmic contacts were formed using a rapid-thermal annealer and a specific contact resistivity of 3×10^{-7} ohm-cm^2. An automatic wafer register technique [But81] has been used for gate-level alignment. The 0.25-µm T-shaped resist cavities were formed using the trilayer resist technique. After the development of gate-level resist, the GaAs gate area was recessed using a wet chemical etch to achieve the desired full channel current. Ti, Pt, and Au layers of 0.6-µm total thickness were then evaporated to form Schottky gates.

8.3 MODELS

8.3.1 The Shockley Model

Historically field-effect transistor was first analyzed by W. Shockley in the early 1950s [Sho52]. A diagram showing the cross section of a metal semiconductor field-effect transistor is presented in Fig. 8.11. The metal gate forms a Schottky barrier with the GaAs epilayer and the depletion region under the gate controls the cross section of the conduction channel under the gate, modulating the drain-to-source current.

Consider an *n*-type channel under the gate. The depletion region is wider closer to the drain because the reverse bias across the channel-to-gate depletion layer is larger

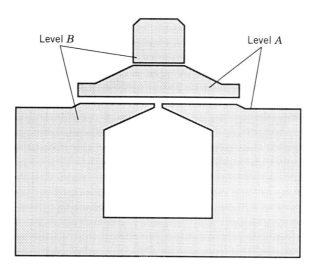

Fig. 8.10 The ohmic pattern is split into level *A* (2-µm-wide border) and level *B* (shaded regions) for both high throughput and fine edge definition. [Chao, P. C., P. M. Smith, S. C. Palmateer, and J. C. M. Huang *IEEE Trans. Electron Devices* **ED-22** (1985): 1042–1046]

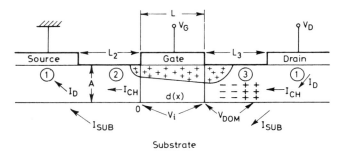

Fig. 8.11 Cross section of a GaAs MESFET. A dipole region (a high-field domain region with a voltage drop V_{dom}) has formed at the drain side of the gate. [Pengelly, R. S., *Microwave Field-Effect Transistors—Theory, Design and Applications*, Research Studies Press (1986)]

there. When the gate voltage V_G is small, the threshold voltage V_T can be expressed by

$$V_T = -V_p + V_{bi}, \tag{8.1}$$

where V_p is the pinch-off voltage and V_{bi} is the built-in voltage. The pinch-off voltage is

$$V_p = \frac{qN_D A^2}{2\varepsilon}. \tag{8.2}$$

Here N_D is the effective donor density and A is the channel thickness width. The effective donor density is assumed to be equal to the electron concentration n_0 in the undepleted portion of the channel, and the doping profile is assumed to be uniform.

When V_G becomes greater than V_T, applying a positive voltage V_{DS} between the source and drain causes electrons to flow. For small values of V_{DS} the layer appears to be a linear resistor, but as larger voltages are applied the electron drift velocity does not rise at the same rate as the electric field E [Run72]. As V_{DS} is increased further, the drain-to-source current starts to saturate. The electric field E reaches a critical value E_c at which the electron reach a saturation velocity v_s, as shown in Fig. 8.11. In short-channel GaAs MESFETs with gate lengths within the order of 0.5 to 2 µm, typical values of the average electric field in the channel are high (5 to 20 kV/cm). Hot-electron effects and the nonlinearity of the electron drift velocity with relation to the electric field curve are important in device operation.

We start from the model that ignores the nonlinear effects and simply implies that the drift velocity v_d is proportional to the longitudinal electric field E up to the point where the channel is pinched off at the drain side of the gate:

$$v_d = \mu E. \tag{8.3}$$

This happens when

$$V_{gs} - V_{ds} \leqslant V_T. \tag{8.4}$$

The proportional constant μ in Eq. (8.3) is the low field mobility. The longitudinal component of the electric field at the drain side becomes very large. Velocity saturation occurs at a finite value of the electric field and, hence, prior to the complete pinch-off of the channel. However, for devices with large pinch-off voltages, long gate lengths, low field mobility, and large electron saturation velocity, the current saturation can be assumed to occur exactly at the channel pinch-off at the drain side of the gate, and Eq. (8.3) is valid up to the pinch-off. This model is named the *Shockley model* after William Shockley [Sho52].

We will now derive the current-voltage characteristics for long-channel FETs based on the Shockley model. Shockley proposed the use of the gradual channel approximation. The gradual channel approximation is based on the assumption that the bias of the gate junction is a slowly varying function of position. That is, the transverse field in the channel (E_x in the x direction) is much larger than the longitudinal field (E_y in the y direction). The following assumptions are made for the conducting channel: The channel is neutral, the region under the gate is totally depleted, and the depletion layer is abrupt; the electric field E in the channel is in the x direction, and the electric field under the gate E_{dep} is in the y direction; and the potential across the channel varies so slowly that at each point the thickness of the depleted area can be found from the solution of the Poisson equation valid for a one-dimensional junction. The incremental change of the channel potential dV is

$$dV = I_{ch}dR = \frac{I_{ch}dx}{q\mu N_D W[A - A_d(x)]}, \qquad (8.5)$$

where I_{ch} is the channel current, dR is the incremental channel resistance, x is the coordinate along the channel, A is the thickness of the active layer, $A_d(x)$ is the thickness of the depletion layer, and W is the gate width. The depletion region width at distance x is given by

$$A_d(x) = \left[\frac{2\varepsilon}{qN_D}[V(x) + V_{bi} - V_G]\right]^{1/2}. \qquad (8.6)$$

Substituting Eq. (8.6) into Eq. (8.5) and integrating with respect to x from 0 (the source side of the gate) to L (the drain side of the gate), we derive the fundamental equation of field-effect transistors:

$$I_{ch} = g_0 \left\{ V_i - \frac{2}{3}\left[\frac{(V_i + V_{bi} - V_G)^{3/2} - (V_{bi} - V_G)^{3/2}}{V_p^{1/2}}\right]\right\}, \qquad (8.7)$$

where V_i is the voltage drop in the channel across the region under the gate,

$$g_0 = q\mu N_D \frac{WA}{L}, \qquad (8.8)$$

is the conductance of the metallurgical channel, L is the gate length, and V_p is the ideal pinch-off voltage.

Equation (8.7) is applicable only up to the point where the neutral channel still exists, even in the narrowest spot at the drain side of the channel:

$$A_d(L) = A_0 = \left[\frac{2\varepsilon}{qN_D}[V_i + V_{bi} - V_G]\right]^{1/2} \leqslant A. \tag{8.9}$$

It is assumed that when $A_d(L) = A$, the current saturation occurs. The saturation voltage $(V_i)_{\text{sat}}^S$ predicted by the Shockley model is given by

$$(V_i)_{\text{sat}}^S = V_p - V_{bi} + V_G. \tag{8.10}$$

Substitution of Eq. (8.10) into Eq. (8.7) leads to the following expression for the saturation current:

$$(I_{ch})_{\text{sat}}^S = g_0\left[\tfrac{1}{3}V_{p0} + \frac{2(V_{bi} - V_G)^{3/2}}{3 V_p^{1/2}} - V_{bi} + V_G\right]. \tag{8.11}$$

The transconductance of the field-effect transistor is

$$g_m = \left.\frac{\partial I}{\partial V_G}\right|_{V_i = \text{const}}. \tag{8.12}$$

From Eq. (8.7) we find in the linear region

$$g_m = g_0\left[\frac{(V_i + V_{bi} - V_G)^{1/2} - (V_{bi} - V_G)^{1/2}}{V_p^{1/2}}\right]. \tag{8.13}$$

Substituting Eq. (8.10) in Eq. (8.13), we find the transconductance in the saturation region:

$$g_m^S = g_0\left\{1 - \left[\frac{(V_{bi} - V_G)}{V_p}\right]^{1/2}\right\}. \tag{8.14}$$

For small drain-to-source voltages

$$V_i \ll V_{bi} - V_G. \tag{8.15}$$

Eqs. (8.7) and (8.13) can be simplified

$$I_D = g_0\left\{1 - \left[\frac{(V_{bi} - V_G)}{V_p}\right]^{1/2}\right\}V_i, \tag{8.16}$$

$$g_m = \frac{g_0 V_i}{2V_p^{1/2}(V_{bi} - V_G)^{1/2}}. \tag{8.17}$$

The total charge in the depletion layer is given by

$$Q = qN_D W \int_0^L h(x)\,dx. \tag{8.18}$$

The derivations of Q with respect to voltages V_{gs} and V_{ds} are related to the capacitances C_{gs} and C_{gd} of the small-signal equivalent circuit:

$$dQ = -C_{gs}dV_g + C_{dg}d(V_i - V_g); \tag{8.19}$$

$$C_{gs} = -\frac{\partial Q}{\partial V_g}\bigg|_{V_i - V_G = \text{const}}, \tag{8.20}$$

$$C_{dg} = \frac{\partial Q}{\partial V_i}\bigg|_{V_G = \text{const}}, \tag{8.21}$$

$$C_{gs} = \frac{C_{gs0}}{(1 - V_{gs}/V_{bi})^{1/2}}, \tag{8.22a}$$

$$C_{gd} = \frac{C_{gs0}}{(1 - V_{dg}/V_{bi})^{1/2}}, \tag{8.22b}$$

$$C_{gs0} = C_{dg0} = \frac{WL}{2}\left(\frac{\varepsilon q N_D}{2V_{bi}}\right)^{1/2}. \tag{8.23}$$

At zero drain-to-source voltage and zero gate voltage, the total gate capacitance is equal to the capacitance of the space-charge region depleted by the built-in voltage:

$$C_{g0} = \frac{\varepsilon WL}{A} = WL\left(\frac{\varepsilon q N_D}{2V_{bi}}\right)^{1/2}. \tag{8.24}$$

This capacitance is equally divided between the source (C_{gs0}) and drain (C_{dg0}) because the space-charge distribution is symmetrical. For nonzero V_{ds} and V_{gs}, the capacitances C_{gs} and C_{dg} behave almost as capacitances of equivalent Schottky diodes connected between the gate and the source, and the gate and the drain, respectively.

The cutoff frequency can be calculated to be

$$f_T = \frac{1}{2\pi}\frac{g_m}{C_{gs}}. \tag{8.25}$$

8.3.2 Analytic Models of GaAs MESFETs

The development of an accurate and simple model for GaAs FET is highly desirable. In this section we consider three models for calculation of the current-voltage characteristics. The first model is the "Raytheon" model; it gives the cubic approximation of the hyperbolic tangent in Curtice model, which accurately describes both the high and low pinch-off voltage devices. However, some effects are not taken into account. The second model considered here is the Jastrzebski model, which includes some new features and keeps the complexity of the model to a minimum without sacrificing its accuracy in microwave applications. The third model is the McCamant model that has been modified by TriQuint and adopted in PSPICE.

Velocity Saturation in the Channel

According to the Shockley model current saturation occurs when a conducting channel is pinched-off at the drain side of the gate. The cross section of the conducting channel predicted by the Shockley model is Zero, and hence the electron velocity has to be infinitely high in order to maintain the finite drain-to-source current. In reality the electron velocity saturates in a high electric field, and this velocity saturation causes the saturation of the current. The importance of the field dependence of the electron mobility for understanding current saturation in field-effect transistors was first mentioned by Dacey and Ross [Dac55]. This concept was later developed in many theoretical models used to describe FET characteristics.

A modified version of Shockley's expression, including velocity saturation, is [Leh70]

$$I_{DS} = I_{DS}^{S_*} \frac{1}{[1 + (V_{DS}/E_c L_g)]}, \tag{8.26}$$

where E_c is the critical field at which electron velocity saturates and L_g the gate length.

Using the modified Shockley expression, it is still difficult to eliminate the discontinuity between the saturates and nonsaturated regimes. For computer-aided design one has to model the current-voltage characteristics in the entire range of the drain-to-source voltages. To overcome this drawback, Curtice [Cur80] has proposed a heuristic functional dependence for the drain-to-source current I_{DS}:

$$I_{DS} = \beta(V_{GS} - V_T)^2 (1 + \lambda V_{DS}) \tanh(\alpha V_{DS}). \tag{8.27}$$

The constant λ in Eq. (8.27) is an empirical constant, and it accounts for the additional output conductance beyond the output conductance.

Raytheon Model (Statz Model)

From the Curtice model, we notice that the dc equations exclusive of parasitic resistances are approximated by [Sus84]

$$I_d = \beta(V_{gs} - V_T)^2 (1 + \mu V_{ds}) \tanh(\alpha V_{ds}). \tag{8.28}$$

In this equation, I_d is the drain current, β is a parameter, V_{gs} is the gate-to-source voltage, V_{ds} is the drain-to-source voltage, V_T is the threshold voltage, μ is a parameter related to the drain conductance, and α determines the voltage at which the drain current saturates at the same drain-to-source voltage irrespective of the gate-to-source voltage. However, when the pinch-off voltage of the transistor is large, the behavior of I_d as a function of V_{gs} is poorly represented. Since the reduction in channel height between the channel entrance and the point where the carrier velocity saturates is usually a negligible fraction of the height at the entrance, the current may be approximately calculated by assuming that all carriers at the channel opening are moving at their saturated velocity. This is a good assumption except for V_{gs} near the pinch-off voltage. For constant channel doping, the saturated drain current I_{ds} can be approximated by

$$I_{ds} = ZV_{sat}(2\varepsilon q N_d)^{1/2}[(-V_T + V_B)^{1/2} - (-V_{gs} + V_B)^{1/2}], \tag{8.29}$$

where Z is the channel width, v_{sat} is the saturated electron velocity, ε is the dielectric constant, V_T is the threshold or pinch-off voltage, and V_B is the built-in potential of the gate junctions.

However, the approximation for the current in Eq. (8.29) breaks down when the voltage drop from the entrance of the channel to the point of velocity saturation of the carriers of comparable to the voltage difference $V_{gs} - V_T$. Under these conditions the assumption of constant channel height breaks down. For a 1-μm channel length FET, the critical voltage drop becomes 0.3 V or less. Hence the approximation in Eq. (8.27) is not valid when $|V_{gs} - V_T| \leqslant 0.3$ V. In the limit of gate voltages near the pinch point, Eq. (8.28) and (8.29) lead to a quadratic form:

$$I_{ds} = \beta(V_{gs} - V_T)^2. \tag{8.30}$$

To smooth connect a law like Eq. (8.30) for small $V_{gs} - V_T$ to an expression like Eq. (8.29) for large $V_{gs} - V_T$, Statz et al. [Sta87] chose the empirical expression

$$I_{ds} = \frac{\beta(V_{gs} - V_T)^2}{1 + b(V_{gs} - V_T)}. \tag{8.31}$$

For small values of $V_{gs} - V_T$, the expression is quadratic, while for larger values, I_{ds} becomes almost linear in $V_{gs} - V_T$. The more gradual doping profiles appear to give a lower value of b. The value of b of a bare transistor is a measure of the doping profile extending into insulating substrate and thus depends on the fabrication process. The value is about 0.3 and can vary from different devices.

The *tanh* function in Eq. (8.28) consumes considerable computer time. A simple polynomial P of the form

$$P = 1 - \left[1 - \frac{\alpha}{n} V_{ds}\right]^n \quad \text{with } n = 2 \text{ or } 3. \tag{8.32}$$

In the saturation region ($V_{ds} > n/\alpha$), the *tanh* function is replaced by unity. The slope at $V_{ds} = 0$ of the polynomial is α and is equal to that of the $\tanh(\alpha V_{ds})$ function.

The preceding GaAs model has been coded in SPICE with the following dc equations:

$$I_d = \frac{\beta(V_{gs} - V_T)^2}{1 + b(V_{gs} - V_T)}\left\{1 - \left(\frac{1 - \alpha V_{ds}}{3}\right)^3\right\}(1 + \mu V_{ds}) \quad \text{for } 0 < V_{ds} < \frac{3}{\alpha}, \tag{8.33a}$$

$$I_d = \frac{\beta(V_{gs} - V_T)^2}{1 + b(V_{gs} - V_T)}(1 + \mu V_{ds}) \quad \text{for } V_{ds} \geqslant \frac{3}{\alpha}. \tag{8.33b}$$

The source and drain capacitance models for the MESFET need to be modified. Previous GaAs device simulations use a diodelike capacitance between source and gate, where the space-charge region's thickness and capacitance is determined by the gate-to-source voltage. A similar diode model is used to describe the gate-to-drain capacitance. However, large errors can be introduced into simulations when low source-to-drain voltages or reverse-biased transistors are encountered.

The behavior for GaAs devices is further complicated by the early onset of carrier-velocity saturation. Van der Ziel [Van63] calculated capacitances for FETs without including the effects of velocity saturation. Figure 8.12 [Sta87] shows the unsaturated velocity values of C_{gs} and C_{gd} as a function of V_{ds} for $V_{gs} = 0$ V. The channel is assumed to have a doping of 1×10^{17} donors/cm^3, and the gate has dimensions of 1×20 μm with $V_B = 0.8$ V. The C_{gs} is approximately constant as a function of V_{ds}. When V_{gs} is not equal to zero, the capacitance curves follow approximately the diode capacitance model as a function of V_{gs}. The gate-drain capacitance C_{gd} starts at the same value as C_{gs} for $V_{ds} = 0$. It then falls continuously with increasing V_{ds} and goes to zero when the drain side of the channel becomes pinched off.

When velocity saturation is taken into account, the situation changes drastically. The results to calculate C_{gs} and C_{gd} can be obtained by taking the partial derivatives of the total gate charge with respect to gate-source and gate-drain potentials. Evaluating the resulting saturated velocity expressions gives a gate-to-source capacitance that rises rather abruptly from the van der Ziel model at the onset of saturation and quickly approaches a nearly constant value as a function of V_{ds}. Similarly the gate-to-drain capacitance drops abruptly to a low value and then stays approximately constant.

To avoid discontinuities that result in convergence problems in simulations, some parameters are added to get the smooth interpolation. The values of C_{gs} and C_{gd} are obtained:

$$C_{gs} = \frac{C_{gs0}(1/2)}{(1 - V_n/V_B)^{1/2}}\left[1 + \frac{V_e - V_T}{[(V_e - V_T)^2 + \delta^2]^{1/2}}\right] \times (1/2)\left[1 + \frac{V_{gs} - V_{gd}}{[(V_{gs} - V_{gd})^2 + (1/\alpha)^2]^{1/2}}\right]$$
$$+ C_{gs0}(1/2)\left[1 - \frac{V_{gs} - V_{gd}}{[(V_{gs} - V_{gd})^2 + (1/\alpha)^2]^{1/2}}\right], \tag{8.34}$$

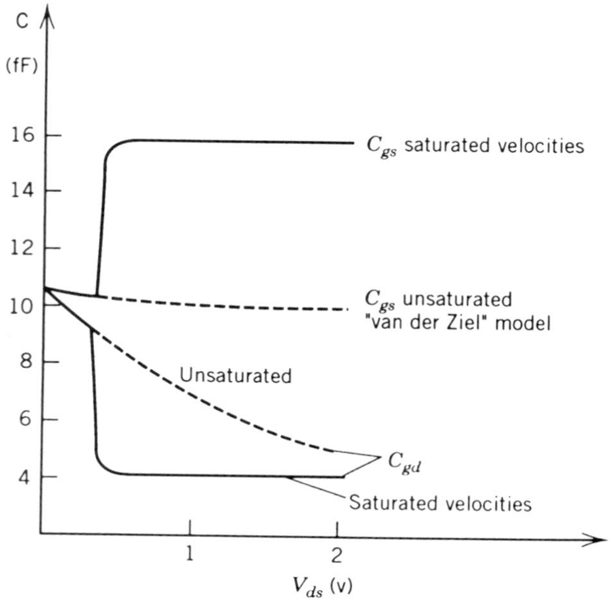

Fig. 8.12 Gate-to-source and gate-to-drain capacitances for a 1×20 μm gate on GaAs with 1×10^{17} donors/cm^3: unsaturated model of van der Ziel (*long dashes*); saturated velocity model (*solid lines*), and simple one-parameter interpolation. [Statz, H., P. Newman, I. W. Smith, R. A. Pucel, and H. A. Haus, *IEEE Trans. Electron Devices* **ED-34**, 2 (1987): 160–168]

$$C_{gd} = \frac{C_{gs0}(1/2)}{(1 - V_n/V_B)^{1/2}} \left[1 + \frac{V_e - V_T}{[(V_e - V_T)^2 + \delta^2]^{1/2}} \right] x(1/2) \left[1 - \frac{V_{gs} - V_{gd}}{[(V_{gs} - V_{gd})^2 + (1/\alpha)^2]^{1/2}} \right]$$
$$+ C_{gs0}(1/2) \left[1 + \frac{V_{gs} - V_{gd}}{[(V_{gs} - V_{gd})^2 + (1/\alpha)^2]^{1/2}} \right], \tag{8.35}$$

where

$$V_e = \frac{V_{gs} + V_{gd} + [(V_{gs} - V_{gd})^2 + \Delta]^{1/2}}{2} \tag{8.36}$$

and

$$V_n = \frac{V_e + V_T + [(V_e - V_T)^2 + \Delta^2]^{1/2}}{2} \tag{8.37a}$$

for

$$\frac{[(V_e - V_T)^2 + \Delta^2]^{1/2}}{2} < V_{max},$$

and

$$V_n = V_{max}$$

for

$$\frac{[(V_e - V_T)^2 + \Delta^2]^{1/2}}{2} \geq V_{max}. \tag{8.37b}$$

In the equations above Δ is between 0 and 0.5. The inclusion of a nonzero Δ produces a smooth transition of width Δ in the value of V_e as a function of V_{gs} or V_{gd}.

Figure 8.13 [Sta87] illustrates the behavior of C_{gs} from Eq. (8.34) as a function of

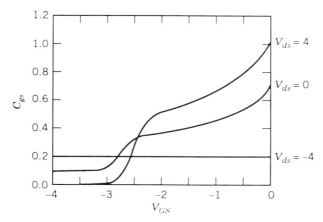

Fig. 8.13 This figure illustrates how, in Statz model, the gate-to-source capacitance behaves as V_{gs} goes through the pinch-off point V_T (= -2.5 V). [Statz, H., P. Newman, I. W. Smith, R. A. Pucel, and H. A. Haus, *IEEE Trans. Electron Devices* **ED-34**, 2 (1987): 160–168]

V_{gs} and for various source-to-drain voltages. For $V_{ds} \gg 0$ (normal bias conditions), C_{gs} follows a diodelike capacitance model as a function of V_{gs}. However, when V_{gs} approaches the pinch-off voltage V_T, C_{gs} falls rapidly to zero within a voltage range δ. For V_{ds} negative, C_{gs} is really a gate-to-drain capacitance because the reverse bias interchanges the roles of source and drain. The capacitance in this range becomes small and independent of V_{gs}. Because of the smooth transition from positive to negative drain-to-source voltages, the situation for $V_{ds} = 0$ is intermediate between the two cases outlines above. Code changes based on the preceding equations have been introduced into the SPICE code, and satisfactory circuit simulations have been obtained.

Materka and Kacprzak Model

The Materka and Kacprzak model uses a very simple but quite accurate formula to describe the dc I–V characteristics of a device:

$$J_{ds}(V_d, V_g) = J_{dss}\left(\frac{1 + V_g}{V_t}\right)^2 \tanh\left(\frac{\alpha V_d}{V_g + V_t}\right), \tag{8.38}$$

$$V_t = V_{t0} + \gamma V_d, \tag{8.39}$$

where

J_{dss} = saturation current,
V_{t0} = threshold voltage of an ideal FET,
V_t = effective threshold voltage,
α, γ = parameters of the model.

The gate-drain breakdown current was simulated by Materka et al. [Mat85] by the reversed diode equation in which the effect of the drain-source current was not included:

$$J_{bgd}(V_{gd}) = -J_{b0}[\exp(-\alpha_b V_{gd}) - 1], \tag{8.40}$$

where J_{b0} and α_b are model parameters.

Jastrzebski Model

The MESFET model proposed by Jastrzebski [Jas88] has less complexity than the Raytheon model. Its parameters can be interpreted on the dc characteristics. This model has the following form:

$$J_{ds}(V_d, V_g) = \begin{cases} J_{dss} V^\varepsilon \left[\tanh\left(\frac{\alpha V_d}{V_{gg}} + V_t\right) + G V_d\right] & \text{for } V_{gg} \geq -V_t, \\ 0 & \text{for } V_{gg} < -V_t, \end{cases} \tag{8.41}$$

$$V = \left(1 + \frac{V_{gg}}{V_t}\right)\left(1 + \frac{V_{g0}}{V_{t0}}\right), \tag{8.42a}$$

$$V_{gg}(t) = V_{gg}(t - \tau) - V_{g0}, \tag{8.42b}$$

$$V_t = V_{t0} + V_{g0} + \gamma V_d, \qquad (8.42c)$$

$$G = \frac{G_0}{J_{dss}(1 + V_{g0}/V_{t0})^\varepsilon}. \qquad (8.42d)$$

The parameters of the above functions have the following interpretation [Jas88] (values for an "ideal" device are given in parentheses):

J_{dss} = saturation current,

ε = exponent of the relationship between current and V_g ($\varepsilon = 2$),

α = parameter controlling saturation voltage V_{sat} [approximately $V_{sat} \sim (V_g + V_{t0})/(\alpha/3 - \gamma), (\alpha = 3)$],

V_{t0} = threshold voltage of an ideal FET [$V_{t0} = V_p - V_{bi}$, where V_p is a pinch-off voltage],

γ = coefficient of the threshold voltage dependence on V_d ($\gamma = 0$),

$V_{g0} = V_g$ voltage at which the $J_{ds}(V_d, V_{g0})$ saturated characteristic has zero slope if $\gamma > 0$ and $G_0 = 0$ [for $V_g < V_{g0}$, the slope is positive, while for $V_g > V_{g0}$ the slope is negative ($V_{g0} = V_{bi}$)],

G_0 = additional component of the conductance [G_0 is equal to the conductance in the saturation at $V_g = V_{g0}$ ($G_0 = 0$)].

Comparison of the Jastrzebski model with the original formula used by Materka and Kacprzak [Eq. (8.38)] indicates that three additional parameters are introduced: ε, V_{g0}, and G_0. The new term $(1 + V_{g0}/V_{t0})$ in Eq. (8.42a) is but a normalizing factor that preserves the physical interpretation of the voltage V_{g0} and simplifies identification of model parameters. Additional conductance control is introduced by the term GV_d, which for most devices may be omitted at dc but is useful for modeling the I–V characteristics in the presence of frequency dispersion.

McCamant Model

McCamant et al. [McC90] have developed a model that includes specific features of MESFET behavior that neither the Curtice nor the Statz equations properly describe. Small-signal parameters (e.g., the S-parameters) are accurately modeled over a wide range of bias conditions. These results were achieved by modifying the Statz model equations to better represent the variation of I_{ds} as a function of the applied voltage.

The $\beta, \alpha, V_t, \lambda$, and b are (constant) model parameters in Eq. (8.33). As listed in Eq. (8.32), there is a polynomial approximation to the *tanh* function devised by Statz to decrease computation time. The first modification to these equations is designed to address the poor fit at near pinch-off values of V_{gs}. This is accomplished by making V_T a function of drain voltage:

$$V_t = V_{t0} - \gamma V_{ds}. \qquad (8.43)$$

This equation improves the drain conductance fit at low drain currents, where the Statz model shows a cutoff independent of drain voltage.

The second modification relates to I_{ds} decreases at higher values of current and voltage, resulting a smaller slope than would be predicted form the Statz model:

$$I_{ds} = \frac{I_{ds0}}{1 + \delta V_{ds} I_{ds0}}, \qquad (8.44)$$

where δ is a (new) model parameter and I_{ds0} is given by the expression

$$I_{ds0} = \beta(V_{gs} - V_t)^Q * P. \qquad (8.44a)$$

Here P is the polynomial function as in Eq. (8.32). This equation resembles the Statz equation (8.33) with b and λ set equal to zero but with new parameters γ and δ introduced. The parameter Q is necessary to model the non-square-law dependence of I_{ds} which is observed for devices with small or positive pinch-off voltage, and has been found useful by other workers [Lar87].

This model provides a better fit for ac characteristics over a wider bias range. Figure 8.14 shows the increased bias range over which the McCamant model provides better RF characterization. The plot gives the contours of the constant S-parameter vector error superimposed on the drain I–V curves. The contours show a 20% vector error between measured S-parameters and S-parameters generated with either the Statz or the McCamant model. The inner curve depicts the Statz model and the outer the McCamant model. The improved model contour encloses the area for the large voltage and low currents as well as for the low drain voltage and high currents. As a result accuracy is improved for gain compression and other nonlinear parameters.

GaAs MESFETs exhibit substantial changes in the value of drain resistance R_{ds} at low frequency [Lar87; Sch88]. The large value at dc decreases with frequency, leveling off around 1 MHz. The parameter most directly influencing R_{ds} in this model is γ. Figure 8.15 illustrates the equivalent circuit model of a MESFET that helps clarify the physical meaning of γ. The current source is controlled by the sum of the gate voltage and an assumed back-gate voltage V_b. The value of γ reflects modulation of V_b by the

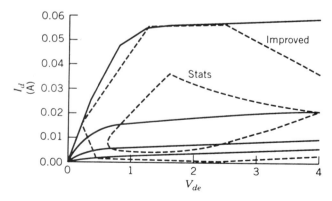

Fig. 8.14 A comparison between the Statz and McCamant models: The inscribed contours represent a maximum error of 0.2 for S_{11}, S_{22}, or S_{12}, or a maximum error of 20% in S_{21} parameters for each of the models. The larger area occupied by the McCamant model is indicative of the improved fit. [McCamant, A. J., G. D. McCormack, and D. H. Smith, *IEEE Trans. Microwave Theory and Tech.* **38**, 6 (1990): 822–823]

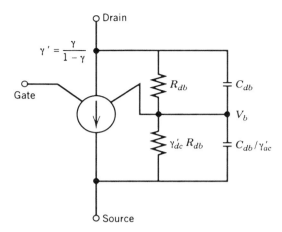

Fig. 8.15 The equivalent circuit model of a GaAs MESFET helps clarify the physical meaning of the parameter γ. The current source in the model is controlled by the sum of the gate voltage and an assumed "back-gate" voltage V_b. [McCamant, A. J., G. D. McCormack, and D. H. Smith, *IEEE Trans. Microwave Theory and Tech.* **38**, 6 (1990): 822–823]

drain electrode. At low frequency this modulation is governed by a redistribution of the (trapped) charge in the substrate and may be represented by resistors, as in the figure. Capacitive coupling dominates at higher frequencies, where γ_{ac} provides a better representation. The shift from ac to dc behavior occurs at frequencies near $1/2\pi R_{bd}C_{bd}$.

By including a voltage dependence of R_{db}, C_{db}, and γ, we can obtain a model that is fully symmetrical with respect to interchange of the source and the drain. Such a model is highly desirable, since, for many circuits, this polarity changes during operation.

8.4 PARAMETER EXTRACTION

8.4.1 Determination of FET Parameters

Determination of Gate Barrier Built-in Voltage and the Ideality Factor

The forward current density J of a Schottky barrier junction for $V > 3kT/q$ can be approximated by [Sze81]

$$J = A^* T^2 \exp\left[\frac{-qV_{bi}}{kT}\right] \exp\left[\frac{qV}{kT}\right], \qquad (8.45)$$

where A^* is the effective Richardson constant, T is the junction temperature, k is Boltzmann constant, n is the ideality factor, and V is the applied forward-bias voltage.

The extrapolated value of current density to zero bias gives the saturation current density J_{sat}. The barrier built-in voltage is then

$$V_{bi} = \frac{kT}{q} \ln\left(\frac{A^* T^2}{J_{sat}}\right), \qquad (8.46)$$

and the ideality factor is

$$n = \frac{q}{kT} \cdot \frac{\partial V}{\partial (\ln J)}. \qquad (8.47)$$

Fukui has given a method of determining various GaAs FET parameters [Fuk79] by a series of dc measurements. The first of these is to measure the forward I–V characteristic of the gate junction at room temperature. A plot of gate forward-bias voltage V_G versus gate current I_G is then plotted. An example of such a plot is shown in Fig. 8.16 [Pen86]. At high values of V_G, the gate current tends to saturate. At low values of V_G, I_G is disturbed by a leakage current around the periphery. In the middle range of V_G, the I_G versus V_G plot on semilog graph paper is linear. From the slope and location of this linear region, it is possible to calculate the ideality factor n of the Schottky junction and the gate built-in voltage V_{bi}, respectively.

Determination of Pinch-off Voltage

By measuring the drain I–V characteristics at low values of I_{ds}, a plot of drain current versus the gate-to-source voltage V_{gs} can be produced. The terminal pinch-off voltage V_p can be determied by extrapolating the plot to the abscissa, as shown in Fig. 8.17.

Determination of Gate-to-Source and Gate-to-Drain Resistance

The drain-to-source resistance R_{DS} at a very low drain-to-source voltage ($\sim 50 \,\text{mV}$) can be plotted as a function of a parameter X defined by [Pen86]

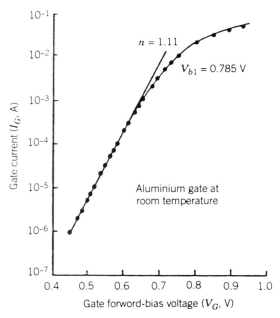

Fig. 8.16 Forward I-V characteristic of an aluminum Schottky gate diode at room temperature. [Pengelly, R. S., *Microwave Field-Effect Transistors—Theory, Design and Applications*, Research Studies Press (1986)]

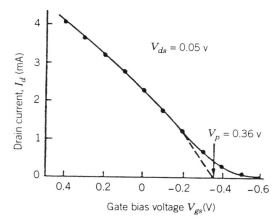

Fig. 8.17 Drain current as a function of gate-bias voltage for forward and reverse drain-source-bias conditions. [Pengelly, R. S., *Microwave Field-Effect Transistors—Theory, Design and Applications*, Research Studies Press (1986)]

$$X = \frac{1}{1 - \sqrt{|(V_{bi} - V_{GS})/(V_{bi} + V_p)|}}. \tag{8.48}$$

R_{DS} is usually frequency dependent. This may due to electron trapping at the active layer-substrate interface [Cam85].

If the plot is a straight line, the value of V_p is accurate. Figure 8.18 [Pen86] shows this plot. By linearly extrapolating the line to the ordinate, a value for the parasitic (gate-to-source and gate-to-drain) resistance, $R_S + R_D$, can be found. The slope of the line is R_0, with $R_0 X$ representing the effective value of the active channel resistance at a given V_{GS}.

Transconductance Determination

The magnitude of the transconductance of a FET device can be assumed to remain constant up to the cutoff frequency f_T. The dc transconductance can, to a first-order approximation, be considered the same as microwave transconductance.

The measured transconductance $g_m(\text{obs})$ is lower than the intrinsic transistor g_m, owing to the gate-to-source resistance R_s. The intrinsic transconductance is given by

$$g_m = \frac{g_m(\text{obs})}{1 - g_m(\text{obs})R_s}. \tag{8.49}$$

8.4.2 Optimization of MESFET Models

It must be stressed that modeling of MESFETs at the design stage is very critical. Requirements set by the microwave industry push the operation of MESFETs up to the limits of their physical capabilities. Microwave circuit designs are usually required to extract maximum performance from the device, so it is extremely difficult, and sometimes impossible, to make them insensitive to changes in the device's parameters.

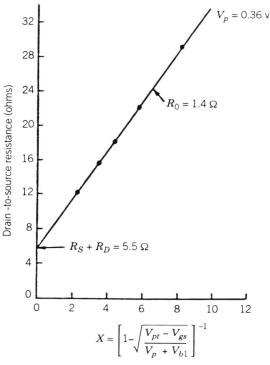

Fig. 8.18 Derivation of he open channel resistance, R_0 and parasitic series resistance R_s and R_D. [Pengelly, R. S., *Microwave Field-Effect Transistors—Theory, Design and Applications*, Research Studies Press (1986)]

In the CAD of such circuits, relatively small errors in the model device can cause large errors in the simulation results [Taj81].

In a typical parameter extraction approach, an optimization method is used. Model elements are varied until the calculated S-parameters fit the measurements. There are well-known difficulties with MESFET model extraction from measured S-parameters [Cur84; Kon86]. First, the results are often inconsistent in that they depend on the starting point and a particular procedure used. Second, the results may not be unique; several almost equally good solutions may exist with completely different element values. Finally, at the point of the best fit, values of some model elements may be nonphysical.

The intrinsic inadequacy of the model has more severe consequences of the S-parameter measurement and dc-embedding errors. Random errors can usually be eliminated in the optimization process. For large de-embedding errors the model may not be able to follow the measured behavior at all or some model parameters may become nonphysical.

Simulated Annealing Optimization

There are several methods that have been applied to the optimization of model parameters. Gradient methods have been successively applied to this optimization problem for some time [Dav59; Fle64; Dav68]. The direct application of such methods can be computationally intensive. Moreover issues such as convergence and entrapment in local minima must be addressed. Variations and improvements of these methods have been proposed [Yan83; Dog83]. The simplex method, proposed by Conway et al. [Con85; Pre88], is independent of gradient methods and claims stable convergence properties. A *simplex* is a geometrical figure consisting, in N dimensions, of $N + 1$ points (vertices) and all their interconnecting line segments, polygonal faces, and so on. However, the problem of local minima has not been directly tackled, the usual recipe being "try a number of initial simplex points."

This section describes the use of simulated annealing (SA) [Kir83], a combinatorial optimization method, to overcome the problem of entrapment in local minima. This method has been successfully used in VLSI design optimization problems [Won88]. It enables one to approach asymptotically the global minima of the objective function [Rom85]. This method has been applied to the optimization of microwave transistor models with good results [Vai89].

Combinatorial Optimization To set up the modeling problem as a combinatorial optimization problem, the limits on each parameter of the device are first specified. Parameter values are then discretized so that a finite (but large) number of solutions is possible. An objective function (or cost function) is defined as

$$F(V) = \sum_{j=1}^{m} (M_j - M'_j)^2, \tag{8.50}$$

where $V = (V_1, \ldots, V_n)$ is the set of model parameters, M_j and M'_j ($j = 1$ to m) are the measured and simulated characteristics of the device, and m is the total number of characteristics to be fitted.

Since the number of possible solutions is extremely large, and an exhaustive search for the optimum solution is impossible, a heuristic approach must be employed. At each step of the iteration the algorithm generates a new solution and tests if it reduces the value of the objective function and if so, accepts the new solution; otherwise, another new solution is generated and tested. One can only hoped that this process leads to the global minimum of $F(V)$.

Simulated Annealing and Model Optimization The simulated annealing (SA) algorithm has the advantage of asymptotically producing the global optimal solution with probability one. A cost-increasing solution may still be accepted, the probability of acceptance depending on (1) a "pseudotemperature" T_k, which it artificially decreased as the iteration proceeds, and (2) the value of $\Delta F(V) = F(V_k) - F(V_{k-1})$, where V_k and V_{k-1} are the values of the parameter vector at iteration steps k and $k - 1$. A Boltzmann-like law is used to determine the probability P of accepting a certain cost-increasing solution V_k at the kth iteration step

$$P(V_k, T_k) = \exp\left[-\frac{\Delta F(V_k)}{T_k}\right]. \tag{8.51}$$

A careful choice of the initial temperature T_0 as well as the rule for decreasing the pseudotemperature T is necessary to save computation time and to escape from local minima. $T_k = \alpha T_{k-1}$ was used for the temperature schedule. The values $\alpha = 0.90$ and $T_0 \geqslant 500$ worked satisfactorily in the chosen examples.

At each iteration new parameter values are generated by first choosing one device parameter V_i at random. A user-defined base value $V_{i\text{-base}}$ is multiplied by a random number R ($0 \leqslant R \leqslant 1$), and a variation $\Delta_i = RV_{i\text{-base}}$ is introduced into parameter V_i. The new parameter value so obtained is used in the next iteration unless it exceeds prescribed limits.

A relative stopping criterion is used, since there is no guarantee that the device model can approximate measured data closely. The optimization process is stopped when the objective function's value has virtually remained unchanged [e.g., $\Delta F(V) \leqslant 0.001$] for ten consecutive iterations.

The optimizer described has been applied to three test cases to demonstrate the optimization process and to evaluate its performance [Vai89].

Future Developments on MESFET Modeling and Parameter Extraction

The design direction of the integrated CAD environment appears to be leading from technology to performance evaluation by integration process, device, and circuit simulation for the development of monolithic microwave-integrated circuits. A physical numerical model [Ghi89] has been developed to generate I–V characterisics and sets of S-parameters for the intrinsic device. The results were used to derive MESFET circuit models. The major advantage of this method is that physical simulation offers much more information than measurements can provide by making available the values of electrical variables inside the device [Jas89].

Other important development includes the forward and reverse modeling techniques reported by Ladbrook [Lad89]. The procedure derives a set of equations that translates the difference in equivalent circuits into differences of doping density, gate length, and gate recess depth so that the technologist can identify those features requiring tighter process control. The task of deducing materials-related and device structural-related values from measured S-parameters and fitted equivalent circuits is very useful in the yield-oriented design approach. The simulated annealing scheme described in the previous section can fit in very well.

Recent development of new measurement-based GaAs FET model provides improved large signal simulation accuracy because the model nonlinearities are explicitly constructed from device data [Roo91a]. This FET model is based directly on processed, measured data, not coefficients of parameters. In addition a new methodology allows the model to predict frequency dispersion of the device currents. This is an extremely fast generation procedure that constructs the model from the data without circuit simulation or optimization.

The new FET model can be represented by essentially the same equivalent-circuit topology as described in Section 8.3. However, the constitutive relations for this model's equivalent circuit elements are not known to take any definite form, such as Eq. (8.7) or Eq. (8.27). Rather, the data for the model functions are obtained by means of algebraic and/or differential equations [Roo91b, Roo91c]. This general approach is the same for any device for which the equivalent circuit is valid. The model is detailed with the individual characteristics of the data which are incorporated directly into the model's nonlinear functions. The circuit-simulation model produced by such a procedure is more accurate than any based on simplified physical or empirical equations.

The process can also admit the use of physics previously ignored in the physical model or even correct a deficiency in the empirical model's equation.

The measured and calculated nonlinear model functions are stored in tabular form as functions of two independent controlling terminal voltages. Two-dimensional spline functions in the nonlinear simulator interpolate the tabulated functions during simulation. Lookup table models use interpolation (or approximation) techniques, rather than analytic equations, to define model-port relations or equivalent-circuit element constitutive relations.

A significant advance in microwave circuit simulation is that measured S-parameter data can now be inserted directly into a linear simulator. Interpolation between the discrete measurement points enables simulations to be done at frequencies for which there is no data. The simulated quantities must still be correlated, however, since the nonlinear model functions are correlated through the device data used in their construction. The new FET model has been compared with the Curtice and Statz models; the data are shown in Fig. 8.19 [Rod91a]. Unlike conventional analytic models, the new model can maintain about the same degree of large-signal accuracy *over a range of bias points* [Roo91a]. In the figure the bias points (and loading conditions) were changed, but all model parameters were fixed. Industry-standard models might be extracted using HP IC-CAP or Silvaco UTMOST III.

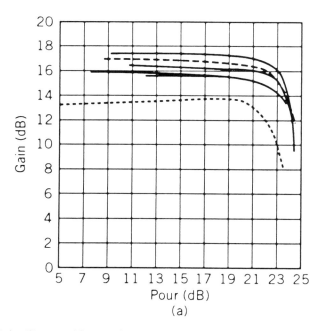

Fig. 8.19 (*a*) Gain; (*b*) second harmonic levels; (*c*) third harmonic levels versus output power for the new model, Curtice cubic, Statz, and three measured 0.4 × 600-μm MESFETs with attenuators; $V_D = 5$, $I_D = 0.65 I_{DSS}$, 50-Ω load. [Root, D. E., *Microwave J.*, September (1991): 126]

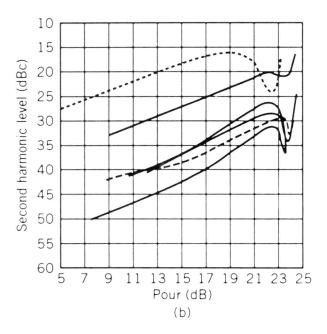

(b)

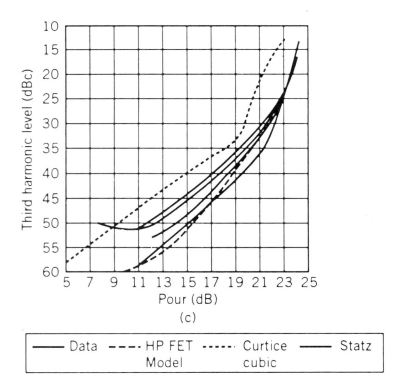

(c)

——— Data   ---- HP FET Model   ····· Curtice cubic   ——— Statz

Fig. 8.19 (*Continued*)

8.5 PARASITIC EFFECTS

The low-frequency anomalies observed in many devices are rather well-known effects in GaAs FETs. This group includes $1/f$ noise [Fol86], low-frequency oscillation [Mil85], backgating [Lee87], and low-frequency oscillation [Cam85]. All can have detrimental effects in sampled-analog and other dc-coupled ICs [Phi88].

8.5.1 $1/f$ Noise Component

Noise with a $1/f$ component is a fundamental problem in GaAs devices. The noise corner in FETs has been known to be high, sometimes even extending to 100 MHz. The variation in noise corner can be extreme, varying from device to device by orders of magnitude, and little progress seems to have been made in tracing the sources of the problem. Some experiments [Fol86] have shown that there may be two sources of $1/f$ noise in GaAs MESFET: the device surface and the device channel regions. The contribution of noise from these two regions is said to be roughly equal. The surface noise results from trapping of electrons in surface states, while the bulk noise is correlated with low field mobility and deep-trap concentration. The cure for broadband $1/f$ noise in GaAs ICs is not yet known. In the case of the GaAs MMIC, this is of little consequence, but for the sampled-analog IC, the dynamic range of signals that can be handled is restricted by this effect.

8.5.2 Backgating Effects

The backgating or sidegating effect occurs when there is imperfect isolation between two devices integrated on the same GaAs SI substrate [Roc87]. It corresponds to the ability of a Schottky or an ohmic contact called a "sidegate" to control the channel conductance of a nearby MESFET or resistor. The conduction mechanisms appear to be inextricably linked with low-frequency oscillation in GaAs FETs and ICs. Observations of backgating effects suggest that the low-level currents flow in semi-insulating or implant-isolated GaAs substrates because of dc bias differences among adjacent devices which have oscillatory components that may be in the hertz or kilohertz range [Phi88]. DLTS studies show that the trapped charge in the substrate, which is responsible for the backgating, is located on the Cr and EL2 levels in chromium-doped substrates and on the EL2 levels in high-purity substrates [Koc82]. These observations suggest that the oscillations are coupled through an IC. When they are coupled to the input stage of a high-gain amplifier, for example, the oscillations are amplified greatly and cause large-amplitude blocking wave forms or other nonlinear behavior.

8.5.3 Low-frequency Variations in Drain Conductance

The drain conductance of FET increases with frequency. In general, several frequencies are observed at which the drain conductance increases, and there is some correlation between the type of material on which the FET is made and the presence or absence of some of these conduction steps [Phi88]. The growth mechanisms of crystals or the methods for preparing active layers may be the sources of the problem.

The major difficulty for the circuit designer is that he drain conductance of a FET may increase with frequencies above a few hertz, with the gain in the FET-based amplifiers usually inversely proportional to this parameter. The use of cascade circuit may counteract this effect. However, recent studies have indicated that drain impedance can be controlled by the so-called self-bootstrapping technique applied to both the amplifying device and the active load [Lee87a]. Each amplifying device and load is made up of two FETs of different threshold voltages. To realize the full benefit of this technique, the threshold voltage of the feedback FET is made more negative than the primary FET to ensure that the latter is biased to operate in the current saturation region.

8.6 NOISE THEORY OF GaAs MESFETs

One of the most important applications of GaAs MESFETs is in low-noise amplifiers used in communications equipment, phase-array radars, space-based electronic detection systems, tracking devices, and so forth. Different MESFETs operate in the frequency range covered from dc to 40 GHz and above. Noise theory with relation to MESFETs remains to be investigated. in this section we will consider the important noise parameters that affect the device performance at microwave frequencies.

8.6.1 Noise Equivalent Circuit of GaAs MESFETs

The noise properties of a linear two-port can be represented by a noiseless two-port with noise current generators connected across the input and output ports. Figure 8.20 [Pen86] illustrates the noise equivalent circuit of GaAs MESFET. The noise current generator at the output of the FET represents the short-circuit channel noise generated in the drain-source path. The mean square of i_{ND} can be expressed [van62; Pen86] as

$$i_{ND}^2 = 4kT_0 \Delta f g_m P, \tag{8.52}$$

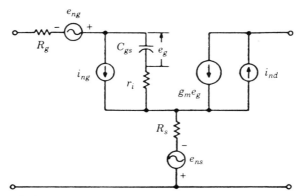

Fig. 8.20 Noise equivalent circuit of GaAs MESFET. [Fukui, H., *IEEE Trans. Microwave Theory and Tech.* **MTT-27** (1979): 643–650]

where

K = Boltzmann constant,
T_0 = lattice temperature,
Δf = bandwidth,
g_m = transconductance,
P = a dimensionless factor depending on the device geometry and the dc bias conditions.

For zero drain bias, i_{ND}^2 represents the thermal noise generated by the drain conductance G_{ds}. It can be shown that $P = G_{ds}/g_m$. For positive drain voltages the noise generated in the channel is larger than the thermal noise generated by G_{ds} [Pen86].

Noise voltages generated in the channel cause fluctuations in the depletion-layer's width. The resulting charge fluctuation in the depletion layer induces a compensating charge variation on the gate electrode. The total induced-gate charge fluctuation is indicated in Fig. 8.20 by the noise generator i_{NG} at the gate terminal where

$$i_{NG}^2 = 4kT_0\Delta f \frac{\omega^2 C_{gs}^2}{g_m} R, \tag{8.53}$$

where R is a factor dependong on FET geometry and the bias conditions. At zero drain voltage, $R = g_m R_i$.

The two noise currents i_{NG} and i_{ND} are caused by the same noise voltages in the channel. A correlation factor C is defined as [Van63]

$$jC = \frac{i_{NG}^* i_{ND}}{(i_{NG}^2 i_{ND}^2)^{1/2}}, \tag{8.54}$$

where j is the imaginary unit and the asterisk denotes the complex conjugate. C is purely imaginary, since i_{NG} is caused by the capacitive coupling of the gate circuit to the noise source in the drain circuit.

Through an exhaustive treatment of a circuit equivalent to that in Fig. 8.20, Pucel et al. [Puc75] obtained an expression for the minimum noise figure:

$$F_{min} = 1 + 2\left(\frac{2\pi f C_{gs}}{g_m}\right)\sqrt{K_g[K_r + g_m(R_s + R_g)]} + 2\left(\frac{2\pi f C_{gs}}{g_m}\right)[K_g g_m(R_g + R_s + K_c R_i)] + \cdots, \tag{8.55}$$

where

$$K_g = P\left[\left(1 - C\frac{\sqrt{R}}{P}\right)^2 + (1 - C^2)\frac{R}{P}\right],$$

$$K_r = \frac{R(1 - C^2)}{(1 - C\sqrt{R/P})^2 + (1 - C)^2 R/P},$$

$$K_c = \frac{1 - C\sqrt{R/P}}{(1 - C\sqrt{R/P})^2 + (1 - C^2)R/P}.$$

In the equations above R_g is the gate resistance, R_s is the source series resistance, and g_m is the transconductance.

8.6.2 Minimum Noise Figure of the GaAs FET

Consider the case where the FET is operating below its cutoff frequency at room temperature. A simple expression for the minimum noise figure F_{min} can be found from the equivalent circuit elements of Eq. (8.55):

$$F_{min} = 1 + 2\pi K_f f C_{gs} \sqrt{\frac{(R_g + R_s)}{g_m}} \times 10^{-3}, \qquad (8.56)$$

where K_f is a fitting factor approximately equal to 2.5 representing the quality of the channel material. Equation (8.56) is a special case of Eq. (8.55), where $R = 0$ and/or $C = 1$ if we neglect all higher-order terms. Since

$$f_T = \frac{g_m}{2\pi C_{gs}},$$

Eq. (8.56) can be rewritten as

$$F_{min} = 1 + K_f f \sqrt{g_m(R_g + R_s)}. \qquad (8.57)$$

Since f_T is also related to the gate length L, we have

$$F_{min} = 1 + K_1 L f \sqrt{g_m(R_g + R_s)}, \qquad (8.58)$$

where $K_1 = 0.27$ when L is in microns.

From the circuit design viewpoint the MESFET can be treated as a blackbox of noisy two ports [Pen86]. The noise properties of this blackbox can be characterized by the use of four noise parameters [Rot56]. A variant on this expression gives

$$F = F_{min} + \frac{R_n}{R_{ss}}\left[\frac{(R_{ss} - R_{OP})^2 + (X_{ss} - X_{OP})^2}{R_{OP}^2 + X_{OP}^2}\right], \qquad (8.59)$$

where

R_n = equivalent noise resistance,
R_{ss} = signal source resistance,
R_{OP} = optimum signal source resistance,
X_{ss} = signal-source reactance,
X_{OP} = optimum signal source reactance.

In this expression F_{min}, R_n, R_{OP}, and X_{OP} are the characteristic noise parameters of the device. Equation (8.59) is be represented on the source impedance Smith chart as a family of circles, each of which corresponds to a constant value of F.

A small R_n is essential for a device to be used in a broadband amplifier where a large tolerance is required in the input match. The smaller the value of R_n, the higher is the gain in the gate structure.

The four noise parameters can be expressed as

$$F_{min} = 1 + K_1 F C_{gs} \sqrt{\frac{(R_g + R_s)}{g_m}},$$

$$R_n = \frac{K_2}{g_m^2},$$

$$R_{OP} = K_3 \left[\frac{1}{4g_m} + R_g + R_s \right],$$

$$X_{OP} = \frac{K_4}{f C_{gs}}, \quad (8.60)$$

where K_1, K_2, K_3, and K_4 are fitting factors.

Figures 8.21 through 8.24 [Fuk79] show the fits obtained for six different FET structures having different channel carrier concentrations, gate lengths, and channel thicknesses. Good fits are obtained for values of

$$K_1 = 0.016, \quad K_2 = 2.2,$$
$$K_3 = 0.03, \quad K_4 = 160,$$

where R_n, R_{OP}, X_{OP}, R_g, and R_s are in ohms, g_m in Siemens, C_{gs} in picofarads and f in gigahertz.

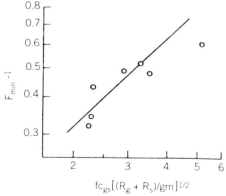

Fig. 8.21 Correlation between the minimum noise figure F_{min} and equivalent circuit elements C_{gs}, g_m, R_s, and R_g. [Fukui, H., *IEEE Trans. Microwave Theory and Tech.* **MTT-27** (1979): 643–650]

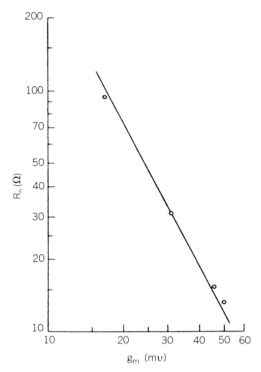

Fig. 8.22 Correlation between the equivalent noise resistance R_n and the transconductance g_m. [Fukui, H., *IEE Trans. Microwave Theory and Tech.* **MTT-27** (1979): 643–650]

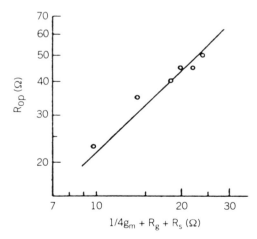

Fig. 8.23 Correlation between the optimum source resistance R_{OP} and equivalent circuit element R_g, R_s, and g_m. [Fukui, H., *IEEE Trans. Microwave Theory and Tech.* **MTT-27** (1979): 643–650]

For the MESFET design it is more convenient to have expressions for g_m, C_{gs}, and f_T. The cutoff frequency f_T can be expressed as

$$f_T = \frac{10^3 g_m}{2\pi C_{gs}} = \frac{9.4}{L} \text{ GHz}, \tag{8.61}$$

$$g_m = k_5 W \left[\frac{N_D}{aL}\right]^{1/3} \text{ S}, \tag{8.62}$$

$$C_{gs} = k_6 W \left[\frac{N_d L^2}{a}\right]^{1/3} \text{ pF}, \tag{8.63}$$

where k_5 and k_6 are found to be 0.02 and 0.34, respectively [Pen86]. Figures 8.25 and 8.26 [Fuk79] show the agreement between the measured values of g_m and C_{gs} and the empirical results of Eqs. (8.62) and (8.63).

Simplified expressions for the gate metallization resistance R_g and the source resistance R_s can be found as

$$R_g = \frac{17 W_1^2}{hWL} \Omega, \tag{8.64}$$

$$R_s = \frac{1}{W}\left[\frac{2.1}{a^{1/2} N_D^{2/3}} + \frac{1.1 L_{sg}}{(a-a_s) N_D^{0.82}}\right] \Omega,$$

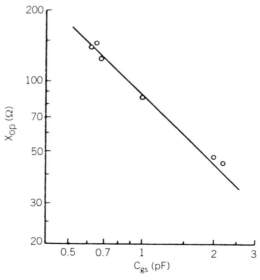

Fig. 8.24 Correlation between the optimum source reactance X_{OP} and the gate-to-source capacitance C_{gs}. [Fukui, H., *IEEE Trans. Microwave Theory and Tech.* **MTT-27** (1979): 643–650]

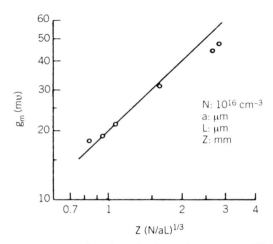

Fig. 8.25 Transconductance g_m as a function of a channel parameters W, L, a, and N_D. [Fukui, H., *IEEE Trans. Microwave Theory and Tech.* **MTT-27** (1979): 643–650]

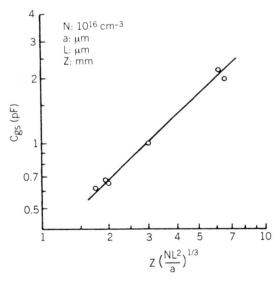

Fig. 8.26 Gate-source capacitance as a function of channel parameters W, L, a, and N_D. [Fukui, H., *IEEE Trans. Microwave Theory and Tech.* **MTT-27** (1979): 643–650]

where

h = gate metallization thickness in microns,
L_{sg} = distance between source and gate in microns,
a_s = depletion layer thickness in microns at the surface in the source-gate gap,
W_1 = unit gate width for a multiparallel gated FET.

8.6.3 Noise Modeling of MESFET

Podell [Pod81] have shown that the noise performance of GaAs MESFETs can be adequately described by two effectively uncorrelated noise sources: one at the input of the FET due to thermal noise generated in the various resistances in the gate-source loop, and the other in the output of the FET due to the Gunn domain between the gate and the drain of the device. The basic procedure is summarized in Fig. 8.27 [Pod81]. The FET input circuit is represented as a simple series RC circuit, with R_1 equal to the sum of R_g, R_i, and R_s and $C_1 = C_{gs}$. The optimum source admittance is derived from the series RC network by transforming this series circuit to its parallel equivalent.

The input conductance g_1 of the FET is

$$g_1 = \frac{1}{(Q_1^2 + 1)R_1},$$

$$b_1 = \frac{\omega C_1}{1 + 1/Q_1^2}, \quad (8.65)$$

where

$$Q_1 = \frac{1}{\omega C_1 R_1}.$$

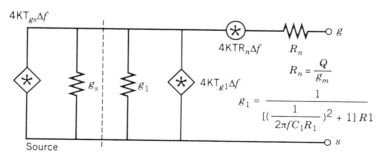

Fig. 8.27 Podell's noise model. [Podell, A. F., *IEEE Trans. Electron Dev.* **ED-28** (1981): 511–517]

Podell has found that R_N in a GaAs FET can be approximated by the expression:

$$R_N = \left(\frac{K_0}{|g_m|}\right) e^{K_2 I}, \qquad (8.66)$$

where $I = I_{DS}/I_{DSS}$ and K_0, K_2 are empirical constants. K_0 is approximately 1 for 0.5 μm FETs and 0.5 for 1 μm FETs, while K_2 has a value of 2.5. Both constants are independent of frequency.

The optimum source admittance ($g_{sOPT} + C_{OPT}$) can be calculated by the following equation:

$$g_{sOPT} = g_1 \left(1 + \frac{1}{A}\right)^{1/2}, \qquad (8.67)$$

$$C_{OPT}^* = \frac{Q_1^2}{Q_1^2 + 1}, \qquad (8.68)$$

where the asterisk denotes the complex conjugate, Q_1 equals $1/2\pi f C_{GS}(R_I + R_G + R_S)$, and $A = g_1 R_N$. In addition

$$F_{min} = 1 + 2A + 2(A + A^2)^{1/2}. \qquad (8.69)$$

Once the equivalent circuit of one FET and its noise figure at one frequency are known from the preceding equations, the optimum noise source impedance and noise figure at any frequency can be calculated.

The equation for F_{min} above relates the noise factor to the parameter A. It is desirable to minimize both the noise and the input Q of the FET, particularly at low microwave frequencies where the input Q is extremely high and losses in the input

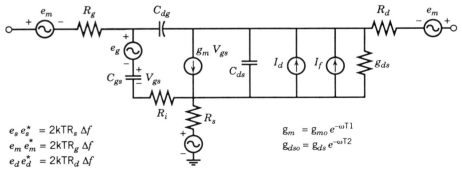

Fig. 8.28 Intrinsic linear FETs, including all noise sources and flicker noise. [Rohde, U., *Microwave J.*, November (1991): 87–99]

matching network can be as important as losses in the input of the FET itself [Pen86]. A low Q will give greater bandwidth potential. This simplified model can produce several other interesting conclusions [Sch87]; among these it can predict the degree of mismatch under noise match conditions. Also from it can be derived an expression relating the associated gain to MAG:

$$|\Gamma| = \frac{1}{F_{min}},$$

$$G_A = \frac{F_{min}^2 - 1}{F_{min}^2}.$$

For very low noise figures ($F_{min} \to 1$), the input reflection coefficient approaches unity, and the ratio of the associated gain to MAG approaches zero. However, these expressions are only approximations, since they neglect source inductance and drain-to-gate feedback [Sch87]. The significance of this theory is that the device noise parameters can be predicted over a broad range of frequencies based only on knowledge of F_{min} at one frequency and the device input equivalent circuit.

A recent study for Rohde [Roh91] has stated that flicker noise should be added as a different source at the output in the noise equivalent circuit (see Fig. 8.28 [Roh91]) for lower-frequency operation (400 MHz to 2 GHz). The noise equivalent circuit model covers the frequency range from low frequencies up to at least 40 GHz. The time-delay factor T_2 shown in Fig. 8.28 is an additional optimizable variable. The optimization process for modeling de-embedding S-parameters should include the stability factor K and the maximum gain function MAG.

REFERENCES

[Tur67] Turner, J., Gallium Arsenide, *Inst. Phys. Ser.* **3** (1967).

[Tur71] Turner, J., A. Waller, R. Bennett, and D. Parker, Inst. Phys. Conf. Serial No. 9, London 1971, pp. 234–239 (1971).

[Pen86] Pengelly, R. S., *Microwave Field-Effect Transistors—Theory, Design and Applications*, Research Studies Press (1986).

[Ruc70] Ruch, J., and W. Fawcette, *J. Appl. Phys.* **41** (Aug. 1970): 3843–3849.

[Hig78] Higgins, J. A., R. L. Kuvas, F. H. Eisen, and D. R. Ch'en, *IEEE Trans. Electron Devices* **ED-25** (1978): 587–596.

[Wel77] Welch, B. M., and R. C. Eden, *Int. Solid State Circuits Conf. Tech. Digest*, pp. 205–208 (1977).

[Yok81] Yokoyama, N., T. Mimura, M. Fukata, and H. Ishikawa, *IEEE ISSCC, Digest of Technical Papers*, pp. 218–219 (1981).

[Yok82] Yokoyama, N., T. Ohnishi, K. Odani, H. Onodera, and M. Abe, *IEEE Trans. Electron Devices* **ED-29** (1982): 1541–1547.

[Yok83] Yokoyama, N., T. Ohnishi, H. Onodera, T. Shinoki, A. Shibatomi, and H. Ishikawa, *IEEE ISSCC, Diegest of Technical Papers*, pp. 44–45 (1983).

[Tse82] Tseng, W. F., and A. Christou, *IEEE IEDM* (1982): 174–176.

[Mat82] Matsumoto, K., N. Hashizume, H. Tanoue, and K. Kanayama, *Jap. J. Appl. Phys.* **21** (1982): L393–L395.

[Lad86] Ladbrook, P. H., in *Gallium Arsenide for Devices and Integrated Circuits*, p. 197, ed. by H. Thomas, IEE Press, (1986).

[Mur77] Murai, F., H. Kurono, and H. Kodera, *Electronics Lett.* **13**, 11 (1977): 316–318.

[Tiw79] Tiwari, S., D. W. Woodard, and L. F. Eastman, *Proc. 7th Biennial Cornell Electrical Eng. Conf.*, pp. 237–248 (1979).

[Can88] Cantos, B. D., and Renba, R. D., *J. Electrochemical Soc.* (May 1988): 1312.

[Ban83] bandy, S. G., Y. G. Chai, R. Chow, C. K. Nishimoto, and G. Zdasiuk, *IEEE Elect. Dev. Lett.* **EDL-4** (1983): 42.

[Kim85] Kim, B., H. Q. Tserng, and H. D. Shih, *IEEE Electron Dev. Lett.* **ELD-6** (1985): 1.

[Can83] Cantos, B. D., D. B. Deal, and D. M. Dobkin, Electrochem. Soc. 163rd Meeting (1983).

[Cha83] Chao, P. C., P. M. Smith, S. Wanuga, W. H. Perkins, R. Tiberio, and E. D. Wolf, *IEEE IEDM* (1983).

[Cha85] Chao, P. C., P. M. Smith, S. C. Palmateer, and J. C. M. Huang, *IEEE Trans. Electron Devices* **ED-22** (1985): 1042–1046.

[Smi88] Smith, P. M., P. C. Chao, L. F. Lester, R. P. Smith, B. R. Lee, D. W. Ferguson, A. A. Jabra, J. M. Ballingall, and K. H. G. Duh, *1988 IEEE MTT-S Int. Microwave Symp. Digest* (1988): 927–930.

[Cha87b] Chao, P. C., P. M. Smith, K. H. G. Duh, J. M. Ballingall, L. F. Lester, B. R. Lee, and A. A. Jabra, *IEDM* (1987): 410–413.

[But81] Butler, M. J., EBMF-6 Product Documentation, Cambridge Instrument Co., UK (1981).

[Sho52] Shockley, W., *Proc. IRE* **40** (1952): 1365–1367.

[Ruc72] Ruch J., *IEEE Trans. Electron Devices* **ED-19** (May 1972): 652–654.

[Shu87] Shur, M., GaAs Devices and Circuits, Plenum (1987).

[Dac55] Dacey, G. C., and I. M. Ross, *Bell Syst. Tech. J.* **34** (1955): 1149–1189.

[Leh70] Lehovec, K., and Zuleeg, R., *Solid State Electron.* **13** (1970): 1415–1426.

[Cur80] Curtice, W. R., *IEEE Trans. Microwave Theory Tech.* **MTT-28**, 5 (1980): 448–456.

[Sus84] Sussman-Fort, S. E., S. Narasimhan, and K. Mayaram, *IEEE Trans. Microwave Theory and Technique* **32**, 4 (1984): 471–473.

[Van63] Van der Ziel, A., *Proc. IEEE* **51** (1963): 461–467.

[Sta87] Statz, H., P. Newman, I. W. Smith, R. A. Pucel, and H. A. Haus, *IEEE Trans. Electron Devices* **ED-34**, 2 (1987): 160–169.

[Mat85] Materka, A., and T. Kacprzak, *IEEE MTT* **MTT-33** (1985): 129–135.

[Jas88] Jastrzebski, A. K., *IEE Colloqium on Large Signal Device Models and Parameter Extraction for Circuit Simulation*, London (1988).

[Mac90] MaCamant, A. J., G. D. McCormack, and D. H. Smith, *IEEE Trans. Microwave Theory and Technique* **38**, 6 (1990): 822–823.

[Lar87] Larson, L. E., *IEEE Int. Symp. Circuits Syst.*, pp. 1–5 (1987).

[Sch88] Scheinberg, N., R. Bayruns, and R. Goyal, *IEEE J. Solid State Circuits* **23** (1988): 605–608.

[Sze81] Sze, S. M., *Physics of Semiconductor Devices*, 2d ed. Wiley (1981).

[Fuk79] Fukui, H., *IEEE Trans. Microwave Theory and Technique* **MTT-27**, 7 (1979): 643–650.

[Cam85] Camacho-Penalosa, C., and C. S. Aitchison, *Electronics Lett.*, **21**, 12 (June 1985): 528–529.

[Taj81] Tajima, Y., B. Wrona, and K. Mishima, *IEEE Trans. Electron Dev.* **ED-28** (1981): 171–175.

[Cur84] Curtice, W. R., and R. L. Camisa, *IEEE Trans. MTT* **MTT-32** (Dec. 1984): 1573–1578.

[Kon86] Kondoh, H., *IEEE MTT-S Digest* (1986): 377–380.

[Dav59] Davidson, W. C., *Research Development Rep.*, ANL-5900, (1959).

[Fle64] Fletcher, R., and M. J. D. Powell, *Computer J.* **6** (1964): 163–168.

[Dav68] Davidon, W. C., *Computer J.* **10** (1968): 406–409.

[Yan83] Yang, P., and P. K. Chatterjee, *IEEE Trans. Electron Devices* **ED-30** (1983): 1214–1219.

[Dog83] Doganis, K., and J. Scharfetter, *IEEE Trans. Electron Devices* **ED-30** (1983): 1219–1228.

[Con85] Conway, P., C. Cahill, W. A. Lane, and S. U. Lidholm *IEEE Trans. Computer-Aided Design* **CAD-4** (1985): 694–698.

[Pre88] Press, W. H., B. P. Flannery, S. A. Teukolsky, and W. T. Vettering, *Numerical Recipes in C*, sec. 10.4, Cambridge University Press (1988).

[Kir83] Kirkpatrick, S., C. D. Gelatt, and M. P. Vecchi, *Science* **220**, 4598 (1983): 671–680.

[Won88] Wong, D. F., H. W. Leong, and C. L. Liu, *Simulated Annealing for VLSI Design*, Kluwer Academic Publishers (1988).

[Vai89] Vai, M. K., S. Prasad, N. C. Li, and F. Kai, *IEEE Trans. Electron Devices* **36**, 4 (1989): 761–762.

[Rob87] Roblin, P., S. Kang, A. Ketterson, and H. Morkoc, *IEEE Trans. Electron Devices* **ED-34**, Sept. (1987): 1919–1927.

[Ghi89] Ghione, G., C. U. Naldi, and F. Filicori, *IEEE Trans.* **MTT-S**, **37**, 3 (1989): 457–468.

[Jas89] Jastrzebski, A. K., in *GaAs Technology and Its Impact on Circuits and Systems*, ed. by D. Haigh and J. Everard, IEE Press (1989).

[Lad89] Ladbrook, P. H., *MMIC Design: GaAs FETs and HEMTs*, Artech House (1989).

[Roo91a] Root, D. E., *Microwave J.* (Sept. 1991): 126.

[Roo91b] Root, D. E., S. Fan, and J. Meyer, *Proc. 21st European Microwave Conf.*, Stuggart, Germany, Sept. (1991).

[Roo91c] Root, D. E., and K. J. Kerwin, in *Microwave Integrated Circuits*, Marcel Deckker (1991).

[Fol86] Folkes, P. A., *Appl. Phys. Lett.* **48**, 5 (Feb. 1986): 344–346.

[Mil85] Miller, D., M. Bujatti, and D. Estreich, IEEE GaAs IC Symp., Monterey, CA (1985).

[Lee87] Lee, W. S. *IEE Electronics Lett.* **23**, 11 (1987): 587–589.

[Cam85] Camcho-Penalosa, C., and C. S. Aitchison, *IEE Electronics Lett.* **21** (1985): 528–529.

[Phi88] Phillips, J. A., and S. J. Harrold, in *GaAs Integrated Circuits*, p. 320, ed. by Joseph Mun, McMillan (1988).

[Roc87] Rocchi, M., B. Gabillard, E. Delhaye, and T. Ducourant, in *GaAs Integrated Circuits*, ed. by Joseph Mun, McMillan (1987).

[Koc82] Kocot, C., and C. A. Stolte, *IEEE Trans. Electron Devices* **ED-29**, 7 (1982): 1059–1064.

[Lee87a] Lee, W. S., and Mun, J., *IEE Electronics Lett.* **23** (1987): 705–707.

[van62] van der Ziel, A., *Proc. IRE* **50** (1962): 1808–1812.

[Puc75] Pucel, R., H. Haus, and H. Statz, *Advances in Electronics and Electron Physics*, vol. 38 pp. 195–265, Academic Press (1975).

[Rot56] Rothe, H., and W. Dahlke, *Proc. IRE* **44** (June 1956): 811–818.

[Pod81] Podell, A. F., *IEEE Trans. Electron Devices* **ED-28**, (May 1981): 511–517.

[Sch87] Schellenberg, J. M., and T. R. Apel, in *GaAs Integrated Circuits*, pp. 204–288, ed. by Joseph Mun, McMillan (1987).

[Roh91] Rohde, U., *Microwave J.* (Nov. 1991): 87–99.

9

HIGH ELECTRON-MOBILITY TRANSISTOR (HEMT)

9.1 INRODUCTION TO HEMT

In today's technical and business world, we need high-speed computers to solve problems calling for high-volume data processing, real-time signal processing, graphics, and remote imaging. Then there are the increasing demands of military and commercial applications on low-noise, high-frequency amplification. This chapter describes the basic principle of the field-effect transistor in terms of selectively doped heterojunctions yielding high electron mobility and velocity, which at the device terminals give rise to a high transconductance FET that can be operated at a millimeter-wave frequency range with ultra-low noise. This device has the superior transport properties of electrons moving along the two-dimensional electron gas (2DEG) formed at the heterojunction interface between two compound semiconductor materials. Various acronyms have been coined for these devices (MODFET, HEMT, TEGFET, SDHT, GAGFET, etc.); all describe either the technology employed in creating the structure or the resultant electronic properties [Col85]. HEMT shows much promise in MMICs, demonstrably outperforming the GaAs MESFET in gain, low noise, and frequency response. Devices with switching delays under 5 ps, current-gain cutoff frequencies of about 250 GHz, and maximum oscillation frequencies in excess of 400 GHz have been developed [Moc91]. Enhancement-mode and depletion-mode HEMTs can also be fabricated on the same wafer for digital integrated circuits applications. HEMT VLSI is currently under development.

The structure of the HEMT will first be described, along with the technology of modulation doping. The operation principle and the device technology will be mentioned. The issues of light sensitivity, threshold voltage variation, threshold voltage shift with temperature, and the persistent photoconductivity (PPC) will be investigated. The recent improvement in performance on pseudomorphic HEMT will be described. We will also discuss HEMT circuit applications in MMIC and high-speed digital ICs.

9.2 THE BASIC HEMT STRUCTURE

Modulation-doped structures consist of single or multiple periods of doped $Al_xGa_{1-x}As$ layers and undoped GaAs layers. The electrons transferring from the $Al_xGa_{1-x}As$ layers into the undoped GaAs layers are confined at the heterointerface (first suggested by Esaki and Tsu in 1969 [Esa69; Esa70] and experimentally observed by Dingle et al. in 1978 [Din78]). Being spatially separated from the donors, the electrons, even at extremely high concentrations, are not subjected to ionized impurity scattering and thus can exhibit very high mobilities. This is especially pronounced at cryogenic temperatures where the ionized impurity scattering would have been dominant.

9.2.1 Principles of Modulation Doping

Charge Transfer

The charge transfer and carrier confinement that form the basis for most modulation-doping effects in heterostructures are produced by the step in the band-edge energies occurring at the heterojunction [Gos85]. Optical determination of confined particle energy levels permits a determination of confining potential barrier heights. The energy can be deduced from the interband optical absorption spectra [Din74], photoluminescence excitation spectra [Wei81] and from intersubband Raman scattering spectra [Abs79]. In the $GaAs/Al_xGa_{1-x}As$ system both electrons and holes can be confined in the GaAs layers, whereas in the InAs/GaSb system they might be confined in separate layers, with the electrons in the InAs layers and the holes in the GaSb layers [Sai77]. In the $GaAs/Al_xGa_{1-x}As$ system about 85% of the difference in band gap between the two species occurs in the conduction band, corresponding to the electron affinity difference between the materials; the remaining 15% occurs in the valence band [Din75]. These discontinuities are illustrated in Fig. 9.1. The band-gap difference between GaAs and $Al_xGa_{1-x}As$ is $1.247x$ electron volts, so for $x = 0.30$ the electrons experience a 0.32-eV conduction band-edge discontinuity, and holes are subject to a 0.06-eV valence band-edge discontinuity.

These band-edge discontinuities form an upper limit for the difference in energy between electrons bound to a donor in the barrier layer near the interface and the electrons in the (GaAs) channel layer. The actual energy difference is reduced by the following effects:

1. *Donor binding energy.* The electrons bound to donors in barrier material will lie below the bottom of the barrier conduction band. This binding energy is greater in $Al_xGa_{1-x}As$ than in GaAs, reaching values on 50 to 100 meV in the range of $x = 0.20$ to 0.40 [Gos85]. It reduces the energy gained on dropping into the channel layer.
2. *Conduction electron quantum energy.* The lowest conduction electron energy in the quantum-confined states in the channel is raised by the size quantization effect. For then wells the quantum state energies E_n are approximated by the energy of a particle in a square box of the same width L with infinite barriers.
3. *Conduction electron Fermi energy.* For a nonzero concentration of electrons in the channel layer, the electron energies will occupy a Fermi distribution. Electron states up to a Fermi energy of $\Pi\hbar^2 n/m^*$ above the quantum energy E_1 will be

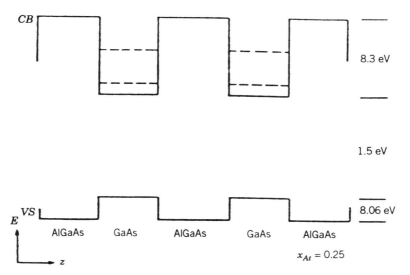

Fig. 9.1 Conduction-band and valence-band edges in undoped GaAs/(Al,Ga)As heterostructure. Electrons introduced into GaAs layers occupy quantum energies by dashed lines. [Gossard, A. C., Modulation doping of semiconductor heterostructures, in *Molecular Beam Epitaxy and Heterostructures*, ed. by L. L. Chang, and K. Ploog, Martinus Nijhoff Publishers (1985); permission from publisher]

occupied for a two-dimensional gas of concentration n carriers/cm^2 with only one transverse quantum state occupied, and a spin degeneracy of two. For $m^* = 0.0665 m_0$ and $n = 1 \times 10^{12}$ cm^{-2}, this energy is 35 meV.

4. *Spatial transfer Coulomb energy.* The charge transfer into the channel creates a negatively charged channel region and leaves behind a positive space charge of the ionized donors. The electric field and potential drop associated with this space charge are also appreciable. The potential drop for separation d of charge density ρ is $\rho d / \varepsilon \varepsilon_0$, which is 138 meV for separation of 10^{12} electron/cm^2 by a distance of 100 Å in GaAs of dielectric constant 13.1. The Coulomb potentials alter the shape of the potential wells and barriers and thus alter the quantum energies. As a result an exact treatment requires a self-consistent calculation. The various effects are illustrated in the band diagram of Fig. 9.2.

Mobility Determining Process

The dominant source of carrier scattering in uniformly doped semiconductors are phonons and impurities, with the phonon processes dominating scattering at high temperatures and the impurity processes predominant at low temperature. A plot of the mobility limits for these processes for very weakly doped GaAs ($\sim 10^{13}$ cm^{-3}) is given in Fig. 9.3 [Wol70]. In modulation-doped layers we may expect the phonon processes and rates to be roughly similar to the phonon scattering in the uniformly doped material. The impurity scattering is drastically altered. When the impurity scatters are separated from the carriers, the impurity scattering decreases because of the falloff in both the Coulomb field and in the short wavelength potential fluctuations

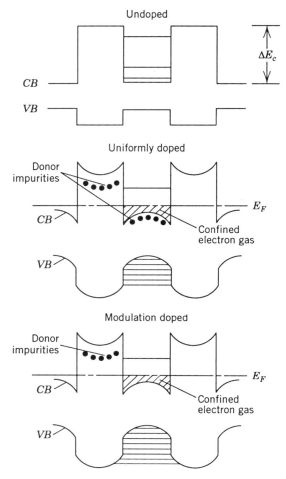

Fig. 9.2 Band edges, donor states, and electron states in undoped, uniformly doped, and modulation-doped multilayered semiconductor systems. [Gossard, A. C., Modulation doping of semiconductor heterostructures, in *Molecular Beam Epitaxy and Heterostructures*, ed by L. L. Chang, and K. Ploog, Martinus Nijhoff Publishers (1985); permission from publisher]

with distance from the impurity distribution. These effects have been calculated extensively by Price [Pri82] and Mori [Mor79]. A simple expression for the resultant mobility is obtained for a degenerate two-dimensional sheet of n carriers/cm^2 separated by d from a two-dimensional sheet of n_I ions/cm^2:

$$\mu = \frac{16\sqrt{\pi}ed^3 n^{3/2}}{\hbar n_I}.$$

Note the increase in mobility with carrier concentration n. The increasing mobility with high n is a result of the increase in the Fermi wavevector with n. With large Fermi wavevectors, small k elastic ion scattering is less effective in relaxing the electron

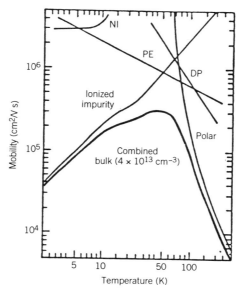

Fig. 9.3 Theoretical mobility limits for main scattering processes in uniformly doped GaAs. Ni refers to neutral-impurity scattering, PE indicates scattering by piezoelectrically active acoustic phonons, and DP indicates deformation-potential scattering by acoustic phonons. In modulation-doped material with no impurity scattering, upper limit on mobility would be $\sim 5 \times 10^6$ cm^2/V-s. [Gossard, A. C., Modulation doping of semiconductor heterostructures, in *Molecular Beam Epitaxy and Heterostructures*, ed. by L. L. Chang, and K. Ploog, Martinus Nijhoff Publishers (1985); permission from publisher]

momentum distribution. The calculated mobility reaches 60,000 cm^2/V-s at $n = n_I = 10^{12}$ cm^{-2} and $d = 100$ Å. The mobility increase continues until a second quantum state becomes occupied, which occurs when the Fermi energy becomes equal to the difference between the first and the second quantum states. The second quantum state forms the basis for a second subband of states and permits scattering of electrons between the subbands, which then produces a decrease in mobility.

In the case of the incomplete electron transfer to the quantum well channels, electron states in both channel and barrier layers will be occupied. The conductivity will then consist of parallel conduction in the channels and barriers. For $Al_xGa_{1-x}As$ barriers, however, donor states in the barriers are sufficiently deep that carriers are typically frozen out at cryogenic temperatures in the barriers, and conductions is dominated entirely by the quantum well channel barriers. This fact will be explained more fully in our discussion of the HEMT device structure.

9.2.2 The Structure of a HEMT

The HEMT is in many ways similar to an Si nMOSFET, but consists of layers of compound semiconductors rather than silicon. Figure 9.4 compares the silicon MOSFET with the conventional HEMT and the inverted HEMT structure. The bulk silicon is replaced by a layer of undoped GaAs, and the silicon dioxide is replaced by a layer of *n*-doped AlGaAs. Advanced techniques (notably MBE or MOCVD)

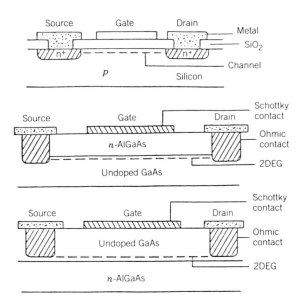

Fig. 9.4. (a) Silicon MOSFET; (b) conventional HEMT; (c) inverted HEMT structure. [Ladbrook, P. H., in *Gallium Arsenide for Devices and Integrated Circuits*, ed. by H. Thomas and D. V. Morgan, IEE Press (1986); permission from publisher]

are used to produce the layers with an atomically smooth heterojunction interface between them. The physical purpose of each layer of HEMT structure is as follows: An active channel is formed on the top surface of the undoped GaAs buffer layer which is grown on a semi-insulating substrate. Typically this layer's thickness is approximately 1 µm. On top of the active channel, a thin (20- to 60-Å) layer of undoped AlGaAs (called the *spacer*) is grown. Above the spacer is the doped AlGaAs layer, which supplies the electrons for channel conduction. The electrons contributed by the *n*-doping in the AlGaAs are free to move through the entire crystal until they fall into the lowest energy states allowed them. In the HEMT the lowest energy states are to be found just to the GaAs side of the heterojunction interface; all the electrons accumulate there in a thin sheet. These electrons are free to move only in the two-dimensional plane of the interface and form a two-dimensional electron gas (2DEG).

The spatial separation of conduction electrons from their parent donor impurities reduces the rate at which the electrons are scattered by the ionized impurities. Consequently the momentum gained from an externally applied field is randomized less quickly. As is silicon MOSFET it is the high quality of the GaAs/AlGaAs interface (equivalent to low Q_{ss} on the Si/SiO$_2$ interface) that makes high-performance HEMT operation possible. The ability to obtain a large sheet concentration of conduction electrons without setting off a large impurity scattering rate is one of the major advantages of the HEMT structure over the standard MESFET. The HEMT structure is capped by a layer of highly doped GaAs that passivates the AlGaAs and facilitates ohmic contact to the 2DEG.

The larger band-gap material used with GaAs to form the heterojunction must be closely lattice-matched to the GaAs. AlGaAs is lattice-matched to within 0.1% of

GaAs, and the growth of extremely abrupt heterojunctions by MBE is possible using these two materials. The band gap of $Al_xGa_{1-x}As$ depends on the Al mole fraction x. the energies of the conduction band minima measured from the top of the valence band ($K = 0, 0, 0$) are shown in Fig. 9.5. The band diagram of a typical GaAs/n-$Al_xGa_{1-x}As$ heterojunction is shown in Fig. 9.6. The sum of the conduction band and valence band discontinuities is equal to the energy gap discontinuity determined using Fig. 9.6.

The $Al_xGa_{1-x}As$ is doped, and the conduction band edge in the GaAs is lower in energy than the donor energy level in the AlGaAs. The electrons have sufficient energy to overcome the donor-binding energy, so they will transfer from the AlGaAs to the GaAs. This process satisfies the equilibrium requirement of a constant Fermi energy through the heterojunction. The electron transfer from AlGaAs to GaAs causes strong electric fields ($\sim 10^5$ V/cm) perpendicular to the interface, which in turn causes bending of the energy bands near the interface. The band bending results in the formation of a quasi-triangular potential well approximately 50 Å wide, to which the electrons are confined, forming a two-dimensional electron gas.

The electron energies in the quantum well are increased because of their quantum-mechanical confinement, and they form discrete energy subbands. In a typical HEMT heterojunction the Fermi level lies higher in energy than the lowest energy subband E_0, which necessitates the use of Fermi-Dirac rather than Maxwell-Boltzmann statistics in calculating the sheet carrier concentration n_{s0}. A thin undoped AlGaAs spacer layer is to separate further the electrons in the 2DEG from the ionized impurity scattering centers, thus enhancing the mobility of the 2DEG electrons. The number

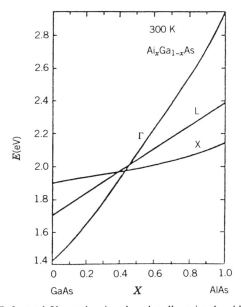

Fig. 9.5. Energies of Γ, L, and X conduction band valleys in the $Al_xGa_{1-x}As$ system as a function of aluminum mole fraction x. Energies are relative to the top of the valence band at $K = (0, 0, 0)$. [Ladbrook, P. H., in *Gallium Arsenide for Devices and Integrated Circuits*, ed. by H. Thomas and D. V. Morgan, IEE Press (1986); permission from publisher]

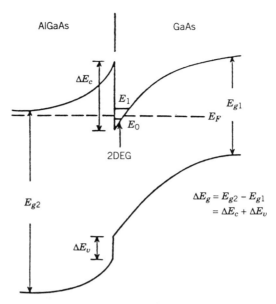

Fig. 9.6. Band diagram of a GaAs/n-Al$_x$Ga$_{1-x}$As heterojunction showing the two-dimensional gas of free electrons formed on the GaAs side of the interface by electrons originating from the n-doping in the Al$_x$Ga$_{1-x}$As. E_0 and E_1 are energies of the subbands in the potential well. E_{g1} and E_{g2} are GaAs and AlGaAs energy band gaps, respectively. ΔE_c and ΔE_v are the conduction and valence-band discontinuities, respectively. [Ladbrook, P. H., in *Gallium Arsenide for Devices and Integrated Circuits*, ed. by H. Thomas and D. V. Morgan, IEE Press (1986), permission from publisher]

of electrons confined in the 2DEG is a function of the thickness of this layer, the donor density in the doped AlGaAs N_D, and the Al mole fraction. Higher-order subbands formed in the narrow quantum well (~ 50 Å) may be populated if the electron concentration exceeds 10^{11} cm^{-2}. In such cases the electron mobility decreases, but improvements in the current carrying capability of the devices more than compensates for the reduction in mobility [Mor85].

9.3 HETEROJUNCTION INTERFACE SHEET CARRIER CONCENTRATION

We have described the electrons that are confined in a quasi-triangular potential well forming 2DEG in the AlGaAs/GaAs heterostructure of a HEMT. Figure 9.6 gives the conduction band-edge diagram of a single-period modulation-doped heterostructure [Mor85]. Consider an electric field E_0 that is quasi-constant. The solution of the longitudinal quantized energy is [Lan77]

$$E_n = \left(\frac{\hbar^2}{2m^*}\right)^{1/3} \left(\frac{3\pi q E_0}{2}\right)^{2/3} \left(\frac{n+3}{4}\right)^{2/3}, \qquad (9.1)$$

where E_n is the energy level and n the index, and E_0 is the electric field at the interface. The energies of the two lowest subbands are thus

$$E_0 = 1.83 \times 10^{-6} \times E_0^{2/3},$$
$$E_1 = 3.23 \times 10^{-6} \times E_0^{2/3},$$

with E_0 in V/m.

The interface carrier concentration can be related to the subband energies if it is expressed in terms of the electric field. It is solved by Poisson's equation. In the depletion approximation, and assuming no impurities in the GaAs layer,

$$\frac{dE_0}{dx} = -\frac{qn(x)}{\varepsilon_2}, \tag{9.2}$$

where ε_2 is the dielectric constant of GaAs. Integration within the depletion region results in

$$\varepsilon_2 E_0 = qn_{s0}, \tag{9.3}$$

where E_0 and n_s are the interface electric field and carrier concentration respectively. From Eqs. (9.1) and (9.3),

$$E_0 = \lambda_0 (n_{s0})^{2/3},$$
$$E_1 = \lambda_1 (n_{s0})^{2/3}, \tag{9.4}$$

where λ_0 and λ_1 are adjustment parameters included to yield a good agreement with experiments.

Consider the AlGaAs/GaAs band diagram with a Schottky barrier deposited on the AlGaAs layer, as shown in Fig. 9.6. The density of states of a two-dimensional system is given by

$$D = \frac{qm^*}{\pi h^2}, \tag{9.5}$$

and the interface carrier concentration can be calculated using the Fermi-Dirac distribution $f(E)$:

$$n_{s0} = D \int_{E_0}^{E_1} f(E) dE + 2D \int_{E_1}^{\infty} f(E) dE. \tag{9.6}$$

From Gauss's law,

$$\varepsilon_1 E_0 d_1 = \varepsilon_2 E_0 d_2 = qn_s,$$

and Poisson's equation,

$$\frac{d^2 V}{dX^2} = \frac{-qN_D}{\varepsilon_1}, \tag{9.7}$$

we obtain

$$n_s = \frac{\varepsilon_1 [V_G - V_{\text{off}}]}{qd}, \tag{9.8}$$

where $d = d_D + d_i$ (d_D is the doped region thickness) and

$$V_{\text{off}} = \phi_B - V_P + \frac{E_f}{q} - \frac{\Delta E_c}{q},$$

$$V_P = \frac{qN_D d_D^2}{2\varepsilon_1}. \tag{9.9}$$

9.4 TRANSPORT IN HEMT STRUCTURES

9.4.1 Low Field Mobility in 2DEGs

The modulation-doped structures grown by MBE were observed for mobility enhancement at low temperatures. In doped bulk GaAs, the Hall mobility is dominated by ionized impurity scattering and is proportional to $T^{1.5}$. The 77 K mobility of bulk material with a low compensation ratio is typically 1.5×10^5 cm^2/V-s for a donor density of 10^{14} cm^{-3} and reduces to 8000 cm^2/V-s for a doping density of 10^{17} cm^{-3} due to ionized impurity scattering. By greatly reducing ionized impurity scattering, modulation doping allows a similar number of carriers in HEMT devices while maintaining high mobility. The background impurity concentration in a practical HEMT is approximately 10^{14} cm^{-3}, but the high density of electrons in the 2DEG (typically 10^{12} cm^{-2}) results in carrier screening, which enhances the mobility. Mobilities as high as 1.95×10^5 cm^2/V-s have been achieved in devices at 77 K.

Below a temperature of 100 K optical phonon scattering is negligible, but above this temperature it is the dominant scattering mechanism in both 2DEGs and bulk GaAs. The 2DEG electron mobility for polar optical scattering is proportional to T^{-k}, where values of k lie between 1.77 and 2.4. The dependence of electron mobility on temperature for a typical HEMT device is shown in Fig. 9.7. The 300 K mobility is limited to approximately 9000 cm^2/V-s for both 2DEGs and bulk GaAs [Lad86].

9.4.2 2DEG Transport in Moderate Electric Fields

The extremely high mobilities of electrons in 2DEG under small applied electric fields at cryogenic temperatures degrade as the field strength is increased to the levels present in the gate-drain region of FET devices subject to typical operating voltages. Electrons in the 2DEG rapidly gain energy and become hot. Above a critical field in the range 50 to 500 V/cm, however, the mobility decreases as a result of spontaneous phonon emission by the hot electrons. The energy separation between the first two subbands in the 2DEG is typically 40 meV, and high mobility electrons in the lowest-energy subband quickly gain sufficient energy from the field to be scattered into the lower-mobility second subband. The higher the initial mobility in the lower subband, the faster the mobility decreases as a function of the applied electric field, as shown

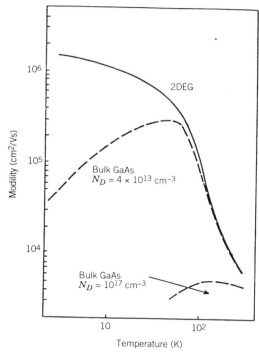

Fig. 9.7 Low-field electron mobility in a two-dimensional electron gas, as a function of lattice temperature, compared with the mobility in bulk GaAs at two different doping levels. [Ladbrook, P. H., in *Gallium Arsenide for Devices and Integrated Circuits*, ed. by H. Thomas and D. V. Morgan, IEE Press (1986); permission from publisher]

in Fig. 9.8. Even in moderate electric fields, the dominant scattering mechanism is polar optical phonon scattering rather than Coulombic scattering, which dominates the cryogenic Hall mobility. At higher temperatures, where polar optical phonon scattering is dominant, the mobility remains as the field is increased.

In MESFETs with gate lengths less than 1 µm, velocity overshoot effects became important in determining device speed. Electrons accelerated by the electric field in the channel or the 2DEG in a HEMT may attain a velocity greater than the static peak velocity before being scattering into a higher-energy subband, as shown in Fig. 9.9. This dynamic over-velocity has an effect on the switching speed of the FET, since it increases the average electron velocity. The amount of overshoot increases as the carrier low-field mobility increases, and hence the enhanced mobility of HEMTs, particularly at cryogenic temperatures, means that this effect is more important in HEMTs than MESFETs.

9.5 CAPACITANCE-VOLTAGE AND CURRENT-VOLTAGE CHARACTERISTICS

The model for C–V characteristics is very important for understanding the operating principle and for developing the model for the small- and large-signal characteristics

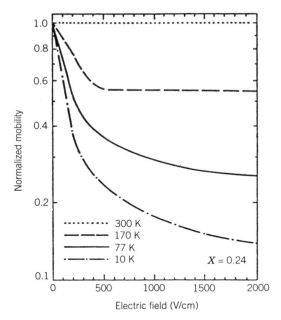

Fig. 9.8 Dependence of electron mobility in a two-dimensional electron gas on electric field strength. [Ladbrook, P. H., in *Gallium Arsenide for Devices and Integrated Circuits*, ed. by H. Thomas and D. V. Morgan, IEE Press (1986); permission from publisher]

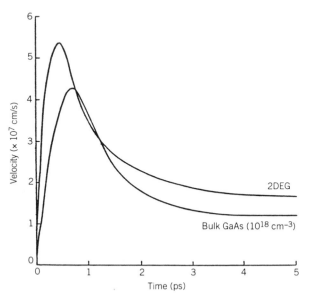

Fig. 9.9 Time response of electrons in a 2DEG and 10^{18} cm^{-3} n-doped GaAs to a 10 kV/cm step function in electric field for a lattice temperature of 77 K (from Monte Carlo simulation). [Ladbrook, P. H., in *Gallium Arsenide for Devices and Integrated Circuits*, ed. by H. Thomas and D. V. Morgan, IEE Press (1986); permission from publisher]

of HEMTs. In this section we describe a charge-control model that takes into account the dependence of the Fermi level on gate voltage and the finite width of the two-dimensional gas.

9.5.1 Charge Control Model

Delagebeaudeuf and Linh developed the charge-control model for HEMT [Del81]. Figure 9.6 shows the band diagram of the studied heterostructure at equilibrium. When a doped AlGaAs layer is grown on an undoped GaAs layers, a two-dimensional electron gas is formed at the interface, since that is a difference in the electron affinity of these layers. The amount of charge transfer across the interface is found by equating the charge depleted from the AlGaAs to the charge accumulated in the potential well. By assuming the total depletion approximation in the space-charge layer, the charge depleted from the AlGaAs is [Del81]

$$n_{s0} = \sqrt{\frac{2\varepsilon N_d(\Delta E_c - E_{F2} - E_{Fi})}{q^2 + N_d^2 d_i^2}} - N_d d_i. \tag{9.10}$$

The charge accumulated in the potential well is given by

$$n_s = \frac{DkT}{q} \ln\left[(1 + e^{q(E_{Fi} - E_0)/kT})(1 + e^{q(E_{Fi} - E_1)/kT})\right], \tag{9.11}$$

where D is the density of states, $E_0 = \lambda_0 n_s^{2/3}$ and $E_1 = \lambda_1 n_s^{2/3}$ are the positions of the first two allowed energy levels in the triangular well. The energy reference in the well is the GaAs conduction band edge at the heterojunction. The constants λ_0, λ_1, and D are derived in the formulation of the triangular well but have been adjusted slightly to obtain closer agreement with measured subband splitting and electron effective mass ($D = qm^*/\pi\hbar^2$). When a Schottky gate is deposited on the AlGaAs layer, this results in a certain amount of depletion beneath the gate. If the AlGaAs layer is thin enough or a sufficiently large negative gate voltage is applied, the gate depletion and junction depletion regions will overlap, in which case Eq. (9.10) would be replaced by

$$n_s = \frac{\varepsilon}{qd}\left[V_g - \left(\phi_b - V_{P2} + \frac{E_{Fi}}{q} - \frac{\Delta E_c}{q}\right)\right], \tag{9.12}$$

where ϕ_b is the Schottky-barrier height, V_g is the gate voltage, and $V_{P2} = qN_d d_D^2/2\varepsilon$, d_D being the thickness of the doped AlGaAs beneath the gate and $d = d_D + d_i$. The simultaneous solution of Eqs. (9.11) and (9.12) then yields the two-dimensional electron gas n_s in the potential well for n_s greater than zero and less than the equilibrium n_{s0}. If the AlGaAs layer is too thick or a sufficiently large positive gate voltage is applied, a parallel conduction path is created.

In the model given in Delagebeaudeuf and Linh [Del82] the total charge in the two-dimensional electron gas is found by neglecting the variation of the Fermi level with the gate voltage:

$$n_s = \frac{\varepsilon}{qd}(V_g - V_{\text{off}}^0), \tag{9.13}$$

where

$$V_{\text{off}}^0 = \phi_b - \Delta E_c/q - V_{P2}$$

is the threshold voltage. The I–V characteristics of a modulation-doped FET can be found from Eq. (9.13) by assuming that the current saturation is reached when the electric field at the drain side of the gate is equal to the saturation field $E_s = v_s/\mu$ [Dru81], where v_s is the saturation velocity and μ the low-field mobility.

Equation (9.11) is a quadratic equation with respect to $\exp(qE_{Fi}/kT)$. The calculated electron density n_s at 300 K is about half of that predicted by a less accurate three-dimensional electron gas model, which neglects the quantization in the potential well and uses the Joyce-Dixon approximation [Joy77]. For the values of n_s between 5×10^{11} cm^{-2} and 1.5×10^{12} cm^{-2}, these dependencies can be approximated as

$$E_{Fi} = \Delta E_{F0}(T) + a n_s, \tag{9.14}$$

where $a = 0.125 \times 10^{-16}$ V-m^{-2}, $\Delta E_{F0} = 0$ at 300 K, and $\Delta F_{F0} = 0.025$ V at 77 K and below. Substituting Eq. (9.14) into Eq. (9.12), we obtain the modified equation of the charge control model:

$$n_s = \frac{\varepsilon}{q(d + \Delta d)}(V_g - V_{\text{off}}), \tag{9.15}$$

where

$$V_{\text{off}} = V_{\text{off}}^0 + \Delta E_{F0}, \tag{9.16}$$

$$\Delta d = \frac{\varepsilon a}{q} = 80 \text{ Å}. \tag{9.17}$$

The estimate for Δd given by Eq. (9.17) is in good agreement with the experimental data reported.

Assuming, as in [Dru81; Yos84], that the current saturation occurs when the electric field at the drain side of the gate exceeds the velocity saturation field $E_s = v_s/\mu$ and using the Shockley model in order to describe the longitudinal field distribution in the channel below the saturation voltage, we find that

$$V_{ds}^s = V_g' + V_{s1} - (V_g'^2 + V_{s1}^2)^{1/2} + I_{ds}^s(R_s + R_d) \tag{9.18}$$

and

$$I_{ds}^s = \frac{\beta V_{s1}^2 * \sqrt{(1 + 2\beta R_s V_g' + V_g'^2/V_{s1}^2) - (1 + \beta R_s V_g')}}{1 - \beta^2 R_s^2 V_{s1}^2}, \tag{9.19}$$

where $V_g' = V_g - V_{\text{off}}$, $V_{s1} = E_s L$, $\beta = \varepsilon \mu W/(d + \Delta d)L$, V_{ds}^s is the drain-to-source saturation voltage, I_{ds}^s is the saturation current, L is the gate length, and W is the gate width.

9.5.2 Equivalent Circuit and Figure of Merits

The relative importance of f_t and f_{max} in millimeter-wave evaluation of HEMT is discussed in this section. Das [Das85] has derived the f_{max} from the equivalent circuit model shown in Fig. 9.10 as

$$f_{max} = \frac{f_t}{\{(4g_0/g_m)[g_m R_i + ((R_s + R_g)/(1/g_m + R_s))] + (4C_{gd}/5C_{gs})[1 + 2.5(C_{gd}/C_{gs})](1 + g_m R_s)^2\}^{1/2}}, \tag{9.20}$$

$$f_t = \frac{g_m}{2\pi C_{gs}}. \tag{9.21}$$

F_t is more important than f_{max} in terms of digital performance. However, f_{max} is preferable to f_t for characterizing high-frequency devices in the reality that f_{max} takes into account the losses associated with gate resistance R_g and output conductance C_{gd} [Les88]. Another important reason for choosing f_{max} over f_t for millimeter-wave evaluation stems for device scaling properties. As frequencies of operation are increased into the millimeter-wave region, device sizes must shrink for easier matching of the input impedance of the device. The bonding pad stays about the same size regardless of the frequency. As a result the gate bonding pad capacitance becomes significant compared to C_{gs} in these small, high-frequency devices. In the design of devices that will operate with reasonable gain at a frequency around 100 GHz, g_m/g_0 and C_{gs}/C_{gd} gain more influence. Hence the relative contribution of f_t to the f_{max} of the device is less.

9.5.3 Transmission-Line Model

A submicron HEMT is modeled as a transmission line for the microwave frequency ac analysis [Hua89]. RC lines can be successfully used to simulate the characteristics of certain active microcircuit elements, in particular, FET devices. It may be considered to have two parts: (1) an active part associated with the 2DEG layer between the

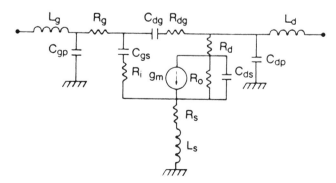

Fig. 9.10 HEMT equivalent circuit model. [Das, M. B., *IEEE Trans. Electron Dev.* **ED-32**, 1 (1985): 11–16; Lester, L. F., P. M. Smith, P. Ho, P. C. Chao, R. C. Tiberio, K. H. G. Duh, and E. D. Wolf, *IEEE IEDM* (1988): 172–175; permission from publisher]

source and drain with transconductance g_m, which is controlled by the gate voltage, and (2) a passive part that includes nonlinear voltage-controlled resistances and capacitances.

The transmission-line model is shown in Fig. 9.11 [Hua89]. For each increment j in the figure, the increment capacitance c_j is modeled as

$$c_j = C_0 l W dx, \tag{9.22}$$

where $C_0 = \varepsilon_{\text{AlGaAs}}/(d_i + d_d + \Delta d)$, which is a constant provided that the AlGaAs layer is completely depleted. The increment resistance r_j and the incremental transconductance g_{mj} are modeled as follows:

$$r_j = \frac{V_{j+1} - V_j}{I_{ds}}, \tag{9.23}$$

$$g_{mj} = \frac{\beta V_{j+1}}{dx/L_g + \alpha V_{j+1}}, \tag{9.24}$$

where V_j is the dc voltage along the channel. At each node of Fig. 9.11, Kirchhoff's current law is used for small-signal ac analysis, as shown in Fig. 9.12. For each node the following equation is derived:

$$-\left(\frac{1}{r_j} + g_{mj}\right)\tilde{v}_{j-1} + \left(\frac{1}{r_j} + \frac{1}{r_{j+1}} + j\omega c_{j+1} + g_{m(j+1)}\right)\tilde{v}_j$$
$$- \frac{1}{r_{j+1}}\tilde{v}_{j+1} = \tilde{v}_{gs}(j\omega c_{j+1} + g_{m(j+1)} - g_{mj}), \tag{9.25}$$

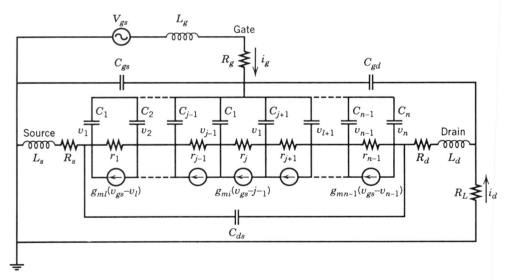

Fig. 9.11 Transmission-line model for HEMT. [Huang, D. H., and H. C. Lin, *IEEE Trans. Microwave Theory and Tech.* **37** (1989): 1361; permission for publisher]

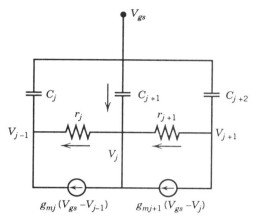

Fig. 9.12 Incremental cells of transmission-line model. [Huang, D. H., and H. C. Lin, *IEEE Trans. Microwave Theory and Tech.* **37** (1989): 1361; permission from publisher]

where $\tilde{v}_1$ is the ac voltage. Equation (9.25) shows that the ac voltage at each node depends on the ac voltage of the previous and the next node only. By applying Kirchhoff's current law to every node along the channel, the following tridiagonal linear equation can be obtained:

$$\begin{cases} d_1\tilde{v}_1 + e_1\tilde{v}_2 &= b_1, \\ a_1\tilde{v}_1 + d_2\tilde{v}_2 + e_2\tilde{v}_3 &= b_2, \\ \quad a_2\tilde{v}_2 + d_3\tilde{v}_3 + e_3\tilde{v}_4 &= b_3, \\ \quad\quad\quad a_{N-1}\tilde{v}_{N-1} + d_N\tilde{v}_N &= b_N, \end{cases}$$

where

$$a_{j-1} = -(1/r_j + g_{mj}),$$
$$b_j = \tilde{v}_{gs}[j\omega c_{j+1} + g_{m(j+1)} - g_{mj}] \quad \text{for} \quad j < N,$$
$$b_j = \tilde{v}_{gs}[j\omega c_{j+1} - g_{mj}] + \tilde{i}_d \quad \text{for} \quad j = N,$$
$$e_j = -1/r_{j+1},$$
$$d_j = 1/r_j + 1/r_{j+1} + j\omega c_{j+1} + g_{m(j+1)}.$$

The small-signal ac current and voltage distribution along the channel can be calculated by solving the above tridiagonal linear equation using the Gaussian elimination method. The terminal small-signal parameters are then determined by mixed parameters:

$$\tilde{i}_g = m_{11}\tilde{v}_{gs} + m_{12}\tilde{i}_d, \tag{9.26}$$

$$\tilde{v}_{ds} = m_{21}\tilde{v}_{gs} + m_{22}\tilde{i}_d, \tag{9.27}$$

where $\tilde{i}_g = \sum(\tilde{v}_{gs} - \tilde{v}_j)j\omega c_j$ and $\tilde{v}_{ds} = \tilde{v}_N$. By shorting the input circuit or opening the output circuit, the parameters m_{12} and m_{22} or m_{11} and m_{21} can be calculated, respectively. The M parameters are then converted to S parameters. The extrinsic S parameters, including parasitics, are calculated for the bias condition of $V_g = -1.77$ V and $V_{ds} = 2.5$ V (saturated region) of a 0.3×100-μm^2 Allied Signal HEMT. The S-parameters are plotted as a function of frequency from 0.5 GHz up to 26 GHz in Fig. 9.13. The calculated curves (solid lines) fit the data (Δ) reasonably well. From Fig. 9.13(b), one can see that the measured S_{21} is larger than the calculated results at high frequencies. This increase of the transconductance can be attributed to an effective decrease of the source access resistance at microwave frequencies.

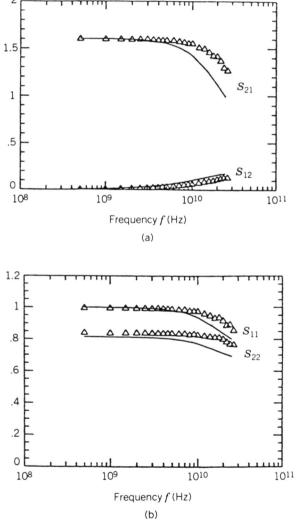

Fig. 9.13 The calculated and experimental S-parameters (magnitude): experimental data (Δ); calculations based on transmission-line model (solid lines). [Huang, D. H., and H. C. Lin, *IEEE Trans. Microwave Theory and Tech.* **37** (1989): 1361; permission from publisher]

9.6 PERSISTENT PHOTOCONDUCTIVITY AND DRAIN I–V COLLAPSE

In the previous sections we mentioned that HEMT devices have better performance than the MESFETs. There are a number of problems that have to be overcome for the HEMT to realize its potential in the commercial IC market. These problems are light sensitivity, threshold voltage variation, threshold voltage shift with temperature, and instability in the I–V characteristics, both at 77 K and 300 K. In this section we will study the causes of persistent photoconductivity (PPC) and drain I–V collapse. The frequence dependence of transconductance g_m on HEMT will also be discussed.

When exposed to light illumination at low temperature (< 140 K), the AlGaAs/GaAs heterostructure HEMT produces a photocurrent that persists even after the light source is removed [Tu85; Sto79]. The origin of this persistent photoconductivity has been studied by numerous authors. For photon energy higher than the AlGaAs band gap, the main cause of PPC seems to be electron photoexcitation from deep traps in the n-type AlGaAs donor layer [Kle84]. Due to a large lattice relaxation after ionization, the photoexcited electrons that have left the traps face an energy barrier for recapture. At low temperature and in complete darkness, the electrons do not have sufficient thermal energy, so the traps remain ionized. Under illumination or at room temperature, the traps are continuously emptied and filled, and an equilibrium state is reached. These deep traps are donor-defect complexes, the DX centers, studied by Lang et al. [Lan74]. For photon energy between the band gaps of GaAs and AlGaAs, PPC in HEMT structures arises from electron-hole generation in the bulk GaAs and a subsequent charge separation at the interface.

The I–V instability at room temperature manifests itself in a hysteresis loop of I_{DS}-versus-V_{SD} curves [Duh84]. The instability can be eliminated or reduced by shining light on the device. At 77 K in complete darkness, the drain I–V curves distort when the drain bias is above a certain threshold. This is the drain I–V collapse problem, as shown in Fig. 9.14. The collapse occurs independent of gate bias and is affected by the DX centers. Electrons frozen in the deep traps can tunnel through the barrier under the influence of the applied electric field; they are collected at a blocking contact on the drain side of the transistor, thereby setting up an internal field between the source and the drain [Tu85]. The electric field pinches off the current flow at the low drain bias. If the n^+ drain region is sufficiently close to the drain end of the gate, the electrons will be collected before the trapping occurs, and the collapse can be reduced or eliminated [Fis84].

9.7 INVERTED HEMT

In a conventional HEMT structure the ternary (AlGaAs) is grown on top of a binary (GaAs) compound. There were difficulties in growing a high-performance inverted structure (binary on top of ternary) [Mor85]. Since in a conventional HEMT the sheet resistance of 2DEG is rather high, some processing techniques are required for the reduction of the source resistance to improve the FET performance at room temperature. On the other hand, the source resistance can be easily lowered by the top n^+-GaAs layer and recessed gate process in the inverted structure. In this section we describe the growth condition and processing technique for the inverted HEMT.

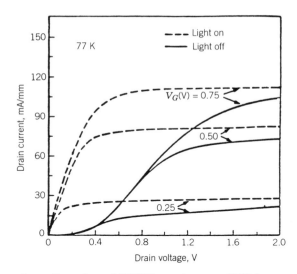

Fig. 9.14 Current-voltage distortion of SDHT transistors at 77 K in complete darkness and under illumination. [Kastalsky, A., and R. A. Kiehl, *IEEE Electron Devices*, **ED-33**, 3, 414–423 (1986); permission from publisher]

9.7.1 Crystal Growth

A high-performance inverted HEMT was reported by Nishi et al. [Nis86]. The inverted heterostructures were grown by the MBE system. The growth rate was about 0.5 µm/h. A 2-in. HB semi-insulating GaAs was used as a substrate. After chemical etching the substrate was immediately fixed to a substrate holder by In solder. A direct heating system without In solder was also used. The schematic cross section of an inverted HEMT is shown in Fig. 9.15. The epitaxial layer consists of undoped GaAs and AlGaAs with thicknesses of 1000 Å, n-AlGaAs (1.1×10^{18} cm^{-3}, 80 Å), an

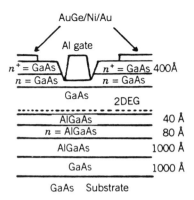

Fig. 9.15 Schematic cross section of the inverted HEMT. [Nishi, S., T. Saito, S. Seki, Y. Sano, H. Inomata, T. Itoh, M. Akimaya, and K. Kaminishi, in *Gallium Arsenide and Related Compounds*, ed. by W. T. Lindley (1986); permission from publisher]

AlGaAs separation layer (40 Å), undoped GaAs (200 Å), n-GaAs (5×10^{17} cm^{-3}), and n^+-GaAs (4×10^{18} cm^{-3}, 400 Å). The maximum carrier density of 2DEG is 8×10^{11} cm^{-2}.

Figure 9.16 illustrates the dependence of 2DEG mobility at 77 K on substrate temperature at the inverted heterointerface. In this experiment the substrate was monitored by a thermocouple behind a Mo substrate holder. The thickness of the AlGaAs separation layer was fixed at 60 Å. To suppress both a depletion of 2DEG by the surface potential and a parallel conduction in n-GaAs, a thick undoped GaAs channel layer of 2000 Å with thin top n-GaAs layer was grown on the AlGaAs layer. The electron mobility was measured by the van der Pauw method. The electron mobility increased gradually with decreasing the substrate temperature, and it decreased rapidly with the substrate temperature below 500°C. The sheet carrier concentrations were also decreased. The fall of mobility and sheet carrier concentration below 500°C were due to the increase of the deep-level concentration in the grown layer. The higher mobilities at fairly low substrate temperatures may be due to the improvement of AlGaAs surface morphology and the reduction of surface segregation of doped Si. When the substrate temperature range was 600° to 650°C, a sufficiently high electron mobility was obtained at room temperature (>6000 cm^2/V-s).

9.7.2 Device Characteristics

Figure 9.17 shows the relation between 2DEG density n_s and the thickness of n-AlGaAs [Yam85]. Two different thicknesses (200 and 400 Å) of the undoped GaAs channel layers are compared in terms of FET performance. In Fig. 9.18, K-values ($K = I_D/[V_{GS} - V_T]^2$) are shown to be at room temperature a function of threshold voltage for two different thicknesses. The gate length and the width are 0.8 and 10 μm, respectively. The K-values greatly improve as the thickness of undoped GaAs is reduced. The K-value at the threshold voltage of 0 V was improved about 40%. Figure 9.19 shows the square roots of the saturation current at room temperature and 77 K as a function of gate voltage. The thickness of the undoped GaAs layer is 200 Å. Extremely

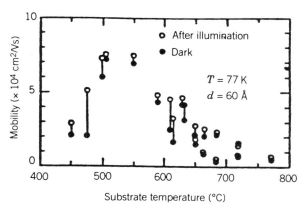

Fig. 9.16 Dependence of 2DEG mobility of 77 K on substrate temperature. [Nishi, S., T. Saito, S. Seki, Y. Sano, H. Inomata, T. Itoh, M. Akimaya, and K. Kaminishi, in *Gallium Arsenide and Related Compounds*, ed. by W. T. Lindley (1986); permission for publisher]

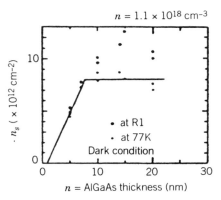

Fig. 9.17 Relation between N_s and thickness of n-AlGaAs. [Nishi, S., T. Saito, S. Seki, Y. Sano, H. Inomata, T. Itoh, M. Akimaya, and K. Kaminishi, in *Gallium Arsenide and Related Compounds*, ed. by W. T. Lindley (1986); permission from publisher]

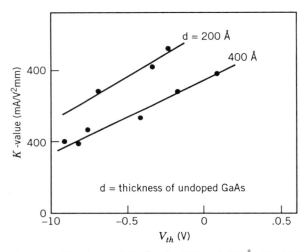

Fig. 9.18 Relation between K-value and V_{th} for $d = 200$ and 400 Å. [Nishi, S., T. Saito, S. Seki, Y. Sano, H. Inomata, T. Itoh, M. Akimaya, and K. Kaminishi, in *Gallium Arsenide and Related Compounds*, ed. by W. T. Lindley (1986); permission from publisher]

high K-values of $480\,mA/V^2$-mm at room temperature and $860\,mA/V^2$-mm at 77 K were obtained. The maximum transconductances at room temperature and 77 K were 400 mS/mm and 550 mS/mm, respectively. Recent results on a 0.2-μm gate pseudomorphic inverted HEMT have demonstrated that a transconductance of 580 mS/mm and a cutoff frequency of 110 GHz are possible [Tsu91].

9.8 PSEUDOMORPHIC HEMT

The AlGaAs/GaAs HEMT performance at low temperature is somewhat complicated by the formation of DX centers, which act as electron traps in the $Al_xGa_{1-x}As$

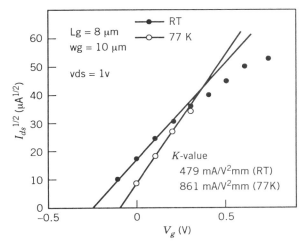

Fig. 9.19 I_{ds} versus V_g relations at room temperature and 77 K. [Nishi, S., T. Saito, S. Seki, Y. Sano, H. Inomata, T. Itoh, M. Akimaya, and K. Kaminishi, in *Gallium Arsenide and Related Compounds*, ed. by W. T. Lindley (1986); permission from publisher]

layers when $x > 0.2$. This leads to the collapse of the drain current-voltage (I–V) characteristics and to persistent photoconductivity because the electrons are captured by the traps. One way to avoid the "collapse" problem is to use InGaAs as the lower band gap or a 2DEG channel material instead of GaAs. This section describes high-electron mobility transistors that utilize a conducting channel consisting of a single $In_{0.15}Ga_{0.85}As$ quantum well grown pseudomorphically on a GaAs substrate. The lattice-mismatched system of heterojunction semiconductors and the requirement of pseudomorphic growth on lattice-mismatched semiconductors will be first described. The initiative of developing pseudomorphic HEMT and the device structure and processing methods will be explained. The AlGaAs/InGaAs/GaAs pseudomorphic HEMT with a gate length of 0.1 μm have been successfully fabricated and tested. The advantages of the pseudomorphic HEMT will be described.

9.8.1 Materials Consideration

Modern optoelectronic and high-speed devices with heterostructures can be divided into two categories:

1. *Lattice-matched system.* Each heterojunction structure is made with materials that are lattice-matched; that is, the unstrained lattice constant of the materials is approximately equal. Examples are GaAs/AlGaAs system for high-speed transistors and the InP/InGaAsP system for double-heterojunction (DH) laser applications.
2. *Lattice-mismatched system.* Each heterojunction structure has a strained-layer superlattice (SL) ensemble that is not lattice-matched to the substrate. A lattice-mismatched layer can be grown sufficiently thin so that the mismatch is accommodated entirely as elastic strain. The interface between the materials is essentially free from the dislocations, and the layer is pseudomorphic.

One research direction for non-lattice-matched systems is to develop a device technology based on the growth of epilayers that are unstrained (as a result of an accommodation of the mismatch at the epilayer/substrate interface by defects such as interface misfit dislocations). Then the epilayer/substrate ensemble becomes a new combined "substrate" on which additional epilayer structures can be grown or on which devices can be directly fabricated. This research is largely driven by the perceived benefits of GaAs devices grown on Si substrates and a possible integration of the two device technologies [Woo87].

For unstrained layers, on the other hand, the misfit between the substrate and epilayer is accommodated by some combination of interface misfit dislocations and other defects such as threading dislocations. It was found that [Mat77; Bie83] the threading dislocations did not propagate into the strained-layer superlattice region. The strained layer thus became a technique in itself that could "filter" threading dislocations. It is possible to use a single nonlatticed matched layer to bend over all the threading dislocations into the interface between the mismatched layers and thus convert the threading dislocations into misfit dislocations that accommodate the lattice mismatch. When the mismatch was confined to less than about 1.5%, 60° misfit dislocations were found to form as the result of the glide of dislocation half-loops nucleated at epilayer surface [Woo83].

The 60° misfit dislocations have been found to quench radiative recombination [Pet79]. These electrically active dislocations have been associated with Fermi-level pinning [Woo83; Bat86]. This is evidence that the edge-type dislocation is not associated with nonradiative recombination [Pet79]. Given that 60° misfit dislocations are electrically and optically active, and that the threading dislocations are optically active with large densities at the epilayer during edge misfit dislocations, it is clear that optoelectronic devices fabricated in non-lattice-matched systems must include a strained-layer threading dislocation filter, or else be so designed as to isolate the adverse effect of the active misfit dislocation. One example of this system is the GaInAs interdigited-metal–semiconductor-metal (IMSM) structure grown on a GaAs substrate [Rog87]. For non-lattice-matched epitaxy, the strain is added to the growth forces in such a way that two-dimensional growth is driven toward three-dimensional growth, or the Stranski-Krastanov mechanism [Lud84]. A rule of thumb is that as the mismatch approaches 1.5 to 2.0% the growth mechanism changes from two to three dimensions. Therefore the structure in Figure 9.20 was step graded in less than 2% mismatch increments. The thickness of each increment was well above h_c, the critical thickness above which, at equilibrium, the lowest energy of the system is for the strained (pseudomorphic) layer to relax back to its unstrained condition through the formation of interface misfit dislocations. This is shown analytically as [Mat70]

$$h_c = b(1 - v\cos^2\theta)\ln\left(\frac{h_c}{b}\right)8f(1+v)\cos\Delta, \qquad (9.28)$$

where b is the Burgers vector, v is Poisson's ratio, f is the misfit factor $\Delta a/a$, and θ and Δ are the angles between the slip planes and the crystal surfaces.

Another research direction is to use the strained layer in non-lattice-matched system. As Matthews and Blakeslee [Mat74] have pointed out, epigrowth of a strained and highly perfect non-latticed-matched layer is possible. The various growth mechanisms associated with non-lattice-matched epitaxy are summarized in Fig. 9.21,

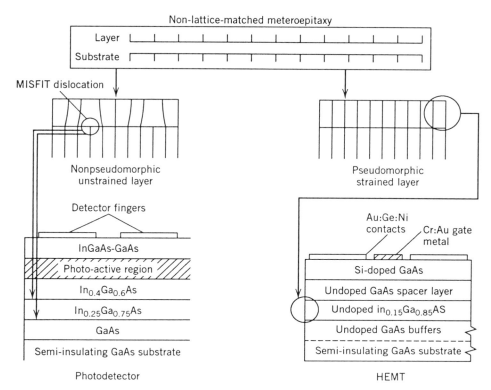

Fig. 9.20 Schematic diagram illustrating two different regimes for fabricating devices in non-lattice-matched systems. *Left side*: a photodetector device fabricated on an unstrained layer formed by step grading in small mismatch increments to relieve strain via 60° glide misfit dislocations. *Right side*: a HEMT device in which the channel is a single layer that is strained and stable (pseudomorphic). [Woodall, J. A., P. D. Kirchner, D. L. Rogers, M. Chisholm, and J. J. Rosenberg, So who needs lattice matched heterojunctions anyway? in *U.S.–Japan Conference, Florida*, IBM (1987)]

which includes a plot of Eq. (9.28). Note that for the mismatch above of about 2%, Eq. (9.28) takes the shape of the dashed line. The dashed curve reflects the fact that 3-D growth generally occurs before the critical thickness is reached.

9.8.2 Fabrication of a 0.1-μm Pseudomorphic HEMT

We now turn to the performance of a 0.1-μm gate-length pseudomorphic HEMT as reported by Chao et al. [Cha87]. The 0.1-μm gate-length MESFET generates short-channel effects [Cha87]. This effect is largely due to the space-charge injection of carriers, which is proportional to 1/(effective gate length)^2, into the buffer layer under the channel. This increases the device output conductance and causes a shift in the pinch-off voltage and transconductance g_m compression near the pinch-off region. To overcome this problem, an AlGaAs/InGaAs/GaAs pseudomorphic HEMT planar-doped structure was developed for the 0.1-μm device. The structure of the device and the band diagram are illustrated in Figs. 9.22(a) and (b), respectively. The pseudo-

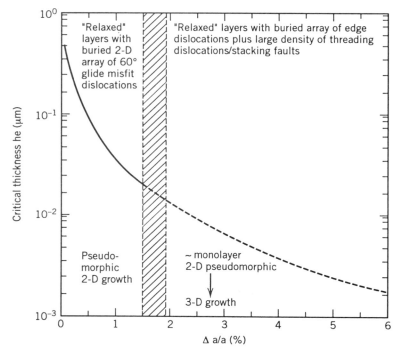

Fig. 9.21 Illustration of types of growth occurring in non-lattice-matched systems. Line plot in Eq. (9.43) from text of critical thickness versus misfit. [Woodall, J. A., P. D. Kirchner, D. L. Rogers, M. Chisholm, and J. J. Rosenberg, So who needs lattice matched heterojunctions anyway? in *U.S.–Japan Conference, Florida*, IBM (1987), permission from IBM]

morphic HEMT has a quantum well channel structure and thus has improved carrier confinement. This will reduced short-channel effect. The pseudomorphic HEMT also provides very high electron velocity and sheet-charge density due to the superior carrier transport properties in the InGaAs channel and the large conduction band discontinuity with AlGaAs. The planar-doped structure (to be discussed in the next section) was used to improve the device's breakdown voltage and also the gate-length/gate-to-channel separation ratio which increases the device's g_m and gain.

The 0.1-μm HEMT devices were fabricated on n^+-GaAs ($10^{18}/cm^3$)/n^+-AlGaAs($10^{18}/cm^3$, 500 Å)/AlGaAs(undoped, 30 Å)/InGaAs (undoped)/GaAs pseudomorphic heterostructures grown by MBE. Hall measurements indicate a sheet-carrier concentration of $2 \times 10^{12}/cm^2$ and an electron mobility of 15,000 cm^2/V-s at 77 K. Devices with 50-μm-wide gates were fabricated on the wafer by direct-write electron-beam lithography. AuGeNi-alloyed ohmic contacts were formed by a rapid thermal annealer, and their specific contact resistance was 10^{-7} Ω-cm^2. An automatic wafer registration technique was used on each device for gate-level alignment. This resulted in a typical gate overlay accuracy with respect to the source-drain level of better than ±0.05 μm. The 0.1-μm gate lines were exposed at an electrode dose of 3.5 nC/cm at 50 kV. After the development of gate-level resist, the exposed area was recessed to achieve the desired full channel current. The Schottky gates were formed by evaporating and lifting off Ti/Pt/Au metal layers.

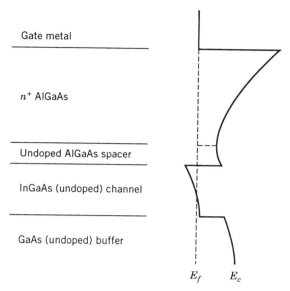

Fig. 9.22 The structure and band diagram based on the pseudomorphic HEMT fabricated by Chao et al. [Chao, P. C., P. M. Smith, K. H. G. Duh, J. M. Ballingall, L. F. Lester, B. R. Lee, A. A. Jabra, and R. C. Tiberio, *IEEE IEDM* (1987): 410–413; permission from publisher]

9.9 PULSED-DOPED HEMT

The pulsed-doping, also called *planar* or *delta doping*, technique can provide high-density, quasi-two-dimensional electron or hole gas systems in compound semiconductors [Plo88; Gil88]. Figure 9.23 shows the cross-sectional profile of a pulsed-doped HEMT [Yue88]. A very thin layer (3 Å) $Al_{0.26}Ga_{0.74}As$ with a concentration of 8×10^{18} cm^{-3} is sandwiched between two AlGaAs layers. Compared with conventional HEMTs, the pulse-doped HEMT has higher breakdown voltage, higher 2DEG density, and higher intrinsic transconductance, and provides better control of the threshold voltage and improved linearity [Hon89].

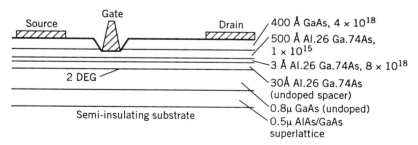

Fig. 9.23 Cross section profile for the pulse-doped HEMT. [Yuen, C., M. Riaziat, S. Bandy, and G. Zdasiuk, *Microwave J.* (August 1988): 87–104; permission from publisher]

9.10 SUBTHRESHOLD CURRENT IN HEMT

The submicrometer gate HEMTs have shown remarkable progress in high-frequency and high-speed operations. However, they suffer the serious drawback of short-channel effects such as the increase of output conductance and subthreshold current [Han88]. The improvement of the output conductance has been reported [Cha89]. It is important to understand the mechanism of the subthreshold conduction in HEMTs so that a set of optimal device parameters can be obtained.

In this section we discuss the study of subthreshold current I_s in conventional (GaAs/AlGaAs)HEMTs and in pseudomorphic (InGaAs/AlGaAs) HEMTs with a short gate length L_g down to 0.12 μm [Jia90]. The subthreshold current in the submicrometer gate HEMT is due to the charge injection from the source limited by the channel potential barrier Π, which is a function of both V_d and V_g. With a higher potential barrier in the buffer, the pseudomorphic quantum well HEMTs can improve I_s by suppressing the current passing through the GaAs buffer region.

Figure 9.24 shows the potential profile in the subthreshold region of long-channel and short-channel HEMTs. For a long-channel HEMT, the potential height barely changes with the applied voltage V_d. For a short-channel device, however, the peak position and the height of the potential very with the applied V_d. Potential Π becomes a function of both V_d and V_g. Any physical parameter change, which increases the role of V_d and decreases the role of V_g, will result in a stronger short-channel effect.

The important problem for the subthreshold current of HEMTs is the location of the charge injection, namely the position of the lowest potential height. It can be either at the channel interface or deep in the bulk, depending on the values of applied V_d and V_s [Tro79].

With an appropriate quantum well device structure we can expect the barrier height in the buffer region to be much higher than that of the conventional HEMT. This well reduce the undesirable I_s. Figure 9.25 compares the I_s and V_g of a conventional GaAs/AlGaAs HEMT with those of a pseudomorphic GaInAs/AlGaAs HEMT whose gate length is 0.12 μm and $V_{th} = 0$ V. The gate swing S of both devices still shows increases with V_d. The pseudomorphic HEMT, however, has a much smaller S than the conventional one for the same V_d. In the case of large V_d and more negative V_g, a pseudomorphic HEMT has a much smaller I_s than the conventional

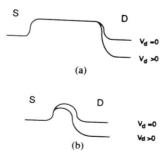

Fig. 9.24 Interface profiles as a function for V_d for (a) long-channel HEMT and (b) short-channel HEMT. [Jiang, C., D. C. Tsui, B. J. F. Lin, H. Lee, A. Lepore, and M. Levy, *IEEE Electron Device Lett.* **11**, 1 (1990): 63–65; permission from publisher]

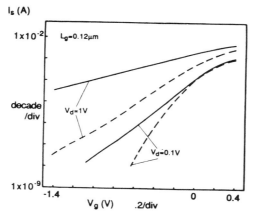

Fig. 9.25 Subthreshold current I_s versus gate bias V_s for gate length $L_g = 0.12\,\mu\text{m}$ and gate width $W_g = 50\,\mu\text{m}$, with drain voltage $V_d = 1$ and 0.1 V. *Solid lines*: data from the conventional GaAs/AlGaAs HEMTs. *Dashed lines*: data from the pseudomorphic InGaAs/AlGaAs HEMTs. [Jiang, C., D. C. Tsui, B. J. F. Lin, H. Lee, A. Lepore, and M. Levy, *IEEE Electron Device Lett.* **11**, 1 (1990): 63–65; permission from publisher]

one. This result indicates that a pseudomorphic HEMT can suppress the charge injection through the buffer region. Figure 9.26 shows the difference seen between the plot of I_s with relation to V_d for these two devices. In the case of the conventional HEMT, I_s increases exponentially with V_d in the small-drain-voltage region, indicating that the charge injection from the source is limited by the channel barrier. In the large-drain-voltage region, I_s increases linearly with the drain voltage. In a short channel of a 0.12-μm-gate device, this behavior indicates that I_s is dominated by the

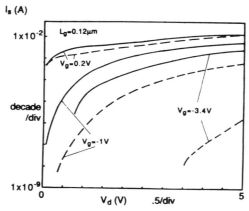

Fig. 9.26 Subthreshold current I_s versus drain voltage V_d for gate length $L_g = 0.12\,\mu\text{m}$ and gate width $W_g = 50\,\mu\text{m}$ with bias $V_g = 0.2$, -1, and -3.4 V. *Solid lines*: data from the conventional GaAs/AlGaAs HEMTs. *Dashed lines*: data from the pseudomorphic InGaAs/AlGaAs HEMTs. [Jiang, C., D. C. Tsui, B. J. F. Lin, H. Lee, A. Lepore, and M. Levy, *IEEE Electron Device Lett.* **11**, 1 (1990): 63–65; permission from publisher]

space-charge-limited current [Han88]. In such a case the gate has less control on I_s. In the region of V_d around 5 V and $V_g = -1$ V and -3.4 V, I_s of the pseudomorphic quantum well HEMT is, respectively, about 10 and 1000 times smaller than that of the conventional HEMT. These results indicate that the pseudomorphic quantum well HEMT suppresses the charge injection through the buffer region. The pseudomorphic HEMT is a better device in high-speed application.

9.11 VLSI GaAs HEMT ICs

Future signal-processing systems will require high data throughput rates as well as low power consumption. GaAs heterostructure FET technology is a leading candidate for such high-performance integrated circuits. Since HEMT ICs have low levels of integration, GaAs technology has not been seriously considered for insertion into high-performance signal-processing systems. It is generally accepted that the level of integration has to approach 5000 gates before system designers can effectively use the technology.

Akinwande [Aki89] described a GaAs heterojunction FET (HFET) integrated circuits technology with a level of integration approaching VLSI. It is a self-aligned, refractory metal gate, ion-implanted process. The heterostructure active layers are grown by MBE on SI LEC substrates. A WSi$_x$ gate metal is sputter deposited and is delineated by optical lithography and reactive ion etching in CF_4/O_2. Heavily doped source and drain contact layers and saturated resistors (ungated FETs) are formed by Si ion-implantation. The implant layers are activated by rapid optical anneal at 850°C. The active devices are isolated by deep oxygen implants followed by another rapid optical anneal. The ohmic contacts are defined by dielectric assisted lift-off (DAL) of Au:Ge-based metallization and alloyed in forming gas at 450°C. The

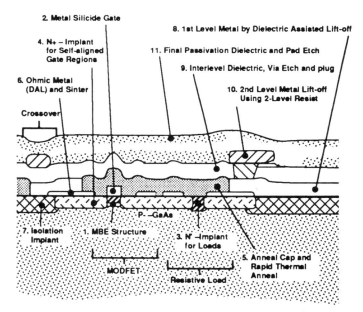

Fig. 9.27 Integrated circuit structure. [Akiwande, A. I., *IEDM 89* (1989): 97; permission from publisher]

active devices are interconnected by two levels of metallization which are separated by PECVD silicon dioxide, using 2 μm lines and spaces on both levels. A via plug layer (2 × 2 μm minimum size) defined by DAL connects the two metal layers. The final metal layer is Cr:Si, and it is delineated by ion milling. Figure 9.27 shows an IC cross section after passivation with a low-stress dielectric.

Three HFET structures are used in the fabrication of the circuits. These are the conventional $Al_xGa_{1-x}As/GaAs$ HEMT, the $Al_xGa_{1-x}As/GaAs$ superlattice HEMT, and the doped $In_yGa_{1-y}As$ channel HFET. The structures are shown in Fig. 9.28. Table 9.1 compares the different devices fabricated by the self-aligned gate process. The V_T standard deviation is 35 mV across a 3-in. wafer, and that the transconductance is between 250 and 350 mS/mm for 1-μm-gate lengths. This process is suitable for high-speed VLSI circuits [Aki89].

Circuit demonstrations include a 500 MHz, 4500-gate 16 × 16 Complex Multiplier chip, a 250 MHz, 3800-gate Butterfly Adder chip, and a four-bit ADC. The level of

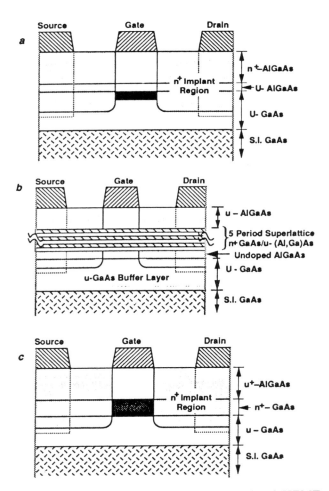

Fig. 9.28 Heterostructure FET device structures (*a*) conventional HEMT, (*b*) superlattice HEMT, and (*c*) doped InGaAs channel HEMT. [Akiwande, A. I., *IEDM 89* (1989): 97; permission from publisher]

TABLE 9.1 Device Parameters for the Three HFET Structures

Device Parameter	Conventional MODFET	Superlattice MODFET	Doped Channel HFET
V_T (Volts)	0.442 ± 0.025	0.355 ± 0.032	0.126 ± 0.065
β (mA/V^2/mm)	377.7 ± 25.0	346.5 ± 23.8	248.3 ± 29.4
I_{dss} (mA/mm)	97.8 ± 7.1	118.5 ± 8.1	173.5 ± 7.2
g_m (mS/mm)	254.6 ± 16.0	254.0 ± 16.3	341.1 ± 20.3
g_{lin} (mS/mm)	481.3 ± 15.4	442.1 ± 16.7	347.0 ± 42.9
g_{ds} (mS/mm)	3.73 ± 1.02	4.21 ± 1.13	6.96 ± 1.28
V_{on} (Volts)	0.820 ± 0.010	0.895 ± 0.014	1.000 ± 0.012
R_s (Ω-mm)	0.564 ± 0.067	0.596 ± 0.065	0.650 ± 0.120

I_{des} was measured at $V_{DS} = 2.0$ V and $V_{GS} = 1.0$ V.
Source: Akiwande, A. I., IEDM 89, (1989): 97.

integration should approach 15,000 to 20,000 gates, while the power dissipated per gate will continue to decrease as the device is scaled to less than 0.5 μm.

REFERENCES

[Col85] Coleman, J. J., in *Gallium Arsenide Technology*, ed. by D. K. Ferry, SAMS (1985).

[Mor91] Morkoc, H., *IEEE Circuits and Devices* (Nov. 1991): 15–20, p. 79.

[Esa69] Esaki, L., and R. Tsu, Internal Report RC 2418, IBM Research, March 26 (1969).

[Esa70] Esaki, L., and Tsu, R., *IBM J. Research* **14** (1970): 61.

[Din78] Dingle, R., H. L. Stormer, A. C. Gossard, and W. Wiegmann, *Appl. Phys. Lett.* **33**, 7 (1978): 665–667.

[Gos85] Gossard, A. C., Modulation doping of semiconductor heterostructures, in *Molecular Beam Epitaxy and Heterostructures*, ed. by L. L. Chang, and K. Ploog, Martinus Nijhoff (1985).

[Din74] Dingle R., W. Wiegmann, and C. H. Henry, *Phys. Rev. Lett.* **33** (1974): 827.

[Wei81] Weisbuch, C., R. C. Miller, R. Dingle, A. C. Gossard, and W. Wiegmann, *Solid State Comm.* **37** (1981): 219.

[Abs79] Abstreiter, G., and K. Ploog, *Phys. Rev. Lett.* **42** (1979): 1308.

[Sai77] Sai-Halasz, G. A., R. Tsu, and L. Esaki, *Appl. Phys. Lett.* **30** (1977): 651.

[Din75] Dingle, R., in *Advances in Solid State Physics*, vol. 15, p. 21, ed. by H. J. Queisser, Pergamon Press (1975).

[Wol70] Wolfe, C. M., G. E. Stillman, and W. T. Lindley, *J. Appl. Phys.* **41** (1970): 3088.

[Pri82] Price, P. J., *Surf. Sci.* **113** (1982): 199.

[Mor79] Mori, S., and T. Ando, *Phys. Rev. B* **19** (1979): 6433.

[Mor85] Morkoc, H., in *Molecular Beam Epitaxy and Heterostructures*, ed. by L. L. Chang, and K. Ploog, Martinus Nijhoff (1985).

[Lan77] Landau, L. D., and E. M. Lifshitz, *Quantum Mechanics*, 3d ed., Pergamon Press (1977).

[Lad86] Ladbrook, P. H., in *Gallium Arsenide for Devices and Integrated Circuits*, ed. by H. Thomas and D. V. Morgan, IEE Press (1986).

[Del81] Delagebeaudeuf, D., and N. T. Linh, *IEEE Trans. Electron Dev.* **ED-28** (1981): 790–795.

[Del82] Delagebeaudeuf, D, and N. Linh, *IEEE Trans. Electron Dev.* **ED-29**, 6 (1982): 955–960.

[Dru81] Drummond, T. J., H. Morkoc, K. Lee, and M. S. Shur, *IEEE Electron Device Lett.* EDL-3(11) (1981): 338–341.

[Joy77] Joyce, W. B., and R. W. Dixon, *Appl. Phys. Lett.* **31** (1977): 354.

[Yos84] Yoshida, J., and M. Kurata, *IEEE Electron Dev. Lett.* **EDL-5**, 12 (1984): 508–510.

[Das85] Das, M. B., *IEEE Trans. Electron Dev.* **ED-32**, 1 (1985): 11–16.

[Les88] Lester, L. F., P. M. Smith, P. Ho, P. C. Chao, R. C. Tiberio, K. H. G. Duh, and E. D. Wolf, *IEEE IEDM* (1988): 172–175.

[Hua89] Huang, D. H., *IEEE Trans. Microwave Theory and Technique* **37** (1989): 1361.

[Tu85] Tu, C. W., R. H. Hendel, and R. Dingle, Molecular beam epitaxy and the technology of selectively doped heterostructure transistors, in *Gallium Arsenide Technology*, ed. by D. K. Ferry, SAMS (1985).

[Sto79] Stormer, H. L., R. Dingle, A. C. Gossard, W. Wiegmann, and M. D. Sturge, *Solid State Commun.* **29** (1979): 705.

[Kle84] Klem, J., T. J. Drummond, R. Fischer, T. Henderson, and H. Morkoc, *J. Electron Mat.* **13** (1984): 741.

[Lan79] Lang, D. V., R. A. Logan, and M. Jaros, *Phys. Rev.* B **19** (1979): 1015.

[Duh84] Duh, K. H., X. C. Zhu, A. vander Ziel, and H. Morkoc, *IEEE Trans. Electron Dev.* **ED-31** (1984): 1345.

[Fis84] Fischer, R., T. J. Drummond, J. Klem, W. Kopp, T. S. Hederson, D. Perrachione, and H. Morkoc, *IEEE Trans. Electron Dev.* **ED-31** (1984): 1028.

[Nis86] Nishi, S., T. Saito, S. Seki, Y. Sano, H. Inomata, T. Itoh, M. Akimaya, and K. Kaminishi, in *Gallium Arsenide and Related Compounds*, ed. by W. T. Lindley, IOP Publishing (1986).

[Yam85] Yamashita, Y., R. Kawazu, K. Kawamura, S. Ohno, T. Asano, K. Kobayashi, and G. Nagamatsu, *J. Vac. Sci. Tech.* **3** (1985): 314.

[Tsu91] Tsuji, H., H. I. Fujishiro, M. Shikata, K. Tanaka, and S. Nishi, *1991 GaAs IC Symp.* pp. 113–116 IEEE (1991).

[Woo87] Woodall, J. A., P. D. Kirchner, D. L. Rogers, M. Chisholm, and J. J. Rosenberg, So who needs lattice matched heterojunctions anyway?, IBM, in U.S.–Japan Conf., Florida (1987).

[Mat77] Matthews, J. M., and A. E. Blakeslee, *J. Vac. Sci. Tech.* **14** (1977): 989.

[Bie83] Biefield, R. M., G. C. Osbourn, P. L. Courley, and I. J. Fritz, *J. Electro Mat.* **12** (1983): 903.

[Woo83] Woodall, J. M., G. D. Pettit, T. N. Jackson, C. Lanza, K. L. Kavanagh, and J. W. Mayer, *Phys. Rev. Lett.* **51** (1983): 1783.

[Pet79] Petroff, P. M., R. A. Logan, and A. Savage, *J. Microscopy* **118** (1979): 225.

[Bat86] Batson, P. E., K. L. Kavanagh, J. M. Woodall, and J. W. Mayer, *Phys. Rev. Lett.* **57** (1986): 2729.

[Rog87] Rogers, D. L., J. M. Woodall, G. D. Pettit, and D. McInturff, 1987 Device Research Conf., Abs. VI-A-8, Santa Barbara (1987).

[Lud84] Ludeck, R., *J. Vac. Sci. Tech.* B **2**, 3 (1984): 400.

[Mat70] Matthews, J. W., S. Mader, and T. B. Light, *J. Appl. Phys.* **41** (1970): 3800.

[Mat74] Matthews, J. W., and A. E. Blakeslee, *J. Cryst. Growth* **27** (1974): 118.

[Cha87] Chao, P. C., R. C. Tiberio, K. H. G. Duh, P. M. Smith, J. M. Ballingall, L. F. Lester, B. R. Lee, A. A. Jabra, and G. G. Gifford, *IEEE Electron Dev. Lett.* **EDL-8** (Oct. 1987): 489.

[Cha87a] Chao P. C., P. M. Smith, K. H. G. Duh, J. M. Ballingall, L. F. Lester, B. R. Lee, A. A. Jabra, and R. C. Tiberio, *IEEE IEDM*, pp. 410–413 (1987).

[Lee88] Lester, L. F., P. M. Smith, P. Ho, P. C. Chao, R. C. Tiberio, K. H. G. Duh, and E. D. Wolf, *IEEE IEDM*, pp. 172–175 (1988).

[Plo88] Ploog, M., M. Hauser, and A. Fisher, *Inst. Phys. Conf. Ser.*, no. 91, ch. 1, pp. 27–32 (1988).

[Gil88] Gillman, G., B. Vinter, E. Barbier, and T. Tardella, *Appl. Phys. Lett.* **52**, 12, pp. 972–974 (1988).

[Yue88] C. Yuen, M. Riaziat, S. Bandy, and G. Zdasiuk, *Microwave J.* (Aug. 1988): 87–104.

[Hon89] Hong, W. P., J. Harbison, and J. H. Abeles, *IEEE Electron Dev. Lett.* **10**, 7 (1989): 310–312.

[Han88] Han, C. J., P. P. Ruden, D. Grider, A. Fraasch, K. Newstrom, P. Joslyn, and M. Shur, in *IEDM Tech. Dig.* (1988): 696–699.

[Cha89] Chao, P. C., *IEEE Trans. Electron Devices* **36** (1989): 46.

[Jia90] Jiang, C., D. C. Tsui, B. J. F. Lin, H. Lee, A. Lepore, and M. Levy, *IEEE Electron Device Lett.* **11**, 1 (1990): 63–65.

[Tro79] Troutman, R. R., *IEEE Trans. Electron Devices* **ED-26** (1979): 461.

[Aki89] Akinwande, A. I., *IEEE IEDM* **89** (1989): 97.

10

HETEROJUNCTION BIPOLAR TRANSISTORS

10.1 INTRODUCTION TO HETEROJUNCTION BIPOLAR TRANSISTORS

Very low logic voltage swings occur when extremely uniform threshold voltage devices experience high nonlinearity. Because of this phenomenon the device structures preferred are those in which the threshold voltage is highly insensitive to normal processing variations [Wel83]. While GaAs MESFET pinch-off voltage is fairly insensitive to horizontal geometry variations (e.g., L_g), V_p is very sensitive to both the thickness (vertical geometry) and doping level in the channel layer. In contrast to the MESFET, an almost ideal device from the standpoint of threshold voltage variations is the bipolar transistor.

Every device has a dimension in the direction of current flow that controls the speed of the device. In MESFETs the current flow is parallel to the surface, and the critical control dimension is established by fine-line lithography. In the bipolar transistor the speed-determining part of the current path is perpendicular to the surface and to the epilayers, and for the first-order, speed is governed by the layer thickness. Since vertical layer thickness can be easily made much smaller than horizontal lithography dimensions, there is, for the horizontal dimensions, an inherently higher-speed potential in bipolar structures than in MESFETs [Kro82]. In this chapter we will discuss the structure of the device and its theoretical performance, different fabrication techniques, the advantages of HBTs, and the figure of merit for high-frequency bipolar transistors.

10.2 THE STRUCTURES OF HETEROJUNCTION BIPOLAR TRANSISTORS

10.2.1 Homojunction and Heterojunction Bipolar Transistors

The concept of using a heterojunction in a bipolar transistor suggested in the early 1950s by W. Shockley. [Sho51]. Kroemer was one of the early pioneers of the heterojunction device, and he wrote an important article on the subject [Kro57]. Today we know that the GaAs homojunction bipolar transistors are not as successful as the Si bipolar transistors. This is due to a number of device design compromises that relate to the fact that the base doping level N_A must be limited to a small percentage of the emitter doping in order to maintain good emitter injection efficiency. Two types of heterojunction structures can generally be distinguished. Figure 10.1(a) shows the band diagram with an abrupt emitter-base junction in which the Al mole fraction is made to change abruptly from the AlGaAs emitter to the GaAs base, and Fig. 10.1(b), the graded heterojunction in which the change is made gradually over a distance from 200 to 600 Å. as pointed out by Kroemer [Kro83] and discussed by Ankri et al. [Ank82], emitter injection efficiency γ is defined as $\gamma = 1 - I_{hE}/I_{eB}$, where I_{eB} is the electron current injected into the base and I_{hE} is the hole current injected back into the emitter. For a homojunction transistor, nondegeneracy doped and with short emitter and base, $1 - \gamma$, is

$$1 - \gamma = \frac{I_{hE}}{I_{eB}} = \frac{N_B w_B}{N_E w_E} \exp\left(\frac{\Delta E_G}{kT}\right), \tag{10.1}$$

where N_E, N_B, w_E, and w_B are the doping and thickness of the emitter and base regions respectively. ΔE_G is the band-gap difference between base and emitter where, in silicon the emitter has a smaller band gap than the base because of heavy doping effects in the emitter. The formula presented above is a simplified and approximate

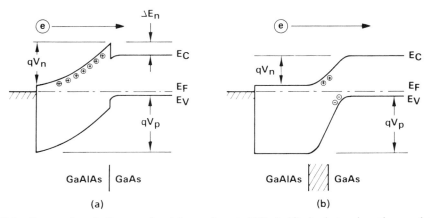

Fig. 10.1 Energy band diagram for (a) an abrupt AlGaAs/GaAs heterojunction and (b) a graded AlGaAs/GaAs heterojunction. [Yuan, H. T., W. V. McLevige, and H. D. Shih, in *VLSI Electronics: Microstructure Science*, vol. 11, pp. 173–213, ed. by N. G. Einspruch, Academic Press (1985); permission from Academic Press, Inc.]

form, partly to illustrate the interplay between the base and emitter doping and the band-gap shrinkage in determining the current gain.

High-speed operation also implies short electron transit times by the base accomplished through the reduced base width. Thin base regions (< 1000 Å) at modest doping levels imply high levels of base resistance (R_B). Hence GaAs homojunction bipolar transistors require both high electron mobility μ_n, for short emitter-collector transit times, and high hole mobility μ_p, for low R_B. The hole mobility in GaAs is unfortunately low ($\mu_p = 250 \, \text{cm}^2/\text{V-s}$). Therefore GaAs homojunction bipolar transistors can not provide better performance than Si bipolar transistors [Wel83].

The availability of high-quality AlGaAs/GaAs heterojunction structures makes it possible to fabricate heterojunction bipolar transistors. In the HBT device the wide-band-gap n-AlGaAs emitter allows the doping level in the p^+-GaAs base region to be made very high without degrading the current gain, since hole injection from the base into the wider-band-gap emitter is virtually impossible. The heavy base doping levels in heterojunction bipolars allow the use of very thin (< 1000 Å) base widths without causing excessive base resistance. High base doping is to avoid punch through. Lower emitter doping in HBT results in lower emitter-base capacitance. Therefore extremely high f_ts can be achieved provided that the low emitter contact resistance can be obtained. The use of GaAs/AlGaAs HBTs in integrated circuits presents a number of advantages such as high threshold uniformity, drive capabilities, and packing densities [Wel83].

10.2.2 The Collector-up Heterojunction Bipolar Structure

The preceding description of the HBT structure is limited to designs where the emitter layer is the topmost layer. It is commonly referred to as the *emitter-up* structure. An alternative design is the *collector-up* structure, where the order of layers is reversed: The collector is the topmost and the emitter is the bottom layer. Figure 10.2 gives simple device structures in emitter-up and collector-up configurations [Kim91]. The major difference between these two structures is the placement of the base layer with respect to the emitter and collector layers. In the case of the emitter-up structure, the extrinsic base layer is located directly above the collector layer. Emitter injection efficiency is high because the emitter-base junction is only turned on above the intrinsic base region.

In the collector-up HBT structure the base layer is directly above the emitter. Most of the carriers entering the base layer are collected at the base contacts. To prevent this from occurring and thus improve the emitter injection efficiency, the external base region boundaries are extended so that a smaller number of carriers are created at the external base boundaries as compared to the intrinsic base region. High-performance AlGaAs/InGaAs/GaAs collector-up HBT fabricated by using a novel self-aligned base/collector process has been developed [Cha89]. Transistors with a collector width down to 2.6 μm and base doping up to $1 \times 10^{20}/\text{cm}^3$ have been fabricated and tested. Based on S-parameters measured up to 26 GHz, an extrapolated current gain bandwidth, f_t, of 65 GHz and a maximum frequency of oscillation, f_{max}, of 102 GHz have been obtained.

The C-up device has been regarded as the optimum transistor for high f_{max} because it offers the smallest value of base/collector capacitance [Kro82] [See Eq. (10.36)]. The technology to fabricate C-up transistors is challenging. Inasmuch as the base-

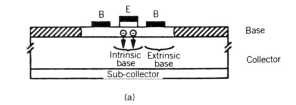

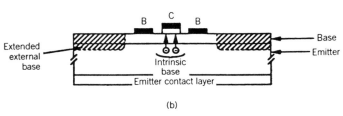

Fig. 10.2 Device cross-sectional views for (a) emitter-up and (b) collector-up structures. [Kim, M. E., B. Bayraktaroglu, and A. Gupta, HBT devices and applications, in *HEMTs and HBTs*, ed. by F. Ali and A. Gupta, Artech House (1991); permission from Artech House]

collector area is less than the base-emitter area, it is critical to minimize the carrier injection from the emitter into the extrinsic base in order to obtain current gain. It is necessary to etch away the collector material in the region above the base contacts in order to reduce C_{BC}. A selective etch technique that can automatically etch the collector layer (6000 Å thick) to stop at the base layer (800 Å thick) is indispensable to successful fabrication of the devices. A self-aligned process is also required to maximize the extrinsic base area to lower the base resistance.

To meet the device requirements, several novel processing techniques have been successfully developed. This includes a self-aligned base/collector process that employs a single layer of photoresist to define the active collector area and the base contact area simultaneously. Figure 10.3 shows the C-up HBT cross section. This photoresist mask allows the selective etching and the ion implantation to be carried out in the extrinsic base area. By using a CCl_2F_2/He plasma in RIE, the thick GaAs collector layer can be etched differentially from the pseudomorphically grown InGaAs base region. The added indium in the base provides the etch selectivity. It also has a built-in quasi-electric field for electron transport (see Fig. 10.4). The carrier injection from the emitter to the extrinsic base was greatly reduced by implanting protons (10^{13} to 10^{14} cm^{-3}) to the extrinsic emitter region. Figure 10.5 shows the I–V characteristics of test base-emitter (B-E) diodes with and without implant.

10.2.3 Double Heterojunction Bipolar Transistors

Double heterojunction bipolar transistors are better suited for integrated circuits applications than single heterojunctions. They offer several advantages: suppression of hole injection from base into collector in digital switching transistors under conditions of saturation, emitter/collector interchangeability in ICs, and reduction of the

THE STRUCTURES OF HETEROJUNCTION BIPOLAR TRANSISTORS 403

- Self-aligned fabrication process
- Etching of collector material in base contact areas
- Suppression of emitter injection under base contacts

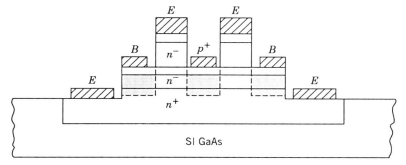

Fig. 10.3 Collector-up fabricational challenges. [Chang, M. F., Rockwell International Science Center (1990); permission from Rockwell International]

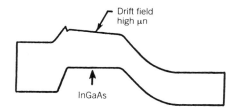

Fig. 10.4 InGaAs pseudomorphic base provides built-in quasi-electric field for electron transport. [Chang, M. F., N. H. Sheng, P. M. Asbeck, G. J. Sullivan, K. C. Wang, R. J. Anderson, and J. A. Higgins, 47th Annual Device Research Conference Proceedings, Cambridge, MA, June 19–21 (1989); permission from Rockwell International]

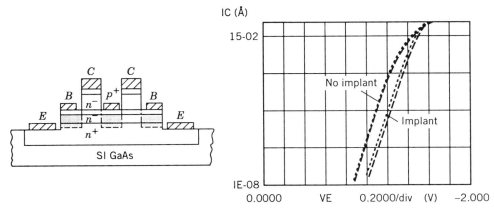

Fig. 10.5 Approach to suppress injection under contacts. [Chang, M. F., Rockwell International Science Center (1990); permission from Rockwell International]

emitter/collector offset voltage [Bai87]. Theoretically the double heterojunction bipolar transistor is considered as a generalization of the single heterojunction transistor. For example, consider the structure whose band diagram is shown in Fig. 10.6. For the current collector the expression

$$\frac{1}{I_c} = \frac{1}{I_1} + \frac{1}{I_2} + \frac{1}{I_3} \tag{10.2}$$

is obtained with currents I_1 and I_2 similar to those used in a single heterojunction

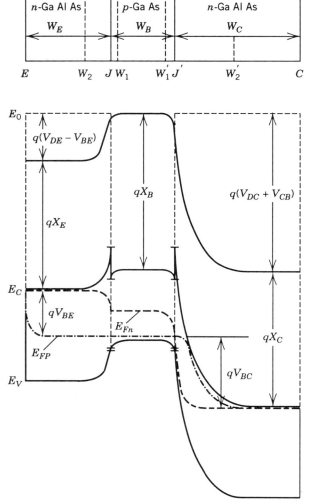

Fig. 10.6 Energy band diagram of a double heterojunction *npn* GaAsAs/GaAs/GaAlAs bipolar transistor. [Bailbe, J. P., A. Marty, G. Rey, J. Tasselli, and A. Bouyahyaoui, *Solid State Electron.* **28** (1985): 627; permission from Pergamon Press]

bipolar transport:

$$I_1 = \frac{Sq^2 n_{iB}^2}{Q_B} D_{nB} \left[\exp\left(\frac{qV_{BE}}{kT}\right) - 1 \right], \quad (10.3\text{a})$$

$$I_2 = \frac{Sq^2 n_{iB}^2}{Q_T} D_{nB} \left[\exp\left(\frac{qV_{BE}}{kT}\right) - 1 \right], \quad (10.3\text{b})$$

where n_{iB} denotes the intrinsic concentration in the base region (GaAs material), D_{nB} represents the diffusion constant of electrons, and S the junction surface. Q_B is the Gummel number: $Q_B = qN_A W_B$, where N_A and W_B refer to the doping and thickness of the base region. Q_T is given by the following expression [Bai87]:

$$Q_T = qN_A \int_{W_2}^{W_1} \frac{D_{nB} N_{CB}}{D_n N_C} \exp\left[\frac{E_c(x) - E_{CB}}{kT}\right] dx,$$

where E_{CB} is the energy level of the bottom part of the conduction band, N_C the effective state density, and W_1 and W_2 the extensions of the depleted zone in the n^- and p^+ regions, respectively.

I_3 is given by

$$I_3 = \frac{Sq^2 n_{iC}^2}{Q'_T} D_{nC} \left[\exp\left(\frac{qV_{BE}}{kT}\right) - 1 \right], \quad (10.3\text{c})$$

with

$$Q'_T = qN_A \int \frac{D_{nB} N_{CB}}{D_n N_C} \exp\left(\frac{E_c(x) - E_{CB}}{kT}\right) dx. \quad (10.4)$$

In normal as well as in reverse working mode, the correct operation of such a device is governed by the absence of barriers [Bai85], that is, discontinuities in the conduction band likely to disrupt the collection of electrons in the reverse-biased junction. The AlGaAs/GaAs transistors must be graded so that the spikes are either deleted or substantially reduced. In Fig. 10.7 the p^+ zones obtained by implantation act as a box, limiting the emitter-collector surfaces which thereby become equal. Such a structure almost completely precludes injection of carriers in the lateral areas, giving

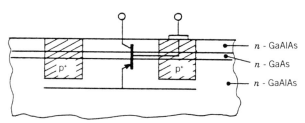

Fig. 10.7 Collector top double heterojunction transistor. [Bailbe, J. P., A. Marty, G. Rey, J. Tasselli, and A. Bouyahyaoui, *Solid State Electron.* **28** (1985): 627; permission from Pergamon Press]

406 HETEROJUNCTION BIPOLAR TRANSISTORS

rise to the high value of the threshold voltage in the (n)AlGaAs/(p)AlGaAs homojunction. This structure is very much like a symmetrical transistor, and consequently the roles or collector and emitter could be exchanged. Such an exchange of roles allows the architectures of digital integrated circuits to be optimized, as illustrated in Fig. 10.8 in the cross-sectional view of an ECL gate envisaged by Kroemer [Kro82].

Double heterojunction bipolar transistors offer an additional advantages associated with the suppression of hole injection in the collector working in a saturated mode. This property, brought about by the extremely high value of the injection efficiency in the collector-based heterojunction, points to a significant improvement of the dynamical performances of some saturated logics, such as I^2L, owing to the dramatic reduction in the transistor's desaturation time. Figure 10.9 illustrates such a structure.

10.2.4 The GaInP/GaAs Heterojunction Bipolar Transistor

Recently the $Ga_{0.51}In_{0.49}P/GaAs$ material system has attracted much attention because it offers larger band gap energy (the $Ga_{0.51}In_{0.49}P$ has a band gap at room temperature of 1.9 eV) and the possibility of growing strain-layer superlattice structures. There is a larger valence-band discontinuity ($\Delta E_v = 0.24$–0.3 eV [Rao87; Wat87; Haf89; Bis90], 0.4–0.43 eV [Kob89; Che91]) as well as a smaller conduction-band discontinuity ($\Delta E_c = 0.19$–0.22 eV [Rao87; Wat87; Haf89; Bis90], 0.03–0.06 eV [Kob89; Che91]) compared with the band discontinuities of the AlGaAs/GaAs material system for the same band-gap difference. The $Ga_{0.51}In_{0.49}P/GaAs$ heterostructure, where most of the energy-gap discontinuity occurs in the valence band, is very suitable for obtaining a high-electron injection efficiency from the emitter to the base and thus

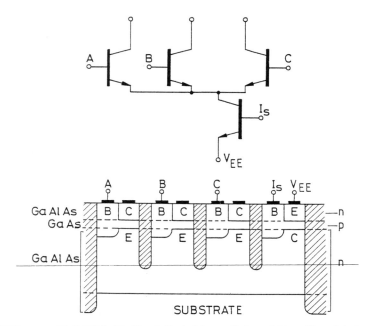

Fig. 10.8 ECL gate with DHBT. [Bailbe, J. P., A. Marty, G. Rey, J. Tasselli, and A. Bouyahyaoui, *Solid State Electron.* **28** (1985): 627; permission from Pergamon Press]

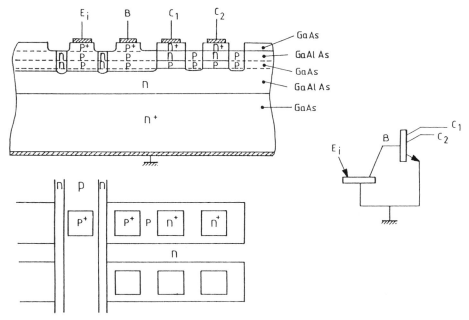

Fig. 10.9 Realization of an I^2L gate with DHBTs. [Bailbe, J. P., A. Marty, G. Rey, J. Tasselli, and A. Bouyahyaoui, *Solid State Electron.* **28** (1985): 627; permission from Pergamon Press]

a high heterojunction bipolar transistor (HBT) current gain [Ale90; Lu92]. Since the introduction of $Ga_{0.51}In_{0.49}P/GaAs$ HBTs by Kroemer [Kro83], several devices have been grown—by MBE [Mon85], MOCVD [Kob89; Raz90; Liu92] and by CBE [Ale90; Lu92]—and fabricated.

The $Ga_{0.51}In_{0.49}P/GaAs$ HBT devices demonstrate near-ideal I–V characteristics with very small magnitudes of the base-emitter junction space-charge recombination current and current gain [Liu92]. Lu et al. applied the emitter edge thinning technique [Lu92] in the fabrication of $Ga_{0.51}In_{0.49}P/GaAs$ HBTs, which resulted in low-surface recombination current, high dc current gain (1580), and small offset voltage. Liu et al. [Liu92] have compared the current gain of the GaInP/GaAs HBTs to an AlGaAs/GaAs HBT. The base current ideality factor is larger for the AlGaAs/GaAs HBT at low I_c levels than the GaInP/GaAs HBTs. The explanation is that the AlGaAs emitter layer is more susceptible to impurity incorporation during epitaxial growth and contains more recombination centers. Consequently the space-charge recombination current is larger in AlGaAs HBTs and completely dominates the base current, resulting in an ideality factor that is very close to 2. The GaInP/GaAs HBTs are useful for applications that require low power operation and/or high linearity.

10.3 DEVICE TECHNOLOGY

HBT technology has progressed rapidly in recent years. Although most of the work has centered on MBE-grown structures, attention is also being focused on MOCVD.

The key characteristics needed for high-yield HBT fabrication include control over layer doping, thickness, and composition; uniformity across wafers; reproducibility from run to run; and freedom from surface defects. The dimension of the transistor influences the power dissipation at which the highest speed is achieved. It is important to reduce the dimensions of HBTs in order to decrease their parasitics and to improve their speed-power performance. This requires self-aligned processing techniques that can accurately produce very small geometry HBTs while employing high-throughput optical lithography.

10.3.1 Material Qualification

To characterize HBT material for epitaxial growth development and to monitor wafers for fabrication, numerous laboratories have found it worthwhile to establish a rapid turnaround process that allows key device characteristics to be measured. Although the process is destructive, it is found that the growth system stability is sufficient, with layer characteristics remaining reproducible over periods of at least one week; this means that the evaluation wafers match the production wafers [Cha88]. Key results may be obtained with as few as two mask steps, aided by the fact that ohmic contacts to surface emitter cap layers of InAs and to base layers of GaAs doped above 2×10^{19} cm^{-3} can be made without the need for alloyed metal. Transistors with relatively large emitters (70×70 μm) are produced. In most wafers the current gain for this size of device is controlled by emitter area-dependent effects rather than by periphery effects. Measurements of Gummel plots on such structures are sensitive indicators of the extent of base dopant diffusion; the value of V_{be} required to attain a given current shifts at a rate of about 1 meV per angstrom of junction diffusion, for a graded emitter-base junction with 25% AlAs graded over 300 Å. Measurements of current gain with relation to current density indicate what is obtainable in subsequently processed wafers, provided that the large-area structures are evaluated in regimes where current crowding and self-heating are not important (below 200 A/cm^{-2}). Emitter-base capacitance measurements provide information on doping of the AlGaAs emitter and dopant diffusion. The interpretation of doping and its position is complicated, however, by the AlGaAs composition variations in the region profiled. The changes in the conduction band energy give rise to an apparent doping concentration that is proportional to the second derivative of the AlAs mole fraction with relation to its position. Figure 10.10 compares the computed apparent doping and its expected position for a constant emitter doping level of 8×10^{17} cm^{-3} and a narrow AlGaAs emitter region that is graded linearly on both sides.

10.3.2 Surface Defects

MBE-grown material tends to have relatively high densities of various morphological defects, known collectively as *oval defects*. It has been shown that for a variety of oval defects types, HBT operation is not impaired unless the defect intersects the emitter area [Cha88]. An estimate of the yield limitation of HBT circuits due to these defects may be obtained, assuming that an average defect has a size of 5×10 μm with a 2-μm-diameter core region and that there is a probability of these defects intersecting with emitters 2×2 μm in size. Figure 10.11 shows the resultant yield estimates. At a defect level of 500 cm^{-2}, the yield is limited to 50% for the circuit

DEVICE TECHNOLOGY   409

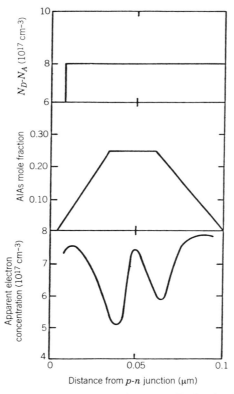

Fig. 10.10 Distance from p–n junction (μm). [Chang, M. F., Rockwell International Science Center (1990); permission from Rockwell International]

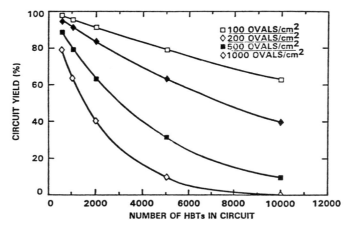

Fig. 10.11 Circuit yield versus number of HBTs in circuit. [Chang, M. F., Rockwell International Science Center (1990); permission from Rockwell International]

containing 3000 transistors. For larger circuits it is important to reduce oval defects accordingly. Defect densities of several hundred per cm^2 are obtainable today from MBE, and surface defects densities below 20 per cm^2 are frequent in MOCVD growth.

10.3.3 Recombination Centers

Deep-level recombination in the base-emitter depletion region dominates the current gain characteristics of most HBTs of large area. The low substrate temperatures that are required for epitaxial growth in order to prevent dopant diffusion tend to increase the number of centers and thus worsen current gain. Recombination centers with a density on the order of 10^{15} cm^{-3} and capture cross section on the order of 10^{15} cm^{-3} and 10^{-14} cm^2 for electrons and holes, respectively, are sufficient to account for the observed base current. For graded base-emitter junctions the depletion region is located in the AlGaAs composition intermediate between the base and the emitter. The deep-level recombination varies according to the band gap E_{gr} at the plane of maximum recombination by $\exp(-E_{gr}/2kT)$ [Lee81]. The recombination current is thus reduced if a limited amount of base dopant diffusion occurs, causing the p–n junction to move toward the emitter. A trade-off must be made between the need for reproducible V_{be} and the need for high current gain, which dictate different relationships between p–n junction and heterojunction. In any case it has been repeatedly shown that epitaxial material consistent with high current gain (>100) at current densities above $1000\,A/cm^2$ can be grown with base acceptor concentrations above $2 \times 10^{19}\,cm^{-3}$.

10.3.4 Etch Control

To access the base in a representative structure, it is necessary to etch through 1500 to 2000 Å of GaAs and AlGaAs, and to stop etching within 150 Å after reaching the base layer. The required control is marginally possible with standard techniques. However, the fabrication task is greatly simplified because of the availability of composition-selective etches. Plasma etching of GaAs, stopping at AlGaAs, is easily accomplished with selectivity greater than 200:1. Etching of AlGaAs with respect to GaAs may also be accomplished with selectivity greater than 10:1. A related task of etching is typically present in accessing the subcollector layer. In this case it is necessary to etch through on the order of 8000 Å of material, and to stop within 2000 Å after reaching the subcollector. The tolerance for this etch step is sufficiently relaxed that control is not a problem.

10.3.5 Nonplanarity

Interconnect metal step coverage is a potential problem when device contacts are present at various depths from the surface. The etched GaAs surfaces are frequently undercut in specific crystallographic directions, creating a geometry with a high risk of metal discontinuity. For edges oriented in the crystallographic direction perpendicular to the undercut direction (the $\langle 0\overline{1}1 \rangle$ direction on $\{001\}$ GaAs wafers), however, relatively gently sloped etched sidewalls are produced (with 55° angles from the wafer surface). It has been found that by requiring the interconnects to the collector to follow the gently sloped crystallographic directions, excellent yield of interconnects can be attained.

10.3.6 Abrupt and Graded Band-gap HBTs

In this section we discuss the effect of heterojunction grading in HBT. The base-emitter turn-on voltage V_{be} required to produce a given collector current I_c is a key characteristic for digital and A/D converter applications. This parameter is exceptionally easy to reproduce in HBTs, provided that structurally the band gap at the base-emitter junction is suitably graded to avoid the formation of a potential barrier. Then V_{be} is given by [Kro85]

$$V_{be} = \left(\frac{kT}{q}\right)\ln\left(\frac{p_b w_b I_c}{q D_n n_i^2 A}\right) + \frac{R_b I_c}{\beta} + R_e I_c\left(\frac{\beta+1}{\beta}\right). \quad (10.6)$$

Here p_b and w_b are the doping and thickness of the base region, respectively; n_i and D_n are its corresponding intrinsic carrier concentration and electron diffusivity, and R_e and R_b are parasitic resistances for emitter and base, respectively. The current gain is β, and A is the emitter area. As a result of the slow dependence of V_{be} on layer characteristics (2 mV change in V_{be} produced by a 10% change in p_b), excellent reproducibility is obtained.

When an abrupt or rapidly graded heterojunction is used, there may occur a potential barrier at the base-emitter junction of an HBT or a diffusion of acceptors from base to emitter; then the collector current is limited at least partially by transport over the barrier, and Eq. (10.6) no longer applies [Mar79, Cha91]. Under these circumstances the turn-on voltage is shifted to higher voltages and becomes sensitive to the details of the potential barrier in the conduction band. As a result significant nonuniformities of V_{be} would be found, as shown by the superimposed Gummel plots of Fig. 10.12. Nonuniformity is found on a local scale (between adjacent devices within a differential pair) as well as on the whole wafer. The results imply that in applications where V_{be} matching is required, it is important in the case of AlGaAs/GaAs HBTs

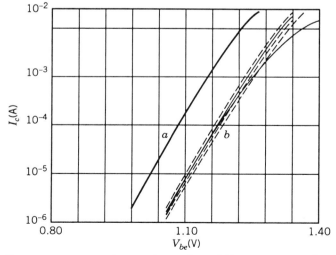

Fig. 10.12 Collector current versus base-emitter voltage. [Chang, M. F., Rockwell International Science Center (1990); permission from Rockwell International]

to utilize graded heterojunction [Tai86]. With an appropriate choice of layer structure, parameter reproducibility requirements for HBT fabrication appear to be well within the capabilities of present MBE-grown technology.

For an abrupt heterojunction there is an additional potential for a spike to occur at the base-emitter junction. The spike would reduce injection into the base, and hence current gain. It may or may not limit current, depending on the rate of emission of electrons over it, their rate of thermalization in the base, and the rate at which the thermalized electrons are carried away by diffusion. The spike will limit the current when

$$I < \frac{e\theta}{kT} + \ln\left(\frac{eDN_C}{fA^*\tau^2 w_B}\right), \quad (10.7)$$

where θ is the height of the potential spike, D the electron diffusion coefficient in the base, w_B the base width, A^* the Richardson constant, and f the fraction of current emitted over the spike which is thermalized into the base. The grading of the emitter-base heterojunction, and the composition and doping in the base, may be used to improve the electron's transit time there. A quasi-electric field will be generated [Kro57] as an Al gradient is created in the base of the AlGaAs/GaAs HBT, decreasing in Al content from emitter to collector. The advantages of grading the emitter are that the offset voltages can be reduced and the injection efficiency can be improved. A parabolically graded layer is preferred to the linearly graded layer for the p^+–n junction when it comes to reducing potential spikes in the conduction band [Cap85]. The base transit time τ_B is $w_B/\mu_n E$, where E is the quasi-electric field proportional to $\Delta\theta$, the band lowering. The whole band-gap change is effective, since the valence band is kept flat by the high p-type doping. The group velocity of electrons in GaAs approaches its limiting value of 1×10^8 cm/s for a potential step in the 0.15 to 0.3 V range. Even a 50-mV step will give a boost in velocity of 5×10^7 cm/s. Variations in epitaxial structure are summarized in Table 10.1.

The collector current is base transport limited, according to the Moll-Ross-Kroemer relation [Kro85]

$$J_C = \frac{qD_n n_{ie}^2 \exp(qV_{BE}/kT)}{\int p(x)dx}, \quad (10.8)$$

when $n_{ie}^2 = 2$ to $3n_i^2$ for $p = 5 \times 10^{18}$ cm^{-3} at 300 K.

TABLE 10.1

Structure	Objective
1. Graded E–B heterojunction	Higher gain
2. Abrupt E–B heterojunction	Faster (ballistic)
3. Graded base region	Higher speed
4. Wide gap collector	Reduce stored charge in saturation
5. Superlattice E, C	Higher gain
6. Inverted transistor	Lower parasitics; easier fabrication (I^2L)

Source: Chang, M. F., Rockwell International Notes, 1988.

10.3.7 Carbon-Doped Base HBTs

In most of the research work as described in the sections above, n–p–n HBT epitaxial structures were produced either by MBE with Be as the acceptor dopants in the base or by MOCVD with Zn as the acceptor dopants. However, the concentration-dependent diffusity of Be and Zn has made it difficult to prepare well-controlled HBT structures with very high base doping (above $4 \times 10^{19}\,\text{cm}^{-3}$) [Wan91]. Uematsu and Wada [Uem91] used Be-doped GaAs tunnel diodes to show that the Be diffusion under forward-bias conditions was enhanced by a factor of $\sim 10^{15}$ at 300 K and that the enhancement mechanism was most likely recombination-enhanced diffusion of Be interstitials. They suggest that electron-hole recombination at a particular site in the lattice liberates Ga interstitials, which migrate to a substitutional Be_{Ga} site and interchange, leaving a Be interstitial that is free to move through the crystal. Ren et al. [Ren91] report that the Be-doped HBT shows a remarkable decrease in current gain from 16 to 1.5 in just two hours. Both the base-emitter and base-collector junction ideality factors rise rapidly during device operation. This current-induced degradation is consistent with recombination-enhanced diffusion of Be interstitials producing graded junctions.

Recently carbon has been shown to be an acceptor in GaAs with high-doping efficiency ($>90\%$ at concentrations up to $1 \times 10^{20}\,\text{cm}^{-3}$ and low diffusivity ($<10^{-16}\,\text{cm}^2\text{-s}^{-1}$ at 950°C) [Abe90]. It has been successfully used in MBE [Mal88], atomic layer epitaxy [Hay88], MOCVD [Sai88], and MOMBE [Ho91]. The use of carbon has the potential for allowing high base doping levels, with the associated lower base resistance and better device reliability. The current-induced degradation of Be-doped HBTs, presumably as a result of recombination-enhanced Be diffusion, is absent in C-doped structures. This high degree of stability is most likely a result of the fact that C occupies the As sublattices, and not the Ga sublattice as in the case of Be, and has higher solubility than Be [Ren91]. Important issues related to C-doped base HBTs are the obtainable current gain, current-gain cutoff frequency f_t, maximum frequency of oscillation f_{max}, and reproducibility of these devices.

Device Fabrication

An example of a carbon-doped base HBT grown by MOCVD is as follows: The structure consists of an n^+ ($4 \times 10^{18}\,\text{cm}^{-3}$) 5000-Å-thick GaAs subcollector, an n ($3 \times 10^{16}\,\text{cm}^{-3}$) 7000-Å-thick Ga collector, an p^+ (1–$4 \times 10^{19}\,\text{cm}^{-3}$) 500 to 1,200-Å-thick GaAs base, an $n(5 \times 10^{17}\,\text{cm}^{-3})$ 1000-Å-thick AlGaAs emitter, and an n^+ ($5 \times 10^{18}\,\text{cm}^{-3}$) 2000-Å-thick GaAs cap. Carbon tetrachloride (CCl_4) is used as the dopant gas for the p-type doping in the base [Cun90]. Evaporated Ti/Pt/Au is used in base for nonalloyed ohmic contact, while emitter and collector contacts are formed by alloyed AuGe/Ni/Au.

Device Characteristics

A large geometry ($70 \times 70\,\mu\text{m}^2$) transistor was made [Wan91] and characterized for the dc gain. Figure 10.13 shows the Gummel plot for an HBT with a base width of 800 Å and a doping of $2 \times 10^{19}\,\text{cm}^{-3}$. Current gain is 303 at a collector current of 100 mA. The base current has an ideality factor in the range of 1.2 to 1.4 and a collector ideality factor of 1. Current gain varies with base doping and thickness. A

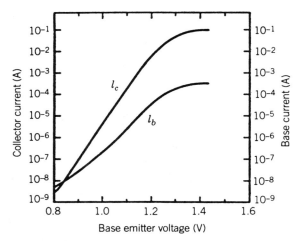

Fig. 10.13 Gummel plot of a C-doped base HBT with a base doping of $2 \times 10^{19}\,\text{cm}^{-3}$. The current gain ($I_c/I_b$) is 303 at a collector current of 100 mA. [Wang, G. W., R. L. Pierson, P. M. Asbeck, K. C. Wang, N. L. Wang, R. Nubling, M. F. Chang, J. Salerno, and S. Sastry, *IEEE Electron Device Lett.* **12**, 6 (1991): 347–349; permission from IEEE]

wide range of parameters for a sequence of 12 different epitaxial growths is identified where current gain is above 30 and the base sheet resistance is below 400 Ω/sq.

The HBT microwave performance was characterized by S-parameters measured from 0.5 to 40 GHz. The device suitable for microwave probing has two emitter fingers. The emitter finger is not self-aligned to the base metal and has a size of $1.4 \times 12\,\mu\text{m}^2$. Proton implant at the extrinsic base was used to reduce the base-collector capacitance [Cha86]. Figure 10.14 shows the current gain and unilateral

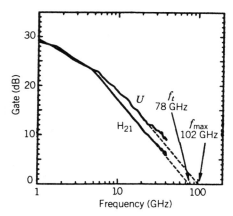

Fig. 10.14 Current gain and unilateral gain of a microwave HBT. The extrapolated f_t and $f_{\max}$ are 76 and 102 GHz, respectively. [Wang, G. W., R. L. Pierson, P. M. Asbeck, K. C. Wang, N. L. Wang, R. Nubling, M. F. Chang, J. Salerno, and S. Sastry, *IEEE Electron Device Lett.* **12**, 6 (1991): 347–349; permission from IEEE]

gain as a function of frequency for a device with a base doping of $4 \times 10^{19}\,cm^{-3}$ and a base thickness of 500 Å. The device is biased at a collector voltage of 1.5 V and a collector current of 35 mA. The extrapolated f_t and f_{max} are 76 and 102 GHz, respectively. This study suggests the potential of the carbon doping technique for HBT structures and for other device structures requiring precise p-type doping profiles. The maximum f_{max} of 236 GHz [Cha93] in InAlGaAs/InGaAs and f_t of 175 GHz in InP/InGaAs HBTs [Son93] were reported.

10.4 CHARACTERISTICS OF HETEROJUNCTION TRANSISTORS

10.4.1 Basic Characteristics

In general, the heterojunction transistor has the general characteristics of a conventional bipolar transistor made out of homojunction materials. However, there are a few important differences that still must be resolved [Yua85]: Transport mechanisms across the heterojunction are complicated by the energy-band discontinuity at the heterojunction interface; it cannot be described adequately by the diffusion model alone. Conventional p-n junction theory does not include such effects as direct band-gap recombination and multiple-level trapping centers that are more pronounced in most heterojunction structures.

The basis for the bipolar transistor model began with the early work of Ebers and Moll [Ebe54]; later it was developed by Gummel and Poon [Gum70]. The Ebers-Moll model was general enough to describe the first order I-V characteristics of any bipolar transistors by a curve-fitting technique. This is particularly valid for transistors that have graded heterojunctions in which the transport mechanism is governed by diffusion. Under such conditions the transport based on the Ebers-Moll model, assuming an n-p-n transistor with terminal voltage and current shown in Fig. 10.15, is defined by [Yua85]

$$I_E = -I_{ES}\left[\exp\left(\frac{qV_{BE}}{kT}\right) - 1\right] + \alpha_R I_{CS}\left[\exp\left(\frac{qV_{BC}}{kT}\right) - 1\right] - I_{ER} \quad (10.9)$$

and

$$I_C = +\alpha_F I_{ES}\left[\exp\left(\frac{qV_{BE}}{kT}\right) - 1\right] - I_{CS}\left[\exp\left(\frac{qV_{BC}}{kT}\right) - 1\right] - I_{CR}. \quad (10.10)$$

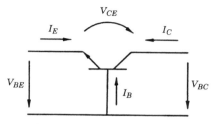

Fig. 10.15 Terminal voltage and current conventions of an n-p-n transistor. [Yuan, H. T., W. V. McLevige, and H. D. Shih, in *VLSI Electronics: Microstructure Science*, vol. 11, pp. 173–213, ed. by N. G. Einspruch, Academic Press (1985); permission from Academic Press, Inc.]

Note the two extra terms, I_{ER} and I_{CR}, added to the familiar terminal current expressions to account for the emitter-base and collector-base junction depletion layer recombination currents. These currents are important in a GaAs transistor because they often dominate the base current even at a moderately high injection level. The magnitude of the current, assuming single-level recombination centers, is determined by

$$I_{ER} = A_E \left(\frac{q n_{ie} W_{BE}}{T_{be}} \right) \left[\exp\left(\frac{q V_{BE}}{2kT} \right) - 1 \right] \tag{10.11}$$

and

$$I_{CR} = A_C \left(\frac{q n_{ic} W_{BC}}{T_{bc}} \right) \left[\exp\left(\frac{q V_{BC}}{2kT} \right) - 1 \right], \tag{10.12}$$

where n_{ie} and n_{ic} are the intrinsic carrier concentration in emitter and collector, respectively.

The parameters I_{ES} and I_{CS} are the emitter-base and collector-base junction reverse-bias saturation currents when the other junction is shorted to the base. In terms of the doping concentrations N_D and N_A and the hole and electron diffusion constants D_h and D_e, the components of the saturated hole and electron current are

$$I_{ES} = I_{ES}(p) + I_{ES}(n), \tag{10.13}$$

or

$$I_{ES} = A_E q n_i^2 \left[\frac{D_h}{W_E N_D} \exp\left(\frac{-\Delta E_g}{kT} \right) + \frac{D_e}{W_B N_A} \right], \tag{10.14}$$

$$\Delta E_g = E_g(\text{AlGaAs}) - E_g(\text{GaAs}), \tag{10.15}$$

and

$$I_{CS} = I_{CS}(p) + I_{ES}(n)$$
$$= A_C q n_i^2 \left(\frac{D_h}{W_C N_D} + \frac{D_e}{W_B N_A} \right), \tag{10.16}$$

where n_i is the intrinsic carrier concentration of GaAs. In these expressions the emitter depth W_E, base width W_B, and collector depth W_C are all assumed to be small compared to their respective carrier diffusion lengths. For a typical AlGaAs-GaAs n-p-n transistor, it is obvious from Eq. (10.14) that the hole component of I_{ES} can be reduced substantially by increasing the mole fraction of Al in an AlGaAs emitter so that

$$\Delta E_g \gg kT. \tag{10.17}$$

Assuming that the depletion-layer recombination current is much larger than the neutral-base recombination current, the common base forward and reverse current gain factors α_F and α_R are defined as

$$\alpha_F = \frac{I_E(n)}{I_E(n) + I_E(p) + A_E \left(\dfrac{q n_{ie} W_{BE}}{\tau_{be}} \right) \exp\left(\dfrac{q V_{BE}}{2kT} \right)}, \tag{10.18}$$

$$\alpha_R = \frac{I_C(n)}{I_C(n) + I_C(p)} + A_C \left(\frac{qn_{ic}W_{BC}}{\tau_{bc}}\right) \exp\left(\frac{qV_{BC}}{2kT}\right), \quad (10.19)$$

where

$$I_E(n) = I_{ES}(n)\exp\left(\frac{qV_{BE}}{kT}\right), \quad (10.20)$$

$$I_E(p) = I_{ES}(p)\exp\left(\frac{qV_{BE}}{kT}\right), \quad (10.21)$$

$$I_C(n) = I_{CS}(n)\exp\left(\frac{qV_{BC}}{kT}\right), \quad (10.22)$$

$$I_C(p) = I_{ES}(p)\exp\left(\frac{qV_{BC}}{kT}\right). \quad (10.23)$$

If depletion-layer recombination current dominates the base current, like many GaAs bipolar transistors, the common emitter current gain will have the following simple expression:

$$h_{FE} = \frac{I_E(n)}{I_{ER}} \quad (10.24)$$

$$= \tau_{be}\left(\frac{D_e}{W_{BE}}\right)^{1/2} \frac{I_E^{1/2}(n)}{A_E q W_B N_A}. \quad (10.25)$$

At sufficient high current level the h_{FE} value will become constant and be defined by

$$h_{FE} = \frac{\alpha_F}{1-\alpha_F} = \frac{I_E(n)}{I_E(p)}. \quad (10.26)$$

The transition from recombination to diffusion-dominated base current usually over a range of 10 to 100 A/cm² in most HBTs. To reduce the recombination current in HBTs must be reduced in order to achieve better device performance.

Higher-level Ebers-Moll models include more complicated effects and parameters such as parasitic resistances and capacitances, base-width modulation, splitting of the base capacitance between internal and external regions, current gain variation with current and voltage, and transit time variation with current. The Gummel-Poon model, whose physics are more closely related to those of bipolar transistors, is difficult to understand, and the user may have difficulty determining appropriate input parameters. Both of these HBT models that can be developed and used in SPICE. This topic is reviewed by Long and Butner [Lon90].

10.4.2 Efficiency versus Collector Layer Thickness

The common-base connection is often preferred in high-power amplifier applications. The breakdown voltage V_{max} may be designed with precision by the selection of the thickness and doping level of the collection layer. A range of values is available up to 50 V with a well-defined trade-off between V_{max} and f_t. The power-added efficiency can be described as [Hig88, Hig91]

$$\text{Power-added efficiency} = E\left(\frac{1-2V_0}{V_{max}}\right)\left[\frac{\sin(x)}{x}\right]\left(\frac{1-1/G}{L_0}\right), \quad (10.27)$$

where G is the power gain, V_0 is the minimum voltage, V_{max} is the breakdown voltage, L_0 is the loss in the output circuit, x is the ratio of f/f_t, and E is the mode efficiency (50% for class A, 78.5% for class B, etc.).

The trade-off between transit time and the maximum voltage is such that for any microwave frequency an optimum collector thickness exists. For the case of 10 GHz, a thickness of between 0.9 and 1.0 μm provides the maximum of Eq. (10.27), a V_{max} of 30 V and an f_t at high voltage ($V_{ce} > 14$ V) of 29 GHz.

10.5 FIGURES OF MERIT FOR HIGH-FREQUENCY TRANSISTORS

The high-frequency gain of the HBTs is controlled by the capacitive elements in the equivalent circuit of Fig. 10.16. In practice, the frequency capability of the transistor is most often specified by determining the frequency where the magnitude of the short-circuit, common-emitter current gain falls to unity. This is called the *transition frequency*, or *cutoff frequency* f_t, and it is a measure of the maximum useful frequency of the transistor when it is used as an amplifier. At high frequency $G_I = f_t/f$. Consider a sample of length d as shown in Fig. 10.17. I_{IN} and I_{OUT} are the input and output current of the device, respectively. The gain G_I can be defined as

$$G_I = \left|\frac{\Delta I_{OUT}}{\Delta I_{IN}}\right| \quad (10.28)$$

and

$$I_{IN} = \frac{\Delta Q_B}{\Delta t} = 2\pi f Q_B; \quad (10.29)$$

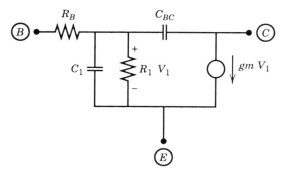

Fig. 10.16 Capacitive elements in the equivalent circuit of HBT. [Chang, M. F., Rockwell International Science Center (1990); permission from Rockwell International]

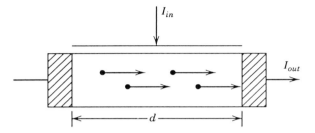

Fig. 10.17 A sample of length d. [Chang, M. F., Rockwell International Science Center (1990); permission from Rockwell International]

hence

$$G_I = \frac{1}{2\pi f} * \frac{\Delta I_{OUT}}{\Delta Q_B} \qquad (10.30)$$

The transition frequency f_T can be obtained as

$$f_T = \frac{\Delta I_{OUT}/\Delta V_{IN}}{\Delta Q_B/\Delta V_{IN}} = \frac{g_m}{2\pi C_{IN}}. \qquad (10.31)$$

Next we consider the maximum frequency of oscillation f_{max}. This is the frequency at which power gain falls to unity. At high frequency we have

$$G_P = \frac{f_{max}^2}{f^2}, \qquad (10.32)$$

where

$$G_P = \frac{1}{4} * G_V * G_I = \frac{1}{4} * G_I^2 * \frac{\text{Re}(Z_{OUT})}{\text{Re}(Z_{IN})}. \qquad (10.33)$$

From the equivalent circuit (Fig. 10.16) we notice that

$$\text{Re}(Z_{IN}) = R_B,$$
$$\text{Re}(Z_{OUT}) = \frac{1}{g_m} * \left(\frac{C_1 + C_{BC}}{C_{BC}}\right). \qquad (10.34)$$

Thus we have

$$G_p = \frac{1}{8\pi R_B C_{BC}} * \frac{f_t}{f^2} \qquad (10.35)$$

and

$$f_{max} = \sqrt{\frac{f_t}{8\pi R_B C_{BC}}}, \qquad (10.36)$$

where R_B is the base resistance and C_{BC} is the base-collector capacitance. Improved expressions are given by Das [Das88] and Prasad [Pra88].

Figures of merit appropriate to digital operation of bipolar transistors have been the subject of considerable attention recently [Tay86; Sto88]. Stork has developed an analytic expression that approximates the gate delay of an ECL gate. The portion of the delay T_d corresponding to the current switch (which can be taken to represent the overall delay for the CML family) can be written (over a limited regime) as [Cha90]

$$T_d^{-1} = \left[\frac{f_t}{K(R_B + R_L/2)(C_{BC} + C_L/2)} \right]^{1/2}, \qquad (10.37)$$

where R_L is the load resistance used, C_L is the load capacitance associated with the output, K is the constant in the order of unity, and R_B and C_{BC} are its base resistance and base-collector capacitance, respectively. Equation (10.37) shows clear parallels with the expression for f_{max}. Optimization of switching speed requires trade-offs to decrease R_B and C_{BC}, although the extent to which these reductions are effective is limited by R_L and C_L. If sufficiently high power dissipation can be allowed in a circuit, R_L can be lowered to match R_B. For circuits with low levels of integration and moderate size transistors, C_L is not dominated by interconnects. Under such conditions digital switching speed can approach $f_{max}/2$. The advantage for HBTs over Si bipolar transistors is clearly evident in this regime, since f_{max} can be five times higher. Equation (10.37) ceases to be valid, however, in the regime where T_d^{-1} approaches $f_t/2$.

10.6 POWER DENSITY IN THE HBT

An HBT has a breakdown voltage V_{max} and a current maximum I_{max}. The voltage maximum is the breakdown voltage of the base-collector diode, and the I_{max} is the limit set by the onset of base-pushout effect. If the HBT is biased at V_{ce} of $0.5V_{max}$ and the quiescent current is $0.5I_{max}$, then the power density in the HBT is at a maximum power level for class A operation. The upper limit is given by the following expression:

$$\text{Power density} = \frac{\varepsilon E_{max}^2 v_s}{8}, \qquad (10.38)$$

where ε is the dielectric constant, E_{max} is the maximum electric field in GaAs ($4.5E5$ V/cm), and v_s is the saturated velocity in the collector. Evaluated for GaAs, this power density is $0.426E6$ W/cm^2. For an emitter of 1×1000 μm, the value is 4 W of power, a level very close to the power levels in working experimental HBTs.

REFERENCES

[Wel83] Welch, B. M., R. C. Eden, and F. S. Lee, *Gallium Arsenide*, pp. 517–573, ed. by M. J. Howes and D. V. Morgan, Wiley (1983).

[Kro82] Kroemer, H., *Proc. IEEE* **70** (1982): 13.

[Sho51] Shockley, W., U.S. Patent No. 2,569,347 (1951).

[Kro57] Kroemer, H., *Proc. IRE* **45** (1957): 1535.
[Kro83] Kroemer, H., *J. Vac. Sci. Tech.* **B 1**, 2 (1983): 126–130.
[Ank82] Ankri, D., and L. F. Eastman, *Electron Lett.* **18** (1982): 750.
[Bai87] Bailbe, J. P., A. Marty, and G. Rey, *Solid State Electron.* **30**, 11 (1987): 1159–1169.
[Bai85] Bailbe, J. P., A. Marty, G. Rey, J. Tasselli, and A. Bouyahyaoui, *Solid State Electron.* **28** (1985): 627.
[Rao87] Rao, M. A., E. J. Caine, H. Kroemer, S. I. Long, and D. I. Babic, *J. Appl. Phys.* **61** (1987): 643–649.
[Wat87] Watanabe, M. O., and Y. Ohba, *Appl. Phys. Lett.* **50** (1987): 906–908.
[Haf89] Hafich, M. J., *Appl. Phys. Lett.* **54**, (1989): 2686–2688.
[Bis90] Biswas, D., *Appl. Phys. Lett.* **56** (1990): 833–835.
[Kob89] Kobayashi, T., K. Taira, F. Nakamura, and H. Kawai, *J. Appl. Phys.* **65** (1989): 4898–4902.
[Che91] Chen, J., J. R. Sites, I. L. Spain, M. J. Haftich, and G. Y. Robinson, *Appl. Phys. Lett.* **58** (1991): 744–746.
[Ale90] Alexandre, F., J. L. Benchimol, J. Dangla, C. Dubon-Chevallier, and V. Amarger, *Electron. Lett.* **26** (1990): 1753–1755.
[Lu92] Lu, S. S., and C. C. Huang, *IEEE Electron Dev. Lett.* **13**, 4 (1992): 214–216.
[Mon85] Mondry, M. J., and H. Kroemer, *IEEE Electron Device Lett.* **EDL-6** (1985): 175–177.
[Raz90] Razeghi, M., et al., *Semicond. Sci. Tech.* **5** (1990): 278–280.
[Lee81] Lee, S. C., and G. L. Pearson, *J. Appl. Phys.* **52** (1981): 275.
[Kim91] Kim, M. E., B. Bayraktaroglu, and A. Gupta, HBT devices and applications, in *HEMTs and HBTs*, ed. by F. Ali and A. Gupta, Artech House (1991).
[Cha89] Chang, M. F., Rockwell International Notes (1989).
[Kro85] Kroemer, H., *Solid State Electron.* **28** (1985): 1101.
[Mar79] Marty, A., G. E. Rey, and J. P. Bailbe, *Solid State Electron.* **22** (1979): 549.
[Cha91] Chang, M. F., and P. M. Asbeck, *Int. J. High-Speed Electron.* (1991).
[Tai86] Taira, K., C. Takano, H. Kawai, and M. Arai, *Appl. Phys. Lett.* **49** (1986): 1278.
[Cap85] Capasso, F., in *Gallium Arsenide Technology*, pp. 303–330, ed. by D. K. Ferry, SAMS (1985).
[Cha88] Chang, M. F., Rockwell International Notes (1988).
[Wan91] Wang, G. W., R. L. Pierson, P. M. Asbeck, K. C. Wang, N. L. Wang, R. Nubling, M. F. Chang, J. Salerno, and S. Sastry, *IEEE Electron Device Lett.* **12**, 6 (1991): 347–349.
[Ren91] Ren, F., T. R. Fullowan, J. Lothian, P. W. Wisk, C. R. Abernathy, R. F. Kopf, A. B. Emerson, S. W. Downey, and S. J. Pearton, *Appl. Phys. Lett.* **59**, 27 (1991): 3613–3615.
[Uem91] Uematsu, M., and K. Wada, *Appl. Phys. Lett.* **58** (1991): 2015.
[Ho91] Ho, W. J., N. L. Wang, R. L. Pierson, M. F. Chang, R. B. Nubling, and J. A. Higgins, *IEDM-91* (1991): 801.
[Cun90] Cunningham, B. T., G. E. Stillman, and G. S. Jackson, *Appl. Phys. Lett.* **56** (1990): 361–363.
[Abe90] Abernathy, C. R., et al., *J. Cryst. Growth* **105** (1990): 375–382.
[Cha86] Chang, M. F., P. M. Asbeck, K. C. Wang, G. J. Sullivan, and D. L. Miller, *Electron. Lett.* **22** (1986): 1173–1174.
[Cha93] Chan, H. F., and Kao, Y. C., *IEEE IEDM Digest* (1993): 783.
[Son93] Song, J. I., and Hong, W. P., *IEEE IEDM Digest* (1993): 787.
[Yua85] Yuan, H. T., W. V. McLevige, and H. D. Shih, in *VLSI Electronics: Microstructure Science*, vol. 11, pp. 173–213, ed. by N. G. Einspruch, Academic Press (1985).

[Ebe54] Ebers, J. J., and J. L. Moll, *Proc. IRE* **42** (1954): 1761–1771.

[Gum70] Gummel, H. K., and H. C. Poon, *Bell Sys. Tech. J.* **49** (1970): 827–852.

[Lon90] Long, S. I., and S. E. Butner, *Gallium Arsenide Digital Integrated Circuit Design*, McGraw-Hill (1990).

[Hig88] Higgins, J. A., GaAs IC Symp. IEEE (1988).

[Hig91] Higgins, J. A., *Microwave J.* (May 1991): 176.

[Das88] Das, M. B., *IEEE Trans. Electron Dev.* **ED-35** (1988): 604.

[Pra88] Prasad S., W. Lee, and C. G. Fonstad, *IEEE Trans. Electron Dev.* **ED-35** (1988): 2288.

[Tay86] Taylor, G. W., and J. G. Simmons, *Solid State Electron.* **29** (1986): 941.

[Sto88] Stork, J. M. C., *IEDM Tech. Dig.* IEEE (1988): 550.

11

RESONANT-TUNNELING TRANSISTORS

11.1 INTRODUCTION TO RESONANT-TUNNELING TRANSISTORS

The first paper on tunneling devices discussed the tunnel diode was written by L. Esaki in 1958 [Esa58]. In 1965 a remarkable, pioneering paper [Mor65] titled "From Physics to Function" introduced the concept of functional device [Cap90]. The key characteristic of such devices is that "they promise to reduce greatly the number of elements and process steps per function when their capabilities are properly matched to an old or new system function." Morton provided a few examples of functional devices, one of which was the tunnel diode.

The strength of Morton's vision was that he foresaw dramatic progress in growth techniques, equipment development, and semiconductor physics [Cap90]. The advent of advanced epitaxial growth techniques, such as MBE and MOCVD as discussed in Chapter 3, and of band-gap engineering [Cap87] has made possible the development of a new class of materials and heterojunction devices with unique optical and electronic properties. Heterojunction superlattices and their transport properties were first investigated by Esaki and Tsu in 1970 [Esa70]. They predicted negative conductance associated with electron transfer into the negative mass regions of the minizone and Bloch oscillations. Pioneering work on resonant tunneling through a heterostructure quantum well was done by Tsu, Esaki, and Chang [Cha74]. The study of the negative differential resistance (NDR) of the resonant-tunneling barrier (RTB) structure has been changed from the AlGaAs/GaAs heterostructure [Gol87] to the InAlAs/InGaAs heterostructure [Sen87a, Lak88]. Before considering the resonant-tunneling transistor, one must study the phenomenon of resonant tunneling. In this chapter we will briefly discuss the resonant-tunneling phenomenon in resonant-tunneling transistors, and their applications to logic circuits.

11.2 WAVE PROPERTY OF ELECTRONS AND RESONANT TUNNELING

The presence of double barriers has been discussed by Bohm [Boh79] using the WKB approximation and by Kane using the method of wave function matching [Kan69]. Bohm considered the case of a particle incident from the left to double barriers that are symmetric about the center of the well, as shown in Fig. 11.1. In his method one assumes that the energy of the incident particle—the potential energy—is a function of position and that the de Broglie wavelength λ of the particle satisfies

$$\frac{\lambda |\partial V/\partial x|}{2(E-V)} \ll 1, \tag{11.1}$$

in other words, that the potential energy changes very slowly as a function of position. Then the WKB method can be used to solve the problem.

Following the procedure shown in Capasso and Kiehl [Cap85], the ratio of transmitted to incident intensities can be obtained as [Boh79]

$$T = 4\left[4\cos^2\frac{1}{2}\left(\pi - \frac{J}{\hbar}\right) + (4\theta^2 + \tfrac{1}{4}\theta^2)^2 * \sin^2\frac{1}{2}\left(\pi - \frac{J}{\hbar}\right)\right]^{-1}. \tag{11.2}$$

Setting $\cos^2 = 1 - \sin^2$, we obtain

$$T = \left[1 + \tfrac{1}{4}(4\theta^2 - \tfrac{1}{4}\theta^2)^2 \sin^2\frac{1}{2}\left(\pi - \frac{J}{\hbar}\right)\right]^{-1}, \tag{11.3}$$

where $J = 2\int_{-b}^{b} P_w dx$, P_w is the absolute value of the momentum inside the well and $\theta = \exp(\int_b^a P_1 dx/\hbar)$, P_1 is the absolute value of the momentum inside the barriers.

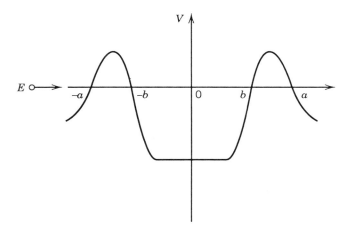

Fig. 11.1 A particle incident from the left to the double barriers which are symmetric about the center of the well. [Wu, J. S., National Cheng Kung University Notes]

Since θ^2, which is $\exp(2\int_b^a p_1 dx/\hbar)$, is usually much bigger, the transmissivity is much smaller. Still there are some points where T is unity, whenever $\pi - J/\hbar = \pm 2N\pi$, or

$$J_N = (N + \tfrac{1}{2})\hbar,$$

or

$$\int_{-b}^{b} P_w dx = \tfrac{1}{2}(N + \tfrac{1}{2})\hbar, \qquad N = 0, 1, 2, \ldots. \tag{11.4}$$

The graph of T versus E will show a value of T that is generally small but that becomes large near $E = E_N$, as shown in Fig. 11.2. Note that each value of E corresponds to a P_w in the well from the relationship of $P_w = [2m(E - V)]^{1/2}$. According to Eq. (11.3) and Fig. 11.2, a particle can tunnel or penetrate two symmetric double barriers without attenuation whenever it owns some values of energy satisfying Eq. (11.4). Such a case is called *resonant tunneling*. Near a transmission resonance the probability of transmission will increase rapidly. The full width at half-maximum, where $T = 1/2$, can be obtained by setting $T = 1/2$.

The large transmissivity at resonance is produced by the fact that the wave is so big inside the well that even if only a small fraction leaks through, it produces a large result. The large amplitude inside also makes possible a large probability of entry into the region between the barriers. This is because the probability current across the barrier is proportional to $\varphi^*\nabla\varphi - \varphi\nabla\varphi^*$, so that if gets large enough, the effect of small barrier transmissivity is canceled. The dependence of the transmissivity on the intensity of the wave inside the barriers is characteristically a wave phenomenon. An analogy of wave penetration near the resonance can be made to the pendulum of simple harmonic motion. If a periodic force is in resonance with the pendulum, the rate of transfer of energy to the pendulum is proportional to the amplitude of vibration already in existence.

Interestingly the condition for a transmissivity resonance in Eq. (11.4) is exactly the same as that for a bound state of a potential well [Boh79]. Resonance occurs when the electron wave function reflected as the first barrier is canceled by the wave that leaks from the well in the same direction or, equivalently, when the energy of

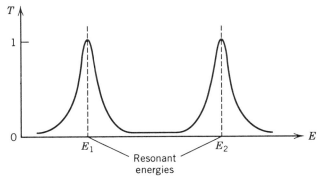

Fig. 11.2 A particle can tunnel through two symmetric double barriers without attenuation whenever it owns some values of energy satisfying Eq. (11.4). This case is called *resonant tunneling*. [Wu, J. S., National Cheng Kung University Notes]

the injected carrier becomes appreciably equal to the energy level of the electrons confined in the well.

The problem of resonant tunneling is very similar to that of the Fabry-Perot interferometer in optics. In the resonance case the wave is reflected at the sharp edges of the potential, which correspond to the edges of a piece of glass in optics where a sharp change in index of reflection takes place. In the treatment of the Fabry-Perot interferometer of two sheets of glass, separated by a distance of $2a$, it is shown that the wave that reflects from the surface at $x = +a$ arrives back at that surface after reflecting from $x = -a$ with a phase shift of $2\pi N$. Then it interfaces constructively with the next wave coming in, and as a result the transmitted coefficient is unity.

11.3 STRUCTURE OF THE RESONANT-TUNNELING DIODE

As a model potential, we consider two identical one-dimensional rectangular barriers of height V_0 and width L_b, separated by a distance L_w (well width), as shown in Fig. 11.3(a). We take one particle with energy E incident from left to represent each wave function in every region, as illustrated in Fig. 11.3(b) where the amplitude of the incident wave function has been set as one. We then solve the effective mass Schrödinger equation for one particle (e.g., an electron) and match the wave functions and their first derivatives at each potential discontinuity [i.e., by using $\varphi_i(x_i) = \varphi_{i+1}(x_i)$

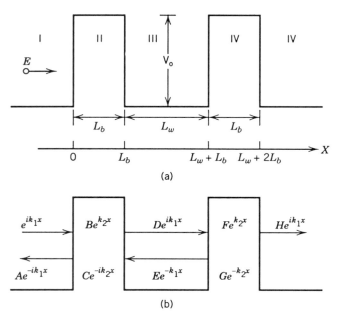

Fig. 11.3 (a) Two identical one-dimensional rectangular barriers of height V_0 and width L_b, separated by a distance L_w (well width). (b) The wave function in each region. [Wu, J. S., National Cheng Kung University Notes]

and $d\varphi_i(x_i)/dx = d\varphi_{i+1}(x_i)/dx$ for $i = 1, 2, 3$, and 4 at the boundary]. After a straightforward calculation we obtain the transmission coefficient T for $E < V_0$.

Next we define the transmission coefficient as $T = |H|^2/1^2 = |H|^2$ and obtain

$$T = \left\{1 + \frac{V_0^2(\sinh^2 K_2 L_b)H^2}{4E^2(V_0 - E)^2}\right\}^{-1}, \tag{11.5}$$

where

$$H = 2[E(V_0 - E)]^{1/2} \cosh K_2 L_b \cos K_1 L_w - (2E - V_0) \sinh K_2 L_b \sin K_1 L_w.$$

We know from Eq. (11.5) that $T = 1$ when $H = 0$; in other words, $H = 0$ is the resonance condition. The resonant-tunneling energy E_n is obtained by solving the equation $H = 0$. Letting $H = 0$ and rearranging the equation, we obtain the resonant condition as

$$\tan K_1 L_w \tanh K_2 L_b = \frac{2[E(V_0 - E)]^{1/2}}{2E - V_0}. \tag{11.6}$$

However, the expression for the discrete energy levels in a rectangular potential well with finite potential height V_0 is given by

$$\left(\frac{V_0 - E}{E}\right)^{1/2} = \begin{cases} \tan \frac{1}{2} K_1 L_w, & N \text{ even}, \\ -\cot \frac{1}{2} K_1 L_w, & N \text{ odd}, \end{cases}$$

where N is a parameter. Combining the two cases, we get a complete relation with a single equation

$$\left[\left(\frac{V_0 - E}{E}\right)^{1/2} - \tan \frac{1}{2} K_1 L_w\right] * \left[\left(\frac{V_0 - E}{E}\right)^{1/2} + \cot \frac{1}{2} K_1 L_w\right] = 0.$$

From this equation we obtain

$$\tan K_1 L_w = \frac{2[E(V_0 - E)]^{1/2}}{2E - V_0}. \tag{11.7}$$

Equation (11.7) can be obtained from Eq. (11.6) by assuming that $\tanh K_2 L_b = 1$.

For the GaAs-AlGaAs structure $\tanh K_2 L_b$ is usually 1, with very small error. When an electron of energy E is incident on a double-barrier structure, the transmission coefficient is unity if E is equal to the bound state energy in the corresponding well with well depth V. Figure 11.4 shows the current-voltage and conductance-voltage characteristics of a double-barrier structure [Esa74]. Conditions at resonance (*a*) and (*c*) and off-resonance (*b*) are indicated in the figure.

11.3.1 Resonant Tunneling through Parabolic Quantum Wells

Parabolic QWs have interesting possibilities for device applications, since the levels in such wells are equally spaced. The I–V characteristics of RT structures with

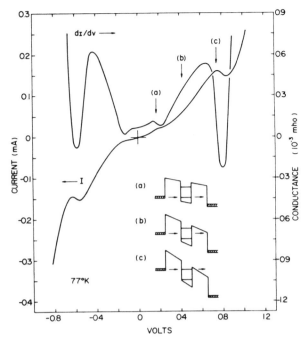

Fig. 11.4 Current-voltage and conductance-voltage characteristics of a double-barrier structure. Conditions at resonance (a), (c), and off-resonance (b). [Esaki, L., and L. L. Chang, *Phys. Rev. Lett.* **33** (1974): 495; Esaki, L., in *Molecular Beam Epitaxy and Heterostructures*, p. 8, ed. by L. L. Chang and K. Ploog, Martinus Nijhoff Publishers (1985)]

parabolic wells are therefore expected to produce nearly equally spaced peaks in voltage. Such resonances have been observed experimentally [Sen87b].

RT samples with parabolic quantum wells have been grown by MBE on n^+ (100) GaAs substrates [Cap90]. The parabolically graded well compositions were produced by growth of short-period (10 Å), variable-duty-cycle, $GaAs/Al_xGa_{1-x}As$ superlattices in which the Al content within each period of the superlattice corresponded to the Al content at the same point in a smooth parabolic well. The energy diagrams at the Γ point are shown in Fig. 11.5 for different bias voltages. Figure 11.6 shows the I–V characteristics and corresponding conductance for this sample for both bias polarities. The group of resonances from the fifth to the eleventh are the most pronounced ones and actually display negative differential resistance. Fourteen resonances are observed in the sample for positive polarity.

The overall features of the I–V characteristics can be interpreted physically by means of the band diagrams in Fig. 11.5. At zero bias the first six energy levels of the well are confined by a parabolic well 225 meV deep, corresponding to the grading from $x = 0$ to $x = 0.30$, and their spacing is 35 meV. When the bias is increased from 0 to 0.3 V, the first four energy levels probed by RT, Fig. 11.5b, remain confined by the parabolic portions of the well, and their spacing is practically independent of the bias. This gives rise to the calculated and observed equal spacing of the first four resonances in the I–V characteristic (Fig. 11.6). When the voltage is raised above 0.3 V, the higher-energy levels become increasingly confined on the emitter side by

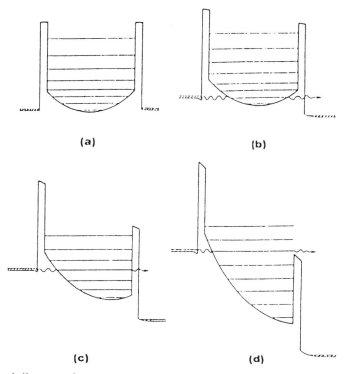

Fig. 11.5 Band diagram of a sample in equilibrium and under different bias conditions. The wells are drawn to scale; for clarity, only half the number of levels in an energy interval are shown. [Capasso, F., S. Sen, and F. Beltram, in *High-Speed Semiconductor Devices*, ed. by S. M. Sze, Wiley (1990)]

the parabolic portion of the well and on the collector side by a rectangular barrier, as shown in Fig. 11.5(c). This leads to the observed gradual increase in the voltage separation of the resonances as the bias is increased from 0.3 to 1.0 V. Above 1 V the electrons injected from the emitter probe the virtual levels in the quasi-continuum above the collector barrier, Fig. 11.5(d). These resonances result from the electron interference effects associated with multiple quantum mechanical reflections at the well-barrier interface for energies above the barrier height. It should be noted that these reflections give rise to the existence of two-dimensional (2-D) quasi-eigenstates in the well region. The observed NDR is due to tunneling into these states. These interference effects produce the four resonances observed above 1 V and must be clearly distinguished from the ones occurring at lower voltages, which are due to RT through the DB.

11.3.2 The Double-Barrier Resonant-Tunneling Structures

The NDR characteristics observed are more pronounced at low temperatures [Hua87; Lak88], but the reason is not very clear. In this section we describe this temperature dependence which is essentially an inherent behavior of the double-barrier resonant-

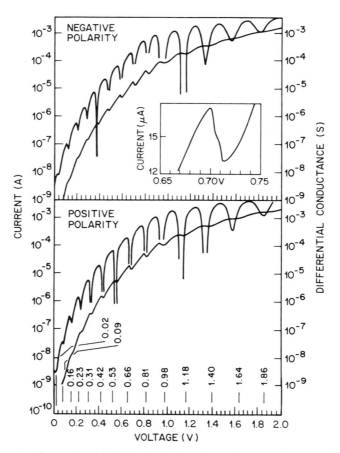

Fig. 11.6 Current-voltage characteristic at 7.1 K and conductance for a parabolic-well RT diode under opposite bias polarity conditions. The inset shows the eighth resonance on a linear scale. The vertical segments near the horizontal axis indicate the calculated positions of the resonances. [Capasso, F., S. Sen, and F. Beltram, in *High-Speed Semiconductor Devices*, ed. by S. M. Sze, Wiley (1990)]

tunneling structure (DBRTS) [Wu89]. The contributions from the thermionic current [Tsu85] and scattering effects [Lak88] are only secondary.

Figure 11.7(a) shows a typical DBRTS. The device has a well width of L_w, symmetrical double barriers each of width L_b, a barrier height V_0, and a Fermi level E_F above the conduction-band edge. The typical current-voltage characteristics at different temperatures are shown in Fig. 11.7(b). Only one bias polarity is presented, and the scale is not indicated. When the temperature is reduced from T_2 to T_1, the peak current increases while the valley current decreases, giving rise to a larger PVCR (peak-to-valley current ratio). This trend has appeared in almost all of the DBRTSs. When the increase in PVCR at low temperatures is caused by a decrease in leakage current because of the suppression of thermionic emission current over the potential barriers, the peak current should not increase [Hua87; Bro88; Lak88], particularly in the low-temperature range of 77 to 4 K. It is clear that there must be something

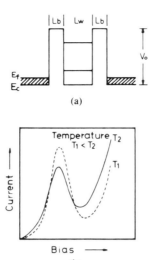

Fig. 11.7 (a) Schematic energy-band diagram of a typical DBRTS. (b) Typical current-voltage characteristics at different temperatures. [Wu, J. S., C. Y. Chang, C. P. Lee, Y. H. Wang, and F. Kai, *IEEE Electron Device Lett.* **10**, 7 (1989): 301–303]

other than thermionic emission to push the peak current up and pull the valley current down when the device is cooled.

The I–V characteristics of resonant-tunneling devices to show the temperature dependence of PVCR ratio and peak/valley currents for a set of structure parameters have been calculated. Accordingly we can take the position of Fermi level to be fixed and not consider the band bending. Then we would use a WKB approximation to estimate the transmission coefficients. The temperature dependence of the I–V characteristics would be accounted for by considering different electron distributions at different temperatures. Figure 11.8 shows the calculated temperature dependences of peak/valley currents and PVCR for the set of structure parameters shown in the insert. When the temperature is reduced, the peak current increases while the valley current decreases, therefore increasing the PVCR. The thermionic current, which is estimated to be about 2.5×10^{-10} A/cm^2, is negligible compared to the valley tunneling current. Even if the band-bending effect is considered, the thermionic current is still insignificant.

Figure 11.9(a) shows the condition when the peak current occurs. At temperature T_1, which is close to 0 K, almost all of the electrons are below the Fermi level. The electron densities per unit energy in this case are higher than those at temperatures above 0 K by a factor of the Fermi distribution function, which is temperature dependent. If the peak transmissivity of the first resonant state occurs at an energy slightly below the quasi-Fermi level of the emitter, the tunneling current, which is proportional to the integral of the product of electron density and transmission coefficient, will be higher than those at higher temperatures. When the temperature rises to T_2, some of the electrons will move upward to higher-energy states, causing a decrease in the amount of electrons having energies below E_F. As the case for peak current is reached, some electrons will tunnel through the barriers, with lower trans-

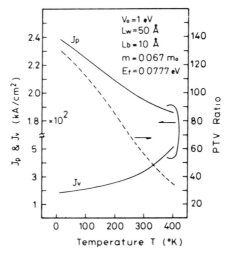

Fig. 11.8 Calculated temperature dependences of peak-to-valley currents and peak-to-valley current ratio (PVCR) for the set of structure parameters shown in the insert. [Wu, J. S., C. Y. Chang, C. P. Le, Y. H. Wang, and F. Kai, *IEEE Electron Device Lett.* **10**, 7 (1989): 301–303].

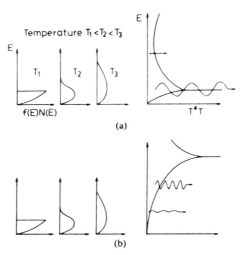

Fig. 11.9 Energy distributions of electrons at different temperatures when (*a*) the peak current and (*b*) the valley current occur. Transmission coefficients shown in the right are not scaled. [Wu, J. S., C. Y. Chang, C. P. Lee, Y. H. Wang, and F. Kai, *IEEE Electron Device Lett.* **10**, 7 (1989): 301–303]

missivities corresponding to higher-energy levels above E_1 and resulting in a smaller current density. If the temperature increases even further, the energy distribution of electrons will extend further upward, with a decrease in peak current. At high temperatures, though a small amount of electrons having high-energy states can resonantly tunnel through the second resonant state, their contribution to the overall current is insignificant. The explanation of the temperature dependence of the valley current is similar to that of the peak current, and it is easy to see if one inverts the discussion of the above effect, as shown in Fig. 11.9(b). Unlike its effect on the peak current, the second resonant state enhances the temperature dependence of the valley current.

Thermionic current over the barriers should not be account for the reduction in the PVCR at high temperatures. It has been shown that [Cap85] at 300 K the thermionic current component is negligible, compared to the measured valley current. In general, one always uses barriers that are as high as possible to achieve a large PVCR, so the thermionic-current-dominated case is unlikely. If the phonon-related scattering at low temperatures is reduced, the PVCR increases, due to the peak current increase rather than the valley current decrease. The scattering process alone cannot explain the experimental results satisfactorily. This unique feature is an inherent characteristic of DBRTS.

11.4 THE REALIZATION OF RESONANT-TUNNELING TRANSISTORS

Several transistors based on the resonant tunneling have been proposed, and some are experimentally realized. In this section the resonant-tunneling transistor (RTT) with different structures will be discussed.

11.4.1 Resonant-Tunneling Hot-Electron Transistor

Figures 11.10 and 11.11 show a cross section and a band diagram of the resonant-tunneling hot-electron transistor (RHET) [Yok85], respectively, where $x = 0.33$ and $y = 0.2$. This device uses n-GaAs layers for emitter, base, and collector with a carrier concentration of 1×10^{18} cm^{-3}. The quantum well, consisting of 56 Å GaAs sandwiched between 50 Å Al$_x$Ga$_{1-x}$As ($x = 0.33$) barriers, is inserted between the emitter and the base. The thicknesses are 1000 Å for the n^+-GaAs base and 3000 Å for the Al$_y$Ga$_{1-y}$As ($y = 0.20$) collector barrier. This sophisticated structure was grown by MBE at 580°C.

Figure 11.11 illustrates the operating principle of the RHET. E_1 and E_2 indicate the energy of the resonant states formed in the quantum well. (A) When the base-emitter voltage is zero, there is no electron injection and no collector current.

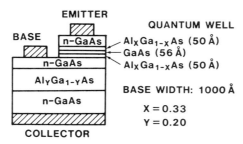

Fig. 11.10 Schematic cross section of the RHET. [Yokoyama, N., K. Imamura, S. Muoto S. Hiyamizu, and H. Nishi, *Jap. J. Appl. Phys.* **24** (1985): L853]

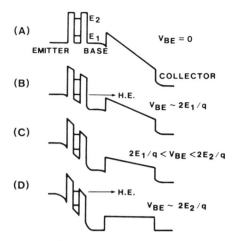

Fig. 11.11 Operating principle of the RHET. [Yokoyama, N., K. Imamura, S. Muto, S. Hiyamizu, and H. Nishi, *Jap. J. Appl. Phys.* **24** (1985): L853]

(B) When base-emitter voltages of around $2E_1/q$ are applied, electrons are injected into the base by resonant tunneling through the first resonant state. Electrons injected into the base are ballistically or near-ballistically transferred to the collector through the base, and the collector current flows. (C) When the base-emitter bias is further increased, the collector current is reduced because the resonant-tunneling current is reduced. (D) The collector current increases again with the base-emitter voltage at around $2E_2/q$, due to resonant tunneling through the second resonant state. The high-speed nature of the HET device is naturally maintained because of the use of resonant-tunneling and hot-electron transport.

Figure 11.12 shows the collector current as a function of the base-emitter voltage in common-emitter configuration with a constant 2 V applied to the collector. The collector current exhibits a peak with respect to the base voltage. Figure 11.13 shows the collector current as a function of the base current with a constant collector voltage of 2 V. As the base current increases, the collector current increases monotonously, and then increases rapidly once the base current reaches 0.84 mA. The common-

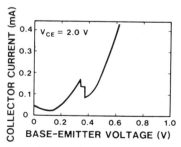

Fig. 11.12 Collector current versus base-emitter voltage with a constant 2 V on the collector. [Yokoyama, N., K. Imamura, S. Muto, S. Hiyamizu, and H. Nishi, *Jap. J. Appl. Phys.* **24** (1985): L853]

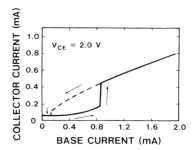

Fig. 11.13 Collector current versus base-emitter current with a constant 2 V on the collector. [Yokoyama, N., K. Imamura, S. Muto, S. Hiyamizu, and H. Nishi, *Jap. J. Appl. Phys.* **24** (1985): L853]

emitter current gain measured at this point can reach 2000. On the other hand, as the base current decreases, the collector current decreases monotonously, rapidly dropping at a base current of 0.1 mA. Thus there is a large hysteresis loop in the collector current with respect to the base current. This is due to the presence of a negative resistance region in the base current with respect to the base voltage.

11.4.2 Asymmetric Barrier Structure

By optimizing the resonant-tunneling structure Fig. 11.14 shows an improvement in device performance [Mor86]. This asymmetric structure can increase the peak current and also decrease the valley current. The increased peak current is due to the increased transmission coefficient through the quantum well. On the other hand, the increased base-side barrier height reduces the thermionic emission over the barrier, thus decreasing the valley current. Figures 11.15(a) and (b) give plots of the collector current

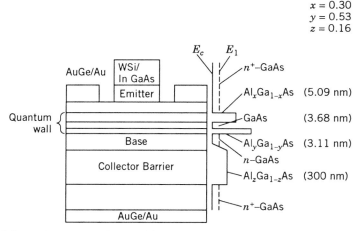

Fig. 11.14 Schematic cross section and band diagram of a resonant tunneling hot-electron transistor. [Mori, T., H. Ohnishi, K. Imamura, S. Muto, and N. Yokoyama, *Appl. Phys. Lett.* **49** (1986): 1779]

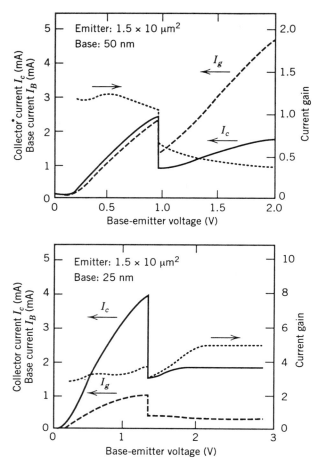

Fig. 11.15 (a) Collector current (*solid line*), base current (*dashed line*), and current gain (*dotted line*) measured at 77 K as functions of the base-emitter voltage for a 50-nm base resonant tunneling hot-electron transistor. (b) Collector current (*solid line*), base current (*dashed line*), and current gain (*dotted line*) measured at 77 K as functions of the base-emitter voltages for a 25-nm base resonant tunneling hot-electron transistor. [Mori, T., H. Ohnishi, K. Imamura, S. Muto, and N. Yokoyama, *Appl. Phys. Lett.* **49** (1986): 1779]

(solid lines), the base current (dashed lines), and the current gain (dotted lines) as functions of the base-emitter voltage for the 500 Å and 250 Å bases, respectively. These were measured at 77 K as the base voltage was increased in the common-emitter configuration with a constant 3 V at the collector.

11.4.3 Resonant-Tunneling Bipolar Transistor

In the RHET a large fraction of the hot electrons lose their energy in the base layer due to hot-electron-phonon and other scattering processes. These electrons cannot surmount the collector potential barrier, resulting in poor current gain. It is also difficult to use RHET at room temperature because of the large leakage current in the base-collector junction caused by the thermionic current surmounting the collector potential barrier. To overcome these drawbacks, the resonant bipolar transistor (RBT) uses a p–n junction instead of the collector potential barrier. Capasso and Kiehl [Cap85] first proposed a RBT having a double barrier in the base region. This section describes the fabrication of the RBT using GaAs/AlGaAs heterostructures grown by MBE and discusses the electrical characteristics of the RBT, focusing on the analysis of the quantum well resonator inserted between the n-type AlGaAs emitter and p-type GaAs base [Cap85].

Figure 11.16 shows the cross section and composition profile of the RBT [Fut86]. A quantum well of 50 Å GaAs, sandwiched between 20 Å AlAs barriers, was inserted between the n-AlGaAs emitter and the p-GaAs base. The base layer was 2000 Å thick, doped with Be to 5×10^{18} cm^{-3}. Emitter doping was 5×10^{17} cm^{-3}, the dopant being Si. The thickness of the collector (1×10^{17} cm^{-3}) was 2500 Å. The AlAs mole fraction of the emitter X was 0.37. The AlAs mole fraction near the emitter side of the quantum well was graded (0.37 to 0.14) over 700 Å. The energy of the conduction band edge of the emitter side of the quantum well was equal to the resonant energy level. This superlattice structure was grown successively on an n^+-GaAs substrate at 580°C by MBE. Emitter electrodes were formed using AuGe/Au/WSi metallization to make shallow ohmic contacts. Base electrodes and collector electrodes were formed by making AuZn/Au and AuGe/Au ohmic contacts, respectively.

The band diagram with the positive base-emitter voltage for the RBT is illustrated in Fig. 11.17. Resonant tunneling occurs when the energy of the injected electrons becomes approximately equal to the energy level of the electrons confined in the well. Hot electrons, injected from the emitter to the base, lose their energy by various scattering processes, as in the case of RHET. But most electrons can reach the collector, their being no collector potential barrier for electrons. The high current gains can thus be achieved. Another merit of the RBT is the small leakage current characteristics of its base-collector p–n junction, enabling it to be used at room temperature.

Figure 11.18 shows the base-emitter I–V characteristics measured at 77 K with the collector open. There is a NDR region at around 1.7 V, due to the resonant-tunneling of electrons. The emitter-base junction measures $5 \times 24 \,\mu\text{m}$. The peak-to-valley ratio

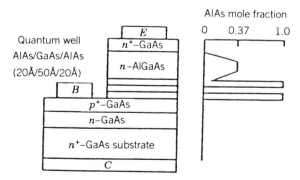

Fig. 11.16 Schematic cross section and composition profile of the RBT. [Futatsugi, T., Y. Yamaguchi, K. Ishii, K. Imamura, S. Muto, N. Yokoyama, and A. Shibatomi, *IEDM 1986*, p. 286 (1986)]

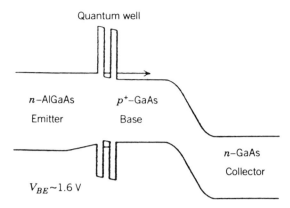

Fig. 11.17 Schematic band diagram for the RBT with the positive base-emitter voltage. [Futatsugi, T., Y. Yamaguchi, K. Ishii, K. Imamura, S. Muto, N. Yokoyama, and A. Shibatomi, *IEDM 1986*, p. 286 (1986)]

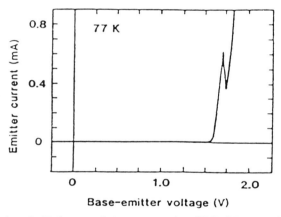

Fig. 11.18 Base-emitter I–V characteristics measured at 77 K. [Futatsugi, T., Y. Yamaguchi, K. Ishii, K. Imamura, S. Muto, N. Yokoyama, and A. Shibatomi, *IEDM 1986*, p. 286 (1986)]

was 1.7. Figure 11.19 shows the collector current of the RBT, measured as a function of the base-emitter voltage in common-emitter configuration, with a constant 3 V on the collector. Figure 11.20 shows the RBT collector current-voltage ($I_C - V_{CE}$) characteristics as measured as 77 K in a common-emitter configuration with the base current I_B as a parameter. It should be noted that there is an abrupt decrease in the collector current when the base current increases from 30 to 40 μA, indicating that the RBT has a negative current gain. The current gain h_{Fe} is typically observed to be 15 in the regular region. The maximum current gain was 20.

11.4.4 RHET Using InGaAs-Based Materials

Yokoyama et al. [Yok88] have developed an RHET using an InAlAs/InGaAs/InAlAs resonant-tunneling barrier, an InGaAs base, and an InAlGaAs collector barrier. This

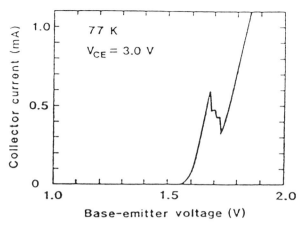

Fig. 11.19 Collector current versus base-emitter voltage at 77 K. [Futatsugi, T., Y. Yamaguchi, K. Ishii, K. Imamura, S. Muto, N. Yokoyama, and A. Shibatomi, *IEDM 1986*, p. 286 (1986)]

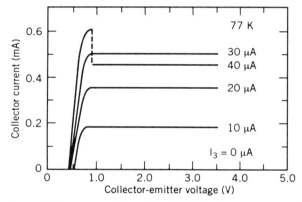

Fig. 11.20 The collector I–V characteristics of the RBT for the common emitter configuration with base current I_B as a parameter. [Futatsugi, T., Y. Yamaguchi, K. Ishii, K. Imamura, S. Muto, N. Yokoyama, and A. Shibatomi, *IEDM 1986*, p. 286 (1986)]

device is shown in Fig. 11.21 [Ohn88]. These alloys have compositions that are lattice-matched with the InP substrate. The emitter common current gain of the 250 Å base RHET was typically in the range from 10 to 25 at 77 K, indicating hot carrier transfer efficiency of 90% to 96%. The transfer efficiency agrees well with theoretical results. The increased transfer efficiency is due to a large Γ–L separation energy as compared to that of GaAs. The collector current peak-to-valley ratio was measured in the range of 20 to 33, larger than that of GaAs by a factor of 4 to 6. The increased ratio is possibly due to the light effective mass and lack of effects of the indirect valleys in the barriers.

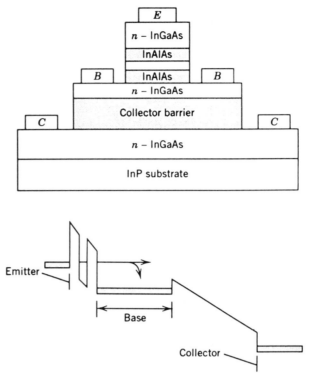

Fig. 11.21 Cross section and band diagram of the $In_{0.53}Ga_{0.47}As$-based RHET. [Ohnishi, H., N. Yokoyama, and A. Shibatomi, *IEDM*, p. 830, IEEE (1988)].

Theoretical and Experimental Analyses of RHET

A Monte Carlo simulation can be used to follow the motion of electrons injected into the GaAs or InGaAs base from the resonant-tunneling barrier. In this calculation the effective masses and nonparabolicities were assumed to be 0.042 and $1.167\,eV^{-1}$ for InGaAs and 0.0755 and $0.586\,eV^{-1}$ for InAlAs, respectively. Conduction-band discontinuity between them was assumed to be 0.53 eV. The carrier density was $1 \times 10^{18}\,cm^{-3}$. The collector barrier was the 200-nm $In_{0.52}(Al_xGa_{1-x})_{0.48}As$ barrier ($x = 0.5$). At the base side of the collector barrier we used the stepwise-graded region over 6 nm ($x = 0.2, 0.25, 0.33$). Emitter-base junction area was $42\,\mu m^2$, and the base-collector junction area was $180\,\mu m^2$.

With this calculation we took into the account of accumulation and depletion layers [Yok88]. Poisson and Schrödinger equations were used to calculate the potential profile of the collector barrier and the collector independently of the resonant-tunneling barrier.

Figure 11.22 shows the calculated emitter I–V characteristics at 77 K. The peak current density is $4.7 \times 10^4\,A/cm^2$ at an emitter-base voltage V_{eb} of $-0.65\,V$, and the peak-to-valley ratio is 33. Figure 11.23 shows the calculated potential profile and the carrier density profile at $V_{eb} = -0.65\,V$. E_0 represents the quasi-bound state. The

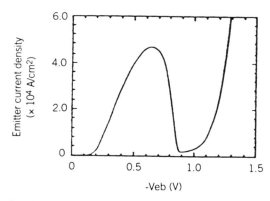

Fig. 11.22 Calculated emitter current density at 77 K as a function of negative emitter-base voltage. [Ohnishi, H., N. Yokoyama, and A. Shibatomi, *IEDM*, p. 830, IEEE (1988)]

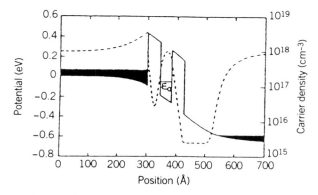

Fig. 11.23 Calculated potential distribution and carrier density at $V_{eb} = -0.65$ V. [Ohnishi, H., N. Yokoyama, and A. Shibatomi, *IEDM*, p. 830, IEEE (1988)]

sheet carrier density integrated over the double-barrier structure is 3.9×10^{11} cm^{-2}, and that of the depletion layer is 1.8×10^{12} cm^{-2}.

Figure 11.24 shows the total energy distribution of hot electrons as they are injected into the base. The energy spread is 130 meV, which is almost equal to the difference between the Fermi energy and the bottom of the quasi-bound level. Figure 11.25 plots the collector I–V characteristics for the 25-nm-base RHET calculated for different emitter currents. The collector current has well-saturated characteristics. The transfer ratio (α) is 0.89 for $I_e = 19.6$ mA and $V_{cb} = 1$ V. The transfer ratio is not high enough. Decrease the collector barrier height can improve α.

11.4.5 Millimeter-Band Oscillations in a Resonant-Tunneling Device

Recent interest in the quantum well resonant tunneling diode can be attributed to the work of Sollner et al. [Sol83], who demonstrated detection in these devices at frequencies up to 2.5 THz. More recently oscillations up to 675 GHz have been

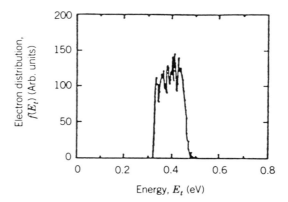

Fig. 11.24 Total energy distribution of electrons injected into a InGaAs base at $V_{eb} = -0.65$ V. [Ohnishi, H., N. Yokoyama, and A. Shibatomi, *IEDM*, p. 830, IEEE (1988)]

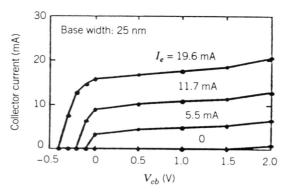

Fig. 11.25 Calculated collector current-voltage characteristics for a 25-nm-base RHET at 77 K. [Ohnishi, H., N. Yokoyama, and A. Shibatomi, *IEDM*, p. 830, IEEE (1988)]

observed in a two-terminal resonant-tunneling device at room temperature [Sol87; Bro89; Bro90]. The device consists of a 4.5-nm layer of GaAs sandwiched between two 3.0-nm layers of $Al_{0.3}Ga_{0.7}As$. Outside of each AlGaAs layer is a 300-nm-thick region of 2×10^{17} cm^{-3} n-GaAs which serves to buffer the double-barrier region from substrate and contact regions. The basis for oscillation is the negative dynamic resistance displayed by the device when the electrons that tunnel through the AlGaAs barriers have energy slightly greater than the energy of the quasi-stationary state of the well. Results for oscillators (InGaAs/AlAs, InAs/AlSb, and GaAs/AlAs) up to 400 GHz are shown in Fig. 11.26 [Sol91], which gives the output power density (output power divided by the diodes area) as a function of the oscillation frequency.

Oscillation was obtained with the device mounted in reduced-height rectangular wave guide, and tuning was performed with a sliding backshort. The observed frequencies and powers of oscillation were 29 to 31 GHz with 1 to 3 μW in the B-band, and 42 to 43 GHz with 0.5 to 1.0 μW in the V-band, wave guide.

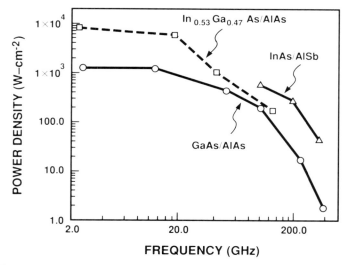

Fig. 11.26 Output power density versus frequency for oscillators employing RTDs from three different materials systems. [Sollner, T. C. L. G., E. R. Brown, J. R. Söderström, T. C. McGill, C. D. Parker, and W. D. Goodhue, in *Resonant Tunneling in Semiconductors*, ed. by L. L. Chang, Plenum Press (1991)]

The high-frequency capability of this particular resonant tunneling device is attributed in large part to its buffer layer doping density. A sample having similar barrier geometry, but much greater buffer layer doping (2×10^{18} cm^{-3}), failed to oscillate in the above frequency bands, while several samples having far less doping (2×10^{16} cm^{-3}) did not even display room-temperature negative resistance. The present device is superior to the sample with greater doping because it has a thicker depletion layer in the buffer region and thus a lower capacitance at the bias voltages required for negative resistance.

REFERENCES

[Esa58] Esaki, L., *Phys. Rev.* **109** (1958): 603.

[Mor65] Morton, J. A., *IEEE Spect.* (Sept. 1965): 62.

[Cap90] Capasso, F., S. Sen, and F. Beltram, in *High-Speed Semiconductor Devices*, ed. by S. M. Sze, Wiley (1990).

[Cap87] Capasso, F., in *Semiconductor and Semimetals*, vol. 24, p. 319, ed. by R. K. Willardson and A. C. Beers, Academic Press (1987).

[Esa70] Esaki, L., and R. Tsu, *IBM J. Res.* **14** (1970): 61.

[Cha74] Chang, L. L., L. Esaki, and R. Tsu, *Appl. Phys. Lett.* **24** (1974): 593–595.

[Gol87] Goldman, V. J., D. C. Tsui, J. E. Cunningham, and W. T. Tsang, *J. Appl. Phys.* **61** (1987): 2693.

[Sen87a] Sen, S., F. Capasso, A. L. Hutchinson, and A. Y. Cho, *Electron Lett.* **23** (1987): 1229.

[Lak88] Lakhani, A. A., R. C. Potter, D. Beyes, H. H. Hier, E. Hempfling, L. Aina, and J. M. O'Connor, *Electron Lett.* **24** (1988): 187.

[Boh79] Bohm, D. *Quantum Theory*, Dover (1979).
[Kan69] Kane, E. O., in *Tunneling Phenomena in Solids*, ed. by E. Burnstein and S. Lundquist, Academic Press (1969).
[Cap85] Capasso, F., and R. A. Kiehl, *J. Appl. Phys.* **58** (1985): 1366.
[Esa74] Esaki, L., and L. L. Chang, *Phys. Rev. Lett.* **33** (1974): 495; Esaki, L., in *Molecular Beam Epitaxy and Heterostructures*, p. 8, ed. by L. L. Chang and K. Ploog, Martinus Nijhoff (1985).
[Sen87b] Sen, S., F. Capasso, A. C. Gossard, R. A. Spah, A. L. Hutchinson, and S. N. G. Chu, *Appl. Phys. Lett.* **51** (1987): 1428.
[Hua87] Huang, C. I., *Appl. Phys. Lett.* **51** (1987): 121–123.
[Lak88] Lakhani, A. A., *Electron Lett.* **24** (1988): 153–155.
[Wu89] Wu, J. S., C. Y. Chang, C. P. Lee, Y. H. Wang, and F. Kai, *IEEE Electron Device Lett.* **10**, 7 (1989): 301–303.
[Tsu85] Tsuchiya, M., H. Sakaki, and J. Yoshino, *Jap. J. Appl. Phys.*, **24** (1985): L466–468.
[Bro88] Brockaert, T. P. E., W. Lee, and C. G. Fonstad, *Appl. Phys. Lett.* **53** (1988): 1545–1547.
[Yok85] Yokoyama, N., K. Imamura, S. Muto, S. Hiyamizu, and H. Nishi, *Jap. J. Appl. Phys.* **24** (1985): L853.
[Mor86] Mori, T., H. Ohnishi, K. Imamura, S. Muto, and N. Yokoyama, *Appl. Phys. Lett.* **49** (1986): 1779.
[Cap85] Capasso, F., and R. A. Kiele, *J. Appl. Phys.* **58** (1985) 1366.
[Fut86] Futatsugi, T., Y. Yamaguchi, K. Ishii, K. Imamura, S. Muto, N. Yokoyama, and A. Shibatomi, *IEDM*, p. 286, IEEE (1986).
[Yok88] Yokoyama, N., K. Imamura, H. Ohnishi, T. Mori, S. Muto, and A. Shibatomi, *Solid State Electron.* **31**, 3/4 (1988): 577–582.
[Ohn88] Ohnishi, H., N. Yokoyama, and A. Shibatomi, *IEDM*, p. 830, IEEE (1988).
[Sol83] Sollner, T. C. L. G., W. D. Goodhue, P. E. Tannenwald, C. D. Parker, and D. D. Peck, *Appl. Phys. Lett.* **43** (1983): 588.
[Sol87] Sollner, T. C. L. G., E. R. Brown, W. D. Goodhue, and H. Q. Le, *Appl. Phys. Lett.* **50** (1987): 332.
[Bro89] Brown, E. R., T. C. L. G. Sollner, C. D. Parker, W. D. Goodhue, and C. L. Chen, *Appl. Phys. Lett.* **55** (1989): 1777.
[Bro90] Brown, E. R., C. D. Parker, L. J. Mahoney, J. R. Söderström, and T. C. McGill, Device Research Conf., Santa Barbara, CA, June 25–27 (1990).
[Sol91] Sollner, T. C. L. G., E. R. Brown, J. R. Söderström, T. C. McGill, C. D. Parker, and W. D. Goodhue, in *Resonant Tunneling in Semiconductors*, ed. by L. L. Chang, Plenum Press (1991).

12

HOT-ELECTRON TRANSISTORS AND NOVEL DEVICES

As semiconductor devices shrink in size and internal fields rise, a large number of carriers in the active regions of the device during its operation are in states of high kinetic energy [Con67]. At a given point in space and time the velocity distribution may be narrowly peaked, which is what is meant when one speaks of "ballistic" electron packets [Lur90]. At other times and locations, the nonequilibrium electron ensemble can have a broad velocity distribution, usually taken to be maxwellian and parametrized by an effective electron temperature $T_e > T$, where T is the lattice temperature. hot-electron phenomena have become important, and they have successfully been applied to semiconductor devices.

The history of hot-electron transistor development can be traced back to the early 1960s "metal-base transistors" [Sze81]. In those days the performance of these devices was severely limited to current gains of less than unity [Sze81], so interest in their development quickly declined. With the continuing advancements in device fabrication, equipment, and processing technology, interest in the metal-base transistors resumed. In the late 1970s a hot-electron transistor (HET) in silicon was developed by Shannon [Sha79]; it had a Schottky barrier emitter and Camel diode collector. The high emitter capacitance of the Schottky barrier led Shannon to fabricate an alternative HET with both a Camel diode emitter and collector, resulting in an all semiconductor HET [Sha81]. Later Malik et al. [Mal81] fabricated a HET in GaAs similar to Shannon's but with out the Camel diodes which were replaced by precisely controlled charge sheets referred to as "planar-doped barriers."

In this chapter we will review planar-doped transistors, quantum well–base transistors, hot-electron spectroscopy, quantum interference devices, and quantum point contacts.

12.1 BALLISTIC-INJECTION DEVICES

12.1.1 Metal-Base Transistors

Ballistic-injection transistors differ by the materials employed and by the physical mechanisms of hot electron injection into the base [Lur90]. The first hot-electron injection device was designed by Mead [Mea 61] based on the concept of electrons tunneling from a metal emitter through a thin oxide barrier into a high-energy state in a metal base, as shown in Fig. 12.1(a). This is the metal–oxide–metal–oxide–metal (MOMOM) transistor. Experimentally Mead found a very small transfer ratio between collector and emitter currents. Only the metal-base transistor has the potential to give a better microwave performance than the bipolar transistor [Sze81]. It was later replaced by the metal–semiconductor junction, resulting in a transistor structure called *MOMS*, as shown in Fig. 12.1(b). The metal-base transistor (MBT), which employs thermionic rather than tunneling injection of hot carriers into the base, was first proposed in the form of a semiconductor–metal–semiconductor (SMS) structure [Ata62; Gep62]. A band diagram of the SMS transistor is illustrated in Fig. 12.1(c). At room

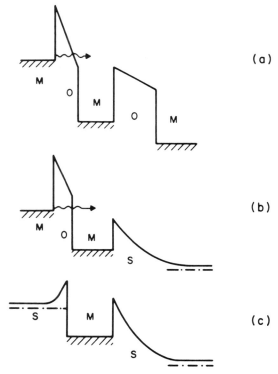

Fig. 12.1 Metal-base transistor: (a) MOMOM; (b) MOMS; (c) SMS. Each of the *M* (metal) of *S* (doped-semiconductor) electrodes is contacted independently and can be biased with respect to the other electrodes. The figure shows the energy-band diagrams of metal-base transistors under operating bias conditions. [Sze, S. M., *Physics of Semiconductor Devices*, Wiley Interscience (1981)]

temperature the early experimental transfer ratio was low, of the order of 0.3, which was obtained from a Si–Au–Ge transistor [Sze66].

The metal-base transistor is the predecessor of the quantum well–base transistor, which has enhanced the transfer ratio (α) to 0.9 and above. The quantum well–base transistor was developed as advances were made in epitaxial growth technology, such as molecular beam epitaxy.

12.1.2 Doped-Base Transistors

The problem with metal-base transistors is that they have a poor transfer ratio α, which is mainly due to the quantum-mechanical (QM) reflection of electrons at the base-collector interface [Cro66]. The doped-base transistor was developed to prevent the QM reflection problem. A number of doped-base devices have been manufactured, including the hot-electron camel transistors [Sha79], planar-doped barrier transistors [Mal81; Hol83], GaAs/AlGaAs heterojunction barrier transistors [Hei85], InGaAs/InAlAs barrier transistor [Red86], and pseudomorphic MBE grown structures with an InGaAs base on GaAs substrates [Has88].

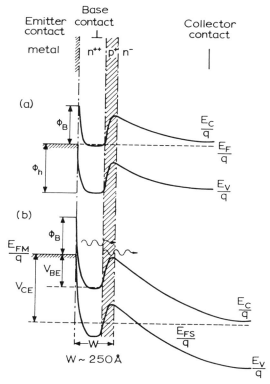

Fig. 12.2 Energy-band diagram for HET with Schottky emitter made by Shannon. (*a*) Without applied bias; (*b*) with applied bias. [Shannon, J. M., *IEE J. Electron Devices* **3** (1979): 144–149]

12.1.3 The Hot-Electron Camel Transistor

In 1979 Shannon [Sha79; Sha84] published the results of his work on the hot-electron transistor formed by shallow implantations of n^+ and p^+ regions in n-type silicon. The energy band diagram for such a device is shown in Fig. 12.2.(a) [Sha79] in which are present both a camel diode collector and a Schottky barrier as the emitter. Fig. 12.2(a) gives the unbiased structure while the Fig. 12.2(b) shows it biased with respect to the base. The high capacitance of the Schottky barrier emitter led Shannon to propose a second HET structure replacing the Schottky barrier with a camel diode [Hay86]. These structures were fabricated by ion implantation which did not allow atomic scale control of doping profiles. A bulk triangular barrier structure similar to that of the camel diode using MBE was demonstrated by Malik et al. [Mal80]. They termed their structure a "planar-doped barrier," and by replacing two of them back-to-back they fabricated the first HET in GaAs [Hay86].

12.1.4 The Tunneling Hot-Electron Transfer Amplifier

Heiblum proposed the concept of tunneling hot-electron transfer amplifier (THETA) devices [Hei81; Hei85] that utilize heterojunctions to form barriers. In the THETA device undoped AlGaAs layers replace the original oxides implemented by Mead. The good lattice match between the conductors and insulators enables the fabrication of single crystal material with perfect interfaces, although lower potential barriers require a low-temperature operation. The THETA device was first realized by Yokoyama et al. [Yok84] in 1984. It has been shown [Hei88] that the existence of ballistic transport of up to 75% of the injected electrons and a high α (0.9) in devices with base widths of 30 nm.

Structure and Operation of the THETA Device

The structure of the THETA is comprised of five epitaxial, MBE-grown layers. Three n-type heavily doped GaAs layers are separated by two undoped AlGaAs barriers. The first GaAs (emitter)–AlGaAs–GaAs (base) structure forms a tunnel injector for hot electrons into the base. The second GaAs (base)–AlGaAs–GaAs (collector) structure forms a barrier with height Φ_c, enabling only hot electrons with $p_x^2/2m > \Phi_c$ to surmount it. p_x is the component of the momentum perpendicular to the one-dimensional collector barrier. The emitter barrier preselects for tunneling predominantly those electrons in the emitter that move perpendicular to the barrier. An energy band diagram for Γ electrons in the heterojunction THETA device in shown in Fig. 12.3. A collimated hot-electron beam, with an energy spread of about 60 meV, is injected into the base. The kinetic energy of the injected electrons is approximated equal to $eV_{BE} + \zeta$, where e is the electron charge, V_{BE} is the applied voltage between base and emitter, and ζ is the Fermi level energy above the bottom of the conduction band in the base. If the base is make thinner than the mean free path (mfp), a substantial fraction of the ballistic electrons will maintain their energy and direction after traversing the base and be collected above the collector barrier.

If hot electrons are injected in the (100) direction into a 300-Å base with kinetic energy of 0.3 eV, they will traverse the base with a drift velocity of about 1×10^8 cm/s in a transit time of 30 fs. If the collector barrier is 500 Å wide and the electrons velocity,

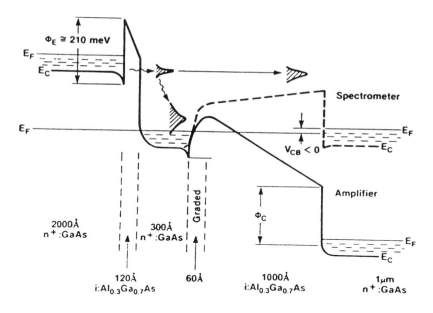

Fig. 12.3 Band diagram of the THETA device. Spectrometer configuration corresponds to $V_{CB} < 0$ when $\delta\Phi_c = e\delta V_{CB}$. [Heiblum, M., *Solid State Electronic* **24** (1981): 343–366; Heiblum, M., D. C. Thomas, C. M. Knoedler, and M. I. Nathan, *Appl. Phys. Lett.* **47** (1985): 1105]

there is about 5×10^7 cm/s; their transit time would be 100 fs. The short transit time is the most attractive feature of this device.

THETA Devices as Amplifiers

When the THETA devices are operated as amplifiers, the emitter is negatively biased and the collector is positively biased with respect to the base (Fig. 12.3). In devices with a 1000-Å-thick base a maximum α of 0.7 was observed at temperatures of 100 K and below. The very weak temperature dependence suggested that optical phonon scattering was negligible. Since the hot electrons relaxed at the bottom of the available conduction bands (Γ, L, and X) before arriving at the collector, increasing the collector voltage V_{CB} will reduce the corresponding collector barrier heights when barriers are graded and aid in the electron collection.

THETA Devices as Energy Spectrometers

Potential steps can be used as launches and detectors of electrons and as instruments for determining the energy distribution of electrons as a function of distance from a step. The latter procedure is called *hot-electron spectroscopy* [Hes90].

Spectroscopy in semiconductors was done initially in 1963 in Si [Sze81] whose surface work function was used as a barrier spectrometer. In 1982 Hesto et al. [Hes82] proposed a way of looking for ballistic electrons in GaAs by fabricating a Schottky barrier as a spectrometer at the end of a short transport region. This was first realized

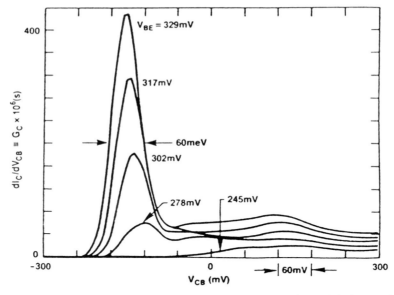

Fig. 12.4 The measured energy spectrum of THETA device. The main peak is due to electrons arriving at the analyzer plane without a single scattering event. [Heiblum, M., D. C. Thomas, C. M Knoedler, and M. I. Nathan, *Appl. Phys. Lett.* **47** (1985): 1105]

in the planar-doped barrier transistor by Hayes et al. [Hay85] and in the THETA device by Yokoyama et al. [Yok84]. The approach calls for scanning the energy distributions of hot electrons by a variable potential barrier Φ_c. Since the collector current is $e\int nv\,dE$, the derivative of the current with respect to Φ_c is the momentum distribution nv of the collected carriers. Over a small range of energy (the distribution width = 60 meV) the electrons velocity is fairly constant, thus the derivatives lead directly to the number distribution of the hot electrons.

To obtain the derivative of the collector current with respect to the collector voltage (in the range $V_{cb} < 0$), we first find the energy distributions, such as those in Fig. 12.4. The peak positions closely coincide with the expected peak energy of the injected ballistic electrons, and the collected distribution widths are as narrow as the injected ones, indicating transfer without collision [Hay85]. When spectroscopy was done in the devices with a base width of 800 Å, similar energy distributions were found, but with a smaller fraction of collected ballistic electrons. In the 300 and 800 Å cases the ballistic fractions were 30% and 15%, respectively, for a doping level of 1×10^{18} cm^{-3} and $\Phi_c = 200$ meV.

The spectrometer's resolution is limited by the quantum mechanical reflections of electrons that are right above the barrier top and by the tunneling of electrons that are right below the barrier top. The spectrometer resolution is about 20 meV [Hei85]. Most scattering events are thermalizing or randomizing events. Hollis et al. [Hol82] and Levi et al. [Lev85] had proposed that coupled modes of optical phonons and plasmons are responsible for the main scattering events. This assumption results in a calculated mfps of 300 to 400 Å in GaAs doped to 1×10^{18} cm^{-3}. However, both the quantization effects and the dominance of surface plasmon over the bulk plasmon modes in the thin quasi two-dimensional base were ignored in their calculations.

Pseudomorphic InGaAs Base THETA Devices

In an effort to widen the ballistic window, low In content (12% to 15% InAs mole fraction) and pseudomorphic InGaAs layers were used as the base layers in recent devices [Seo88; Has88]. In these strained base layers, the L–Γ separation is 380 to 410 meV [Seo88], and the conduction band discontinuity to GaAs is near 100 meV. The collector barrier with a much lower AlAs mole fraction (10%) can be fabricated, while maintaining sufficiently small leakage currents. In these structures the ballistic window is increased by at least 150 meV. In the devices with base widths of 21 nm, base dopings of 7×10^{17} cm^{-3}, InAs mole fractions of 12% in the base, and AlAs mole fractions of 10% in the collector barrier, a current gain $\beta = 27$ was measured at 77 K, corresponding to ballistic fractions higher than 97% and ballistic mfp in excess of 500 nm [Seo88]. This gain is high enough to be suitable for low-temperature circuit operation.

12.1.5 Quantum Well Base Transistors

Induced-Base Transistor

The induced-base transistor is illustrated in Fig. 12.5 [Lur85a]. Base conductivity is provided by a 2-dimensional [2-D] electron gas induced by the collector field at an undoped heterointerface. The density of the induced charge is limited by a dielectric breakdown in the collector barrier [Lur85b]. For a GaAs/AlGaAs system this means that $\sigma/q \leqslant 2 \times 10^{12}$ cm^{-2}. In the IBT operation the lateral electric field in the base is low, and hence the device can take direct advantage of the high electron mobility in a 2-D electron gas at an undoped heterojunction interface. At room temperature μ is limited by phonon scattering, $\mu \leqslant 8000$ cm^2/V-s, yielding $(\mu\sigma)^{-1} = 400 \, \Omega$/sq at the highest sheet concentrations in the base.

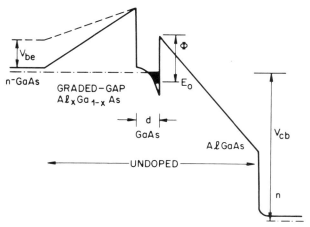

Fig. 12.5 Unipolar ballistic transistors with a monolithic all-semiconductor structure: induced-base (IBT) transistor. [Luryi, S., *IEEE Electron Dev. Lett.* **EDL-6** (1985): 178]

The induced-base conductivity is virtually independent of its thickness down to $d \leqslant 100$ Å. At such short distances the loss of hot electrons due to scattering is small. Injected hot electrons, traveling across the base with a ballistic velocity on the order of 10^8 cm/s, lose their energy mainly through the emission of polar optical phonons. For $d = 100$ Å the attendant decrease in α is estimated to be about 1%. Energy losses to the collective and single-electron excitations of the two-dimensional electron gas are negligible. The next section describes a device based on the IBT concept.

The Doped Quantum Well Gate FET

The conventional HEMT suffers from instabilities in persistent photoconductivity (PPC), threshold voltage shift, and drain current collapse at low temperatures caused by the DX center in n-$Al_xGa_{1-x}As$, especially in those alloys with $x > 0.24$ [Miz85]. The DX center is due to the coexistence of Al and donor atoms. The inferior characteristics of the HEMTs can be eliminated by the spatial separation of Al and donor atoms [Bar84]. In this section we discuss the doped GaAs quantum well FETs with $Al_{0.5}Ga_{0.5}As/GaAs$ heterostructures grown by low-pressure MOCVD. The 2DEG density increases with the increasing doping level in the well, though it starts to increase only slightly at a doping level of about 5×10^{18} cm^{-3}. The PPC effect of the device is negligible, and the gate-drain breakdown voltage was increased because of the undoped $Al_{0.5}Ga_{0.5}As$ layers.

Experimental Procedure and Results

The layer structures were grown on (100)-oriented Cr-O-doped GaAs substrates by the low-pressure MOCVD system. All layer structures were grown at 700°C with a chamber pressure 40 torr. Triethylgallium (TEG) and trimethylaluminum (TMA) were used as the gallium and aluminum sources, respectively. The growth of undoped GaAs has been optimized and characterized as having an electron mobility of 60,000 cm^2/V-s with a background impurity of 1×10^{15} cm^{-3} (77 K). Growth of high-quality Al_xGa_{1-x}-As has been achieved by bubbling the AsH_3 (15% diluted in H_2) into a Ga–In–Al eutectic melt to remove the oxygen and moisture. Figure 12.6 gives a schematic of the device and the band structure [Lin88]. It consists of a 0.5-μm GaAs buffer layer, a 50-Å $Al_{0.5}Ga_{0.5}As$ spacer layer, a 50-Å doped GaAs quantum well, a 350-Å undoped $Al_{0.5}Ga_{0.5}As$, and a 300-Å n^+-GaAs contact layer. The electron mobility and sheet-carrier density at room temperature and 77 K were estimated by the Hall measurement method for wafers in which the n^+-GaAs top layer was selectively etched away by a critic acid solution.

The prepared samples were measured by the van der Pauw measurement method to be under 5000 gauss at room temperature and 77 K. The light sensitivity of these samples was measured under tungsten lamp illumination. The results of the Hall mobility and the 2DEG density measurements on doped quantum well [DQW) modulation-doped samples are shown in Fig. 12.7. Both sets of data obtained at 300 K and 77 K show an increasing 2DEG density with an increasing doping level in the GaAs quantum well, but the increase starts at a doping level above 5×10^{18} cm^{-3}. This saturation effect is more clearly observed at room temperature. The tendency for mobilities to decrease with increasing doping level is obviously due to the increased impurity scattering by Coulomb forces of the ionized Si atoms in the quantum well.

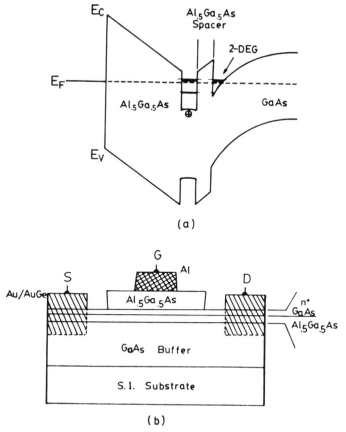

Fig. 12.6 (a) Energy-band diagram of the $Al_{0.5}Ga_{0.5}As$/GaAs heterostructure which is heavily doped in the GaAs quantum well; (b) cross-sectional view of the structure. [Lin, W., M. D. Lei, C. Y. Chang, W. C. Hsu, L. B. Di, and F. Kai, *Jap. J. Appl. Phys.* (December 1988): L2431–L2433].

Figure 12.7 also shows the sensitivity of mobilities and 2DEG densities to light. The light had no substantial effect on electron mobilities at 77 K, though an increase in the 2DEG density was clearly observed. Any increase in percentage of the 2DEG density due to illumination is negligible when the doping level is fixed and the temperature varies. The PPC effect is also negligible, as shown in the figure. As a result the density of the DX center is negligible in this structure.

The DQW FET devices have good characteristics. The drain current collapse at 77 K associated with conventional HEMTs is not observed. There is a threefold increase in transconductance, and the increase in the drain current are attributed to the increase in the electron drift velocity. The DQW FETs show their superior characteristics more than conventional HEMTs by removing the DX centers from the n-$Al_xGa_{1-x}As$ material. The threshold voltage shift between 300 K and 77 K in the dark has been measured as 0.26 V. However, when the devices were illuminated, a large increase in both the transconductance and drain current was observed, and the threshold voltage shift was quite large ($\Delta V = 0.8$ V) at 300 K and 77 K. When the light was shut off at

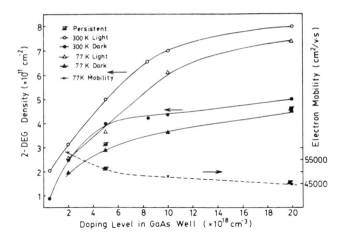

Fig. 12.7 2DEG density and electron mobilities versus doping level in the GaAs well. The PPC effects were measured under 77 K and 2 min after light illumination was shut off. [Lin, W., M. D. Lei, C. Y. Chang, W. C. Hsu, L. B. Di, and F. Kai, *Jap. J. Appl. Phys.* (December 1988): L2431–L2433].

cryogenic temperature, the devices returned to their shapes before illumination. This suggests that the large increase in transconductance and drain current may be due to photon-generated electron-hole pairs in the well in which there are deep donors and a slight diffusion into the AlGaAs layers. Electrons drift to the active channel and then contribute to the increase in the 2DEG density. The generated holes are collected by the gate electrode. The threshold voltage is then shifted to a more positive direction than in the dark. The measured breakdown voltage of this device was about 25 V. This was due to the few total space charges under the gate, and thus a powerful FET may be feasible.

12.2 REAL-SPACE TRANSFER DEVICES

The concept of real-space transfer (RST) [Hes85] describes the process in which electrons in a narrow semiconductor layer, accelerated by an electric field parallel to the layer, acquire a high average energy (become "hot") and then spill over an energy barrier into the adjacent layer. This principle underlies the operation of a three-terminal hot-electron devices, called the *charge-injection transistor* (CHINT) [Lur85]. The basic structure of CHINT is illustrated in Fig. 12.8. The emitter is a conducting layer that has source and drain contacts and acts as a hot-electron cathode. The other conducting layer, the collector, is separated by a potential barrier. When the emitter electrons are heated by the source-drain field, most of them do not reach the drain but are injected over the barrier into collector layer; a strong negative differential resistance (NDR) develops in the drain circuit.

A number of functional applications have been contemplated, based on the unique characteristics of three-terminal RST devices. In this section we describe two device structures, one is the NORAND logic circuits and the other is the light-emitting charge-injection transistor.

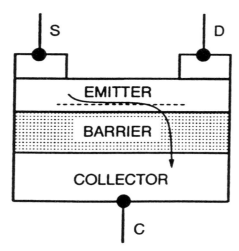

Fig. 12.8 Schematic of a charge-injection transistor. The arrow shows the direction of electron flow. [Luryi, S., *Appl. Phys. Lett.* **58**, 16 (1991): 1727–1729]

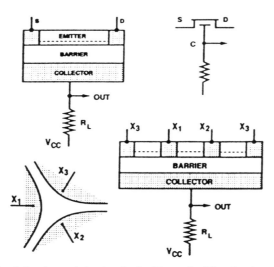

Fig. 12.9 Schematic of the charge injection logic. *Top right*: The circuit symbol of the CHINT; *bottom*: the NORAND element. The figure on the bottom left gives the symmetric arrangement of three identical channels, X_{3-1}, X_{1-2}, and X_{2-3}. By symmetry, one has the same amount of hot-electron injection for any of the six states of binary input in which at least one of the three $\mathbf{X}_j$ is different from the other two. The bottom right-hand figure gives the asymmetric NORAND layout. One of the electrodes is physically split, resulting in a "periodic" boundary condition, equivalent to the threefold rotational symmetry of the figure on the left. [S. Luryi and M. R. Pinto, *Semicond. Sci. Tech.* **7** (1992): B523–B526]

12.2.1 Charge-Injection Logic

The fact that the RST current I_C does not depend on which of the two surface terminals, S or D, is chosen to be the source allows the implementation of devices in which the role of a particular terminal in the circuit is not defined by the layout. Consider the circuit illustrated in the top panel of Fig. 12.9 as a logic element with input **S** and **D** corresponding to the biases on the S and D electrodes, respectively. The point to note is that the logic function **OUT(S, D)** .. 1 represents an exclusive NOR, since **OUT** is high when I_C is low (no RST), and **OUT** is low when the injection current is flowing.

The bottom panel of Fig. 12.9 gives diagram of the proposed logic element NORAND, which has three logic inputs $\mathbf{X}_j$ ($j = 1, 2, 3$) and one output **OUT**. Which of the $\mathbf{X}_j$ will serve as a source and which as a drain is determined only at the time when a particular logic operation is performed. The value of **OUT** is high only when all three $\mathbf{X}_j$s have the same value. All the other six possible logic input configurations lead to the same high injection current resulting from hot-electron emitters formed in two of the three channels, X_{3-1}, X_{1-2}, and X_{2-3}. Three of these configurations correspond to the presence of two sources and one drain, the other three to one source and two drains.

The logic function **OUT**($\{\mathbf{X}_j\}$) is given by

$$\mathbf{OUT}(\{\mathbf{X}_j\}) = (\mathbf{X}_1 \cap \mathbf{X}_2 \cap \mathbf{X}_3) \cup (\bar{\mathbf{X}}_1 \cap \bar{\mathbf{X}}_2 \cap \bar{\mathbf{X}}_3), \tag{12.1}$$

where the symbols $\cap$, $\cup$, and $\bar{A}$ stand for logic functions **AND**, **OR**, and **NOT A**, respectively. We see that the NORAND operates as a $\mathbf{NOR}(\mathbf{X}_1, \mathbf{X}_2) = \overline{X_1 \cup X_2}$ when the input to $\mathbf{X}_3$ is low, $\mathbf{X}_3 = $ logic **0**, and as an $\mathbf{AND}(\mathbf{X}_1, \mathbf{X}_2) = \mathbf{X}_1 \cap \mathbf{X}_2$ when $\mathbf{X}_3 = $ logic **1**. It is clear that the threefold symmetry of the device ensures that the injection current has the same value in all the six states corresponding to **OUT** = logic **0**. However, one can achieve a similar effect without an exact threefold symmetry, as illustrated in the bottom right-hand panel of Fig. 12.9.

The NORAND functional element represents a natural embodiment of the essence of hot-electron injection by the RST. Compared to all existing logic families, the NORAND offers a considerable economy in the layout of basic functional elements. Moreover it promises faster operation of these elements, since the entire function is implemented within a one-gate delay of the high-speed transistor.

12.2.2 Light-Emitting Charge-Injection Transistor

The charge injection described in the previous section occurs between layers of the same conductivity type. Injection of minority carriers by RST holds promise for optoelectronic device applications.

Mastrapasqua et al. [Mas92] have demonstrated the effectiveness of a CHINT with *n*-type emitter channel and a *p*-type collector. The lattice-matched InGaAs/InAlAs/InGaAs heterostructure and grown by MBE on a semi-insulating InP substrate. Figure 12.10(*a*) shows a cross section of the device. The n^+ cap layer is removed in the trench area. The channel length, defined by the trench, is 3 µm and the width 50 µm.

The device was characterized electrically. Figure 12.10(*b*) shows the collector leakage current, with both source and drain grounded, as a function of the collector bias V_C for different temperatures. The collector-channel diode represents a forward-biased *p-n* junction, with a barrier in between. Because of the different discontinuities in the valence and conduction bands ($\Delta E_V = 0.2$ eV and $\Delta E_C = 0.5$ eV), the leakage current

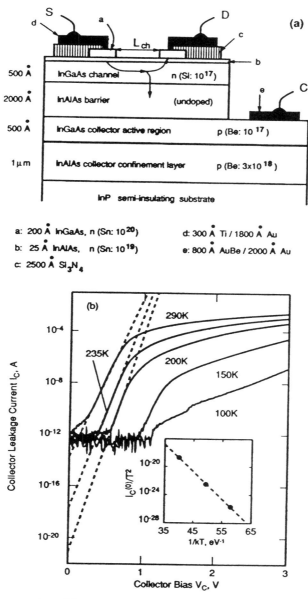

Fig. 12.10 (*a*) Cross section of the sample structure. The real-space transfer current is indicated by the downward arrow. (*b*) Collector current-voltage characteristics (source and drain grounded at different temperatures. Dashed lines indicate the linear extrapolation of I_C to zero forward bias. The intercepts of the dashed lines with the ordinate axis correspond to $I_C^{(0)}$. Inset shows an Arrhenius plot of $I_C^{(0)}/T^2$. [Mastrapasqua, M., F. Capasso, S. Luryi, A. L. Hutchinson, D. L. Sivco, and A. Y. Cho, *Appl. Phys. Lett.* **60**, 19 (1992): 2415–2417]

is mainly due to the transport of holes from the collector to the channel. At high temperatures, $200 \leqslant T \leqslant 300$ K, and a relative low bias, the current obeys the thermionic model $I_C = I_C^0 \exp(qV_C/nkT)$, with an ideality factor $n = 1.4$ and a saturation current $I_C^0 = SA^*T^2 \exp(-\Phi/kT)$, where $S = 10^{-5}$ cm^2 is the total emitter area including the source and drain contacts, Φ is the barrier height at zero bias, and A^* is the effective Richardson constant. The inset to Fig. 12.10(a) shows an Arrhenius plot of $I_C^{(0)}/T^2$ versus T^{-1} in the high-temperature range. The slope of this plot gives $\Phi = 0.90$ eV, which is in agreement with the calculation [Mas92].

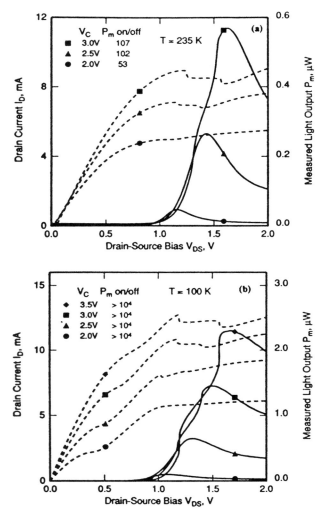

Fig. 12.11 The total measured light power P_m (*solid line*). versus the drain-source bias V_{DS} at $T = 235$ K^m(a) and $T = 100$ K (b). Dashed lines indicate the drain current I_D. The light power on/off ratio is calculated by dividing the maximum value of P^m by its minimum value before the onset of real-space transfer. [Mastrapasqua, M., F. Capasso, S. Luryi, A. L. Hutchinson, D. L. Sivco, and A. Y. Cho, *Appl. Phys. Lett.* **60**, 19 (1992): 2415–2417]

At high V_C the top of the barrier for holes is at the collector interface. This is the operation regime of a complementary CHINT. The barrier height further decreases with V_C because of the accumulation of holes and due to thermally assisted tunneling. At low temperatures, $T \leqslant 150\,K$, thermally assisted tunneling of holes is the dominant leakage mechanism.

The luminescence arising from the recombination of injected electrons in the p-type collector was detected from the back of the polished substrate using a liquid nitrogen-cooled Ge detector and a 0.75-nm spectrometer. Figure 12.11 shows the bias dependent of both the drain current I_D and the measured light power P_m. Due to the symmetry property of the CHINT, the output optical power depends only on the magnitude of the heating bias V_{DS}—and not on the polarity. The device exhibits an exclusive **OR** dependence of the emitted light power on the input voltages regarded as logic signals, $P_m = \text{XOR}(V_S, V_D)$. The device has an extremely high light-power on/off ratio, which confirms that the collector leakage is mostly due to the injection of holes into the channel [Mas92]. To maximize the optical on/off ratio, it is essential to suppress the leakage of electrons into the active region; the oppositely directed flux of holes can be tolerated.

12.3 QUANTUM DEVICES

Advances in nanolithography using electron-beam techniques have allowed a great variety of quantum devices to be fabricated. Fine-line patterning with dimensions in the nanometer scale is essential for these devices because their critical feature size has to be comparable to certain characteristic dimensions such as the Fermi wavelength λ_F (de Broglie wavelength of the electrons at the Fermi level $\lambda_F = \sqrt{2\pi/n_s}$, n_s is the carrier concentration) [Sch88], the elastic scattering length l, and inelastic scattering length L_Φ. Figure 12.12 shows some typical values of the Fermi wavelength, the elastic and inelastic scattering length for metal, Si-MOSFETs, and GaAs HEMTs at 1 K [Sch88].

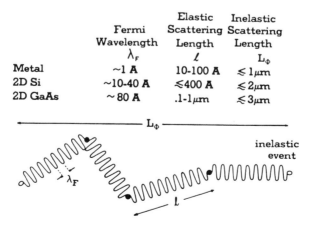

	Fermi Wavelength λ_F	Elastic Scattering Length l	Inelastic Scattering Length L_Φ
Metal	~1 Å	10-100 Å	$\leqslant 1\,\mu m$
2D Si	~10-40 Å	$\leqslant 400$ Å	$\leqslant 2\,\mu m$
2D GaAs	~80 Å	$.1$-$1\,\mu m$	$\leqslant 3\,\mu m$

Fig. 12.12 Typical values for the Fermi wavelength and for elastic and inelastic scattering lengths. [Schmid, H., S. A. Rishton, D. P. Kern, S. Washburn, R. A. Webb, A. Kleinsasser, T. H. P. Chang, and A. Fowler, *J. Vac. Sci. Tech.* B **6**, 1 (1988): 122]

The type of experiment to be carried out determines which of these dimensions is important for the critical features of the design. For ballistic experiments the drift length should be much smaller than the elastic scattering length. For the investigation of quantum transport in single wires, the most important dimension is the length of the wire, which has to be smaller the L_Φ. Resonant scattering can only be observed if the period of the lattice is between the Fermi wavelength λ_F and the elastic scattering length l. For quantum-scattering experiments at a single potential barrier, the width of the barrier ought to be in the range of the Fermi wavelength. For Aharonov-Bohm studies one critical dimension is the perimeter of the ring structures, which should be less than the inelastic scattering length. The ratio of the linewidth to ring diameter is another important factor that has to be considered in the design of Aharonov-Bohm structures. The field enclosed by the ring produces periodic oscillations in the resistance, while the field piercing the wires produces aperiodic fluctuations. The characteristic field scales for the two effects are determined by the areas of the hole and the wires.

A series of experiments on quantum effects in metals and semiconductors have been conducted using a high-resolution electron-beam systems [Coa82; Ker83]. This section highlights some of these experiments in Aharonov-Bohm studies. Superconducting weak links and quantum effects in GaAs will be discussed with special attention given to the fabrication processes employed.

The study of microstructures as they relate to the Aharonov-Bohm [Aha59] effect has been gathering momentum over recent years. Consider a ring structure of the test structure shown in Fig. 12.13 [Sch88]. Suppose that the loop is small enough for electrons to preserve their quantum phase memory when they take one side or the other. Then one might expect quantum interference if a magnetic field threads the loop. Accordingly an extra phase factor $\exp(\pm i\pi HS/\Phi_0)$ can be introduced to transit around the loop, with H the magnetic field, S the loop area, and $\Phi_0 = hc/2e$ the quantum of magnetic flux.

Investigations of the impact of a magnetic vector potential $\mathbf{A}$ on the phase of the electron-wave function have been conducted in a number of experiments with electrons in vacuum [Cha59; Möl62; Möl82; Ton86], in superconductors [Dea61; Dol61; Par64], and in normal metals [Was86]. In these experiments the path of electrons is commonly split into two branches and then recombined, thus forming a closed-loop structure. If

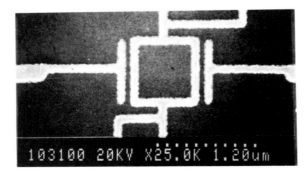

Fig. 12.13 Micrograph of the test structure used for the investigation of the electrostatic Aharonov-Bohm effect. [Schmid, H., S. A. Rishton D. P. Kern, S. Washburn, R. A. Webb, A. Kleinsasser, T. H. P. Chang, and A. Fowler, *J. J. Vac. Sci. Tech.* B, **6**, 1 (1988): 122]

coherence is maintained between the two paths, interference effects can be observed. The magnetic flux enclosed by the loop shifts the relative phase of the partial waves according to $\Delta\Phi = e/h \int \mathbf{A} d\mathbf{s}$. A phase shift of 2π is achieved by an enclosed flux of h/e (or $h/2e$ for superconductors). In metal loops this phase shift results in conductance fluctuations of the device. This experiment proves to be a powerful tool for the study of electron transport in solids.

The impact of electrostatic potentials on the phase of the wavefunction (electrostatic Aharonov-Bohm effect) has been studied with electrons in vacuum [Sch84]. An electron potential contributes to the fourth term of the fourth vector product $A_\mu dx_\mu$. This term contains the scalar potential V associated with electric fields and time. The effect of the electrostatic potential is to cause a phase shift in the wave function of $\Delta\Phi = e/h \int V dt$. A potential difference between the two paths causes the partial waves following those two paths to arrive at the recombination point with different phase.

To investigate the impact of electric fields on the phase of electrons in metal a test structure was fabricated, as shown in Fig. 12.13. It consists of a square closed loop with dimensions of 0.8×0.8 μm and a linewidth of 70 nm at close proximity ($= 50$ nm) to the two opposite legs of the loop. A voltage of opposite polarities can be applied to these electrodes to introduce the electrostatic effects. In addition a magnetic field can be applied normal to the loop to produce magnetically induced conductance fluctuations (h/e oscillations). This structure is fabricated on a plane silicon substrate by electron-beam lithography at 50 kV. The exposure was done in a double-layer polymethylmethacrylate (PMMA) resist [Bea81] with a total thickness of about 160 nm, followed by a liftoff of 60-nm-thick Sb.

Experiments have been carried out with this test structure whereby both a magnetic and an electric field were applied at different combinations of intensities. In the absence of electric field, the well-established h/e oscillation of conductance is observed, as shown in Fig. 12.14, when the external magnetic field is increased. With the application of the electric field, a phase shift of the h/e oscillation was observed which increases with the increasing electric field. As can be seen in the figure, a phase shift of π was obtained with a voltage difference of 0.75 V applied at the two electrodes. Removing the electric field reverts the phase to the initial position. The expected voltage difference

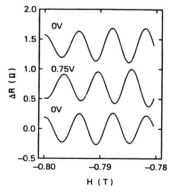

Fig. 12.14 Electrostatically induced phase shifts of the h/e oscillations. [Schmid, H., S. A. Rishton, D. P. Kern, S. Washburn, R. A. Webb, A. Kleinsasser, T. H. P. Chang, and A. Fowler, *J. Vac. Sci. Tech.* B, **6**, 1 (1988): 122]

required to produce mutual phase shift of 2π is $U = 2\pi\hbar/e\tau = 50\,\mu\text{V}$, which is much smaller than the value observed. This large discrepancy between the expected value for the potential difference that would cause a phase shift of 2π and the observed one is not surprising. The potential difference at the two electrodes is highly screened within the metal loop. The electrostatic potential cannot extend into the metal more than the screening length ($l_s = 2\,\text{Å}$), over which distance is decays exponentially. The phase shift of the electron wave function would then be accumulated only when the electrons move within a screening length of the edge of the arms of the loop.

The effect of the classical electrostatic force on the electron trajectories is also

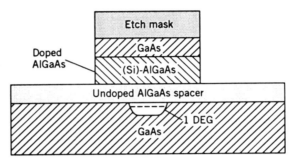

Fig. 12.15 Schematic cross section through the conducting channel of the GaAs wire. The one-dimensional electron gas (1DEG) is formed in the GaAs substrate by electrons transferred from the doped AlGaAs overlayer. Lateral confinement is from the potential at the exposed AlGaAs surfaces at the bottom and sides of the etched mesa. [Mankiewich, P. M., *J. Vac. Sci. Tech* **6**, 1 (1988): 131]

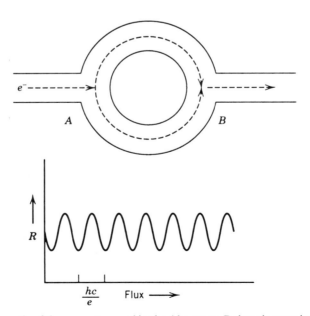

Fig. 12.16 Schematic of the geometry used in the Aharanov-Bohm electron interference experiment. [Mankiewich, P. M., *J. Vac. Sci. Tech.* **6**, 1 (1988): 131]

expected to have a role. The voltage from the capacitive electrodes rearranges the carrier trajectories inside the arms of the loop. These changes in the carrier distribution can cause random phase shifts as the trajectories are shifted.

The ring-shaped structure fabricated on GaAs substrate can also exhibit the Aharonov-Bohm effect [Man88]. A cross section through the conducting channel of the GaAs wire is shown in Fig. 12.15. The geometry used in the Aharonov-Bohm electron interference experiment is shown in Fig. 12.16. The relative phase between the two channels can be changed by applying an external magnetic field through the ring. Then the oscillations in the magnetoresistance that occur as the relative phase changes by 2π for each flux quantum (hc/e) included within the area of the ring can be measured.

12.4 QUASI-ONE-DIMENSIONAL CHANNEL DEVICES

In a one-dimensional electronic system the motion of electrons normal to the channel is quantized, resulting in size-quantization effects in the transport properties of electrons. In 1980 Sakai [Sak80] theoretically analyzed the electron transport in such a system and showed the possibility of suppressed elastic scattering, thus increasing carrier mobility. More recently Yamada [Yam88] simulated the high-field electron transport in such a system by solving the Boltzmann equation and made some observations on electron energy confinement and the possibility of electron velocity runaway [Leb84].

In this section we discuss the results of conductance experiments on a corrugated

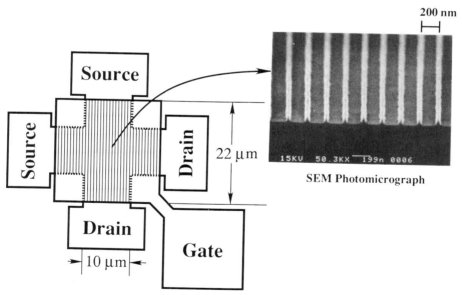

Fig. 12.17 Top view of the corrugated-gate GaAs/AlGaAs HEMT in which the electron gas confinement has been changed from a conventional two-dimensional electron gas channel to an array of one dimensional electron gas channels. [Okada, M., T. Ohshima, M. Masuda, N. Yokoyama, and A. Shibatomi, *Ext. Abs. 20th Conf. Solid State Dev. and Mat.*, pp. 503–506, Tokyo (1988)]

gate GaAs/AlGaAs modulation-doped FET in which the electron gas confinement has been changed from a conventional two-dimensional electron gas channel to an array of one-dimensional electron gas channels [Oka88]. Figure 12.17 shows a top view of the proposed device structure. It consists of a corrugated gate and two source-drain pairs. The corrugated gate structure was formed on a GaAs/AlGaAs modulation-doped heterostructure grown by MBE.

Figure 12.18 shows the epilayer structure of the device. It contains no spacer layer. The Hall mobility of the two-dimensional electron gas confined in the structure was measured as a 2×10^4 cm^2/V-s at 77 K, with a sheet-carrier concentration of 1×10^{12} cm^{-2}.

Figure 12.19 gives a cross section of the corrugated gate structure. When a negative gate voltage is applied, the two-dimensional electron gas changes into quasi-one-

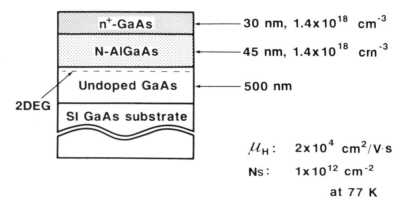

Fig. 12.18 The epilayer structure of the corrigated-gage device. [Okada, M., T. Ohshima, M. Matsuda, N. Yokoyama, and A. Shibatomi, *Ext. Abs. 20th Conf. Solid State Dev. and Mat.*, pp. 503–506, Tokyo (1988)]

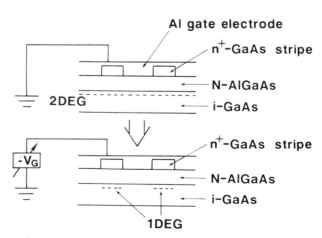

Fig. 12.19 Schematic cross section of corrugated-gate device. [Okada, M., T. Ohshima, M. Matsuda, N. Yokoyama, and A. Shibatomo, *Ext. Abs. 20th Conf. Solid State Dev. and Mat.*, pp. 503–506, Tokyo (1988)]

QUASI-ONE-DIMENSIONAL CHANNEL DEVICES 465

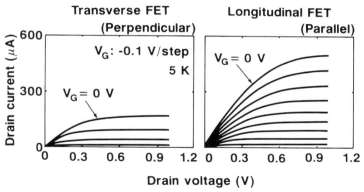

Fig. 12.20 I–V characteristic of transverse and longitudinal FETs at 5 K. [Okada, M., T. Ohshima, M. Matsuda, N. Yokoyama, and A. Shibatomi, *Ext. Abs. 20th Conf. Solid State Dev. and Mat.*, pp. 503–506, Tokyo (1988)]

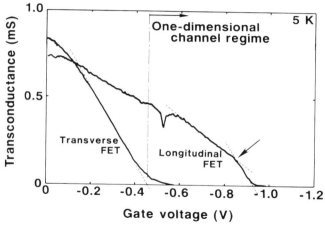

Fig. 12.21 Transconductance versus gate voltage at 5 K. [Okada, M., T. Ohshima, M. Matsuda, N. Yokoyama, and A. Shibatomi, *Ext. Abs. 20th Conf. Solid State Dev. and Mat.*, pp. 503–506, Tokyo (1988)]

dimensional electron gases, because the two-dimensional electron gas is depleted at those areas without n^+-GaAs cap stripes.

Figure 12.20 shows the current-voltage characteristics of a transverse (having the channel perpendicular to the n^+-GaAs stripes) and a longitudinal (having the channel parallel to the n^+-GaAs stripes) FETs. The data were measured at 5 K. The drain currents for these FETs completely saturate for drain voltages greater than 0.9 V. The gate length of these devices is 22 μ, as shown in Fig. 12.17.

Figure 12.21 plots the transconductance measured as a function of the gate voltage for drain voltage of 0.9 V. The transconductance of the transverse FET decreases with increasing negative gate voltage and becomes zero at around −0.45 V. The transconductance of the longitudinal FET has finite values for negative gate voltages

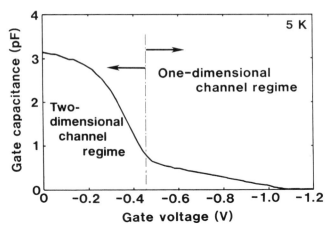

Fig. 12.22 Gate capacitance versus gate voltage at 5 K. [Okada, M., T. Ohshima, M. Matsuda, N. Yokoyama, and A. Shibatomi, *Ext Abs. 20th Conf. Solid State Dev. and Mat.*, pp. 503–506, Tokyo (1988)]

at around -0.45 V. This indicates that the channel changes from two-dimensional to quasi-one-dimensional at around -0.45 V. This result was confirmed by gate capacitance measurements, which are discussed below.

Figure 12.22 plots the gate capacitance against the gate voltage for this device measured at a frequency of 1 MHz. The capacitance decreases sharply with increasing negative gate voltage from -0.2 to -0.45 V, then monotonically decreases with increasing negative gate voltage. This further suggests that the transition from a two-dimensional channel to a one-dimensional channel occurs at around -0.45 V, and that the width of the one-dimensional channel decrease monotonically from -0.45 V to -1.0 V as the negative gate voltage varies.

In the transconductance of a longitudinal FET in a longitudinal FET in the one-dimensional channel regime, there is a critical change in the slope of the transconductance curve (K value) at around -0.85 V. The capacitance monotonically decreased from -0.45 to -1.0 V, as shown in Fig. 12.22.

Since the gate length of the device is 22 μm, a gradual channel approximation can be employed. The transconductance g_m in the saturation region (the source-drain voltage was 0.9 V) is thus approximated by

$$g_m = 2K(V_G - V_T),$$

$$K = \frac{\mu C_G}{2L_G^2}, \tag{12.2}$$

where V_G is the gate voltage, V_T the threshold voltage, C_G the gate capacitance of this device, and μ_F the field-effect mobility. The slope of the transconductance (K value) is proportional to the product of the gate capacitance and the field-effect mobility. The slope of the transconductance (K value) is proportional to the product of the gate capacitance and the field-effect mobility. Considering the monotonic decrease of the gate capacitance around -0.85 V, the critical change in the longitudinal transconduct-

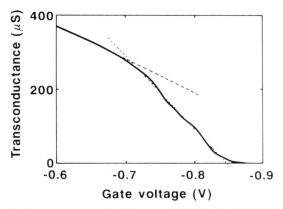

Fig. 12.23 Transconductance versus gate voltage at 5 K. [Okada, M., T. Ohshima, M. Matsuda, N. Yokoyama, and A. Shibatomi, *Ext. Abs. 20th Conf. Solid Sate Dev. and Mat.*, pp. 503–506, Tokyo (1988)]

ance slope (K value) is caused by the enhancement of the field-effect mobility. The field-effect mobility after the change in slope is estimated from the graph to be about three times that before the change in slope.

Figure 12.23 shows the transconductance of the longitudinal FET in the one-dimensional regime for a different sample. The similar critical change in slope is

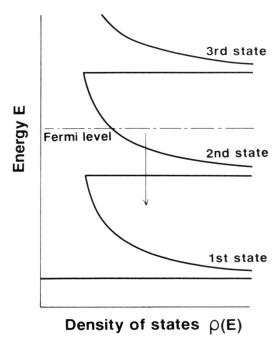

Fig. 12.24 Density of states in a strictly defined one-dimensional electronic system. [Okada, M., T. Ohshima, M. Matsuda, N. Yokoyama, and A. Shibatomi, *Ext. Abs. 20th Conf. Solid State Dev. and Mat.*, pp. 503–506, Tokyo (1988)]

observed at around -0.7 V for this sample. Staircaselike oscillations are clearly visible for the negative gate voltage region from -0.7 to -0.9 V.

The oscillatory behavior can be understood as caused by the one-dimensional quantization of the channels as follows: For one-dimensional electronic system, the density-of-states (DOS) is given by

$$\rho(E) = \left[\frac{m^*}{2\hbar^2\pi^2}\right]^{1/2} \sum_{l,m} \frac{1}{\sqrt{(E - E_l - E_m)}}, \qquad (12.3)$$

as shown in Fig. 12.24. When a negative gate voltage is applied, the Fermi level descends due to decreased carrier density, while the one-dimensional energy levels and intervals between them increase due to the decreased width of the one-dimensional channels. These indicate that the Fermi level passes through successive one-dimensional levels as the negative gate voltage increases. When the Fermi level passes the minimum DOS, the available final states are more limited. Therefore the scattering rates change substantially, and the field-effect mobility and transconductance become oscillatory.

In the oscillatory regime the electrons must be strictly defined into one-dimensional channels. The enhancement of the field-effect mobility is thought to be caused by the transitions from loosely defined one-dimensional channels to strictly defined one-dimensional channels.

12.5 QUANTUM INTERFERENCE DEVICES

In the field of optics, there exist several representative systems that are based on the interference of two or more wave components [Sak89]. As in the Fabry-Perot (FP) interferometer, a pair of parallel mirrors reflect optical waves back and forth and give rise to resonant states and a novel transmission characteristics. This situation is similar to the formation of standing-wave states in quantum well structures and the resonant tunneling of electrons through double-barrier structures (see Chapter 11). In addition the reflection or diffraction of optical waves in multilayered mirrors (MM) or other periodic structures (holograms and grating) is conceptually similar to the Bragg reflection of electron waves in superlattices.

One could form an electron wave analogue of the Michelson interferometer or the Mach-Zehnder interferometer (the electron phase in one is changed by an applied

TABLE 12.1 Representative Systems

Multilayered Superstructures	Lateral Superstructures
1. Superlattices 2. Resonant tunneling through quantum well	1. Lateral planar superlattices 2. Single-mode quantum wires 3. Multiple-mode quantum wires 4. Quantum wire ring (multiple mode) 5. Multiple branched quantum wires of Michelson type 6. Coupled quantum box arrays 7. Quantum boxes

Source: Sakaki, H., *Ext. Abs. 21st. Conf. on Solid State Dev. and Mat.*, Tokyo (1989): 537–540.

electric field) by utilizing some kind of electron waveguide structure that allows electron propagation along two or more different paths in order to generate an appropriate phase difference. However, one must be aware of the conceptual difference between quantum interference devices and lateral surface superlattice structures, such as quantum wires and boxes. Table 12.1 summarizes several representative systems, where various interference phenomena play some role [Sak89]:

The quantum interference device inspired by recent advances in the physics of mesoscopic structures presents a new class of quantum semiconductor devices [Cap90]. This Section discusses the mesoscopic structure and the examples of the quantum interference devices.

12.5.1 Novel Properties and Geometrical Properties of LSSL

Sakai [Sak89] pointed out that quantum effects associated with lateral confinements in quantum wires and quantum boxes can be observed as long as the following condition is satisfied:

[C-1] The lateral dimensional L of the lateral surface superlattice (LSSL) should be small enough that the energy-level separation

$$E(2) - E(1) = 3\left(\frac{h^2}{2m}\right)\left(\frac{\pi}{L}\right)^2$$

is far larger than the level broadening DE, which is determined either by the carrier scattering or by the inhomogeneities of the LSSL system.

This condition is nearly identical to the condition that the coherence length of electrons l_c be longer than L, and can be easily satisfied as long as the scattering is suppressed by cooling the sample at low temperatures, where l_c is known to exceed 0.1 μm.

However, the condition [C-1] will not be sufficient if we want to observe or utilize some novel features of LSSL that can be exploited in device applications. Those features appear only when the following two conditions are additionally satisfied:

[C-2] The level that separates $E(2) - E(1)$ should be greater than the average Fermi energy E_F (or thermal energy kT_e) of carriers so that nearly all carriers are accommodated at the ground level.

[C-3] The quantum level must be sharp with $DE \ll$ carrier energy E_f or kT_e so that the energy distribution of carriers $n(E)$ gets much sharper in quantum wire/box systems.

In most device applications, all three conditions need to be satisfied.

12.5.2 Mesoscopic Structures

The inelastic mean-free path, which is the distance electrons move on average without losing energy, is determined in semiconductors by electron-phonon interaction and is typically on the order of 10 to 100 [Hes90]. A mesoscopic structure is one whose dimensions are small compared to the electron mean-free path L_ϕ for phase-destroying scattering. A phase-destroying scattering process is one in which the scattering changes its state.

Since 1985 numerous experiments have demonstrated that electron transport in mesoscopic structures is influenced by quantum interference effects similar to phenomena well known in microwave or optical networks [Tim87]. In large devices such interference effects are generally washed out by phase-destroying processes. Advances in nanofabrication [Smi90] have made it possible to build devices with lateral dimensions much smaller than one micron; the smaller dimensions reduce scattering within the device at low temperatures.

In the absence of phase-destroying process, the steady state transmission of electrons through a device is described by the one-electron Schrödinger equation

$$\nabla^2 \Phi(\mathbf{r}) = -\frac{2m}{\hbar^2}[E - V(\mathbf{r})]\Phi(\mathbf{r}). \tag{12.4}$$

Equation (12.4) is similar to the Helmholz wave equation from optics:

$$\nabla^2 E = \omega^2 \mu \varepsilon(\mathbf{r}) F, \tag{12.5}$$

where E is the electric field intensity of the electromagnetic wave, ω is the angular frequency, μ is the permeability, and ε is the spatially varying dielectric constant. A comparison of Eqs. (12.4) and (12.5) shows that electron waves move through a medium with a spatially varying potential in a manner analogous to the movement of light waves through a medium of varying refractive index. The electric field in the Helmholtz equation can be viewed as the wavefunction of a single photon, analogous to the electronic wavefunction in the Schrödinger equation.

To obtain large interference effects, it is usually necessary to restrict the range of wavelengths. In electromagnetics one commonly uses single-mode wave guides and single-frequency sources. Electron waves in solids, by contrast, commonly have a large spread in energy. The energy spread can be restricted by employing low temperatures and low voltages, so that one electron near the Fermi energy level contributes to the conductance. Quantum wires with few modes have become a reality [Wha88].

The mesoscopic structures resemble an electron waveguide. Unlike metal walls of the electromagnetic wave guides, it is the potentials that define the electron wave guides. The potentials can be easily controlled and changed, and so the new nanostructures represent "flexible" wave guides. A device with electronic wave guides was proposed independently by Fowler [Fow85] and Datta [Dat85]. Consider the T-shaped structure shown in Fig. 12.25. Suppose that the structure has metal boundaries and an electromagnetic wave is propagating in the horizontal arm. Then the transmission would be a function of the length L^* of the side stub, since these would be an interference of the waves in the horizontal and perpendicular parts of the structure. For a nanostructure—an electron waveguide—the effective length L^* of the sidearm can be changed by changing a potential. This would modulate, for example, the depletion width L_d of a Schottky barrier. In such an interferometer the gate potential shifts the subband energy E_n^0 in one arm relative to the other, causing a phase difference

$$\Delta \lambda = \Delta k L = \frac{\Delta E_n^0 \tau_T}{\hbar}. \tag{12.6}$$

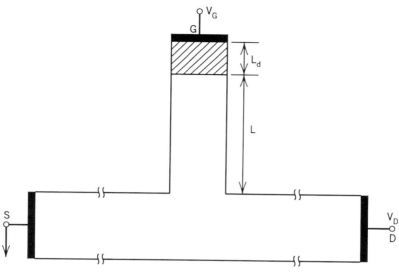

Fig. 12.25 Mesoscopic or quantum interference transistor. The transmission coefficient T between the source S and drain D of the device is a sensitive function of the electron energy E and the transverse stub length L^*, which may be adjusted by changing the depletion length L_d. [Hess, K., *Physics Today* **43**, 2 (1990): 34–42].

Here τ_T is the transit time of an electron across the region of length L. The expected periodic conductance modulation has not been observed. Because the phase shift is proportional to the transit time, the electrostatic effect is more difficult than its magnetic analogue—the Aharonov-Bohm effect, where the phase shift depends only on the flux enclosed by the two alternative paths.

12.5.3 Lateral-Surface-Superlattice Devices

The lateral-surface-superlattice (LSSL) HEMT is a variant of the ordinary single-gate HEMT. The periodic structures in LSSL devices were produced using X-ray nanolithography [Ism89]. In the X-ray system, C_K X-rays with a wavelength of 4.5 nm were used.

Figure 12.26 gives a cross section of the grating-gate device [Ism88]. The layers were prepared by molecular beam epitaxy. Figure 12.27 gives the schematic top view of another grating HEMT; the gate length is 15 µm and the gate width is 20 µm, defined by mesa etching [Ism89]. In the LSSL HEMT, electrons traveling from source to drain fall under the influence of periodic potential barriers, caused by the extension of the depletion region under the gate fingers. The effect of this potential is to create an array of quantum wells having quantized energy levels separated by energy gaps. As the potential on the gate is raised, the electron Fermi wavelength decreases. Back diffraction from the periodic potential occurs whenever an integral number of half-wavelengths is equal to the period of the grid. The plot of drain current as a function of gate voltage is shown in Fig. 12.28. Each region of the negative slope corresponds to the Fermi level passing through a minigap.

The gate potential controls the depth of modulation. At sufficiently negative gate

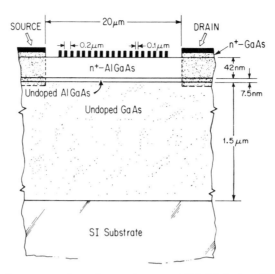

Fig. 12.26 Schematic cross section of a grating-gate HEMT device. The channel width and length are both 20 μm. An undoped GaAs/AlGaAs superlattice was used to trap impurities' diffusion out of the substrate. [Ismail, K., W. Chu, D. A. Antoniadis, and H. I. Smith, *Appl. Phys. Lett.* **52**, 13 (1988): 1071]

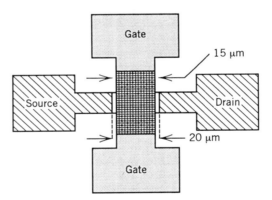

Fig. 12.27 Schematic top view of a LSSL HEMT. The gate length is 15 μm as indicated, and the gate width is 20 μm, defined by the mesa etching. [Ismail, K., W. K. Chu, A. Yen, D. A. Antoniadis, and H. I. Smith, *Appl. Phys. Lett.* **54** (1989) 460]

voltage the two-dimensional electron gas is depleted under the gate metal, leaving an array of isolated, parabolic potential wells, or "quantum dots," each containing only a few electrons. The electronic states of these quantum dots have been probed by far-infrared magnetoabsorption [Liu89] and magnetocapacitance [Ism89b] techniques that make use of the dependence of absorption and capacitance on the magnetic field.

Kotthaus et al. [Kot87] have used holographic lithography to make grating structures with periods of a few hundred nanometers directly on semiconductor substrates and have studied the optical and transport properties of the electrons confined in the

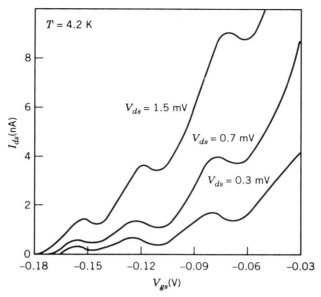

Fig. 12.28 I_{DS} as a function of V_{GS} for three different V_{DS} values. [Ismail, K., W. K. Chu, A. Yen, D. A. Antoniadis, and H. I. Smith, *Appl. Phys. Lett.* **54** (1989): 460]

resulting quantum wires. Reed et al. [Ree88] have fabricated arrays of quantum dots by etching columns less than 100 nm in diameter down through a two-dimensional electron gas in a multilayer compound semiconductor. The notion of arrays of quantum dots interacting in a predictable way with their near neighbors may find application in massively parallel computation.

12.6 QUANTUM POINT CONTACTS

In mesoscopic or nanofabrication regions, the electron moves ballistically and coherently, behaving both as a particle and a wave. With regard to the wave properties of the electron for electron devices, we must ask whether the electron wave source injects a wave with a controlled phase or mode. Quantum point contacts [Wee88, Wha88] are suitable for the wave source injector.

In the case of a point contact as the injector, Molenkamp et al. [Mol90] demonstrated the angular distribution of electrons and explained their single-peak profile using the collimation of electrons having a ballistic trajectory. Kriman et al. [Kri89] calculated the electron diffraction using a narrow slit and demonstrated the Fraunhofer diffraction.

Okada et al. [Oka92] presented the experimental results and theoretical analyses of the angular distribution of electrons injected through a single quantum point contact (QPC). Using a modified Fraunhofer diffraction approximation, they explained the double peaks in the distribution as a point contact quantized in two modes. Theoretical calculations by way of Green's function and mirror images in weak magnetic fields show that the distribution for an injector with two modes agrees well with experiment.

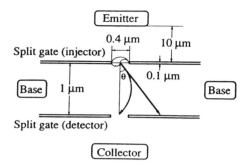

Fig. 12.29 The top view of the grid-gate device. [M. Okada, M. Saito, M. Takatsu, P. E. Schmidt, K. Kosemura, and N. Yokoyama, *Semicond. Sci. Tech.* 7 (1972): B223–B227]

Regarding double slits, Okada et al. fabricated submicron air bridges for the slits and showed the controlled additivity of quantum point contacts. This result suggests that air bridges are suitable for controlling the widths of quantum point contacts independently and are useful for measuring interference by double slits.

12.6.1 Angular Distribution for a Single Quantum Contacts

Figure 12.29 shows the top view of the device. It consists of an emitter, base, and collector with two split Schottky gates, each 0.1 µm long. The split is 0.4 µm wide. One split gate, between the emitter and base, acts as an electron injector. The other, between the base and collector, acts as an electron detector. The injector and detector are 1 µm apart. These structures are formed on an MBE-grown GaAs/AlGaAs modulation-doped heterostructure. The mean-free path of electrons is 2.9 µm, and a Fermi wavelength of 42 nm was obtained by Hall measurement data. Since the mean-free path exceeds the 1 µm between the injector and detector, electrons can move ballistically and behave as coherent waves. The injector point contact is quantized by applying negative voltage to the gate. When the injector has one (two) occupied energy level(s), there is/are one (two) mode(s) for electron waves in the contact, respectively. Electrons are injected into the base region by applying an emitter-base voltage.

The collector current was measured as a function of the magentic field applied perpendicularly to the sample. The magnetic field is less than 0.18 T, and the cyclotron diameter between the injector and detector exceeds 1 µm. Figure 12.30 shows the quantized conductance of the electron injector at 0.35 K. Conductance was measured for an emitter-base voltage of less than 30 µV without a magnetic field. It increases with the gate voltage in $2e^2/h$ steps. The injector has one occupied energy level at about -0.28 V and two occupied energy levels at about -0.195 V. When the magnetic field is less than 0.18 T, conductance steps do not change, indicating that the number of modes in the injector does not change. Figure 12.31 gives the measured angular distribution at 0.35 K for an injector with one mode at a gate voltage of -0.28 V and an injector with two modes at a gate voltage of -0.195 V. The distribution for one mode as a single peak and that for two modes has double peaks.

Okada et al. [Oka92] calculated the angular distribution of electrons through a single slit using a modified version of Fraunhofer diffraction, as shown in Fig. 12.32.

QUANTUM POINT CONTACTS   475

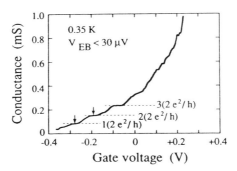

Fig. 12.30 Quantized conductance of the electron injector. Arrow show the measurement gate voltages of angular distribution. [M. Okada, M. Saito, M. Takatsu, P. E. Schmidt, K. Kosemura, and N. Yokoyama, *Semicond. Sci. Tech.* **7** (1992): B223–B227]

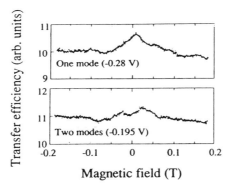

Fig. 12.31 Measured angular distribution at 0.35 K for the injector with one mode at a gate voltage of −0.28 V and the injector with two modes at a gate voltage of −0.195 V. [M. Okada, M. Saito, M. Takatsu, P. E. Schmidt, K. Kosemura, and N. Yokoyama, *Semicond. Sci. Tech.* **7** (1992): B223–B227]

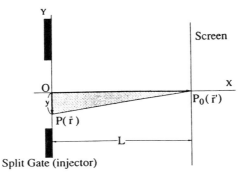

Fig. 12.32 Coordinate system for Green's function. [M. Okada, M. Saito, M. Takatsu, P. E. Schmidt, K. Kosemura, and N. Yokoyama, *Semicond. Sci. Tech.* **7** (1922): B223–B227]

Since the wave is quantized in the slit, the amplitude of the wave was obtained at the injector exit with a cosine function for the first mode and a sine function for the second mode. For homogeneous and weak magnetic fields, we compute the relation between the weak magnetic Green's function G^+ and nonmagnetic field G^0:

$$G^+ = \exp(i\varphi(\mathbf{r}',\mathbf{r}))G^0. \tag{12.7}$$

A weak magnetic field means that the cyclotron diameter is longer than the distance. In Eq. (12.7), φ satisfies

$$\varphi(\mathbf{r}',\mathbf{r}) = -\int_r^{r'} \frac{2\pi e}{h} \mathbf{A}(\mathbf{R})\mathbf{u}_t(\mathbf{R})dt, \tag{12.8}$$

$$\mathbf{u}_t(\mathbf{R}) = \frac{\nabla_{r'} G^0(\mathbf{r}',\mathbf{r})}{|\nabla_{r'} G^0(\mathbf{r}',\mathbf{r})|}. \tag{12.9}$$

In Eq. (12.9), the line integral path is taken in the direction of the gradient of G^0. When we consider the phase difference between P to P_0 and 0 to P_0, the effect of weak magnetic fields is expressed by the magnetic flux through the triangle enclosed by P, P_0, and 0. The wavefunction is derived at an arbitrary point by combining the wavefunction at the boundaries and Dirichlet's boundary condition of Green's function,

$$\psi(\mathbf{r}') = \frac{\pi h}{m^*}\int_s d S \mathbf{n} \Psi(\mathbf{r}_S) \cdot \nabla G_D^+(\mathbf{r}',\mathbf{r}_S), \tag{12.10}$$

where $\mathbf{n}$ is the normal vector at the boundary S. The injector split gate and the exit of the injector QPC are boundaries. For a nonmagnetic field a suitable Green's function is easily obtained using mirror images. However, in weak magnetic fields we need to determine φ, which includes the vector potential $\mathbf{A}$ and the path of the integral. Since reciprocity should still hold in weak magnetic fields, the path of the integral should be taken as shown in Fig. 12.33 and the Green's function is expressed as follows:

$$G_D^+(\mathbf{r}',\mathbf{r}) = \exp(i\varphi_1(\mathbf{r}',\mathbf{r}))G^{\text{Free}}(\mathbf{r}',\mathbf{r}),$$
$$- \exp(i\varphi_2(\mathbf{r}',\mathbf{r}))G^{\text{Free}}(\mathbf{r}',\mathbf{r}'') \tag{12.11}$$

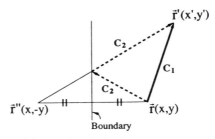

Fig. 12.33 Mirror images and integral paths in homogeneous and weak magnetic fields. [M. Okada, M. Saito, M. Takatsu, P. E. Schmidt, K. Kosemura, and N. Yokoyama, *Semicond. Sci. Tech.* 7 (1992): B223–B227]

with φ_1 and φ_2 expressed as follows:

$$\varphi_1(\mathbf{r}',\mathbf{r}) = -\int_{C_1} \frac{2\pi e}{h} \mathbf{A}(\mathbf{R})\cdot\mathbf{u}_t(\mathbf{R})dt, \qquad (12.12)$$

$$\varphi_2(\mathbf{r}',\mathbf{r}) = -\int_{C_2} \frac{2\pi e}{h} \mathbf{A}(\mathbf{R})\cdot\mathbf{u}_t(\mathbf{R})dt, \qquad (12.13)$$

where $\mathbf{r}''$ is the mirror image of $\mathbf{r}$, and paths C_1 and C_2 are as shown in Fig. 12.33. Using these paths guarantees reciprocity.

Figure 12.34 shows the distribution for an injector with one mode and an injector with two modes. The calculated distribution for the two modes agrees well with the experiment, assuming a width of 150 nm. This suggests the calculations support the

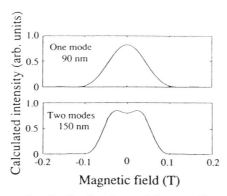

Fig. 12.34 Calculated angular distribution for an injector with two modes. [M. Okada, M. Saito, M. Takatsu, P. E. Schmidt, K. Kosemura, and N. Yokoyama, *Semicond. Sci. Tech.* 7 (1992): B223–B227]

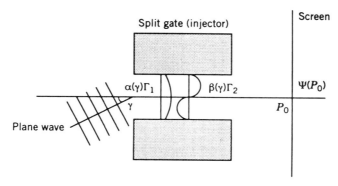

Fig. 12.35 Schematic for the incident wave, wave function in the injector. [M. Okada, M. Saito, M. Takatsu, P. E. Schmidt, K. Kosemura, and N. Yokoyama, *Semicond. Sci. Tech.* 7 (1992): B223–B227]

12.6.2 Electron Wave Interference

The variations in phase differ between first and second modes, since the x direction elements of the wavevector differ. The interference can be cancelled if the system is symmetrical. As shown in Fig. 12.35, the electron plane wave

$$\Phi_\gamma(\mathbf{r}) = C_\gamma \exp(i\mathbf{k}_\gamma \mathbf{r}) \tag{12.14}$$

comes from the γ direction and enters the slit, where it is divided into the first and the second modes,

$$\Omega = \alpha(\gamma)\Gamma_1(y)\exp(ik_x x) + \beta(\gamma)\Gamma_2(y)\exp(ik'_x x). \tag{12.15}$$

In this equation $\Gamma_1(y)$ and $\Gamma_2(y)$ are the y direction elements of the wavefunction, and $\alpha(\gamma)$ and $\beta(\gamma)$ the expansion coefficients; $\Gamma_1(y)$ is an even function, and $\Gamma_2(y)$ is an odd function from system symmetry. Higher-order wave elements decreases near the entrance of the slit because they have the negative energy. The intensity at detector P_0 is

$$|\psi(P_0)|^2 = |\alpha(\gamma)|^2|G_0[\Gamma_1]|^2 + |\beta(\gamma)|^2|G_0[\Gamma_2]|^2$$
$$+ \{\alpha^*(\gamma)\beta(\gamma)\exp[i(k_x - k'_x)d]G_0[\Gamma_1]^*G_0[\Gamma_2] + cc\}, \tag{12.16}$$

where $G_0[\Gamma]$ means the integration of Eq. (12.10). For the opposite angle $-\gamma$, given system, symmetry, $\alpha(-\gamma)$ and $\beta(-\gamma)$ are

$$\alpha(-\gamma) = \alpha(\gamma)\exp(i\theta),$$
$$\beta(-\gamma) = -\beta(\gamma)\exp(i\theta), \tag{12.17}$$

where θ is a constant given as follows:

$$C_{-\gamma} = C_\gamma \exp(i\theta). \tag{12.18}$$

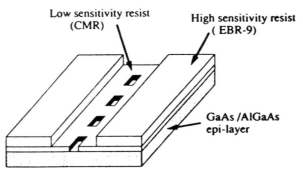

Fig. 12.36 Exposed and developed biresist and air bridges. [M. Okada, M. Saito, M. Takatsu, P. E. Schmidt, K. Kosemura, and N. Yokoyama, *Semicond. Sci. Tech.* **7** (1992): B223–B227]

The term of interference is canceled by adding the contributions from γ and $-\gamma$. This conclusion was derived by the symmetry of the system.

Another possibility of electron interference is caused by an electron injected through double slits. The electron phase is also controlled by the widths of the slits. If we control the widths of double slits independently by applied gate voltages, we can observe shifts in the peaks of diffraction. A process for submicron Schottky contacts interconnected by a submicron air bridge was developed [Sch91]. The air bridge has the advantage that isolated islands between the split gates do not float electrically and do not decrease the sheet-carrier density near the point contact. The process is based on the use of different sensitivity resists and different electron-beam densities. Figure 12.36 shows the exposed and developed biresist and air bridges. In this figure CMR is a low-sensitivity positive electron-beam resist, and the EBR-9 is a higher-sensitivity positive electron-beam resist. Exposure was made first in the rectangular holes, and another exposure was made on a line interconnecting all holes. Aluminum was then evaporated and liftoff done. Figure 12.37 shows the quantized conductance with four parallel point contacts. Figure 12.37(a) shows the quantized conductance by $2e^2/h$, Fig. 12.37(b) that by three times $2e^2/h$, and Fig. 12.37(c) that by four times $2e^2/h$. This result means that

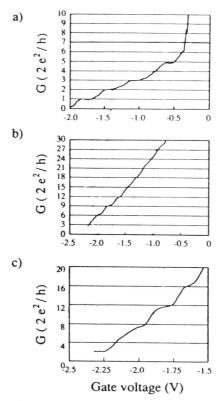

Fig. 12.37 Controlled quantized conductance with four parallel point contacts by gate voltages: (a) one-point contact; (b) three-point contacts; (c) four-point contacts. [M. Okada, M. Saito, M. Takatsu, P. E. Schmidt, K. Kosemura, and N. Yokoyama, *Semicond. Sci. Tech.* 7 (1992): B223–B227]

air bridges are effective in controlling the widths of double slits independently and help in mesoscopic fabrication.

REFERENCES

[Con67] Conwell, E. M., *High Field Transport in Semiconductors*, Academic Press (1967).

[Lur90] Luryi, S., in *High Speed Semiconductor Devices*, ed. by S. M. Sze, Wiley-Interscience, (1990).

[Sze81] Sze, S. M., *Physics of Semiconductor Devices*, Wiley-Interscience (1981).

[Sha79] Shannon, J. M., *IEE J. Electron Devices* **3** (1979): 144–149.

[Sha81] Shannon, J. M., and A. Gill, *Elect. Lett.* **17** (1981): 620–621.

[Hay85] Hayes, J. R., A. F. J. Levi, and W. Wiegmann, *Phys. Rev. Lett.* **54** (1985): 1570–1572.

[Mal81] Malik, R. J., M. A. Hollis, L. F. Eastman, C. E. C. Wood, D. W. Woodward, and T. R. AuCoin, *Proc. 8th Biennial Cornell Conf. on Active Microwave Semicond. Devices and Circuits*, August (1981).

[Mea61] Mead, C. A., *J. Appl. Phys.* **32** (1961): 646.

[Ata62] Atalla, M. M., and D. Kahng, IRE-AIEE Solid State DRC, University of New Hampshire, Durham (1962).

[Gep62] Geppert, D. V., *Proc. IRE* **50** (1962): 1527.

[Sze66] Sze, S. M., and H. K. Gummel, *Solid State Electron.* **9** (1966): 751.

[Cro66] Crowell, C. R., and S. M. Sze, *J. Appl. Phys.* **7** (1966): 2683.

[Hei85] Heiblum, M., D. C. Thomas, C. M. Knoedler, and M. I. Nathan, *Appl. Phys. Lett.* **47** (1985): 1105.

[Hol83] Hollis, M. A., S. C. Palmater, L. F. Eastman, N. V. Dandekar, and P. M. Smith, *IEEE Electron Dev. Lett.* **EDL-4** (1983): 440.

[Red86] Reddy, U. K., J. Chen, C. K. Peng, and H. Morkoc, *Appl. Phys. Lett.* **48** (1986): 1799.

[Has88] Hase, I., K. Taira, H. Kawai, T. Watanabe, K. Kaneko, and N. Watanabe, *Electron. Lett.* **24** (1988): 279.

[Sha79] Shannon, J. M., *IEE J. Electron Devices* **3** (1979): 144–149.

[Sha84] Shannon, J. M., *Inst. Phys. Conf. Ser.* **69** (1984): 45.

[Hay86] Hayes, J. R., and A. F. J. Levi, *IEEE J. Quantum Electronics* **QE-22**, 9 (1986): 1744.

[Hay86b] Hayes, J. R., *18th Conf. on Solid State Devices and Materials* (1986): 331–334.

[Hei81] Heiblum, M., *Solid State Electron.* **24** (1981): 343–366.

[Hei85b] Heiblum, M., M. I. Nathan, D. C. Thomas, and C. M. Knoedler, *Phys. Rev. Lett.* **55** (1985): 2200.

[Yok84] Yokoyama, N., K. Imamura, T. Ohshima, N. Nishi, S. Muto, K. Kondo, and S. Hiyamizu, *Jap. J. Appl. Phys.* **23** (1984): L311.

[Hei89] Heiblum, M., *1989 IEDM* (1989): 822–825.

[Hes90] Hess, K., *Physics Today* **43** (1990): p. 56

[Hes82] Hesto, P., J. F. Pone, and R. Castagne, *Appl. Phys. Lett.* **40** (1982): 405.

[Hay85] Hayes, J. R., *Phys. Rev. Lett.* **54** (1985): 1570–1572.

[Lev85] Levi, A. F. J., J. R. Hayes, P. M. Platzman, and W. Wiegmann, *Phys. Rev. Lett.* **55** (1985): 2071–2073.

[Seo88] Seo, K., M. Heiblum, C. M. Knoedler, W. P. Hong, and P. Bhattacharya, *Appl. Phys. Lett.* (1988).

[Has88] Hase, I., K. Taira, H. Kawai, T. Watanabe, K. Kaneko, and N. Watanabe, *Electron Lett.* **24** (1988): 279.

[Lur85a] Luryi, S., *IEEE Electron Dev. Lett.* **EDL-6** (1985): 178.

[Lur85b] Luryi, S., *Physica* B **134** (1985): 466.

[Lin88] Lin, W., M. D. Lei, C. Y. Chang, W. C. Hsu, L. B. Di, and F. Kai, *Jap. J. Appl. Phys.* (Dec. 1988): L2431–L2433.

[Miz85] Mizuta, M., M. Tachikawa, H. Kukimoto, and S. Minomura, *Jap. J. Appl. Phys.* **24** (1985): L143.

[Bab84] Baba, T., T. Mizutani, M. Ogawa, and K. Ohata, *Jap. J. Appl. Phys.* **23** (1984): L654.

[Hes85] Hess, K., *Festkörperprobleme* **25** (1985): 321.

[Lur85] Luryi, S., and A. Kastalsky, *Superlatt. Microstructures* **1** (1985): 389.

[Mas92] Mastrapasqua, M., F. Capasso, S. Luryi, A. L. Hutchinson, D. L. Sivco, and A. Y. Cho, *Appl. Phys. Lett.* **60**, 19 (1992): 2415–2417.

[Sch88] Schmid, H., S. A. Rishton, D. P. Kern, S. Washburn, R. A. Webb, A. Kleinsasser, T. H. P. Chang, and A. Fowler, *J. Vac. Sci. Tech.* B **6** (1988): 122.

[Coa82] Coane, P. J., D. P. Kern, A. J. Speth, and T. H. P. Chang, in *Proc. 10th Int. Conf. on Electron and Ion Beam Science and Technology*, p. 2, ed. by R. Bakish, Electrochemical Society (1982).

[Ker83] Kern, D. P., P. J. Houzego, P. J. Coane, and T. H. P. Chang, *J. Vac. Sci. Tech.* B **1** (1983): 1096.

[Aha59] Aharonov, Y., and D. Bohm, *Phys. Rev.* **115** (1959): 485.

[Cha59] Chambers, R. G., *Phys. Rev. Lett.* **5** (1959): 3.

[Möl62] Möllenstedt, G., and W. Bayh, *Physics* B **1**, 18 (1962): 299.

[Möl82] Möllenstedt, G., H. Schmid, and H. Lichte, In *Proc. 10th Int. Congress on Electron Microscopy*, vol. 1, p. 433, Deutsche Gesschäft für Electronenmikroskopie e.V., Frankfurt (1982).

[Ton86] Tonomura, A., T. Matsuda, R. Suzuki, A. Fukuhara, N. Osakabene, H. Umezaki, J. Endo, K. Shinagawa, Y. Sugita, and H. Fujiwara, *Phys. Rev. Lett.* **48** (1986): 1443.

[Was86] Washburn, S., and R. A. Webb, *Adv. Phys.* **35** (1986): 375.

[Bea81] Beaumont, S. P., P. G. Bower, T. Tamamara, and C. D. W. Wilkinson, *Appl. Phys. Lett.* **38** (1981): 436.

[Man88] Mankiewich, P. M., *J. Vac. Sci. Tech.* **6**, 1 (1988): 131.

[Lik76] Likharev, K. K., *Sov. Tech. Phys. Lett.* **2** (1976): 12.

[Sak80] Sakaki, H., *Jap. J. Appl. Phys.* **19** (1980): L735.

[Yam88] Yamada, T., *Proc. ICEM* p. 96, Tokyo (1988).

[Leb84] Leburton, J. P., *J. Appl. Phys.* **56** (1984): 2850.

[Oka88] Okada, M., T. Ohshima, M. Matsuda, N. Yokoyama, and A. Shibatomi, *Ext. Abs. 20th Conf. Solid State Dev. and Mat.*, pp. 503–506, Tokyo (1988).

[Sak89] Sakaki, H., *Ext. Abs. 21st. Conf. on Solid State Dev. and Mat.*, pp. 537–540, Tokyo (1989).

[Cap90] Capasso, F., and S. Datta, *Physics Today* **43**, 2 (1990): 74–82.

[Hes90] Hess, K., *Physics Today*, **43**, 2 (1990): 34–42.

[Ism89] Ismail, K., W. K. Chu, A. Yen, D. A. Antoniadis, and W. I. Smith *Appl. Phys. Lett.* **54** (1989): 460.

[Tim87] Timp, G., A. M. Chang, J. E. Cunningham, T. Y. Chang, R. Mankiewich, and R. E. Howard, *Phys. Rev. Lett.* **58** (1987): 2814.

[Tim87b] Timp, G., A. M. Chang, P. Mankiewich, R. Behringer, J. E. Cunningham, T. Y. Chang, and R. E. Howard, *Phys. Rev. Lett.* **59** (1987): 732.

[Smi90] Smith H. I., and H. G. Craighead, *Physics Today* **43**, 2 (1990): 24–30.

[Wha88] Wharam, D. A., T. J. Thornton, R. Newbury, M. Pepper, H. Ahmed, J. E. F. Frost, D. G. Hasko, D. C. Peacock, D. A. Ritchie, and G. A. C. Jones, *J. Phys.* C **21** (1988): L209.

[Fow85] Fowler, A. B., U.S. Patent 45503320 (1985).

[Dat85] Datta, S., *Phys. Rev. Lett.* **55** (1985): 2344.

[Ism89b] Ismail, K., W. Chu, A. Yen, D. A. Antoniadis, and H. I. Smith *Appl. Phys. Lett.* **54** (1989): 460.

[Ism88] Ismail, K., W. Chu, D. A. Antoniadis, and H. I. Smith, *Appl. Phys. Lett.* **52**, 13 (1988): 1071.

[Liu89] Liu, C. T., K. Nakamura, D. C. Tsui, K. Ismail, D. A. Antoniadis, and H. I. Smith, *Appl. Phys. Lett.* **55** (1989): 168.

[Kot87] Kotthaus, J. P., *Phys. Scr.* T **19** (1987): 120.

[Ree88] Reed, M. A., J. N. Randall, R. J. Aggarwall, T. Matyi, T. M. Moore, and A. E. Wetsel, *Phys. Rev. Lett.* **60** (1988): 535.

[Wee88] van Wees, B. J., H. van Houten, C. W. J. Beenakkar, J. G. Williamson, L. P. Kouwenhoven, D. van der Marel, and C. T. Foxon, *Phys. Rev. Lett.* **60** (1988): 848.

[Wha88] Wharam, D. A., T. J. Thornton, R. Newbury, M. Pepper, H. Ahmed, J. E. F. Frost, D. G. Hasko, D. C. Peacock, D. A. Ritchie, and G. C. A. Jones, *J. Phys.* C:, *Solid State Phys.* **21** (1988): L209.

[Mol90] Molenkamp, L. W., A. A. M. Staring, C. W. J. Beenakker, R. Eppenga, C. E. Timmering, J. G. Williamson, C. J. P. M. Harmens, and C. T. Foxon, *Phys. Rev.* B **41** (1990): 1274.

[Kri89] Kriman, A. M., G. H. Bernstein, B. S. Haukness, and D. K. Ferry, *Superlett. Microstruct.* **6** (1989): 381.

[Oka92] Okada, M., M. Saito, M. Takatsu, P. E. Schmidt, K. Kosemura, and N. Yokoyama, *Semicond. Sci. Tech.* **7** (1992): B223–B227.

[Sch91] Schmidt, P. E., K. Kosemura, M. Okada, and N. Yokoyama, *J. Vac. Sci. Tech.* B **9** (1991): 1598.

13

GaAs FET AMPLIFIERS AND MONOLITHIC MICROWAVE INTEGRATED CIRCUITS

13.1 INTRODUCTION TO MONOLITHIC MICROWAVE INTEGRATED CIRCUITS

The superior physical properties of compound semiconductors such as gallium arsenide and indium phosphide over silicon has led researchers to use the material as a semiconducting and semi-insulating substrate for both high-frequency analog and high-speed digital integrated circuits. The monolithic microwave integrated circuit (MMIC) consists of both active and passive circuit elements integrated on a single semi-insulating GaAs substrate or chip forming a component or subsystem, as shown in Fig. 13.1 [Sch88]. The active devices consist typically of MESFETs, HEMTs, and Schottky barrier diodes, while the passive elements include both lumped and distributed matching elements, bias networks, and interconnections between active and passive elements and ground. While currently at the multistage amplifier level of complexity, this technology holds the promise of integrating whole subsystems, such as a transmitter/receiver (T/R) module, on a single chip of GaAs. The advantages of MMIC are as follows:

1. Low cost circuits due to batch processing capability.
2. Enhanced reproducibility due to uniform processing and integration of all parts of the circuits.
3. Improved reliability due to the elimination of wire bonds and discrete components.
4. Size and weight reduction.
5. Circuit design flexibility and multifunction performance on a single chip.
6. Broad-band performance owing to the reduction of parasitic elements and optimization of device size.

Analog ICs which contain complete receiver front ends, phase- and amplitude-

484 MONOLITHIC MICROWAVE INTEGRATED CIRCUITS

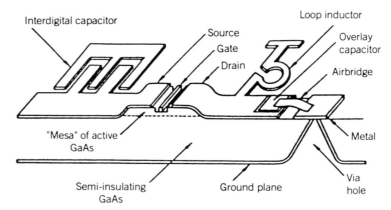

Fig. 13.1 Schematic of a monolithic microwave integrated circuit (MMIC). [Schellenberg, J. M., and T. R. Apel, in *GaAs Integrated Circuits, Design and Technology*, ch. 4, p. 204, ed. by J. Mun, Macmillan (1988)]

coded transmitters, analog-to-digital converters and synthesizers, are presently being developed. Many activities are now concentrating on the ability to integrate large numbers of FETs and HBTs in logic and memory applications. High-power watt-level MMIC are also under development.

This chapter will deal with the design of GaAs FET amplifiers and MMIC. Passive (planar lumped and distributed) elements will not be included in this chapter; readers can refer to Pucel's paper [Puc85]. However, low-noise amplifier design parameters and examples will be given, and recent developments in high-power watt-level MMIC will be discussed. The large signal circuit model for nonlinear analysis and sensitivity is included for optimization studies. Commercial applications and application-specific MMIC will be explored.

13.2 COMPARISON OF THE HYBRID AND MONOLITHIC APPROACHES

MMIC technology depends on the exploitation of the high-resistivity properties of semi-insulating semiconductor substrates for passive components and the good microwave device properties of the semiconductor layers grown on the SI substrate. The active devices commonly used in MMICs are the MESFET and the Schottky barrier diode on GaAs substrate.

Hybrid integrated circuit (HIC) technology, in contrast, depends on the insulating properties of the substrate [Puc85], primarily, as a low-loss microwave medium that supports planar transmission-line interconnect patterns, as well as passive components bonded onto these interconnects or deposited in situ. The function of the substrate is as a "holder" for all of the passive and active components. Good thermal conductivity is also important in high-power applications.

MMICs are fabricated by batch processing which results in low cost, high performance, small size, and reproducibility. The MMIC approach can create new markets where volume production is required, or where the limitations of hybrid technology simply do not allow performance and/or cost objectives to be met. Two examples of applications where technological objectives are difficult to meet by the hybrid approach are [Puc85] multioctave bandwidth amplifiers and millimeter-wave circuits. In both of these the limitations imposed by bond-wire inductances and other parasitics, and also by discrete elements, preclude wide-band or high-frequency performance.

However, there are certain components that are difficult to realize with the same performance as their hybrid counterparts. One example is the narrow-band (high-Q) filter. Because of the higher transmission line losses associated with the thinner GaAs substrates, coupled with the higher parasitic capacitances to the ground plane, lower-Q performance can be expected from MMICs. This low-Q limitation is a problem in some circuit applications, such as low-noise stable VCOs.

For FET power amplifiers a trade-off must be made between good thermal performance and good RF design. For example, to minimize the thermal resistance through the substrate, one must use a wafer as thin as practical. However, a thin wafer increases the skin-effect losses of microstrip lines and the attenuation within the circuit. Furthermore, since the use of heat sinks requires the metallization of the bottom side of the chip, additional parasitic capacitance to ground is introduced, and corrections must be made to planar inductors and other components to account for "image" currents in the ground plane and/or for increased capacitive parasitics.

The thermal limitation associated with the substrate in the MMIC approach is a disadvantage when compared to hybrid circuits. In hybrid circuits the thermal rise in the FET (or IMPATT diode) can be minimized by the correct thermal design of

TABLE 13.1 Attributes of the HIC and MMIC Approaches

Attribute	HICs	MMICs
Batch processing	−	+
Labor intensive	−	+
Low cost/circuit	−	+
On-chip wire bonds	−	+
Potential reliability	−	+
Small size/weight	−	+
Circuit design flexibility	−	+
Bandwidth/frequency range	−	+
Controlled parasitics	−	+
Substrate cost	+	−
On-chip RF testing	+	−
"Tweakability"	+	−
Repairability	+	−
Variety of active elements	+	−
Wafer handling	+	−
Heat sinking applications	+	−
Circuit-Q-factors	+	−
MS impedance range	+	−
Capital equipment costs	+	−

Source: Pucel, R. A., in *Gallium Arsenide Technology*, ed. by D. K. Ferry, SAMS (1985): 201.

the FET chip itself, in combination with an optimum method for channeling the heat generated to the appropriate heat sink. No limitations are imposed by the dielectric substrate, since metal heat sinks, consisting of posts passing through the substrate, can be incorporated rather easily. No such option exists in the MMIC approach.

Small size and volume, light weight, and high reliability are intrinsic properties of the MMIC approach. Small size allows the batch processing of hundreds of circuits per wafer. Since the essence of batch processing is that the cost of fabrication is determined by the cost of processing the entire wafer, it follows that the processing cost per chip is proportional to the area of the chip. This implies that the higher the circuit count is per wafer, the lower will be the circuit cost.

The elimination of wire bonding and the integration of active components within a printed circuit will remove many of the undesired parasitics that limit the broad-band performance of circuits that employ packaged discrete devices. The MMIC approach will ease the difficulty of attaining multioctave performance. Such broad-banding approaches as distributed amplifiers become feasible. Nevertheless, in cases where very narrow-band performance is required, the HIC approach is better.

Table 13.1 summarizes the attributes of both the HIC and MMIC approaches [Puc85]. The "+" sign implies the advantage, the "−" sign a disadvantage.

13.3 GENERAL DESIGN CONSIDERATIONS

The substrate sizes are 4 in. in diameter. Suppose that the maximum linear dimension per circuit falls between $\lambda_g/10$ and $\lambda_g/4$, where λ_g is the wavelength of the propagation mode in GaAs [Puc85]. The lower limit takes into account the approximate maximum size of the lumped elements; the upper limit, the typical maximum size of the distributed elements. In the vicinity of 10 to 20 GHz, some distributed elements as large as $\lambda_g/4$ (e.g., couplers) will be used. Therefore, above this frequency range, linear circuit dimensions on the order of $\lambda_g/4$ will be the rule. We can postulate a "linear" admixture of lumped- and distributed-element weighting so that we can obtain $\lambda_g/10$ at 1 GHz and $\lambda_g/4$ at 16 GHz as the probable linear dimension of a circuit-function chip.

The constraints imposed on the substrate thickness are as follows [Puc85]:

1. Volume of material used.
2. Fragility of wafer.
3. Thermal resistance
4. Propagation losses.
5. Higher-order mode propagation.
6. Impedance/linewidth consideration.
7. Thickness tolerance versus impedance tolerance.

To keep material cost down, one should use as thin a substrate as can be handled without compromising the fragility. Thermal considerations also require a thin wafer. However, a thin wafer emphasizes the effect of the ground plane. Microstrip propagation losses increase inversely with substrate thickness. Furthermore the Q-factor and inductance of thin film inductors decrease with decreasing substrate thickness. In

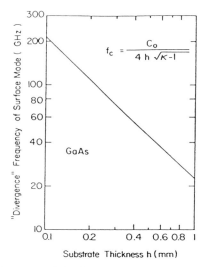

Fig. 13.2 Frequency of the onset of the lowest-order TE surface wave on a GaAs substrate as a function of substrate thickness. [Pucel, R. A., in *Gallium Arsenide Technology*, pp. 189–248, ed. by D. K. Ferry, SAMS (1985)]

contrast, undesired higher-order surface mode excitation is inhibited for thinner substrates.

Figure 13.2 shows a graph of frequency denoting the onset of the lowest-order (TE) surface mode as a function of the substrate thickness. For a substrate thickness of 0.1 mm (4 mils), the "safe" operating frequency is below 200 GHz. It appears that surface mode propagation is not a limiting factor in the choice of substrate thickness.

The linewidth dimensions for a given impedance level of some propagation modes, such as microstrip, are proportional to the substrate thickness. Therefore thicker substrates alleviate the effect of thickness and linewidth tolerances.

The choice of substrate thickness is a trade-off of many factors, since it is strongly dependent on the frequency of operation and the power dissipation of the circuit. The most important of the factors is perhaps the thermal consideration. In the frequency range up to 30 GHz, a substrate thickness on the order of 0.1 to 0.15 mm is appropriate for power amplifier circuits, with thicknesses up to 0.6 mm acceptable for low-noise amplifiers and similar circuits provided that a satisfactory means of dicing the thicker wafers can be found [Puc85].

Metal films are used in MMICs to define transmission-line interconnects, as electrodes for thin film capacitors, as conductors for planar inductors, and as a means to connect the chip to the external system. The choice of metal that is to be used is based on a number of factors, namely (1) good adherence to the substrate, (2) a technology compatible with semiconductor device fabrication, and (3) low RF losses.

Conductors can be divided into three categories [Cau70]. The first category contains good conductors (e.g., Ag, Cu, Au, Al) that have low resistivity but poor adhesion to dielectrics. Aluminum is an anomaly, since it has fair adhesive properties. The second category has poorer conductors (e.g., Cr, Ta, Ti), but they have good adherence

488 MONOLITHIC MICROWAVE INTEGRATED CIRCUITS

to semiconductors. The more popular ones are listed—chromium, tantalum, and titanium. The metals of these first two categories are generally deposited by vacuum evaporation, electroplating, resistance boat, or electron-beam heating. The third category of metals contains some fair conductors (e.g., Mo, W) that have a good adherence to dielectrics. Tungsten and molybdenum are refractory metals, and vacuum evaporation using electron-beam heating is required for deposition.

Most MMICs use chromium or titanium for the bonding layer and gold for the thicker cover layer. Usually the gold is deposited in two steps: A thin (0.5 μm) evaporated layer is formed over the bonding layer to provide a good electrical conductor for the subsequent electroplating of a much thicker layer, usually on the order of 3 to 6 μm.

13.4 LOW-NOISE AMPLIFIER

Modern GaAs FET devices are capable of producing noise figures of 1 dB at 20 GHz with almost 10-dB associated gain. This section deals with the theory and some design examples of amplifiers using GaAS MESFETs and HEMTs.

13.4.1 Low-Noise Design

S-Parameters

S-parameters are measured using traveling waves. It is essential to specify the plane of reference of the measurements with respect to the FETs. A brief review of S-parameters is given here. The readers can refer to microwave books for detail discussions on the theory of S-parameters [Mon48; Hew72].

Assume a microwave junction have n ports, each of which is a lossless uniform transmission line, as shown in Fig. 13.3. a_j is the incident traveling wave coming

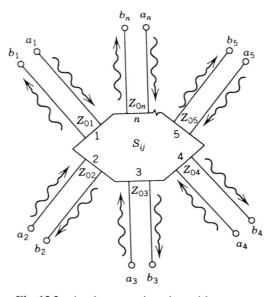

Fig. 13.3 A microwave junction with n-ports.

toward the junction, and b_i is the reflected traveling wave coming outward from the junction. From transmission-line theory, the incident and reflected waves are related by [Lia85]

$$b_i = \sum_{j}^{n} S_{ij} a_j \quad \text{for } i = 1, 2, 3, \ldots, n, \tag{13.1}$$

where

$S_{ij} = \Gamma_{ij}$ is the reflection coefficient of the ith port if $i = j$ with all other ports matched,

$S_{ij} = T_{ij}$ is the forward transmission coefficient of the jth port if $i > j$ with all other ports matched,

$S_{ij} = T_{ij}$ is the reverse transmission coefficient of the jth port if $i < j$ with all other ports matched.

In general, Eq. (13.1) can be written

$$b_1 = S_{11}a_1 + S_{12}a_2 + S_{13}a_3 + \cdots + S_{1n}a_n$$
$$b_2 = S_{21}a_1 + S_{22}a_2 + S_{23}a_3 + \cdots + S_{2n}a_n$$
$$\ldots\ldots\ldots\ldots\ldots\ldots\ldots\ldots\ldots\ldots\ldots\ldots\ldots\ldots\ldots$$
$$b_n = S_{n1}a_1 + S_{n2}a_2 + S_{n3}a_3 + \cdots + S_{nn}a_n. \tag{13.2}$$

In matrix notation, boldface letters are used to represent matrix quantities:

$$\mathbf{b} = \mathbf{Sa}, \tag{13.3}$$

where both **b** and **a** are column matrices. The $n \times n$ matrix **S** is called the *scattering matrix*:

$$\mathbf{S} = \begin{bmatrix} S_{11} & S_{12} & \cdots & S_{1n} \\ S_{21} & S_{22} & \cdots & S_{2n} \\ \ldots & \ldots & \ldots & \ldots \\ S_{n1} & S_{n2} & \cdots & S_{nn} \end{bmatrix} \tag{13.4}$$

The coefficients $S_{11}, S_{12}, \ldots, S_{nn}$ are the scattering parameters or scattering coefficients.

Stability of a Two-Port

One important parameter in the design of microwave amplifiers is that of stability. An amplifier can be either unconditionally or conditionally stable. A circuit is unconditionally stable if its input and output resistances are positive for passive terminations. If the amplifier is conditionally stable, then either the input or output resistances are negative.

The measured S-parameters of a transistor enable the maximum available gain (MAG) of the device to be determined.

The transducer power gain G_T of a transistor can be calculated as

$$G_T = \frac{|S_{21}|(1-|\Gamma_s|^2)(1-|\Gamma_{L_2}|^2)}{|(1-S_{11}\Gamma_s)(1-S_{22}\Gamma_L) - S_{12}S_{21}\Gamma_s\Gamma_L|^2}. \tag{13.5}$$

When the generator and load Γ_s and Γ_L are conjugately matched to the two FET ports, the gain can be maximized. Assuming that $S_{12} = 0$, we have

$$G_{max} = \frac{|S_{12}|}{(1-|S_{11}|^2)(1-|S_{22}|^2)}, \tag{13.6}$$

where $\Gamma_s = s_{11}^*$, $\Gamma_L = s_{22}^*$, and s_{22}^* are the complex conjugate reflection coefficients, and the device is assumed to be unilateral. The S-parameters also determine Rollett's stability factor K [Rol62]:

$$K = \frac{1 + |S_{11}S_{22} - S_{12}S_{21}|^2 - |S_{11}|^2 - |S_{22}|^2}{2|S_{12}S_{21}|}. \tag{13.7}$$

If K is larger than unity (i.e., when the FET is unconditionally stable), an optimum combination of Γ_s and Γ_L can simultaneously match the two FET ports to maximize the gain. If K is smaller than unity, the FET is only conditionally stable, and Γ_s and Γ_L must be carefully chosen to operate the device in a stable region. Since the FET's stability factor K will change with frequency, it is possible to plot the regions of stability onto a Smith chart at each frequency. Another method involves negative feedback with either purely reactive or lossy networks to achieve $K > 1$. Reactive feedback is particularly attractive for low-noise applications, where it is possible to achieve a simultaneous noise and gain match [Eng74].

The input and output reflection coefficients with arbitrary source and load terminations are given by

$$\Gamma_{IN} = S_{11} + \frac{S_{21}S_{12}\Gamma_L}{1 - S_{22}\Gamma_L}, \tag{13.8}$$

$$\Gamma_{OUT} = S_{22} + \frac{S_{21}S_{12}\Gamma_s}{1 - S_{11}\Gamma_s}. \tag{13.9}$$

If we set $|\Gamma_{IN}|$ equal to unity, a boundary is established beyond which the device is unstable and

$$\frac{|S_{11} + |S_{21}S_{12}\Gamma_L||}{1 - S_{22}\Gamma_L} = 1. \tag{13.10}$$

This equation can be changed in form to give the solution as a circle whose radius is

$$\gamma_L = \frac{S_{12}S_{21}}{|S_{22}|^2 - |\Delta|^2} \tag{13.11}$$

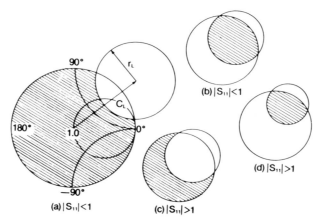

Fig. 13.4 The four different regions of stability in the Γ plane. [Pengelly, R. S., *Microwave Field-Effect Transistors—Theory, Design and Applications*, 2d ed., Research Studies Press (1986)]

and whose center C_L is

$$C_L = \frac{(S_{22} - \Delta S_{11}^*)^*}{|S_{22}|^2 - |\Delta|^2}, \tag{13.12}$$

where $\Delta = S_{11}S_{22} - S_{12}S_{21}$ and the origin of the Smith chart is at $\Gamma_L = 0$.

The area either inside or outside the circle represents a stable operating condition. If $|S_{11}| < 1$ and $|\Gamma_{IN}| < 1$, then the shaded region in Figs. 13.4(a) and (b) will enable the stable gain to be realized. In Fig. 13.4(b) the stability circle encloses the origin of the Smith chart, and hence its interior represents the stable region. In Fig. 13.4(a) the origin is outside the circle, and the exterior of the circle represents the stable region.

For $|S_{11}| > 1$, the origin represents an unstable point. Figure 13.4(c) is the case where the stability encircles the origin and its exterior is the stable region. For the case where the circle does not encircle the origin, its interior represents the stable region as in Fig. 13.4(d). For unconditional stability we must ensure that the magnitude of the vector C_L minus the radius of the stability circle r_L is greater than one.

For the case where $K < 1$ the maximum stable gain (MSG) of the device can be calculated as

$$\text{MSG} = \frac{|S_{21}|}{|S_{12}|}. \tag{13.13}$$

In general,

$$\text{MAG} = |S_{21}|(K - \sqrt{K^2 - 1}) \tag{13.14}$$

The MAG equals MSG when K is 1, and the MSG is the gain that can be obtained from the GaAs FET when the input and output reflection coefficients fall on the boundaries of the instability regions.

Transducer Power Gain

The transducer power gain of an amplifier from port i to port k is given by $G_T = |S_{ki}|^2$. We can write

$$\Gamma_s = \frac{Z_s - Z_o}{Z_s + Z_o}, \tag{13.15}$$

$$\Gamma_L = \frac{Z_L - Z_o}{Z_L + Z_o}. \tag{13.16}$$

When S_{12} is sufficiently small to be neglected, the device is defined to have a unilateral transducer gain

$$G_U = S_{21}^2 \frac{(1 - \Gamma_s^2)(1 - \Gamma_L^2)}{(1 - S_{11}\Gamma_s)^2(1 - S_{22}\Gamma_L)^2}, \tag{13.17}$$

$$G_U = G_s G_o G_L. \tag{13.18}$$

TABLE 13.2 Nine Power Gains

Transducer power gain in 50-ohm system	$G_T = |S_{21}|^2$	
Transducer power gain for arbitrary Γ_G and Γ_L	$G_T = \dfrac{(1 - |\Gamma_G|^2)|S_{21}|^2(1 - |\Gamma_L|^2)}{|(1 - S_{11}\Gamma_G)(1 - S_{22}\Gamma_L) - S_{12}S_{21}\Gamma_G\Gamma_L|^2}$	
Untilateral transducer power gain	$G_{TU} = \dfrac{|S_{21}|^2(1 - |\Gamma_G|^2)(1 - |\Gamma_L|^2)}{|1 - S_{11}\Gamma_G|^2|1 - S_{22}\Gamma_L|^2}$	
Power gain with input conjugate matched	$G = \dfrac{|S_{21}|^2(1 - |\Gamma_L|^2)}{|1 - S_{22}\Gamma_L|(1 - |S'_{11}|^2)} = \dfrac{|S_{21}|^2}{1 - |S_{11}|^2}$	(for $\Gamma_L = 0$)
Available power gain with output conjugate matched	$G_A = \dfrac{|S_{21}|^2(1 - |\Gamma_G|^2)}{|1 - S_{11}\Gamma_G|^2(1 - |S'_{22}|^2)} = \dfrac{|S_{21}|^2}{1 - |S_{22}|^2}$	(for $\Gamma_G = 0$)
Maximum available power gain	$G_{ma} = \left|\dfrac{S_{21}}{S_{12}}\right|(k - \sqrt{k^2 - 1})$	
Maximum unilateral transducer power gain	$G_{TU\,max} = \dfrac{|S_{21}|^2}{(1 - |S_{11}|^2)(1 - |S_{22}|^2)}$	
Maximum stable power gain	$G_{ms} = \dfrac{|S_{21}|}{|S_{12}|}$	
Unilateral power gain	$U = \dfrac{1/2|S_{21}/S_{12} - 1|^2}{k|S_{21}/S_{12}| - \mathrm{Re}(S_{21}/S_{12})}$	

Source: Vendelin, G. D., *Design of Amplifiers and Oscillators by the S-Parameter Method*, Wiley (1982).

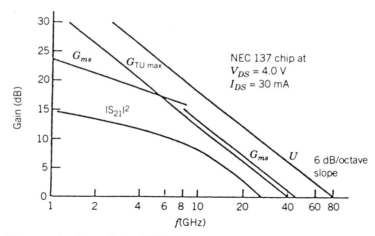

Fig. 13.5 Power gains for a GaAs MESFET versus frequency. [Vendelin, G. D., *Design of Amplifiers and Oscillators by the S-Parameter Method*, Wiley-Interscience (1982)]

The first term is related to the transistor used and remains invariant throughout the amplifier design. The other two terms, however, are not only related to the S-parameters S_{11} and S_{22} but also to the load and source coefficients. The latter two quantities are used to control the design of the amplifier in terms of gain slope compensation, and so forth.

The unilateral transducer gain can be divided into three independent gain blocks, where

$$G_s = \frac{1 - \Gamma_s^2}{(1 - S_{11}\Gamma_s)^2}, \qquad (13.19a)$$

$$G_0 = S_{21}^2, \qquad (13.19b)$$

$$G_L = \frac{1 - \Gamma_L^2}{(1 - S_{22}\Gamma_L)^2}. \qquad (13.19c)$$

Table 13.2 lists the nine power gains used by microwave engineers [Ven82]. The power gains for a typical common-source GaAs MESFET are plotted versus frequency in Fig. 13.5. This figure indicates that the largest gain is always U and that MAG (G_{ma}) will exist only for $k > 1$.

Circles of Constant Unilateral Gain

Considering Eq. (13.19a), we have that for $\Gamma_s = S_{11}^*$, $G_s = G_{s_{max}}$ and that for $\Gamma_s = 1$, $G_s = 0$. For any arbitrary value of G_s there is a value of Γ_s that lies on a circle. The output gain term in Eq. (13.19c) is of a similar form, and thus a similar set of circles can be generated. Such circles are called *constant gain circles* and can be generated from the FET S-parameters, since

$$d_i = \frac{g_i S_{ii}}{1 - S_{ii}^2(1 - g_i)},$$

$$R_i = \frac{\sqrt{1 - g_i(1 - S_{ii}^2)}}{1 - S_{ii}^2(1 - g_i)}, \qquad (13.20)$$

where

$$g_i = G_i(1 - S_{ii}^2) = \frac{G_i}{G_{i_{max}}}.$$

Here G_i is the constant gain represented by the circle, d_i is the distance from the center of the Smith chart to the center of the constant gain circle along the vector S_{11}^*, and R_i is the radius of a particular circle. The gain circle enables the design engineer to choose a matching topology which, for example, will allow matching over a range of frequencies to produce constant gain [Pen86].

Unilateral Figure of Merit

The assumption that $S_{12} = 0$ is often made in initial amplifier design. It is desirable to quantify the error involved in making this assumption. The transducer power gain G_T can be written

$$G_T = G_U \frac{1}{(1 - X)^2}, \qquad (13.21)$$

where

$$X = \frac{\Gamma_s \Gamma_L S_{12} S_{21}}{(1 - \Gamma_s S_{11})(1 - \Gamma_L S_{22})}. \qquad (13.22)$$

The ratio of the true gain to the unilateral gain is bounded by

$$\frac{1}{(1 + X)^2} < \frac{G_T}{G_U} < \frac{1}{(1 - X)^2}. \qquad (13.23)$$

For complex conjugate matching $S_{11} < 1$ and $S_{22} < 1$,

$$\frac{1}{(1 + X')^2} < \frac{G_T}{G_U} < \frac{1}{(1 - X')^2}, \qquad (13.24)$$

where

$$X' = \frac{S_{11} S_{22} S_{12} S_{21}}{(1 - S_{11}^2)(1 - S_{22}^2)}.$$

For a value X' at 10 GHz equal to 0.1, we have $1/(1 + X')^2$ equivalent to $-0.8\,\text{dB}$ and $1/(1 - X')^2$ equivalent to $+0.9\,\text{dB}$. The error in gain by assuming that the device is unilateral is approximately $\pm 0.8\,\text{dB}$.

Bias Network

The design of the bias circuits for monolithic ICs is as important as the design of the matching networks. The bias circuit determines the device operating point, amplifier stability particularly at low frequencies, temperature stability, and gain.

Depending on the application—low noise, high gain, class A power, class AB or B high efficiency—an optimum dc operating point exists. Low-noise amplifiers operate at a relatively low drain-source voltage V_{DS} and current I_{DS}, typically $I_{DS} = 0.15 I_{DSS}$ and $V_{DS} = 3$ V. For higher gain the bias point is adjusted upward to a higher I_{DS} level, often all the way to I_{DSS} for maximum gain. For power applications the bias point must be shifted to a high voltage (typically 8 to 10 V) and an I_{DS} level of approximately $0.5 I_{DSS}$. For high efficiency class AB or B operation, I_{DS} and V_{DS} must both be reduced from their high-power values. Examples of commonly used bias networks can be found in [Ven82].

13.4.2 Noise

The noise performance of a low-noise FET amplifier is determined by both the device noise parameters and the input/output matching networks. The device noise parameters are described in Chapter 8. Since the performance limits are set by the FET device, the matching networks are to extract potential performance over the frequency band of interest. As such, the matching networks determine the amplifier bandwidth, control the gain magnitude and flatness, and minimize the noise figure over the desired bandwidth. In this section the interaction of the device noise parameters with the input/output matching networks is examined.

Noise Match

A typical low-noise FET amplifier stage is shown in Fig. 13.6. The roles of the input and the output matching networks are quite different. The input matching network M_{IN} is required to transform the system input impedance (usually 50 ohms) to an impedance at the device input that minimizes the device noise contributions, while the output network must extract the maximum device gain (i.e., network M_{OUT} must conjugately match the resulting device output impedance). The effect of the input reflection coefficient on the stage noise figure can be expressed as

$$F(\Gamma_s) = F_{\min} + 4 r_n \frac{|\Gamma_s - \Gamma_o|^2}{(1 - |\Gamma_s|^2)|1 + \Gamma_o^2|}, \qquad (13.25)$$

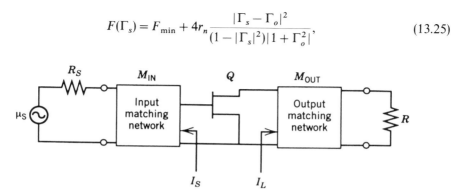

Fig. 13.6 Low-noise amplifier stage.

where

F_{min} = minimum noise figure,
r_n = normalized equivalent noise resistance $R_n/50$,
Γ_s = source reflection coefficient,
Γ_o = optimum source reflection coefficient for F_{min}.

Equation (13.25) expresses the relationship between the stage noise figure and the input reflection coefficient Γ_s. The source reflection coefficient can be varied until a minimum noise figure is read in a noise figure meter. The value of F_{min}, which occurs when $\Gamma_s = \Gamma_o$, can be read from the meter, and the source reflection coefficient that produces F_{min} can be determined accurately using a network analyzer. The noise resistance r_n can be measured by reading the noise figure when $\Gamma_s = 0$, called $F(\Gamma_s)$. From Eq. (13.25), we obtain

$$r_n = (F(\Gamma_s) - F_{min}) \frac{(1+\Gamma_o)^2}{4\Gamma_o^2}. \tag{13.26}$$

For the common-source configuration the reflection coefficient for minimum noise is not the same as the reflection coefficient for maximum gain. This implies that when the device is matched for minimum noise, the amplifier input reflection coefficient will not be zero. This input mismatch is characteristic of low-noise, common-source, FET amplifiers matched for minimum noise.

While the input network provides the optimum impedance for minimum noise figure, the role of the output network M_{OUT} is to extract the maximum device gain. To accomplish this, F_2 must conjugately match the output impedance of the FET over the band of interest. The required load reflection can be expressed in terms of the input reflection coefficient and the device S-parameters as

$$\Gamma_L = (S'_{22})^*, \tag{13.27}$$

where

$$S'_{22} = S_{22} + \frac{S_{12}S_{21}\Gamma_s}{1 - S_{11}\Gamma_s}. \tag{13.28}$$

The resulting device gain, termed the *available power gain*, is given by

$$G_A = \frac{|S_{21}|^2(1-\Gamma_s^2)}{|1 - S_{11}\Gamma_s^2|(1-|S'_{22}|^2)}, \tag{13.29}$$

and it is a function only of the input reflection coefficient and the device S-parameters.

13.4.3 Low-Noise Amplifier Using HEMT

HEMT devices are capable of achieving higher gains, lower noise and broader noise bandwidths than MESFET devices of similar gate lengths. Therefore MMICs using HEMTs as active devices will have reduced chip sizes, improved receiver performances,

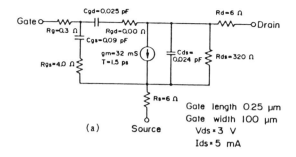

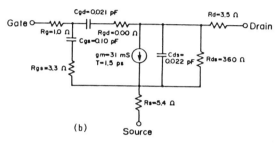

Fig. 13.7 Equivalent circuit of the HEMT chip: (a) HEMT equivalent circuit and element values obtained from the S-parameter measurement; (b) HEMT equivalent circuit and element values. [Ishizaki, M., T. Hamabe, Y. Oohashi, and S. Asai, *1988 MTT-S Digest*, p. 461, IEEE (1988)]

and extended frequency operations into the mm-wave region. The InP-based HEMT amplifier has been developed in a V-band application with a 9-dB gain at 62.5 GHz [Sov90]. This section will give a description of the low-noise amplifier using HEMT ICs. For the device characteristics of HEMT, the reader should refer to Chapter 9.

Ishizaki et al. [Ish88] reported designing a 43 GHz-band-balanced low-noise amplifier using HEMTs with a gate length of 0.25 µm. The optimum noise figure obtained was 1.0 dB with an associated gain of 8.2 dB at 20 GHz, and the noise at room temperature figure was 1.7 dB with a gain of 6.1 dB at 30 GHz. This section describes the design example of a 43 GHz low-noise amplifier using HEMT technology.

The devices used were low-noise HEMTs with a T-shape of 0.25 µm and a gate width of 100 µm. The S-parameters of the device were measured with a wafer prober over 2 to 18 GHz. Figure 13.7(a) shows the HEMT equivalent circuit and element values obtained from the S-parameter measurement.

A prototype is made of a 50-GHz-band one-stage amplifier. Figure 13.8 shows the gain and noise figure characteristics of the device up to 50 GHz. The noise figure is 5.5 dB for the optimum gain of 5.5 dB, and the gain is 3 dB for the optimum noise figure of 3.6 dB at 50 GHz. Figure 13.7(b) shows the results of the HEMT equivalent circuit and element values revised from the characteristics of the one-stage amplifier at 50 GHz. Figure 13.9 shows a comparison between the values calculated using the equivalent-circuit element values shown in Fig. 13.7(b) and the measured characteristics at 43 GHz.

A circulator is usually connected to the input/output sections of a low-noise

498 MONOLITHIC MICROWAVE INTEGRATED CIRCUITS

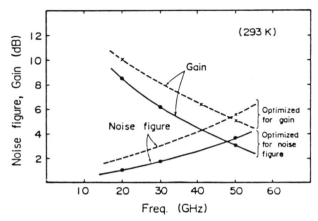

Fig. 13.8 Characteristics of a 0.25-μm gate-length HEMT. [Ishizaki, M., T. Hamabe, Y. Oohashi, and S. Asai, *1988 MTT-S Digest*, p. 461, IEEE (1988)]

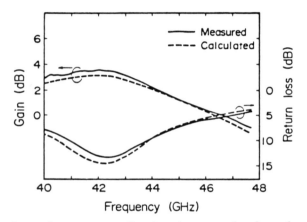

Fig. 13.9 Comparison of computer-predicted and measured gain and return loss of the 43 GHz-band amplifier. [Ishizaki, M., T. Hamabe, Y. Oohashi, and S. Asai, *1988 MTT-S Digest*, p. 461, IEEE (1988)]

amplifier so that the amplifier impedance matches that of other circuits. However, the circulator's characteristics at ultra low temperatures are not known. Ishizaki et al. [Ish88] adopted a balanced amplifier configuration in which a waveguide-type hybrid was used as the input/output section of the amplifier. Figure 13.10 gives a schematic of the configuration.

This amplifier consists of hybrid circuits, waveguide to microwave IC (MIC) transition circuits, and carrier-mounted amplifiers. The hybrid circuit was formed by waveguide branch lines using WRI-400 (WR-22) waveguides [Pat59; Mar48]. The coupling loss was 3.1 ± 0.1 dB, and the return loss and isolation were 18 dB or more from 40 to 45 GHz.

Figure 13.11 shows the gain, noise figure, and input return loss characteristics of the unit amplifier. The amplifier had a gain of 11 dB and a noise figure of 4.2 dB or

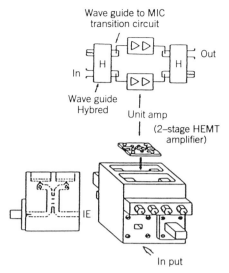

Fig. 13.10 Configuration of the HEMT amplifier. [Ishizaki, M., T. Hamabe, Y. Oohashi, and S. Asai, *1988 MTT-S Digest*, p. 461, IEEE (1988)]

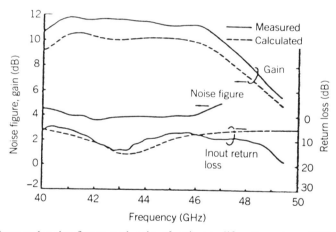

Fig. 13.11 Measured noise figure and gain of unit amplifier (two-stage HEMTs amplifier). [Ishizaki, M., T. Hamabe, Y. Oohashi, and S. Asai, *1988 MTT-S Digest*, p. 461, IEEE (1988)]

less over 40 to 46 GHz when biased with a drain voltage of 3 V, and a drain current of 8 mA for the front stage and 10 mA for the rear stage. The dotted lines in Fig. 13.11 represent the gain and input return loss characteristics obtained by computer simulation.

Two unit amplifiers with the same characteristics were built in this way. Figure 13.12 shows the characteristics of the amplifier consisting of these amplifiers. A gain of 9 dB or more, a noise figure of 5 dB or less, and input/output return losses of 15 dB or more were obtained over 40 to 50 GHz. The satisfactory voltage standing wave ratios (VSWRs) of the amplifier enable the multistages of the amplifiers to be directly connected.

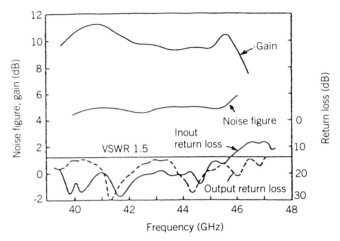

Fig. 13.12 Gain, noise figure, and return loss performance of the 43-GHz-band amplifier. [Ishizaki, M., T. Hamabe, Y. Oohashi, and S. Asai, *1988 MTT-S Digest*, p. 461, IEEE (1988)]

13.5 HIGH-POWER GaAs MMIC

Since 1979 there has been given a considerable amount of emphasis given to the realization of high-power MMIC amplifiers. High-power, high-efficiency amplification of microwave power using GaAs FETs have been demonstrated over the past few years at frequencies in C, X, and K_u band. A power FET is essentially a multicell structure where the GaAs material and the fabrication techniques used are optimized for higher-breakdown voltage than a small-signal FET. Power amplifier design is considerably more complex than small-signal linear amplifier design due to the following factors:

1. To realize high output power levels, a large gate periphery device is required. The large device results in low input/output impedance levels (typically an input impedance of 1 Ω/watt of output power). This in turn produces broad-band matching difficulties and excessive matching circuit losses.
2. The gain performance of a power amplifier is often prescribed for a range of power levels—not just the maximum output power level. However, device data as a function of power level is usually quite limited, which makes it difficult to predict the subsequent gain variation.
3. The FET device is operated in the nonlinear region, which results in the S-parameters being a function of the power level. The small-signal S-parameters are not useful for power amplifier design. The small-signal S-parameters can be used in large-signal amplifiers operating in class A. However, for classes AB, B, or C the small-signal S-parameters are not suitable.

This section describes the dc and RF characteristics of a power amplifier. High-power, high-efficiency amplifiers with wide bandwidths have been developed [LeS90; Ava90; Cam88]. A few design examples are given.

13.5.1 Nonlinearity in Class A Operation

The major difference between devices used for low-level amplification and those used for high-power operation is linearity. A set of large-signal S-parameters is needed to characterize the transistor for power amplifications. However, the measurement of large-signal S-parameters is difficult and is not properly defined. Therefore an alternate set of large-signal parameters is needed to characterize the transistor [Gon84]. This can be done by providing information of source and load reflection coefficients as a function of output power and gain, especially the measurement of the source and load reflection coefficients, together with the output power, when the transistor is operated at its 1-dB-gain compression point. The listing of the 1-dB compression point data is used to specify the power-handling capabilities of the transistor.

The 1-dB-gain compression point (called G_{1dB}) is defined as the power gain where the nonlinearities of the transistor reduces the power gain by 1 dB over the small-signal linear power gain:

$$G_{1dB}(dB) = G_0(dB) - 1, \qquad (13.30)$$

where $G_0(dB)$ is the small-signal linear power gain in decibles. Since the power gain is defined as

$$G_p = \frac{P_{OUT}}{P_{IN}},$$

or

$$P_{OUT}(dBm) = G_p(dB) + P_{IN}(dBm),$$

we can write the output power at the 1-dB-gain compression point, called P_{1dB}, as

$$P_{1dB}(dBm) = G_{1dB}(dB) + P_{IN}(dBm). \qquad (13.31)$$

Substituting Eq. (13.30) into (13.31) gives

$$P_{1dB}(dBm) - P_{IN}(dBm) = G_o(dB) - 1. \qquad (13.32)$$

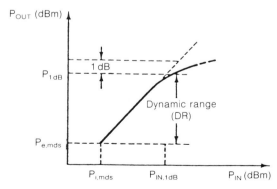

Fig. 13.13 The 1-dB-gain compression point and the dynamic range of microwave amplifier. [Gonzalez, G., Microwave Transistor Amplifiers, Prentice Hall (1984)]

Equation (13.32) shows that the 1-dB-gain compression point is that point at which the output power minus the input power in dBm is equal to the small-signal power gain minus 1 dB.

A typical plot of P_{OUT} versus P_{IN} which illustrates the 1-dB-gain compression point is shown in Fig. 13.13. Observe the linear output characteristics for power levels between the minimum detectable signal output ($P_{o,mds}$) and P_{1dB}. The dynamic range (DR), shown in Fig. 13.13, is that range where the amplifier has a linear power gain. The dynamic range is limited at low-power levels by the noise level. An input signal ($P_{i,mds}$) is detectable only if its output power level ($P_{o,mds}$) is above the noise power level.

13.5.2 Dynamic Load Line and Thermal Effects

The power-added efficiency (PAE) can be defined as [Soa83]

$$\text{PAE} = \frac{P_o - P_i}{P_{dc}}. \tag{13.33}$$

This establishes the efficiency of the device as a power converter, changing dc energy into microwave RF energy. Equation (13.33) may be rewritten as

$$\text{PAE} = \frac{1 - P_d}{P_{dc}}, \tag{13.34}$$

which indicates the importance of dissipated power on power-added efficiency. As dissipated power decreases with the RF drive, the power-added efficiency under a large-signal operation becomes substantially higher for the same dc power, unlike under small-signal conditions.

In the class A operation the maximum power-added efficiency is 50%. This is the case for an ideal voltage or current gain device that is linear in the first quadrant of the V-I plane, Fig. 13.14, with the idealized characteristics extending to infinity on the positive voltage and current axes.

The bias point for the device is set at (V_Q, I_Q) and the dynamic load line, corresponding to a load resistor R_L, is set at 45°. Under these conditions the peak-to-peak signal voltage and signal current swings from $2V_Q$ and $2I_Q$, respectively. The maximum RF power is written as

$$P_{om} = \frac{V_m I_m}{8} = \frac{V_Q I_Q}{2}. \tag{13.35}$$

The RF power corresponds to the area of the triangle $I_Q O V_Q$, and thus the maximum power added efficiency is

$$(\eta_{add})_m = \frac{P_{om} - P_{im}}{P_{dc}}$$

$$= \frac{1}{2}\left(1 - \frac{1}{G}\right). \tag{13.36}$$

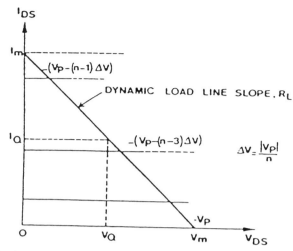

Fig. 13.14 Idealized active device I–V characteristics. [Soares, R. A., in *Applications of GaAs MESFETs*, p. 160, ed. by R. Soares, J. Graffeuil, and J. Obregon, Artech House (1983)]

The maximum power efficiency of 50% is only attainable for a device with infinite RF power gain. To maximize power-added efficiency for a device with a given output power, the dynamic load resistance should be set to $R_L = V_Q/I_Q$ to maximize the RF power gain.

In a practical GaAs MESFET, physical limitations in the device limit maximum permissible current excursion to $(I_f - I_m)$, Fig. 13.15, and maximum voltage excursion to $(V_{SD}^L - V_K)$. The use of gate recess structures has eliminated failure due to drain avalanche burnout under normal operating conditions and permitted the power

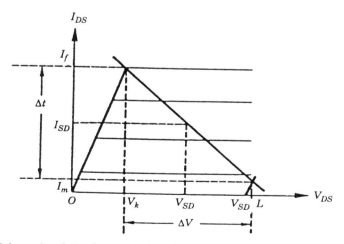

Fig. 13.15 Schematic of dc characteristics of a power GaAs MESFET. [Soares, R. A., in *Applications of GaAs MESFETs*, p. 160, ed. by R. Soares, J. Graffeuil, and J. Obregon, Artech House (1983)]

MESFET to be biased at a higher drain voltages [Fuk78]. Gate-drain avalanche breakdown is now the major factor limiting RF output power [DiL79; Wro83].

The gate-drain avalanche breakdown voltage V_{GD}, for a gate biased to its pinch-off voltage V_p, limits V_{so} by the following relationship:

$$V_{SD}^L = V_{GD} - V_P. \qquad (13.37)$$

Beyond this point [DiL79] excess current appears in the drain circuit, which is not modulated by the RF signal and therefore will not contribute to the fundamental frequency output power of the device. From Fig. 13.15, the maximum quasi-linear RF output power from the transistor is

$$P_m = \frac{I_f}{8}[(V_{GD} - V_P) - V_K] \qquad \text{for } \frac{I_m}{I_f} \ll 1. \qquad (13.38)$$

The parameter I_f, drain current flow with a forward gate bias offset voltage, is a function of the GaAs MESFET gate width z. An increase in I_f will lead to decrease in V_{GD} and an increase in V_P. Therefore the power transistors' designs have to be optimized.

The optimum load resistance for a practical device is given by the load line in Fig. 13.15:

$$(R_L)_{opt} = \frac{(V_{GD} - V_P) - V_K}{I_f}. \qquad (13.39)$$

The optimum bias points are thus established as $(V_{GD} - V_P - V_k)/2$ and $I_f/2$.

13.5.3 RF Characteristics of Power GaAs FETs

The small-signal scattering parameters of power FETs are usually supplied on manufacturers' data sheets [Pen86]. This enables the design engineer to obtain a device bandwidth capability and also indicates the regions of instability of the device. The low-frequency S-parameters of a power FET usually indicate, below 2 GHz, that the regions of instability are large particularly for the output of such a device, so it is necessary to use bias networks. It is apparent that the gain of a power FET is substantially lower than that of a small-signal device, a result of the device being optimized for its power-handling performance.

The small-signal S-parameters of a power FET are not so useful for accurate prediction of the large-signal gain and power output of a device either at a spot frequency or over a bandwidth. The source and load impedances for the device operating at its 1-dB-gain compression point are more easily measured, and together with the small-signal S-parameters, give sufficiently accurately information for design. These impedances are measured by matching the device using, for example, stub tuners at the RF power levels of interest and then separately measuring the tuner impedance using a conventional small-signal network analyzer. The required impedance is the conjugate of the measured stub impedance.

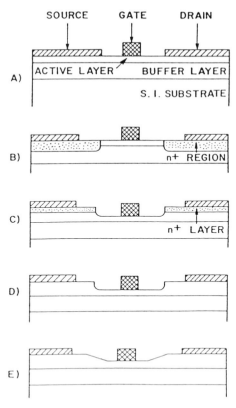

Fig. 13.16 Basic FET structures used in power GaAs FETs. [Hesegawa, F., in *GaAs FET Principles and Technology*, pp. 219–255, ed. by J. V. DiLorenzo and D. D. Khandelwal, Artech House (1982)]

13.5.4 Design and Performance of GaAs Power FET Amplifiers

Basic GaAs FET structures utilized in power amplifiers are shown in Fig. 13.16 [Hes82]. They include devices with and without gate recess and with and without ion-implanted n^+ region under the ohmic contacts, which reduces the series source resistance. As can be seen from Fig. 13.17, the presence and shape of the gate recess strongly influence the drain breakdown voltage. The drain-to-source breakdown voltage is also dependent on the drain-to-gate spacing. The breakdown voltage may substantially increase if the drain-to-gate spacing is large enough to house a stationary high-field domain forming at the drain side of the gate [Shu87]. For this reason there may be some advantage in increasing the gate-to-drain spacing. As shown in [Eas80], the high-field domain size may be estimated as

$$d_{\text{dom}} = 2.06 \left(\frac{\varepsilon V_{dbr}}{q N_d^{1/2} n_{cr}^{1/2}} \right)^{1/2}, \qquad (13.40)$$

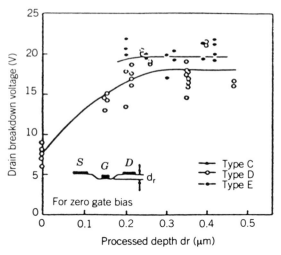

Fig. 13.17 Dependence of the drain-to-source breakdown voltage on the recessed depth of the gate region at zero gate bias for different structures shown in Fig. 13.15. [Hesegawa, F., in *GaAs FET Principles and Technology*, pp. 219–255, ed. by J. V. DiLorenzo and D. D. Khandelwal, Artech House (1982)]

where V_{dbr} is the drain-to-source breakdown voltage, N_d is the doping density in the channel, and n_{cr} is the characteristic doping density (3×10^{15} cm^{-3} for GaAs). The device structure shown in Fig. 13.16(b) may have the smallest source series resistance. The device structure of Fig. 13.16(e) has the highest breakdown voltage.

The total gate periphery of the GaAs amplifier should be divided into several "unit gates" or gate fingers in order to minimize the series gate resistance and related gain degradation [Shu87]. Depending on how the unit gates are connected, the power FET designs may be divided into two groups—with the crossover of the connecting metal over the contacts and without the crossover. The isolation between the connecting metal and the device contacts that are crossed in devices with a crossover design may be achieved by using either suitable dielectric (typically CVD deposited SiO_2) or air-bridge technology. Air bridges may be made by using a thick gold plating on top of a thick photoresist. The photoresist is then removed to form an air bridge. This structure has a slight advantage over dielectric isolation in a smaller parasitic capacitance. Power FET structures without crossovers include a wire-bonding construction, the flip chip configuration, and the via-hole connection [Has82].

The thermal design of GaAs power FETs was reviewed in [Wem82]. In the analytical treatment the heat sources introduced by FETs were approximated by planar line heat sources on an infinite dielectric slab. This approach led to the following simple expression for the thermal impedance θ:

$$\theta = (\pi k)^{-1} \ln \left[\frac{8C}{\pi d} \right], \tag{13.41}$$

where θ is in °C mm/W, k is the substrate thermal conductivity in W/°C mm, C is the substrate thickness, and d is the width of the thermal source. This expression is valid when $C \gg d$ and in agreement with a computer solution when parameter d is adjusted to provide a good fit.

13.5.5 Ka-Band Monolithic GaAs FET Power Amplifier Modules

Recent advances in GaAs power FET technology have made it possible to build power amplifiers at Ka-band [Kim86; Kim87]. This section describes a monolithic power combining scheme that uses several of the small-gate-width devices (400 μm) in a six-way traveling-wave monolithic divider combiner to make a 0.6-W amplifier at 34 GHz with 2.8-dB gain [Cam88]. Modules with several cascaded stages producing 0.55-W with 27-dB gain at 34 GHz will be discussed. A two-way hybrid combining scheme making use of the 0.6-W monolithic chips producing 1 W of output power will also be described.

Unit Cell Monolithic Design

The design of the power amplifiers described in this section includes a 400-μm monolithic unit cell. An equivalent circuit of the monolithically matched unit cell is shown in Fig. 13.18. This unit cell uses two 200-μm FETs with overlay source grounding. Two smaller FETs are used rather than a single 400-μm FET to decrease the source lead inductance and therefore increase the gain. The input uses open circuited stubs rather than MIM capacitors for matching purposes. Matching is achieved on the input side by using an inductor to ground at the device and an open circuited

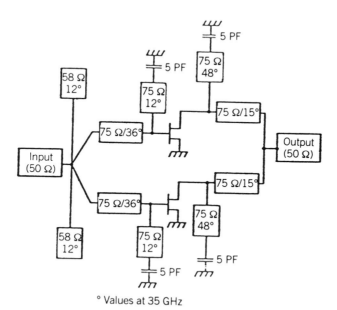

° Values at 35 GHz

Fig. 13.18 Schematic showing the circuit elements for the Ka-band monolithic unit cell consisting of two 200-μm FETs. [Camilleri, N., B. Kim, H. Q. Tserng, and H. D. Shih, in *IEEE 1988 Microwave and Millimeter-Wave Monolithic Circuits Symp. Digest of Papers*, pp. 129–132, IEEE (1988)]

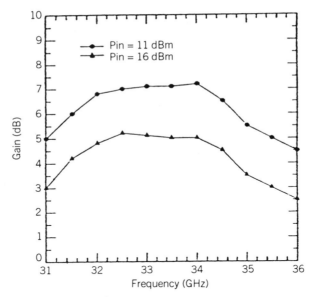

Fig. 13.19 Typical gain curve for the average Ka-band monolithic unit cell. [Camilleri, N., B. Kim, H. Q. Tserng, and H. D. Shih, in *IEEE 1988 Microwave and Millimeter-Wave Monolithic Circuits Symp. Digest of Papers*, pp. 129–132, IEEE (1988)]

stub 300 µm away from the device. The output is matched by using an inductor to ground at the device. The respective line lengths and impedances are given in the equivalent circuit of Fig. 13.18.

The maximum power performance achieved under compression for this unit cell is 200 mW, with 3.8-dB gain and 21% efficiency. At the 1-dB compression point the best cells produce 170 mW (22 dBm) with 5-dB gain. The gain curve for the average unit cells at different power levels and bias conditions is given in Fig. 13.19.

Six-Way Combined Monolithic Chip Design

The monolithically combined chip consists of six 400-µm unit cells that are power combined using a six-way traveling-wave divider/combiner, as shown in Fig. 13.20. The traveling-wave divider/combiner [Ber80; Tse85] is ideal for monolithic design, since it is compact and has broad-band and low-loss characteristics.

The design was implemented on 100-µm-thick semi-insulating GaAs substrate. The total chip size was 19 mm^2; the total loss for the divider/combiner has been measured to be 2 dB at 34 GHz. The performance of the six-way combined chip is shown in Fig. 13.21. The chip can deliver 0.6 W (27.8 dBm) with 2.8-dB gain and 8.5% power-added efficiency.

One-Watt Power Amplifier Design

The 1-W amplifier was implemented using two levels of power combining. The first level was implemented in monolithic form using a six-way divider/combiner. The second level was in hybrid form where four of the monolithically combined chips

Fig. 13.20 Picture for the monolithic GaAs six-way combined Ka-band amplifier chip. [Camilleri, N., B. Kim, H. Q. Tserng, and H. D. Shih, in *IEEE 1988 Microwave and Millimeter-Wave Monolithic Circuits Symp. Digest of Papers*, pp. 129–132, IEEE (1988)]

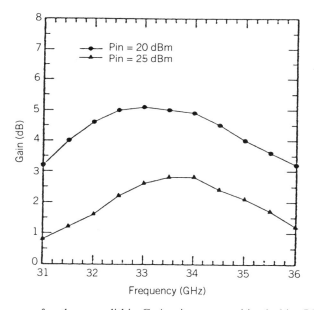

Fig. 13.21 Gain curve for the monolithic GaAs six-way combined chip. [Camilleri, N., B. Kim, H. Q. Tserng, and H. D. Shih, in *IEEE 1988 Microwave and Millimeter-Wave Monolithic Circuits Symp. Digest of Papers*, pp. 129–132, IEEE (1988)]

were used to make a two-stage, two-way-power-combined amplifier. The design used four of the monolithically combined chips. The two-stage configuration was used to increase the available gain of the combined amplifiers to compensate for the loss of the two-way divider/combiner.

The two-way divider/combiner consisted of a compensated in-line quarter-wave resistive power splitter [How74]. The power splitter was constructed on a 100-μm-thick semi-insulating GaAs substrate. The measured loss of this power splitter was

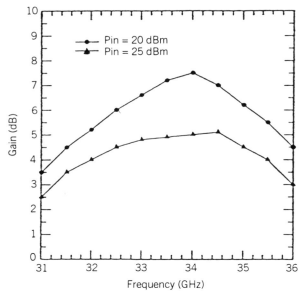

Fig. 13.22 Gain curve for the two stage two-way combined power amplifier module. [Camilleri, N., B. Kim, H. Q. Tserng, and H. D. Shih, in *IEEE 1988 Microwave and Millimeter-Wave Monolithic Circuits Symp. Digest of Papers*, pp. 129–132, IEEE (1988)]

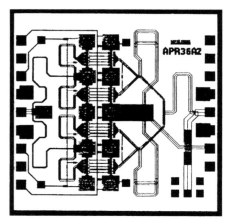

Fig. 13.23 Layout of the HBT MMIC power amplifier. [Ramachandran, R., M. Nijjar, A. Podell, E. Stoneham, and S. Mitchell, in *GaAs IC Symp. Digest*, p. 357 (1990)]

0.5 dB for the divider and 0.5 dB for the combiner. A 90° offset between the two arms was included to offset any mismatch and obtain a better input and output VSWR. Due to biasing considerations a dc blocking capacitor was used to isolate the first stage from the second. The first and second stages were biased separately to maximize the gain, since they operate at different power levels. The performance of this two stage two-way-combined amplifier is shown in Fig. 13.22. Under power conditions the chip can deliver 1 W (30 dBm) with 5-dB gain and 5.8% power-added efficiency.

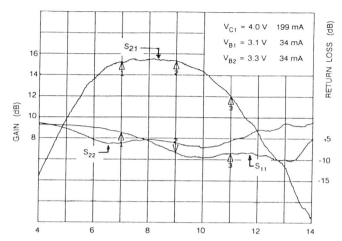

Fig. 13.24 Measured gain of HBT MMIC power amplifier. [Ramachandran, R., M. Nijjar, A. Podell, E. Stoneham, and S. Mitchell, in *GaAs IC Symp. Digest*, p. 357 (1990)]

Under low-power conditions, and with the bias adjusted for maximum gain, the chip has 7.5-dB gain with 0.56 W (27.5 dBm) power output.

13.5.6 A HBT MMIC Power Amplifier

The heterojunction bipolar transistor (HBT) has received a lot of attention because of its high-power density (4 W/mm) and power-added efficiency [She87; Wan89; Bay87; Ali89]. This transistor is especially suitable for applications requiring high efficiency at high output levels. The high efficiency is achieved due to higher operating voltages combined with a small knee voltage whereas high-power output is realized due to the high current densities possible. Figure 13.23 shows a layout of the chip with chip size 80 × 80 mils [Ram90].

The AlGaAs/GaAs HBTs used in the design were produced from MBE-grown wafers with a base doping from $1 \times 10^{19}\,\text{cm}^{-3}$ to $5 \times 10^{19}\,\text{cm}^{-3}$ [Ram90]. The collector doping concentration was chosen to achieve the necessary breakdown voltage (> 20 V). A dual-liftoff self-aligned process was employed for device fabrication in order to minimize base resistance and collector-base capacitance. The emitter finger width used was 2 μm, resulting in an f_t of 40 GHz and an f_{max} of 80 GHz. A complete MMIC process with two layers of metal interconnects, nitride resistors, and MIM capacitors was integrated with the device processing to enable the HBT monolithic circuit implementation.

Over 60 chips from six wafers were measured using wafer-probing techniques. Figure 13.24 is a graph of the measured gain of an average MMIC with $V_{cc} = 4\,\text{V}$ and $I_c = 200\,\text{mA}$. The peak gain is 15 dB at 8 GHz and gain is greater than 14 dB over the 6 to 10 GHz band. The gain flatness within the band is less than ± 1 dB. After small-signal evaluation the MMICs were mounted on a molybdenum carrier for power output and efficiency measurements. Figure 13.25 shows the output power and power-added efficiency measured from 6 to 10 GHz for chips from two different wafers. At 7.5 GHz a power output of over 31 dBm was achieved with 13-dB gain

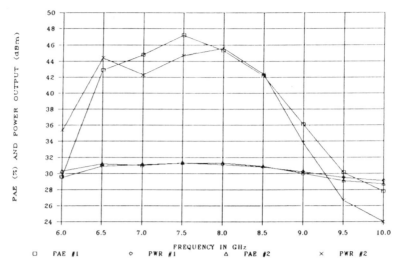

Fig. 13.25 PAE and power output of HBT MMIC (two wafers). [Ramachandran, R., M. Nijjar, A. Podell, E. Stoneham, and S. Mitchell, in *GaAs IC Symp. Digest*, p. 357 (1990)]

and 47% power added efficiency (PAE) while operating in a class AB mode. This corresponds to 3 Watts per millimeter of emitter length. Over the 6.5 to 8.5 GHz band, the PAE is greater than 42%, and the power output is greater than 31 dBm. These results demonstrate the great potential of HBT MMICs for high-efficiency power applications.

13.6 LARGE-SIGNAL CIRCUIT MODEL FOR NONLINEAR ANALYSIS

The general-purpose nonlinear circuit analysis programs that exist were designed primarily for transient (time-domain) analysis of silicon integrated circuits [Nag75]. Time-domain analysis requires all circuit parameters to be input in the time dimension; for example, transmission line lengths are usually given in picoseconds of transit time [Pen85]. Fast Fourier transformation is employed to display results in the frequency dimension. A good example of time-domain nonlinear analysis is the ANAMIC program by Sobhy et al. [Sob85] which employs state-space variable techniques.

However, time-domain analysis is in many cases inefficient [Cur85]. Rizzoli et al. [Riz83] described a general-purpose nonlinear microwave circuit design technique that efficiently analyzes the device-circuit interaction by application of the "harmonic-balance technique [Nak76]. Camacho-Penalosa [Cam83] described the application of this technique to the microwave FET. Later Camacho-Penalosa and Aithison [Cam87] stimulated work on harmonic balance in the microwave community. Harmonic-balance techniques have been successfully applied to nonlinear distributed FET model as well [Ong90].

In this section we will describe the commonly used harmonic-balance method and its application to large-signal MESFET characterization. The load-pull method [Tak76] will be briefly described.

13.6.1 Harmonic-Balance Method

Harmonic balance [Cun58; Mee81] can be seen as the extension of phasor analysis [Des69] from linear to nonlinear differential equations. In phasor analysis the steady state solution to an ordinary linear equation whose stimulus is sinusoidal is found by assuming that the solution has the form $x(t) = \text{Re}(Xe^{jw_0 t})$, substituting it into the differential equation, evaluating the derivatives, and solving the resulting algebraic equation for X. When the differential equation is not linear, the solution is rarely a simple harmonic function of time but can often be approximated to the first order by such a function. With harmonic balance, an approximate solution is found by assuming the solution to be purely sinusoidal and choosing its magnitude and phase to satisfy the differential equation at the fundamental only [Kun86]. The approximate solution $x(t) = \text{Re}(Xe^{jw_0 t})$ is substituted into the differential equation, all frequency components generated other than the fundamental are ignored, and the resulting algebraic equation is solved for X.

The assumed solution when using harmonic balance need not be purely sinusoidal. It could consist of a linear combination of sinusoids that are harmonically related,

TABLE 13.3 The Method of Harmonic Balance

Given a differential equation of the form

$$f(x, \dot{x}, u) = 0, \tag{1}$$

where $u \in P(T_0)$ is the stimulus wave form, x is the unknown wave form to be found, and f is continuous and real.

Step 0 Assume that the solution x exists, is real, and belongs to $P(T_0)$. Then

$$x(t) = \sum X(k)e^{jw_0 t},$$

where

$$w_0 = 2\pi/T_0.$$

Step 1 Substitute the assumed solution and its derivative into f. Note that $x \in P(T_0)$ implies that $dx/dt \in P(T_0)$, and since $u \in P(T_0)$ as well, $f(x, \dot{x}, u) \in P(T_0)$. Write the resulting equation as a Fourier series:

$$f(x(t), \dot{x}(t), u(t)) = \sum F(X, U, k)e^{jkw_0 t}, \tag{2}$$

where

$$X = [\ldots, X(-1), X(0), X(1), \ldots]^T$$
$$U = [\ldots, U(-1), U(0), U(1), \ldots]^T$$
$$u(t) = \sum U(k)e^{jw_0 t}.$$

Step 2 Solve the system of nonlinear algebraic equations

$$F(X, U, k) = 0 \quad \text{for all } k \in \mathbf{Z}, \tag{3}$$

the integer numbers for X.

Source: Bandler, J. W., Q. J. Zhang, and R. M. Biernacki, *IEEE MTT-S Digest* (1988): 1041–1044.

making the solution periodic. If the differential equation is such that, once the solution is substituted in, the resulting equation can be factored into a sum of purely sinusoidal terms, then superposition and the orthogonality of sinusoids at different harmonics can be exploited to break the resulting algebraic equation up into a collection of simpler equations, one for each harmonic. The equations are solved by finding the coefficients of the sinusoids in the assumed solution that result in the balancing of the algebraic equation at each harmonic. The harmonic-balance method is very attractive if the steady state solution is the only required result [Obr85].

The application of harmonic balance can be stated as a simple procedure referred to as the *method of harmonic balance* [Kun86], as shown in Table 13.3.

In the table the statement (3) is satisfied if and only if (1) is satisfied is called the *principle of harmonic balance*. It can easily be proved by applying Parsevals theorem to (2) [Kun86].

13.6.2 Harmonic-Balance Simulation and Sensitivity Analysis

The harmonic-balance (HB) method has become an important tool for the analysis of nonlinear circuits. In this section a novel theory for exact sensitivity analysis of nonlinear circuits based on harmonic-balance simulation is introduced [Ban88]. The linear part of the circuit can be large and can be hierarchically decomposed, highly suited to modern microwave CAD. Analysis of the nonlinear part is performed in the time domain, and the large-signal steady state periodic analysis of the overall circuit is carried out by the HB method.

The adjoint network concept involves solving a set of linear equations whose coefficient matrix is available in many existing HB programs [Gil86; Kun86; Kun88]. The solution of a singly adjoint system is sufficient for the computation of sensitivities with respect to all parameters in both the linear and nonlinear subnetworks, as well as in bias, driving sources and terminations.

The sensitivities were proposed to be exact in terms of the harmonic balance method itself [Ban88]. Computational effort includes solving the adjoint linear equations and calculating the Fourier transforms of all time-domain derivatives at the nonlinear element level. Significant central processing unit (CPU) time savings are achieved over the perturbation method [Ban88].

Notation

In this section we will describe the notation used by Bandler et al. [Ban88]. Real vectors containing voltages and currents at time t are denoted by $\mathbf{v}(t)$ and $\mathbf{i}(t)$. Capitals $\mathbf{V}(k)$ and $\mathbf{I}(k)$ are used to indicate complex vectors of voltages and currents at harmonic k. A subscript t at $\mathbf{V}_t(k)$ indicates that the vector contains the nodal voltages at all N_t nodes (both internal and external) of a linear subnetwork. If there is no subscript, then the vector corresponds to the port voltages (currents) at all N ports of the reduced linear subnetwork. A bar denotes the split real and imaginary parts of a complex vector. In particular, $\underline{V}$ or $\underline{I}$ are real vectors containing the real and the imaginary parts of $\mathbf{V}(k)$ or $\mathbf{I}(k)$ for all harmonics $k, k = 0, 1, \ldots, H - 1$. The total number of harmonics taken into consideration, including dc, is H. The caret distinguishes quantities of the adjoint system.

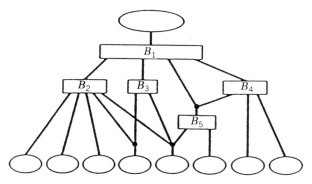

Fig. 13.26 An arbitrary circuit hierarchy. Each thick line represents a group of nodes. Each rectangular box represents a connection block for a subcircuit. Each bottom circular box represents a circuit element, and the top circular box represents the sources and loads. [Bandler, J. W., Q. J. Zhang, and R. M. Biernacki, *1988 MTT-S Digest*, pp. 1041–1044, IEEE (1988)]

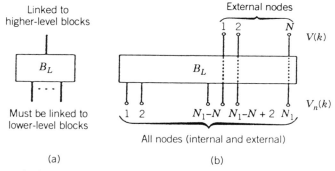

Fig. 13.27 A typical subcircuit connection block: (*a*) As seen in Fig. 13.26; (*b*) detailed representation of all the nodes of the subnetwork. [Bandler, J. W., Q. J. Zhang, and R. M. Biernacki, *1988 MTT-S Digest*, pp. 1041–1044, IEEE (1988)]

Linear and Nonlinear Simulation

Consider the arbitrary circuit of Fig. 13.26. A typical subnetwork containing internal and external nodes is shown in Fig. 13.27. An unpartitioned or nonhierarchical approach is a special case of Fig. 13.26 when only one level exists. Consider an unterminated N-port circuit at the highest level of hierarchy because of the importance of the reference plane in microwave circuits. The N-port description is needed for the harmonic-balance equations.

As an example, all nodal voltages $\mathbf{V}_t(k)$ of a subnetwork were computed using the results of a higher-level simulation, using the external voltages $\mathbf{V}(k)$. We solve

$$\mathbf{A}(k)\begin{bmatrix}\mathbf{V}_t(k)\\ \mathbf{I}(k)\end{bmatrix} = \begin{bmatrix}0\\ \mathbf{V}(k)\end{bmatrix}, \qquad (13.42)$$

where the matrix $\mathbf{A}(k)$ is a modified nodal admittance matrix of the subnetwork.

516   MONOLITHIC MICROWAVE INTEGRATED CIRCUITS

The simulation of the overall nonlinear circuit is to find a $\underline{V}$ such that

$$F(\underline{V}) = \underline{I}_{NL}(\underline{V}) + \underline{I}_L(\underline{V}) = 0, \tag{13.43}$$

where the vectors $\underline{I}_L$ and $\underline{I}_{NL}$ are defined as the currents into the linear and nonlinear parts at the nodes of their connection. The Newton update for solving Eq. (13.43) is

$$\underline{V}_{new} = \underline{V}_{old} - \underline{J}^{-1} \underline{F}(\underline{V}_{old}), \tag{13.44}$$

where $\underline{J}$ is the Jacobian matrix.

Adjoint System Simulation

Suppose $\underline{V}_{out}$ is the real or imaginary part of output voltage V_{out} and can be selected from the voltage vector $\underline{V}$ be a vector $\underline{e}$ as

$$\underline{V}_{out} = \underline{e}^T \underline{V}. \tag{13.45}$$

The adjoint system is the linear equation

$$\underline{J}^T \hat{\mathbf{V}} = \underline{e}, \tag{13.46}$$

where $\underline{J}$ is the Jacobian at the solution of Eq. (13.43). Notice that the LU factors of $\underline{J}$ is available from the last iteration of Eq. (13.44).

The adjoint voltages can be computed even if the output port is suppressed from the harmonic equation (13.43). The adjoint voltages at the external nodes of the linear subnetwork can be computed first. The resulting vector, denoted by $\hat{\mathbf{V}}_L$, is then transformed to the actual adjoint excitations of the overall circuit (including both the linear and nonlinear parts) to be incorporated in Eq. (13.46) instead of $\underline{e}$. The final equation takes the form

$$\underline{J}^T \hat{\mathbf{V}} = \underline{Y} \hat{\mathbf{V}}_L, \tag{13.47}$$

where $\underline{Y}$ is the split real/imaginary nodal Y matrix for all harmonics. The solution of Eq. (13.46) or (13.47) provides the adjoint voltages at the external ports at the highest level of the hierarchy. The internal adjoint voltages can be determined from the equation

$$\mathbf{A}^T(k) \begin{bmatrix} \hat{\mathbf{V}}_t(k) \\ -\hat{\mathbf{I}}(k) \end{bmatrix} = \begin{bmatrix} 0 \\ -\hat{\mathbf{V}}(k) \end{bmatrix}. \tag{13.48}$$

Equation (13.48) is used iteratively down the levels of the hierarchy until all desired adjoint voltages are found.

Sensitivity Expressions

Sensitivity analysis plays an important role in optimization. Suppose that a variable x belongs to branch b. The following formula for computing the exact sensitivity of

V_{OUT} w.r.t. x can be derived [Ban88]:

$$\frac{\partial V_{\text{OUT}}}{\partial x} = \begin{cases} -\sum_k \text{Real}[\hat{\mathbf{V}}_b(k)V_b^*(k)G_b^*(k)], & (13.49\text{a}) \\ -\sum_k \text{Real}[\hat{\mathbf{V}}_b(k)G_b^*(k)], & (13.49\text{b}) \\ -\sum_k \text{Imag}[\hat{\mathbf{V}}_b(k)G_b^*(k)], & (13.49\text{c}) \end{cases}$$

where

(13.49a) if $x \in$ linear subnetwork,

(13.49b) if $x \in$ nonlinear VCCS (voltage-controlled current source) or nonlinear resistor or real part of a complex driving source,

(13.49c) if $x \in$ nonlinear capacitor or imaginary part of a complex driving source.

Here the asterisk denotes the complex conjugate. Complex quantities $\mathbf{V}_b(k)$ and $\hat{\mathbf{V}}_b(k)$ are the voltages of branch b at harmonic k and are obtained from vectors $\underline{V}$ and $\hat{\mathbf{V}}$, respectively. $G_b(k)$ denotes the sensitivity expression of the element containing variable x. A list of various cases of $G_b(k)$ is given in Table 13.4.

To approximate the sensitivities using the traditional perturbation method, one needs a circuit simulation for each variable. The best possible situation for this method is that all simulations finish in one iteration. For this adjoint sensitivity analysis, the major computation (i.e., solving the adjoint equations) is done only once for all variables. The worst case for this approach takes less computation than the best situation of the perturbation method [Ban88].

Equations (13.49) can be used to formulate the gradient vectors for design optimi-

TABLE 13.4 Sensitivity Expressions

Type of Element	Expression for $G_b(k)$
Linear G	1
Linear R	$-1/R^2$
Linear C	$j\omega_k$
Linear L	$-1/(j\omega_k L^2)$
Nonlinear VCCS or resistor $i = i(v(t), x)$	[kth Fourier coefficient of $\partial i/\partial x$]
Nonlinear capacitor $q = q(v(t), x)$	ω_k [kth Fourier coefficient of $\partial q/\partial x$]
Current driving source	1
Voltage driving source	1/source impedance

ω_k is the kth harmonic angular frequency

Source: Bandler, J. W., Q. J. Zhang, and R. M. Biernacki, *IEEE MTT-S Digest* (1988): 1041–1044.

zation and yield maximization of nonlinear circuits. Table 13.5 lists the gradients of a FET mixer conversion gain with respect to various variables, expressed as simple functions of $\partial V_{OUT}/\partial x$.

13.6.3 Load-Pull Method

The load-pull method can be used to determine the output impedance under large-signal conditions. The device is tuned with variable tuners to the desired performance level. The optimum load can be determined by measuring the output network containing the tuners. An improved method by Takayama [Tak76], shown in Fig. 13.28, generates the tuner electronically by feeding some of the input power back into the output. This incident signal can be varied in phase and amplitude, and thus arbitrary load impedances can be generated electronically. The advantage is the elimination of mechanical tuners that do not allow orthogonal tuning and create reproducibility problems. However, this method has a basic problem of accurately determining the output power for output reflection coefficients close to unity, so it is suitable for large unmatched devices. This method is particularly well-suited for computer control, since it does allow orthogonal tuning.

13.6.4 Parameter-Extraction Program

Choosing a large-signal model that provides sufficient accuracy in the prediction of device characteristics can be a problem for the microwave engineer. In this section we describe a parameter-extraction program developed at Motorola by Miller et al. [Mil90]. Six empirical MESFET models and one physically based model are

TABLE 13.5 Gradients of Mixer Conversion Gain

Variable x	Gradient Expression
RF power	$c\,\text{Real}\{(\partial V_{out}/\partial x)/V_{out}\} - 1$
$R_g(f_{RF})$	$c\,\text{Real}\{(\partial V_{out}/\partial x)/V_{out}\} + c/(2R_g(f_{RF}))$
$R_d(f_{IF})$	$c\,\text{Real}\{(\partial V_{out}/\partial x)/V_{out} - 1/(R_d(f_{IF}) + jX_d(f_{IF}))\} + c/(2R_d(f_{IF}))$
$X_d(f_{IF})$	$c\,\text{Real}\{(\partial V_{out}/\partial x)/V_{out} - j/(R_d(f_{IF}) + jX_d(f_{IF}))\}$
Any other parameter	$c\,\text{Real}\{(\partial V_{out}/\partial x)/V_{out}\}$

$c = 20/\ln 10$.

R and X represent the real and the imaginary parts of the impedance terminations, respectively. Subscripts g and d represent the gate and the drain terminations, respectively.

Complex quality $\partial V_{out}/\partial x$ is obtained by solving (5)–(8) twice, once for the real part and the other for the imaginary part. The LU factors of $\bar{J}$ and the Fourier transforms of element sensitivities are common between the two operations.

Source: Bandler, J. W., Q. J. Zhang, and R. M. Biernacki, *IEEE MTT-S Digest* (1988): 1041–1044.

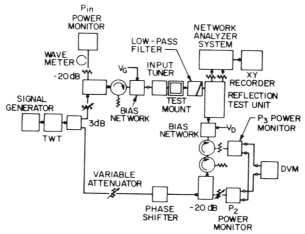

Fig. 13.28 Electronic load-pull system. [Takayama, Y., *1976 IEEE MTT-S International Microwave Symp. Digest*, pp. 218–220, IEEE (1976)]

incorporated into the program together with three new empirical HEMT models. These extraction routines determine the model parameters with which optimum performance predictions can be made by each model.

Compared are the following MESFET models:

1. The model proposed by Curtice and Ettenberg [Cur85].
2. The hyperbolic tangent model presented by Curtice, modified by Meta-Software [Cur80], [Met89].
3. The physically based model originally proposed by Lehovec and Zuleeg, modified by Hartgring and Golio [Leh70; Har81; Gol85].
4. The model proposed by Materka and Kacprzak [Kac83].
5. The model proposed by Statz, et al. [Sta87].
6. The model proposed by Curtice [Cur80].
7. The model proposed by Statz, et al., modified by Triquint Semiconductor, Inc. [McC89].

The three HEMT models are obtained by applying a general modification procedure to existing MESFET models. The following models are used for this portion of the study:

8. Model 6 with additional HEMT parameters.
9. Model 4 with additional HEMT parameters.
10. Model 2 with additional HEMT parameters.

The parameter-extraction program uses two methods to optimize the parameters of the model equations. A constrained random optimization is used first, followed by the Levenberg-Marquardt method [Lev44; Mar63]. The use of a random technique enables the Levenberg-Marquardt optimization to begin from a new starting point each time the extraction program is executed, thus addressing the problem of entrapment of local minima. A better algorithm using simulated annealing to avoid entrapment of local minimum [Vai89] is described in Chapter 8.

The model equations for drain current are incorporated into the extraction program for each model. The derivatives for I_{ds} with respect to V_{gs} and V_{ds} are expressed analytically to determine the equations for transconductance and output conductance, respectively. These expressions are also incorporated into the extraction program. The model equations for the device capacitance are not included in this comparison. To obtain capacitance parameters for the full-range signal prediction comparisons, a random optimization routine was used.

Optimization of Model Parameters

Optimization problem is to adjust the model parameters in the device model equation to improve the fit between measured and modeled data. The objective function chosen in this program measures the error in both the dc and ac characteristics of the device. This objective function is termed the *error magnitude unit* (EMU) and is defined as

$$\text{EMU} = \frac{\text{EMU}'}{3n}. \tag{13.50}$$

Here n is the number of bias points, $\text{EMU}' = \sum (E_{ids}(i)WF_{ids} + E_{gm}(i)WF_{gm} + E_{Gds}(i)WF_{Gds})]/E_{aaa}(i)$, where

$$E_{aaa}(i) = \left[\frac{\text{Meas}(i) - \text{Mod}(i)}{\text{Meas}(i)^2}\right]^2,$$

and WF_{ids}, WF_{gm}, and WF_{Gds} are normalized weighting factors.

Figures 13.29, 13.30, and 13.31 show the measured and modeled results of this optimization approach for dc I_{ds}, RF G_{ds}, and RF g_m; the weighting factors are all set to 1. The device used in these measurements was a depletion mode RF probeable 0.5×300-μm MESFET manufactured by Triquint Semiconductor. The parameter extraction was performed for model 2. The limitation of this approach was that it could not model the low-frequency dispersion of G_{ds} and g_m observed in GaAs MESFETs [Gol88].

13.7 APPLICATIONS OF GaAs ICs

U.S. government-funded research and development contracts have led to early progress in the development of GaAs ICs for military applications, namely in radar, electronic warfare, and communication systems [Gla88]. Commercial applications of GaAs IC technology took a little longer to realize—the first commercial GaAs IC product is believed to have been Pacific Monolithics' TVRO module introduced in March 1985. Today GaAs ICs are the key devices in test and measurement instruments, fiber optics, and radio communication systems. In this section we will briefly mention the electron warfare applications of MMIC and the commercial applications.

APPLICATIONS OF GaAs ICs   521

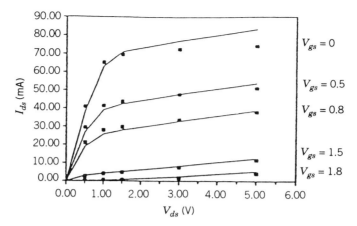

Fig. 13.29 A comparison of dc measured and modeled drain-source current versus drain-source voltage. Blocks are measured for a 0.5 × 300-μm device, while the solid lines are predictions from model 2 using optimized parameters. [Miller, M., M. Golio, B. Beckwith, E. Arnold, D. Halchin, S. Ageno, and S. Dorn, *1990 MTT-S Digest*, pp. 1279–1282, IEEE (1990)]

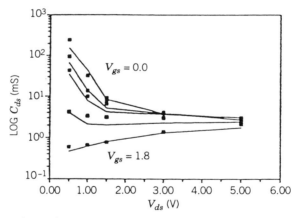

Fig. 13.30 A comparison of RF measured and modeled output conductance. Blocks are measured for a 0.5 × 300-l_m device, while the solid lines are predictions from model 2 using optimized parameters. [Miller, M., M. Golio, B. Beckwith, E. Arnold, D. Halchin, S. Ageno, and S. Dorn, *1990 MTT-S Digest*, pp. 1279–1282, IEEE (1990)]

13.7.1 Electronic Warfare Applications

The electronic warfare systems have experienced rapid technical growth. The microwave environment is becoming ever more cluttered by increasing emitter/beam densities and higher frequencies of operation. Advanced radar systems are using spread spectrum techniques and complex intrapulse modulations. These features place new performance demands on electron warfare processing [Roo88].

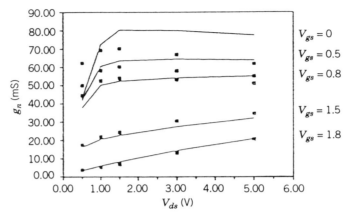

Fig. 13.31 A comparison of RF measured and modeled transconductance. Blocks are measured for a 0.5×300-μm device, while the solid lines are predictions from model 2 using optimized parameters. [Miller, M., M. Golio, B. Beckwith, E. Arnold, D. Halchin, S. Ageno, and S. Dorn, *1990 MTT-S Digest*, pp. 1279–1282, IEEE (1990)]

Currently there is a joint venture between Raytheon and Texas Instruments to produce GaAs devices under phase 2 of the tri-service Microwave/Millimeter Wave Monolithic IC (MIMIC) program with a contract of $83.5 million. As a result GaAs products could turn up in such weaponry as the Advanced Air-to-Air Missile (AARAAM) and Sense-and-Destroy Armor (SARARM) [Isc92].

13.7.2 Commercial Applications

The commercial market for GaAs ICs can be divided into four segments: industrial, communications, computers, and consumer. Each of these market segments can be split into three groups according to circuit type: analog, digital, and mixed analog-digital [Nog90]. Two market models are shown in this section. The first represents known current production GaAs ICs that are being shipped in their end use product. The second model is a combination of applications under development that are being supported with either custom GaAs ICs or production circuits from the merchant suppliers.

Industrial

The primary industrial application for GaAs ICs is in test and measurement instrumentation. All three types of circuits—analog, digital, and mixed analog-digital—are currently being used in industrial instruments. Anritsu and Advantest have more than ten instruments that are equipped with GaAs ICs [Nog90].

Pulse pattern generators use four such kinds of GaAs ICs: T-type and D-type flip-flops, line receiver ICs, and switch array ICs. The T-type flip-flop chain generates subharmonic timing pulses from 10 Gb/s to a few hundred Mb/s pulse. The line receiver IC delivers complimentary output from a single input, and these output offset levels and logic swings are adjustable for 50-ohm loads. The line receiver IC is used as a pulse date divider. The GaAs FET switch is suitable for pulsing circuits because of its low-pass dc performance. A delay time control unit contains switch array ICs and delay lines, which adjust data delay electrically at the multiplexer input.

The spectrum analyzer uses wide-band amplifiers and prescaler ICs. The prescaler IC is a main component of frequency counter block. The wide amplifier ICs are used in two blocks, as a receiver amplifier and as an oscillator buffer. These applications require different wideband amplifiers. Low harmonic distortion and gain flatness are necessary for receiver amplifiers to prevent an input signal from generating error signals. A stable input impedance is required for the YIG-tuned oscillator buffer amplifier, since the impedance of an oscillator load is sensitive to a pure oscillator spectrum.

The HP8780A sector signal generator—a synthesizer capable of complex vector modulation—utilizes an analog circuit that can be described as "precision" analog [Kel87]. This amplifier operates from 50 MHz to 3 GHz with a 30-dB gain and a 20-dB possible gain adjustment. The input is single ended, but it is converted to a differential topology at the first stage. The output is differential, and is considered a precision analog circuit since waveform symmetry, total gain, gain flatness, and distortion are fully controlled. A key design parameter is low distortion at 10-dBm output power levels. This chip utilizes 38 FETs, and the size is 1×2 mm.

Communications

The first commercial use of GaAs ICs in communications was in AT&T Communications' 1.7 Gb/s lightwave system, which began production in November 1986 [Gla88]. AT&T utilizes three different GaAs ICs in their fiber optic system: The first circuit is a transimpedance preamplifier of about 30 transistors complexity that has a 4-GHz bandwidth at 1.5 kΩ impedance level. The chip includes a transimpedance stage (which converts input photodiode current to output voltage), two voltage gain stages, and an output buffer stage for driving low-impedance transmission loads. The second IC is a decision and regeneration circuit that has a bit error rate of 3×10^{-11} at -32-dBm signal levels. This circuit utilizes enhancement- and depletion-mode devices and consists of a high-gain bandwidth amplifier, D-type flip-flop, and the complementary 50-ohm output buffers. The final IC is a 50-mA laser driver circuit with a 20–80% rise and fall time of 150 ps. The circuit converts an input voltage signal at rates up to 3 Gb/s to an output pulse current that drives the laser.

Digital multiplexers and demultiplexers are the basic components in serial-to-parallel and parallel-to-serial conversion for time division multiplexed communication systems. The high speed of GaAs digital integrated circuits can accommodate several silicon channels with no compromise in the individual channel bandwidth. In integrated electrooptical devices, the lasers, detectors, and drive electronics can all be monolithically integrated with multiplexers and demultiplexers to provide multi-gigahertz transceiver capability. NEC utilizes two kinds of digital ICs in their first fiber optic system. The first circuit is a 1:4 demultiplexer of MSI that has a 2-Gb/s maximum operating speed. The IC consists of SCFL circuits and an ECL-compatible output data level, so the IC can be used with Si-bipolar ICs on the same HIC in the fiber optic system. The second IC is a 4:1 multiplexer that has the same features as the demultiplexer. These digital ICs are the key devices of the fiber optic terminal. The development is now moving toward 10 Gb/s [Nog90].

The future for GaAs ICs is extremely promising in the communications market, with continued applications in fiber optic communication systems. Under development are a NASA Search and Rescue program for navigation and rescue, MLS (microwave landing system) for advanced commercial aircraft, and phased array antennas for communications. NTT has been working on a 30-GHz-band transmitter/receiver for satellites. Other future applications include the GPS receiver, MLS radars, and phased array antennas for communications satellites.

Computers

These are many promising GaAs digital IC applications in production for computers and networking products. GaAs will enhance the performance of the next generation supercomputers. The CRAY™-3 is expected to be the first major computer system to be implemented with GaAs logic [Kei87]. A GaAs 32-bit asynchronous serial data transmitter and receiver chip set has been reported by the Semiconductor Technology Laboratory at Oki [Nog90]. This chip set can be used in a 32-bit data transmission I/O board between several high-speed processors [IEC89]. The chips are sealed in 80-pin packages and have 500-Mb/s transmission speed. The I/O ICs with their required high-speed performance will find suitable applications in the computer market.

GaAs ICs may turn out to be used in microprocessors, for instance, the RISC or SPARC microprocessor. Another potential implementation may be GaAs microprocessors with multiple-chip designs or cache controllers.

Consumer

For the past few years Matsushita wideband amplifiers packaged in plastic have been used in the IF circuits of DBS (direct broadcast satellite) receivers [Nog90]. About two hundred thousand DBS receivers were shipped per month in Japan in 1989. This amplifier system has been also used in cellular radio receivers because of its low-noise figure (2 dB at 1 GHz) and compact size.

The largest microwave market in Japan in TV tuners. In 1988 Hitachi announced that it would start production of a UHF/VHF one-chip tuner GaAs IC. With this IC tuner high-quality, low-noise tuning will be possible for big-screen TV. The GaAs IC tuner will be all-band (include UHF/VHF) and will use an up-conversion system of up to 1.6-GHz or higher-IF frequency.

In 1991 AT&T Microelectronics and the Mitsubishi Electric Corp. signed a unique agreement to develop high-performance GaAs ICs for wireless products, such as cellular phones. One of these ICs will be a dual-module prescaler. A prescaler is a key component of the frequency synthesizer of a radio transmitter/receiver. The frequency synthesizer will use the prescaler, combined with digitally controlled counters in a PLL scheme, to produce 100 or more selectable channels.

In the future the consumer market could include RF circuits of DBS receivers and more highly integrated receivers of CATV and cellular radio. Mobile applications such as in collision avoidance systems, radar detectors and engine speed sensing ICs are to be expected. The key to success in the future consumer market is reducing price, improving reliability, and lowering consumption.

Outlook for the Future

The development of low-cost MMIC in the near future is much needed. There are several factors that may contribute to the production of low-cost MMICs [Ker88]:

1. Development of a stable and standardized process of automation and quality control of GaAs.
2. Reduction of the development costs by improving design accuracy and producing interactive CAD software.
3. Improvement of integration density for a more efficient use in GaAs.
4. Development of fast and accurate testing procedures.
5. Development of low-cost high-performance packages.

The shipping of GaAs ICs to end users is produced with depletion-mode MESFET technology. Some facilities are readying E/D&D processes offering two different pinch-off MESFETs. Now technologies based on HEMTs and HBTs are also being explored.

REFERENCES

[Sch88] Schellenberg, J. M., and T. R. Apel, in *GaAs Integrated Circuits*, ed. by J. Mun, Macmillan (1988).
[Puc85] Pucel, R. A., in *Gallium Assenide Technology*, ed. by D. K. Ferry, SAMS (1985).
[Cau70] Caulton, M., and H. Sobol, *IEEE J. Solid State Circuits*, **SC-5** (1970): 292–303.
[Mon48] Montegomery, C. G., *Principles of Microwave Circuits*, McGraw-Hill (1948).
[Hew72] Hewlett-Packard, S-parameter design, *Hewlett-Packard Application Notes* **154** (Apr. 1972).
[Lia85] Liao, S. Y., *Microwave Devices and Circuits*, 2d ed., Prentice Hall (1985).
[Eng74] Engberg, J., *Proc. 1974 European Microwave Conf.* pp. 385–389 (1974).
[Ven82] Vendelin, G. D., *Design of Amplifiers and Oscillators by the S-Parameter Method*, Wiley-Interscience (1982).
[Pen86] Pengelly, R. S., *Microwave Field-Effect Transistors—Theory, Design and Applications*, 2d ed., Research Studies Press (1986).
[Sov90] Sovero, E., D. Deakin, W. J. Ho, G. D. Robinson, C. W. Farley, J. A. Higgins, and M. F. Chang, *GaAs IC Symp. Digest* (1990): 169.
[Ish88] Ishizaki, M., T. Hamabe, Y. Oohashi, and S. Asai, *1988 MTT-S Digest* (1988): 461.
[Pat59] Patterson, K. G., *IRE Trans. Microwave Theory and Tech.* **MTT-7** (Oct. 1959): 466–473.
[Mar48] Marcuvitz, N., *Waveguide Handbook*, pp. 336–350, McGraw-Hill (1948).
[LeS90] LeSage, S. R., M. J. Mendes, and F. J. Sullivan, *GaAs IC Symp. Digest* (1990): 353.
[Ava90] Avasarala, M., D. S. Day, S. Chan, C. Hua, and J. R. Basset, *GaAs IC Symp. Digest* (1990): 353.
[Cam88] Camilleri, N., B. Kim, H. Q. Tserng, and H. D. Shih, in *IEEE 1988 Microwave and Millimeter-Wave Monolithic Circuits Symp. Digest*, pp. 129–132, IEEE (1988).
[Gon84] Gonzalez, G., *Microwave Transistor Amplifiers*, Prentice Hall, (1984).
[Soa83] Soares, R. A., in *Applications of GaAs MESFETs*, p. 160, ed. by R. Soares, J. Craffeuil, and J. Obregon, Artech House (1983).
[Fuk78] Fukui, H., *Bell Syst. Tech. J.* **57** (1978): 771.
[DiL79] DiLorenzo, J., and W. Wisseman, *IEEE Trans. Microwave Theory Tech.* **MTT-27** (1979): 367.
[DiL82] DiLorenzo, J., *GaAs FET Principles and Technology*, Artech House (1982).
[Wro83] Wroblewski, R., G. Salmer, and Y. Crosnier, *IEEE Trans. Electron Devices* **ED-30** (1983): 154.

[Pen86] Pengelly, R. S. *Microwave Field-Effect Transistors—Theory, Design and Applications*, 2d ed., Research Studies Press (1986).

[Hes82] Hesegawa, F., in *GaAs FET Principles and Technology*, pp. 219–255, ed. by J. V. DiLorenzo and D. D. Khandelwal, Artech House (1982).

[Shu87] Shur, M., *GaAs Devices and Circuits*, Plenum Press (1987).

[Eas80] Eastman, L. F., S. Tiwari, and M. S. Shur, *Solid State Electron.* **23** (1980): 383–389.

[Wem82] Wemple, S. H., and H. Huang, in *GaAs FET Principles and Technology*, pp. 312–347, ed. by J. V. DiLorenzo and D. D. Khandelwal, Artech House (1982).

[Kim86] Kim, B., H. M. Macksey, H. Q. Tserng, H. D. Shih, and N. Camilleri, *GaAs IC Symp. Tech. Digest*, pp. 61–63, IEEE (1986).

[Kim87] Kim, B., H. M. Macksey, H. Q. Tserng, H. D. Shih, and N. Camilleri, *Microwave J.*, March 1987, pp. 153–164 (1986).

[Ber80] Bert, A. G., and D. Kaminsky, *IEEE MTT-Symp. Digest* (1980): 487–489.

[Tse85] Tserng, H. Q., and P. Saunier, *Electron. Lett.* **21** (1985): 950–951.

[How74] Howe, H., *Stripline Circuit Design*, pp. 94–95, Artech House (1974).

[She87] Sheng, N. H., M. F. Chang, and P. Asbeck, *IEEE IEDM Digest*, pp. 619–622, IEEE (1987).

[Wan89] Wang, N. L., *Proc. 12th Biennial Conf. Adv. Concepts in High-Speed Semiconductor Devices and Circuits*, Cornell University, Ithaca (1989).

[Bay87] Bayraktaroglu, B., N. Camilleri, H. D. Shih, and H. Q. Tserng, *IEEE MTT-S Symp. Digest* (1987): 969–972.

[Ali89] Ali, F., I. Bahl, and A. Gupta, eds., *Microwave and Millimeter-Wave Heterostructure Transistors and Their Applications*, Artech House (1989).

[Ram90] Ramachandran, R., M. Nijjar, A. Podell, E. Stoneham, and S. Mitchell, *GaAs IC Symp. Digest* (1990): 357.

[Nag75] Nagel, L. W. SPICE2: A computer program to simulate semiconductor circuits, Electronics Res. Lab., Univ. of California, Berkeley, em. ERL-M520, May 9 (1975).

[Sob85] Sobhy, M. I., A. K. Jastrzebski, R. S. Pengelly, J. Jenkins, and J. Swift, in *Proc. 15th European Microwave Conf.*, Paris, pp. 925–930 (1985).

[Cur85] Curtice, W. R., and M. Ettenberg, *IEEE Trans. Microwave Theory Tech.* **33**, 12 (1985): 1383–1394.

[Riz83] Rizzoli, V., A. Lipparini, and E. Marazzi, *IEEE Trans. Microwave Theory Tech.* **MTT-31** (Sept. 1983): 762–769.

[Nak76] Nakhla, M. S., and J. Vlach, *IEEE Trans. Circuits Syst.* **23** (Feb. 1976).

[Cam83] Camacho-Penalosa, C., *IEEE Trans. Microwave Theory Tech.* **31** (Sept. 1983): 724–730.

[Cam87] Camacho-Penalosa, C., and C. S. Aitchison, *IEEE Trans. Microwave Theory Tech.* **35** (1987): 643–652.

[Ong90] Ongareau, E., M. Aubourg, M. Gayral, and J. Obregon, *IEEE MTT-S Tech. Digest* (1990): 323.

[Tak76] Takayama, Y., *1976 IEEE MTT-S International Microwave Symp. Digest*, pp. 218–220, IEEE (1976).

[Cun58] Cunningham, W. J., *Introduction to Nonlinear Analysis*, McGraw-Hill (1958).

[Mee81] Mees, A. I., *Dynamics of Feedback Systems*, Wiley (1981).

[Des69] Desoer, C. A., and E. S. Kuh, *Basic Circuit Theory*, McGraw-Hill (1969).

[Kun86] Kundert, K. S., and A. Sangiovanni-Vincentelli, *IEEE Trans. Computer-Aided Design* **5** 4 (1986): 521–535.

[Obr85] Obregon, J., in *Proc. 15th European Microwave Conf. Paris*, pp. 1089–1093 (1985).

[Ban88] Bandler, J. W., Q. J. Zhang, and R. M. Biernacki, *1988 IEEE MTT-S Digest*, pp. 1041–1044, IEEE (1988).

[Gil86] Gilmore, R., *IEEE Trans. Microwave Theory Tech.* **34** (1986): 1294–1307.

[Kun86] Kundert, K. S., and A. Sangiovanni-Vincentelli, *IEEE Trans. Computer-Aided Design* **5** (1986): 521–535.

[Kun88] Kundert, K. S., G. B. Sorkin, and A. Sagiovanni-Vincentelli, *IEEE Trans. Microwave Theory Tech.* **36** (1988): 366–378.

[Tak76] Takayama, Y., *1976 IEEE MTT-S International Microwave Symp. Digest* (1976): 218–220.

[Mil90] Miller, M., M. Golio, B. Beckwith, E. Arnold, D. Halchin, S. Ageno, and S. Dorn, *1990 IEEE MTT-S Digest* (1990): 1279–1282.

[Cur85] Curtice, W. R., and M. Ettenberg, *IEEE Trans. Microwave Theory Tech.* **33** (1985): 1383–1394.

[Cur80] Curtice, W. R., *IEEE Trans. Microwave Theory Tech.* **28** (1980): 448–456.

[Met89] *HSPICE User's Manual*, Meta-Software, Campbell, CA (1989).

[Leh70] Lehovec, K., and R. Zuleeg, *Solid State Electronics* **13** (1970): 1415–1429.

[Har81] Hartgring, C., Silicon MESFETs, Ph.D. dissertation, UC Berkeley (1981).

[Gol85] Golio, J. M., *IEEE Circuits and Devices* (1985): 21–30.

[Kac83] Kacprzak, T., and A. Materka, *IEEE J. Solid-State Circuits* **18** (1983): 211–213.

[Sta87] Statz, H., *IEEE Trans. on Elec. Devices* **34** (1987): 160–169.

[McC89] McCamant, A., D. Smith, and G. McCormack, *An Improved GaAs MESFET Model for SPICE*, TriQuint Semiconductor Inc., Beaverton, OR (1989).

[Lev44] Levenberg, K., *Quart. Appl. Math.* 2 (1944): 164–168.

[Mar63] Marquardt, D., *SIAM J. Appl. Math.* **11** (1963): 431–441.

[Vai89] Vai, M., S. Prasad, N. C. Li, and F. Kai, *IEEE Trans. Electron Devices* (Apr. 1989).

[Gol89] Golio, J. M., and D. Warren, *IEEE Trans. Microwave Theory Tech.* (Nov. 1988): 1535–1539.

[Gla88] Gladstone, J., *1988 IEEE-MTT-S Digest* (1988): 93.

[Roo88] Roosild, S., and A. Firstenberg, in *GaAs Integrated Circuits*, ed. by J. Mun, Macmillan (1988).

[Isc92] Iscoff, R., *Semiconductor Int.* (Mar. 1992): 60.

[Nog90] Noguchi, T., *GaAs IC Symp.* (1990): 263.

[Kel87] Kelly, W. M., M. J. Woodward, E. B. Rodal, P. A. Szente, and J. D. McVey, *Hewlett-Packard J.* **39**, 11 (Dec. 1987): 48–52.

[Kei87] Keifer, D., and J. Heightly, *Tech. Digest GaAs IC Symp.* (Oct. 1987): 3–6.

[IEC89] *Digest of Technical Papers of IEC Electron Meeting*, **ED89**, pp. 130–162 (Jan. 1990).

[Ker88] Kermarrec, C., and C. Rumelhard, *GaAs MESFET Circuit Design*, ed. by R. Soares Artech House (1988).

14

GaAs DIGITAL INTEGRATED CIRCUITS

14.1 INTRODUCTION

One of the most prominent applications of GaAs technology is in ultra-high-speed digital integrated circuits design for supercomputers. Circuits as complicated as 16 × 16 multipliers [Lev82; Lev83], 1K SRAM, 4K SRAM, 16 K SRAM, 64K gate arrays have been built, and GaAs medium-scale integration (MSI) ICs have been fabricated with reasonably good yields.

GaAs MESFETs were first introduced as discrete devices for microwave applications in the late 1960s. It was not until 1974 that Rory Van Tuyl and Charles Liechti of Hewlett Packard first reported the use of MESFETs for digital applications [Tuy74]. The first integrated circuit was a simple NAND/NOR gate that was configured as buffered FET logic (BFL). This circuit used about five transistors and a few level shifting diodes and achieved performances down to about 100 ps of propagation delay. In 1977 Van Tuyl et al. reported a frequency divider circuit that operated at 4 GHz [Tuy77]. This integrated circuit was based on liquid phase epitaxy technology, and it consumed a large amount of power, about 20 mW per logic gate. However, its speed performance stirred a worldwide interest in GaAs digital circuits [Mun88].

In 1978 Richard Eden and coworkers at Rockwell International reported an improved circuit known as Schottky diode FET logic (SDFL) in which the power hungry level shifting circuit in the BFL output level shifter was replaced by very small diodes at the SDFL gate input. These diodes performed the dual function of level shifting and logic implementation and considerably reduced the power consumption per gate. The power consumption came down by an order of magnitude to a few milliwatts without sacrificing speed, and the simpler circuit had the added advantage of packing density. The epitaxial growth process was replaced by ion implantation which offered much better uniformity and reproducibility [Zuc80]. By 1980 the Rockwell team had taken the SDFL approach on ion-implanted material and had demonstrated the first

GaAs LSI circuit in the form of an 8 × 8 parallel multiplier with a complexity of just over 1000 logic gates [Lee80].

In the early 1980s the depletion-mode MESFETs used in the logic circuit were gradually replaced by enhancement-mode MESFETs. Logic gates built from the enhancement mode MESFET have compatible input and output logic levels without the need for level shifting diodes and so can be directly coupled to each other. Hence the term direct coupled FET logic (DCFL) was developed. The elimination of level-shifting circuits plus the virtue of the very low logic swing of DCFL reduced the power consumption by a few hundred microwatts per gate [Miz80]. The elimination of level-shifting circuits also helped to significantly simplify the circuit layout and consequently further increased the packing density. The development of self-aligned technology improved the GaAs device's performance.

With enhancement-mode and depletion-mode transistors placed on the same chip, circuit designers could now use DCFL to fabricate up to 30K gate array [Vit89]. This circuit structure reduces the chip area needed for a gate to about one-third that for an equivalent ECL gate. It has higher-speed capabilities and much less heat dissipation (about one-quarter) than ECL devices. Figure 14.1 compares the speeds and complexities of CMOS, ECL, and GaAs integrated circuits [Dey92]. Four-layer metal DCFL gallium arsenide processes use fourteen to sixteen mask steps, where comparable CMOS processes use 22 to 26 mask steps, and BiCMOS processes use 26 to 31. The smaller number of masking steps in GaAs tend to equalize the cost [Rou94].

The majority of near-term applications include high-speed acquisition and perhaps storage of very wide-bandwidth pulsed, pseudorandom, or continuous stream data and its processing in real time. Generic areas for the application of these types of digital systems will be in military radar signal processing and signature analysis, electronic countermeasures and electronic support measures, and spread spectrum communications [Gil86].

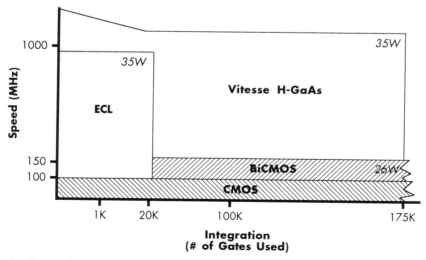

Fig. 14.1 Comparisons of speeds and complexities among CMOS, ECL, and GaAs integrated circuits demonstrate the benefits of GaAs for Sonet applications, which demand fast speed, low power, low cost, and a large number of gates. [Vitesse Semiconductor Corp. (1992)]

Gallium arsenide components can be profitably employed in the region of the data input (i.e., at the front end) of a signal processor confronted with a continuous stream of digital data. Such cases arise in computed tomography [Gil81]. These single, high-speed streams must be partitioned into a set of lower-rate parallel substreams, which in turn can be further processed by silicon devices operating at much slower clock rates. Another application for GaAs integrated circuits operating at high clock rates arises in the implementation of certain types of algorithms that cannot easily be parallelized. The class of problems that can be executed in a straightforward manner with a single uniprocessor or with a small number of coprocessors, all operating at relatively high speed, includes certain types of iterative algorithms.

The design tools are currently under development. Predicting a renaissance for gallium arsenide integrated circuits, Cascade Design Automation announced the first physical design tool, Epoch/GaAs [Goe93]. This tool set includes parameterized module generators, timing analysis, and automatic placement and routing. It support the captive foundry at Motorola and Vitesse H-GaAs-III processing at the present time. It offers interfaces to commercial logic-syntheses and simulation tools from vendors like Cadence, Synopsys, and Viewlogic.

In this chapter we will discuss the GaAs logic families, digital logic circuitry and static random access memory, noise margin limitations on GaAs VLSI, GaAs microprocessors, and packaging technologies.

14.2 HIGH-SPEED GaAs DEVICES AND INTEGRATED CIRCUITS

One of the most important applications of GaAs technology is in ultrafast digital integrated circuit designs for supercomputers. There are several device choices for high-speed GaAs ICs; among them are JFET, MESFET, HEMT, and HBT. Of these device technologies the depletion-mode FET (DFET), shown in Fig. 14.2, is the best developed. The DFET has the largest current drive capacity per unit device width for an all-GaAs FET device. This contributes to its high speed, low fan-out sensitivity, and high-power dissipation. The pinch-off voltage of the DFET, and thus the logic swing of logic gates, is determined by the channel doping and thickness under the Schottky barrier gate. This voltage can be made quite large (-2.5 V) in order to improve the noise immunity of the logic gate [Gre86]. By increasing the pinch-off voltage to zero or above, a low-current, low-power, enhancement-mode FET (EFET) is realized, as shown in Fig. 14.3. The logic swing for the EFET is limited to the difference between the pinch-off voltage (0 V) and the forward turn-on voltage of the

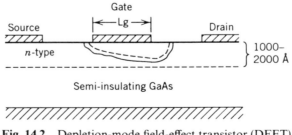

Fig. 14.2 Depletion-mode field-effect transistor (DFET).

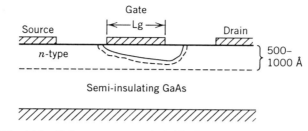

Fig. 14.3 Enhancement-mode field-effect transistor (EFET).

Schottky barrier gate (+0.5 V), thus providing a significantly lower noise immunity for the logic gate.

The third GaAs FET is a junction FET (E-JFET) in which the Schottky barrier of the EFET is replaced with an implanted p^+-region in the active n-channel that forms a p–n junction for the gate, as shown in Fig. 14.4 [Tro79; Shu7; Gre85]. The normally off JFETs are used as drivers. The depletion-mode ion-implanted two-terminal devices may be used as loads in DCFL circuits that are similar to GaAs MESFET DCFL circuits. GaAs JFET logic exhibits a good radiation hardness with the ring oscillator operation not affected by doses as high as 10^7 rad [Zul80]. An additional advantage that has significant impact in the design of large (>16 K) memories is that complementary devices (i.e., both n- and p-channel) can be readily fabricated in JFET technology [Zul84]. Another advantage of GaAs JFET technology is a larger built-in voltage that corresponds to a larger voltage swing and a better noise margin [Shu87]. The ultimate speed of the E-JFET will be less than an EFET of similar dimensions because the added sidewall gate capacitance of the p-n junction gate is a significant fraction of the total gate capacitance at submicrometer gate lengths [Gre86].

High-electron-mobility transistor (HEMT) technology has opened the door to new possibilities for LSI/VLSI with regard to ultrahigh speed and low-power dissipation, especially at low temperature. Complex logic circuits such as a single-clocked divided-by-two circuits based on the master-slave flip-flop consists of eight DCFL (direct-coupled field-effect transistor logic) NOR gates, one inverter, and four output buffers was fabricated back in 1983 [Abe83, Mim82; Abe86]. Static RAM based on HEMT

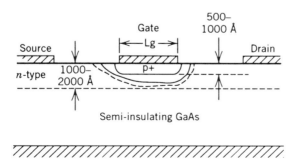

Fig. 14.4 Junction enhancement-mode field-effect transistor (E-JFET). [Greiling, P. T., and C. F. Krumm, in *VLSI Electronics*, vol. 11, ed. by N. G. Einspruch and W. R. Wisseman, Academic Press (1986).

technology has also been fabricated. The logic families and memory circuits will be discussed in the following sections.

Another high-speed device that has been implemented in IC form is the heterojunction bipolar transistor (HBT). In comparison with GaAs ICs implemented with MESFETs, the GaAs HBT technology is expected to benefit from the relative ease of making small structures intrinsic to vertical devices, from the lower sensitivity to capacitive loading and lower voltage swings that are typical of bipolar logic, and from the increased availability of computer-aided design tools that already exist for Si bipolar technology [Hug88]. In addition digital circuits based on HBTs have better threshold voltage control than the pinch-off voltage of MESFETs because the threshold voltage of an HBT is primarily dependent upon the band gaps of the emitter and base material.

These advantages will allow the HBT to play a dominant role in future digital applications. Nonthreshold logic (NTL) ring oscillators have been reported to operate at 16.5 ps/gate and current-mode logic (CML) ring oscillators at 27.6 ps/gate [Cha86].

14.3 GaAs LOGIC FAMILIES

This section discusses different logic families implemented with GaAs MESFETs, HEMTs, and HBTs.

14.3.1 Buffered FET Logic

Logic circuits using depletion-mode MESFETs (D-MESFET) as switches are difficult to implement [Mun88]. In a simple inverter arrangement, the output logic levels of the inverter are incompatible with the input required by the subsequent gate. This limitation can be overcome by using a level-shifting circuit at the output. The level-shifting circuit, which is provided by Schottky diodes incorporated into the output buffer, acts as a buffer between the output of the inverter and the inputs of other gates it is driving, hence termed "buffered FET logic" (BFL). The first significant GaAs ICs reported [Tuy74] utilized D-MESFETs is called "buffered FET logic" [Tuy77]. NAND and NOR implementations of BFL are illustrated in Fig. 14.5 [How85]. This circuit typically uses $-2.5 < V_p < -1$ V depletion-mode MESFETs and requires two power suppliers. Since it requires a negative gate voltage to turn off an n-channel D-MESFET, while its drain voltage is positive, level shifting must be introduced so that the output logic levels match the input levels. In the BFL approach the choice was made to operate with negative logic swings by level shifting the positive drain voltages at the gate output. This is normally accomplished by placing level-shifting diodes in the source follower output stage of the gate.

Placing the level-shifting diodes at the output has a drawback. Since the current through the output driver is relatively high, the power dissipation caused by the level-shifting diodes is also relatively high. However, a key attraction of BFL is that the circuit can be easily implemented with a fabrication process compatible with a conventional GaAs MESFET. Power dissipation of a BFL gate may be reduced by removing the source follower [Shu87]

A large number of circuits have been made using this buffered FET logic circuit approach—starting from simple ring oscillators to more complex sequential logic

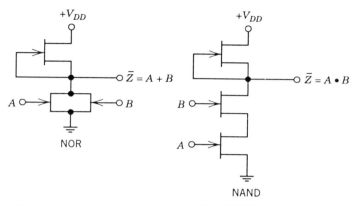

Fig. 14.5 Various circuit configurations for buffered FET logic: basic inverter circuit.

circuits, and including complementary clocked master-slave flip-flop stages used in divided-by-two and divided-by-eight binary ripple counter configurations—and MSI level logic circuits—including a word generator containing 600 active devices developed by Hewlett Packard [Lie82] and a 32-bit adder containing 2500 devices by NEC [Yam83].

14.3.2 Schottky Diode FET Logic

Logic Circuits

In 1978 Eden et al. [Ede78] at Rockwell reported an improved circuit known as Schottky diode FET logic (SDFL), whereby the power level-shifting circuit in the BFL output level shifter was replaced by very small diodes at the SDFL gate input. These diodes performed the dual function of level shifting and logic implementation and considerably reduced the power consumption per gate. The power consumption came down by an order of magnitude to a few milliwatts without sacrificing speed, and the simpler, smaller circuit had the added advantage of an improved packing density.

Two power supplies are needed for SDFL circuits (Fig. 14.6): one negative power supply voltage V_{ss} that is smaller than the threshold voltage of the normally on switching transistor and the regular power supply voltage V_{DD} [Shu87]. A pull-up (PU) transistor serves as a load; a pull-down (PD) transistor connects the gate of the switching transistor to the negative power supply. A level-shifting diode D_s decreases the voltage on the gate of the switching transistor $Q1$ so that, when the input voltage V_{IN} is low, the switching transistor $Q1$ is turned off. A logic **OR** function can be implemented using Schottky logic diodes.

The normally on switching transistors are commonly used in the SDFL circuits. The SDFL circuit approach offers savings in circuit area with a larger voltage swing and noise margins. However, a larger number of transistors and diodes per gate may lead to a somewhat lower speed compared to direct coupled field-effect transistor logic, which we will discuss in next section.

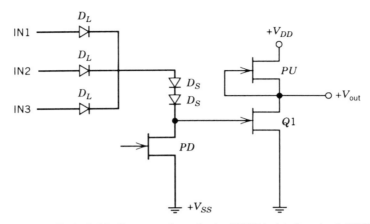

Fig. 14.6 Schottky diode field-effect transistor logic (SDFL). (a) Standard SDFL gate: Q, switching transistors; PU, pull-up transistor, PD, pull-down transistor, D_L, logic diode; D_S, level-shifting diode; **OR** function is implemented using logic Schottky diodes. [Shur, M., *GaAs Devices and Circuits*, p. 437, Plenum Press (1987)).

14.3.3 Direct Coupled Field-Effect Transistor Logic

The depletion-mode MESFET logic circuits has a higher-power consumption for integrating complex circuitry. Depletion-mode MESFETs were gradually replaced by enhancement-mode MESFETs. Logic gates built from the enhancement-mode MESFET have compatible input and output levels without the need for level-shifting diodes and so can be directly coupled to each other. Enhancement-mode logic needs only a single power supply and has lower-power consumption. This logic family is termed *direct coupled field-effect transistor logic* (DCFL). Enhancement-mode logic integrated circuits have been fabricated using E-MESFETs, E-JFETs, and HEMTs.

DCFL gates with a FET load, an ungated FET load, and a resistive load have an advantage of circuit simplicity and very few circuit elements per gate [Shu87], as shown in Fig. 14.7. Figure 14.8 shows the DCFL NOR gate with a resistor load [How85].

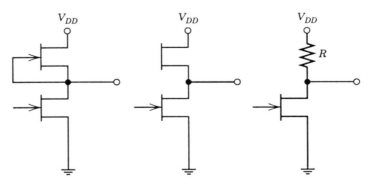

Fig. 14.7 Direct-coupled field-effect transistor logic (DCFL). Basic inverter with a FET load, an ungated FET load, and a resistive load.

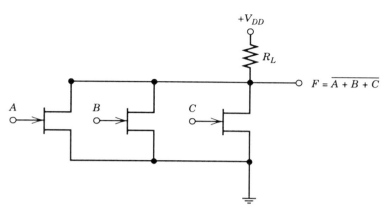

Fig. 14.8 Enhancement-mode JFET or MESFET circuits: simple direct-coupled FET logic (DCFL) NOR gate with resistor load.

The elimination of level-shifting circuits, plus the virtue of the very low logic swing of DCFL, reduces the power consumption to a few hundred microwatts per gate [Miz80]. The reduced logic swing corresponds to a smaller noise margin. The very low power consumption and circuit simplicity leads to high packing density. It has a lower interconnect parasitics and higher speed than for other GaAs logic families.

In DCFL circuit design, typical values of V_{pe} are 0.1 to 0.2 V. The device must be in cutoff when $V_{IN} = V_{oL}$, where $V_{oL} = 0.1$ V. If $V_{pe} > 0.2$ V, since not enough current is available [$I_D = \beta(V_{GS} - V_p)^2$] to discharge load capacitance. Hence W_L must be made larger, leading to more capacitance and lower density. The choice of V_{pd} is from -0.5 to -1.0 V as typical values. If V_{pd} is too small, large W_L is needed at the expense of extra capacitance. If V_{pd} is too large, small W_L is required. For $W_L < 2$ µm, process nonuniformity and scaling nonidealities may degrade the chip's noise margin.

The processing technology for DCFL circuits is difficult. The reduced noise margin placed a heavy demand on material and processing control. To maintain the threshold voltage uniformity, the active-layer doping and thickness are under stringent requirements. Multiple-ion implants are needed for both enhancement-mode drivers and depletion-mode loads [Shu87]. DCFL circuits require a recessed-gate or self-aligned process.

14.3.4 Source Coupled FET Logic

The spread in the threshold voltages of GaAs FETs in different logic gates of the same circuit may limit the integration scale and device yield. This factor is important even in SDFL circuits where the FET pinch-off voltages are larger than in DCFL circuits and the relative variation of pinch-off voltages is smaller. In the source-coupled field-effect transistor logic (SCFL) [Kat82; Shi83; Shi84] the circuitry of the logic gate is such that only relative variation of the threshold voltages is important.

The SCFL circuit includes a FET differential amplifier and a pair of buffer stages, as shown in Fig. 14.9. Here we discuss two cases of the operation of the SCFL gate [Kat82; Shu87]: (1) The transistor·FET1 is on and the transistor FET2 is off. Then

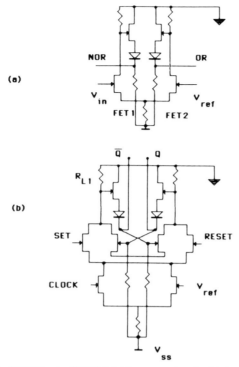

Fig. 14.9 Source coupled FET logic: (SCFL). (a) Inverter circuit; (b) clocked SCFL R-S flip-flop. [Katsu, S., S. Mambu, A. Shimano, and G. Kano, *IEEE Electron Dev. Lett.* **EDL-3**, 8 (1982): 197–199]

the following conditions must be satisfied:

$$V_{INa} - V_s > V_{t1} + V_{ON1}(0), \tag{14.1}$$

$$V_{ref} - V_s < V_{t2}. \tag{14.2}$$

(2) The transistor FET1 is off and transistor FET2 is on. Now we have

$$V_{INb} - V_s < V_{t1}, \tag{14.3}$$

$$V_{ref} - V_s > V_{t2} + V_{ON2}(0). \tag{14.4}$$

Here V_{INa} and V_{INb} are the input voltages required for switching in cases (1) and (2), respectively; V_{t1} and V_{t2} are the threshold voltages of FET1 and FET2, V_s is the source voltage common to both FET1 and FET2, and voltages $V_{ONi}(0)$ ($i = 1, 2$) are defined as

$$V_{ONi}(0) = V_{gsi} - V_{ti}, \quad i = 1, 2, \tag{14.5}$$

where V_{gsi} ($i = 1, 2$) is the gate-to-source voltage required to turn the transistor on.

Substituting Eq. (14.2) into Eq. (14.1), and Eq. (14.4) into Eq. (14.3), we obtain conditions required for a proper operation of an SCFL gate:

$$V_{INa} > V_{ref} + V_{ON1}(0) + V_{t1} - V_{t2}, \qquad (14.6)$$

$$V_{INb} < V_{ref} + V_{ON2}(0) + V_{t1} - V_{t2}. \qquad (14.7)$$

From this analysis we notice that the input level required for switching is only dependent on the difference between the threshold voltages of the FETs. In most cases the threshold voltage difference, $V_{t1} - V_{t2}$, between neighboring FETs on a chip is small.

The power consumption of an SCFL inverter is considerably higher than the power consumption of inverters implemented with other logic families. The SCFL inverter has a more complicated circuit configuration and larger values of V_{ss} required for an optimum output swing. However, total power consumption for the SCFL circuit may compare favorably with those for other logic families because a flip-flop may be implemented using a single SCFL inverter with two complementary outputs. SCFL circuits seem to be more appropriate for very high speed and moderate power SSI and MSI circuits [Lee84].

SCFL circuits are compatible with bipolar ECL logic. FET differential amplifiers usually exhibit smaller voltage gain than bipolar amplifiers. Nevertheless, SCFL circuits may be quite fast because the FETs operate in the saturation region of the current-voltage characteristic where the drain-to-gate capacitance is smaller. A comparison of DCFL and SCFL circuits is given in Table 14.1.

TABLE 14.1 Comparison of DCFL and SCFL Circuits

Advantages	Disadvantages
DCFL	
Simple NMOS-like circuit style	Small noise margins
No level shifters	Impacts ability to main large die
Low-power/medium performance circuit technique	NOR gate is the fundamental logic unit
	Inefficient for large macrocells
	OR/NOR gate requires 5 FETs, has 1 gate delay of intrinsic skew and rising versus falling edge skew
SCFL	
Different circuit technique, therefore independent of threshold variations	Requires level shifters
Excellent noise margins over temperature, voltage, and process variations	Requires twice the number of signals to be routed
High equivalent gate count per "current tree"	
TRUE and COMPLEMENT signals available	
Low switching noise	

14.3.5 Capacitive Coupled Logic

An obvious approach to the level-shifting problem is to use capacitive coupling between stages. Capacitive coupled logic (CCL) was proposed by Livingstone and Mellor [Liv80a; Liv80b]. In this approach a reverse-biased Schottky diode is used as a capacitor, which provides a dc isolation between the states, as shown in Fig. 14.10 [Hai89; Lar89]. The power dissipation is considerably less than for the BFL. However, this inverter must be initialized for proper operation, and it does not work at dc, which is not usually acceptable in logic applications. The CCL was pioneered by British Telecom.

A better version of the CCL was developed by R. C. Eden [Ede84]. A combination of reverse- and forward-biased diodes is used, as shown in Fig. 14.11. The forward-biased diodes (D1-3) act as level shifters, while the reverse-biased diode (D4) is used as a "speed-up" capacitor to couple the switching waveform edges. Since the speed of operation does not depend on the level-shifting diodes, the current through them can be made very small, allowing low power consumption per stage. The capacitor diode-coupled FET logic (CDFL) gate approach obtains dc to very high speed operations with typically 90% to 97% current efficiency and does not require the complicated source-follower output stage.

Generally the CDFL circuit is not usable in chip input and output interface applications. In the GigaBit 10G PicoLogic™ series of GaAs ICs, the outputs are

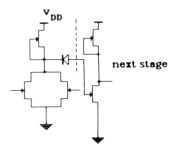

Fig. 14.10 Capacitive coupled logic (CCL). [Eden, R., *IEEE GaAs IC Symp. Tech. Digest* pp. 11–144, IEEE (1984)]

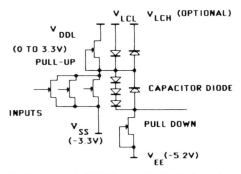

Fig. 14.11 Capacitive diode-coupled FET logic (CDFL) three-input NOR logic gate with output level shifted. [Eden, R., *IEEE GaAs IC Symp. Tech. Digest*, pp. 11–14, IEEE (1984)]

required to drive transmission-line impedances to standard ECL signal levels, while the inputs are to be ECL signal compatible.

14.3.6 Low Pinch-off Voltage FET Logic

A serious limitation of "normally off" logic are yield problems arising from poor device threshold voltage control due to nonuniform enhancement-mode active layers [Wel85]. Alternate circuit designs have been created to circumvent the limitations of "normally off" layers. Low pinch-off voltage FET logic (LPFL) is also called *quasi-normally off logic* [Nuz80; Nuz81]. This logic family occupies an intermediate position between the SDFL and DCFL circuits. The switching transistor has a threshold voltage close to zero. Because of large pinch-off voltages (compared to DCFL) LPFL circuits are more tolerant to the variations in the threshold voltages than DCFL circuits. An example of the LPFL circuit is shown in Fig. 14.12. LPFL circuits may be expected to have larger power dissipation, larger delay time, and higher fan-out capabilities than DCFL circuits.

14.3.7 Heterojunction Bipolar Logic

GaAs/AlGaAs HBT devices demonstrate high-frequency performance, moderate lithography requirements, high transconductance, high current driving capabilities, and threshold voltage uniformity. Logic circuits based on current-mode logic (CML) have been reported [Asb84]. Current-mode logic, which closely resembles ECL, can also be built in GaAs. In this logic the current from an appropriate source is steered between alternate paths depending on the input voltages. Figure 14.13 gives an example of the CML gate. The CML provides complementary outputs [Asb84].

GaAs CML structures are much more complicated than DCFL gates, so a GaAs-based CML chip has a far lower gate count than either DCFL or ECL. The CML is more commonly used in communication circuits; however, it may prove viable in the future for the highest-performance supercomputers. In the recent past the density of CML chips has been too slow, too many CML chips are needed, and the resulting interconnection delay offsets the the logic's tremendous speed—already about 1 GHz. Yet, as the DCFL GaAs approaches BiCMOS densities, the CML GaAs should draw

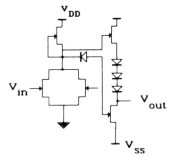

Fig. 14.12 Quasi-normally off logic gate (fan-in 2). [Shur, M., *GaAs Devices and Circuits*, p. 437, Plenum Press (1987)]

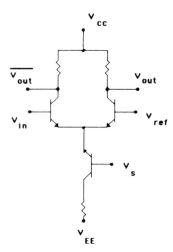

Fig. 14.13 Current-mode logic (CML) gate. [Asbeck, P. M., D. L. Miller, R. J. Anderson, R. N. Deming, R. T. Chen, C. A. Liechti, and F. H. Eisen, *IEEE GaAs IC Symp.*, pp. 133–136, IEEE (1984)]

closer to ECL density. When this happens, the CML's speed may well compensate for its high-power dissipation in top of the line supercomputers [Doz92].

Small-scale complexity digital, analog/digital, and RF analog integrated circuits were fabricated using the 2-μm emitter HBT mesa process [Kim86]. The ICs include divide-by-two prescalers, voltage comparators, a RF analog amplifier, and multiplier circuits. The prescaler and the comparator were based on the CML bi-level latch with emitter followers, implemented into a master-slave combination with an output buffer to facilitate testing.

Ring oscillators and frequency dividers are most commonly used to demonstrate the capabilities of HBT technology. The ring oscillators are of the nonthreshold logic (NTL), current-mode logic (CML), and classical emitter-coupled logic (ECL) types. The NTL logic gates have limited fan-out capability. In the NTL circuit approach the transistor saturation is avoided by the appropriate choice of the emitter and collector resistors as well as by the power supply voltage. CML ring oscillator propagation delays close to 40 ps per stage were obtained [Cha88]. A four-bit pattern generator operating with a clock frequency up to 4 Gbit/s was also demonstrated. An ECL ring oscillator has also been fabricated using graded-band-gap base GaAs/AlGaAs HBTs.

14.3.8 GaAs Gate Array

During the past few years several design styles for semicustom arrays have been developed and used in a wide spectrum of applications. The gate array approach became popular with Si technologies, and it is being widely used for low-cost rapid turnaround development of ICs with no stringent performance requirements. This trend is also being used for GaAs technologies. GaAs technology has been considered a prime candidate for high-speed array designs; previous gate arrays were limited to the 3K–4K range because of difficulties in compromising speed, power, and noise

GaAs LOGIC FAMILIES

margin [Sat87]. GaAs technology has been unable to seriously challenge emitter-couple logic (ECL) in applications needing high-speed, low-power logic at integration levels exceeding a few thousand gates. However, Vitesse Semiconductor has developed a new family of GaAs gate arrays that has finally caught up to ECL technology on performance and cost.

The first member of Vitesse's Fury family of gate arrays, the VSC10000, packs over 14,000 gates on a 280-by-335-mil GaAs die. Using a 0.8-µm, four-layer metal nMOS-like process, Vitesse has designed its gate array to attack the weaknesses of ECL technology while offsetting its advantages. For example, the VSC10000 requires only 11 masking steps to implement the logic array, comprised of rows of single-sized enhancement-mode and depletion-mode GaAs MESFETs and metal interconnect. However, the ECL structures can require up to 25 masking steps needed to fabricate resistors and bipolar transistors of various sizes.

Another difference is in the gate circuitry (Fig. 14.14). A typical ECL NOR gate (the basic logic building block) contains six bipolar transistors and three resistors, and operates on two power supplies. The same function in the Vitesse design uses just three transistors and one 2-V Power supply. The economies of space and power are evident when comparing the size of the VSC10000 with an ECL counterpart, such as the Motorola MCA10000 family. While the MCA10000 is 148,225 square mils in area and typically consumes from 15 to 30 W, the VSC10000 is 93,800 square mils and dissipates 5 W. Low-power dissipation allows the part to be packaged in a 211-leaded pin-grid array or 256-pin leaded chip carrier that requires no exotic cooling design.

The FURY family VSC15K GaAs gate array has been developed to be used in the

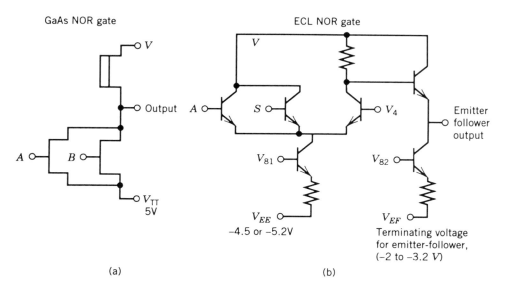

Fig. 14.14 The low-power, high-speed performance of Vitesse's Fury gate array family stems from the use of just three NMOS transistors in a basic NOR gate operating with one power supply (*a*). In contrast, a typical ECL array using NOR gates with six transistors three resistors, two power supplies, and several reference voltages, consumes four times the power and twice the chip area (*b*). Deyhimer, I., *View on Vitesse* **2**, 3 (July 1991)]

Convex computer's memory subsystem as an error checking and correction chip (ECC). The Convex C2 Series represents the first time GaAs has been used in a volume commercial computer system. In the Convex C3800 supercomputer, air cooling, rather than liquid-refrigerant cooling, was a system requirement. GaAs circuitry has lower-power dissipation than Si ECL; therefore it is used extensively in this computer. Other target applications for FURY Series gate arrays include telecommunications, office automation networks, military/aerospace, and test equipment.

The FX gate array family, built with Vitesse's new H-GaAs III technology, offers integration levels similar to the highest-performance CMOS/BiCMOS devices, with significantly higher gate speed. FX offers up to 350 K raw gates (0.6 μm H-GaAs III process technology) and up to 100 K bits of embedded high-speed SRAM. Benchmarks comparing BiCMOS system patitions to H-GaAs III partitions show H-GaAs III to achieve a $2 \times$ or greater performance factor [Dey91].

FX is competitive in power dissipation. The VGFX350K has 378 input/output signal buffers and is packaged in a plastic 557 pin grid array. The device has a maximum power dissipation of 44 W but will typically dissipate less than 30 W [Vit92]. At clock rates between 50 and 100 MHz, CMOS/BiCMOS power dissipation becomes comparable to or exceeds that of H-GaAs.

14.4 GALLIUM ARSENIDE CIRCUITS

As described in the previous section, the design and operation of basic digital GaAs circuits can be implemented in different logic families. The design of these GaAs logic circuits are most efficiently implemented in BFL or DCFL which use most AND-NOR gates, while circuits implemented in SDFL use OR-NAND gates. Both combinational and sequential logic circuits can be used as building blocks in designing a monolithic integrated circuits. Several of these circuits are available in circuit libraries used by GaAs foundries. In this section we discuss the memory circuit designs using GaAs technology.

14.4.1 GaAs Static Random Access Memory

Memory circuits are an integral part of any computer system. There are three basic characteristics of a memory circuit—size, organization, and control [Kan89]. *Organization* refers to the way bits are stored in the memory. If a memory has an n-bit address, and the full address space is used, then there are 2^n memory locations, each of which can store a k-bit word, leading to a total memory size of $2^n \cdot k$ bits. Frequently memory circuits are referred to by a rounded quotation, indicating their capacity. Whenever a memory is referred to as $A \times B$ bits, the A indicates the number of memory locations while B indicates the number of bits that can be stored in each location. *Control* refers to the way data are read or written at any time, then the memory is classified as random access.

A major use of such high-speed RAMs is in cache memories. Unlike Si RAMs, all GaAs RAM designs reported to data are static. This is due to the refresh requirements for dynamic memories and their inherent low-noise margins. To realize practically usable GaAs SRAM, circuit technologies that affect the following key characteristics need to be developed: (1) higher speed than Si bipolar SRAMs, (2) alpha-particle immunity, (3) stable operation over a wide temperature range, and (4) an interface circuit that is compatible with Si bipolar ECL [Tan87].

Several types of GaAs SRAMs are now available: 1K, 4K, and 16 Kb. Special attention has been paid to threshold voltage control, alpha-particle immunity, and thermal stability. To achieve high speed, it is necessary to shorten the relatively long delay time from the memory cell to the output buffer in a conventional circuit, as shown in Fig. 14.15 [Tan87]. To reduce the delay time, a current sense circuit has been developed and is shown in Fig. 14.16. The read operation is performed through the column switch FET (i.e., $F1$ and $F2$) and the common data line load circuit. As shown in the timing diagram of Fig. 14.16, a signal swing for the data line (V_{d1}) that is smaller than the one for the common data line (V_{c1}) can be achieved. This enables a 25% more drivability of the memory cells than with the conventional circuit. By the additional application of the 0.7-μm gate length device technology used for the FET in Fig. 14.17, a total access time reduction of 60% has been achieved.

As for the alpha-particle immunity of GaAs LSIs, the charge multiplication phenomena and the collected charge improvements with a buried P-layer have been reported [Ume86]. Experiments conducted with an accelerated rare of 2.5×10^7 give a FIT of 4×10^5, even for a 1-Kb cell array, as shown in Fig. 14.18. This value is four orders of magnitude larger than that of Si LSIs. However, by increasing the dose of the buried P-type layer from $1 \times 10^{12}\,\text{cm}^{-2}$ to $2 \times 10^{12}\,\text{cm}^{-2}$, the error rate is reduced by one order of magnitude. Furthermore it has been found by calculation that an error rate comparable to that of Si LSIs (less than 100 FIT) can be achieved by utilizing the capacitance that is added to flip-flot nodes of cells, as shown in Fig. 14.18.

For practical applications stable power dissipation over a wide temperature range is required. In GaAs LSIs, an increase in power dissipation occurs when the device temperature goes up. This phenomenon is caused by an increase in the FETs drain-source currents (I_{ds}). To overcome this problem, the device characteristics have been

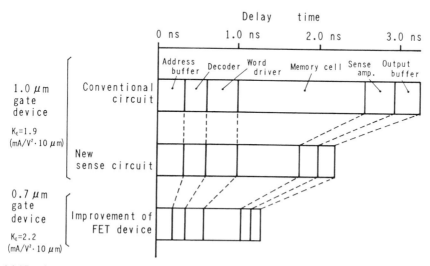

Fig. 14.15 Access time improvements of 4-Kb SRAM using circuit and device technologies: average for a chip. [Tanaka H., H. Yamashita, N. Masuda, N. Matsunaga, M. Miyazaki, H. Yanazawa, A. Masaki, and A. Hashimoto, *ISSCC 87*, pp. 138–139, IEEE (1987)]

544 GaAs DIGITAL INTEGRATED CIRCUITS

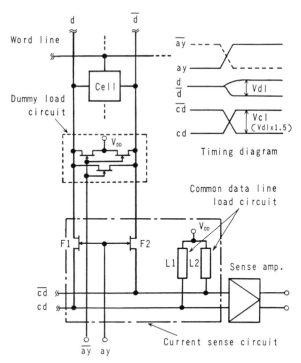

Fig. 14.16 Current sense circuit configuration [Tanaka, H., H. Yamashita, N. Masuda, N. Matsunaga, M. Miyazaki, H. Yanazawa, A. Masaki, and A. Hashimoto, *ISSCC 87*, pp. 138–139, IEEE (1987)]

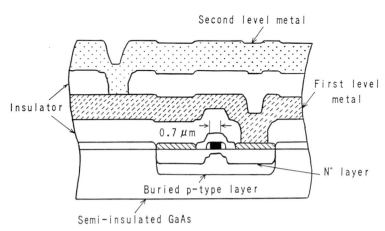

Fig. 14.17 Cross section of a FET device with a buried *P*-type layer. [Tanaka, H., H. Yamashita, N. Masuda, N. Matsunaga, M. Miyazaki, H. Yanazawa, A. Masaki, and A. Hashimoto, *ISSCC 87*, pp. 138–139, IEEE (1987)]

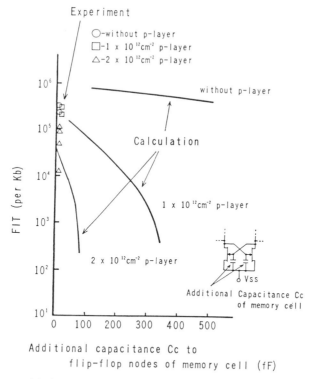

Fig. 14.18 Alpha-particle immunity of GaAs SRAM. [Tanaka, H., H. Yamashita, N. Masuda, N. Matsunaga, M. Miyazaki, H. Yanazawa, A. Masaki, and A. Hashimoto, *ISSCC 87*, pp. 138–139, IEEE (1987)]

investigated. It has been found that a decrease in the I_{ds} is possible by using FETs with a larger negative V_{th} than -1.3, as shown in Fig. 14.19(a). In addition, by using load FETs with a large negative V_{th} for all DCFL circuits included in SRAM, stable power dissipation can be realized as can be seen in Fig. 14.19(b).

The sense amplifier is critical for the operation of the memory. This circuit detects a small bit line voltage difference, which constitutes the logic level stored in the memory cell, and subsequently amplifies this small voltage difference to appropriate levels required for driving the output buffers. There are different sense amplifier designs that can be used to satisfy design requirements in terms of performance, power dissipation, area, and noise margins. Figure 14.20 illustrates four sense amplifier circuit configurations [Ino82]. In configuration (a), the (W/L) ratio of the FETs is such that the load current is smaller than the current driving capability of the switching FETs. the common-source voltage in this case is almost zero independent of the input voltage, and therefore the circuit in this configuration operates as a source-coupled inverter pair. Configuration (b) and (c) have loaded and switching FETs with equal (W/L) ratios and thus operate as source-coupled differential amplifiers. In these configurations the output of the sense amplifier is pulled down, to appropriate levels for driving the output buffer, by biasing the current sink FET to $-V_{ss}$, configuration (b), and by adding

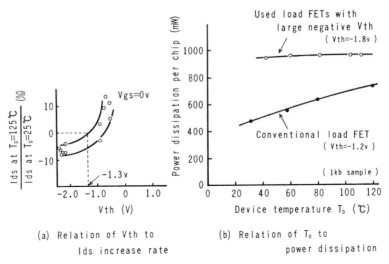

Fig. 14.19 Stable power dissipation over wide temperature range using deep negative threshold voltage (V_{th}). [Tanaka, H., H. Yamashita, N. Masuda, N. Matsunaga, M. Miyazaki, H. Yanazawa, A. Masaki, and A. Hashimoto, *ISSCC 87*, pp. 138–139, IEEE (1987)]

output voltage shift diodes, configuration (c). In configuration (d) the output voltage is fed back to obtain an optimum sensing condition. The last configuration is well known from its utilization in Si nMOS SRAMs. The voltage levels required for driving the output buffer determine the minimum voltage difference that the sense amplifier must be able to sense for correct circuit operation.

For the 4-Kb SRAM with the sense circuit, a minimum access time of 1.0 ns and a maximum of 1.5 ns have been obtained in a chip. The chip size, cell size, and power dissipation for the 4-Kb SRAM are 3.7 × 4.7 mm, 47 × 25 µm, and 1.6 W, respectively.

A few 16-Kb GaAs SRAM designs have been reported [Hir86; Tak87]. For the 16-Kb SRAM, the E/D DCFL has been adopted [Tak87]. In the cascade DCFL circuits the fan-out number, or the driver-size ratio between the succeeding and the preceding stages, is optimized to four so as to minimize the total propagation delay time from the first to the final stages.

A GaAs SRAM designed for manufacturability is reported using conventional D-MESFET processing with sufficient temperature and power supply tolerance [Tse87]. The 16K-bit RAM was processed using a non-self-aligned depletion mode LSI process. The channel doping regions are formed using ion implantation. Device isolation is achieved by mask proton implantation into semi-insulating GaAs. Direct step-on-wafer 10 × lithography is used throughout the process. The ohmic contacts are alloyed Au:Ge:Ni, and the gate metal is Ti:Pt:Au. The ohmic and gate metals are defined using lift-off techniques. Vias are made by a reactive ion etch process in a SiO_xN_y interlevel dielectric. Second-level metal and cermet resistor patterns are defined using ion milling. The minimum feature size is 1.0 µm, and the first- and second-metal interconnect pitch is 4.0 µm. The typical threshold voltage is -0.5 V, with an average standard deviation of 46 mV across a 3-in. wafer. The average K-value is 80 µa/V²-µm for a 20 µm-wide FET.

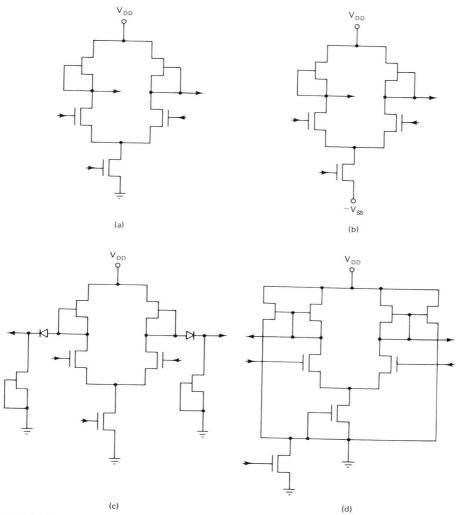

Fig. 14.20 Sense amplifier circuits. [Ino, M., *Proc. GaAs IC Symp.*, p. 2, IEEE (1982)]

The RAM circuits designed and implemented using HEMT technology are being developed. Although fabrication of these devices, process control, and operating conditions are still in the experimental stage and a multitude of issues and problems have to be solved, the circuits produced demonstrate superior performance. N. H. Sheng et al. [She86] has reported a high-speed 1K × 1-bit SRAM using HEMT devices in DCFL configuraion exhibited an access time of 800 psec with an operating power of 450 mW. New designs are underway.

14.4.2 Data Conversion Circuits

Data converters are one of the largest sectors of linear-integrated circuits. Analog-to-digital (ADC) and digital-to-analog (DAC) converters have become fundamental blocks in data acquisition systems, operating as a peripheral to a data-processing computer.

Although much work has been performed in the design and implementation of Si data converters, the trend is inevitably toward higher sampling rates while keeping a sufficient resolution. Many systems applications in the area of instrumentation, signal processing, and telecommunications require performance levels even higher than what is available in Si. Especially in military applications, such as real-time radar signal processing or FW systems, require medium/high resolution in the Gb/s range.

Here we discuss only linear converters in which the digital signal is directly proportional to the amplitude of the analog signal. The data conversion process can be described by the expression

$$V_a = V_{ref}(b_1 2^{-1} + b_2 2^{-2} + \cdots\cdots + b_{n-1} 2^{n-1} + b_n 2^n), \qquad (14.8)$$

which can be applied for both analog-to-digital (A-D) and digital-to-analog (D-A) conversions. In the former case an analog voltage V_a is converted into an n-bit binary number. V_{ref} is the "high" logic level of the system.

For D-A conversion the input $b_1 b_2 \cdots b_n$ is converted into an analog voltage by the scaling factor V_{ref}, which represents a reference voltage. The accuracy of V_{ref} is crucial in obtaining the required resolution for the conversion. According to Eq. (14.8), the conversion resolution is $\pm 1/2$ LSB (least significant bit), resulting in a quantization error $\pm \delta V_a$, where

$$\delta V_a = V_{ref} \cdot 2^{-(n+1)}. \qquad (14.9)$$

The sampling of V_a in A-D conversion introduces another error due to the time involved in performing the sampling. This error $\delta_t V_a$ is given by

$$\delta_t V_a = \int_0^{t_a} \frac{\delta V_a}{\delta_t} dt, \qquad (14.10)$$

where t_a is the sampling time of measurement.

The quantization error, often called *quantization noise*, computed by Eq. (14.10) can be used to define the signal-to-quantization noise (S/QN) ratio. The S/QN is given by

$$\frac{S}{QN} = V_a \cdot (V_{ref})^{-1} \cdot 2^{n+1}. \qquad (14.11)$$

As the expression above indicates, S/QN increases linearly with V_a, having its minimum value when $V_a = V_{ref} \cdot 2^{-n}$ or S/QN = 2.

The prime parameters characterizating the performance of data converters are resolution, accuracy, and dynamic response [Kan89]. *Integral linearity* or *relative accuracy* is defined as the straightness of the transfer function of the converter. It expresses the deviation of points of the converted signal from a reference straight line which is an empirical fit to a result in the best linearity specification. The integral linearity expresses the maximum deviation in terms of the LSB fraction.

Differential linearity or *differential nonlinearity* expresses the deviation of the analog output from the ideal for a unit change in input. Differential nonlinearity is measured in LSBs, and when it is ± 1 LSB or less, monotonic behavior is guaranteed. Generally DACs must be monotonic, especially when used for control systems, because non-monotonicity can lead to positive feedback and subsequently to loop instability.

Integral and differential linearity are related in the sense that the maximum differential is smaller than or equal to two times the integral linearity. If a DAC specification contains information for one of these parameters, then this rule can be used to assume the worst-case specification for the other.

DAC settling time is defined as the total time interval between application of a new input code and the point where the analog output settles to within a specified error band around its final value. The most common specification for this value is ± 1/2 LSB. The maximum throughput rate is the maximum number of conversions per second that the DAC can perform.

The definition of A-D conversion characteristic parameters is very similar to the definitions given above for D-A converters. Integral and differential linearities are the same as previously defined, and most circuit techniques for ADCs guarantee monotonic operation. However, there is a possibility of missing codes. This phenomenon manifests itself as a change at the output code of more than one bit for a unit increase (or decrease) at the input. An integral linearity of ± 1/2 LSB guarantees no missing codes for ADCs.

14.4.3 Data Communication Chip Set

TriQuint Semiconductor's FC-265 serial data-communications chip set supports fiber channel (FC), asynchronous transfer mode (ATM), and other network-communication schemes. The chip set operates at serial data rates to 265.625 Mbaud [Ken92].

The three-chip FC-265 set includes a GaAs transmitter, a receiver, and an encoder/decoder. The GA9101 transmitter serializes an encoded ten-bit, TTL-compatible word into a differential, positive ECL signal. It has an on-chip, phase-locked loop that synthesizes a reference bit clock of 19.44, 20, or 26.5625 MHz. It multiplies the reference clock to produce a bit clock of 194.4, 200, or 265.625 MHz, which is used to provide the bit timing for the transmit path.

The GA9102 receiver has an integrated clock-and-data-recovery (CDR) circuit that recovers the clock information from the input data. The GA9103 CMOS encoder has an eight-bit interface to the host and a ten-bit interface to the transmitter and receiver.

14.5 GaAs MICROPROCESSOR

In 1982 the Mayo Foundation was assigned by the U.S. Strategic Technology Office of the Defense Advanced Research Projects Agency (DARPA) the task of identifying a machine architecture that could be implemented with fewer custom ICs fabricated in GaAs rather than in Si [Hen87]. Although the availability of large numbers of gates on VLSI chips had enabled silicon microprocessor designers to create relatively powerful architecture, based upon parallelism and rich set of assembly language instruction types, the architectural complexity and the need for low-power consumption constrained the microcycle clock rate of these silicon-based processors to the range of 2 to 25 MHz.

A microprocessor implemented in GaAs would have to contain less than 10 K to 20 K equivalent gates on a chip in order to obtain reasonable fabrication yield. A microprocessor architectural approach that seems adequate for a GaAs microprocessor is represented by the family of reduced instruction set computers (RISC). A RISC architecture has the following characteristics [Pat80a]:

1. A small number of reduced instructions.
2. A fixed instruction format.
3. Hardware rather than microcode architecture.
4. Simple cycle execution for most instructions.
5. A load and store architecture.

RISC machines are designed beginning with features that all machines have, and with the basic operations from which all others are constructed: arithmatic functions, logic functions, shift operations, and test and branch operations. To these are added operations that handle the normal flow of program execution [Pat80b].

GaAs chips are used in supercomputers and personal computers. CONVEX Computer as designed an entire super-minicomputer with Vitesse GaAs ASICs and standard products. COMPAQ Computer has used a Vitesse ASIC chip to handle microprocessor support logic in its 50-MHz 486-based Systempro PC.

Recently, Intel announces Pentium chip for PC. The new processor is expected to add a new dimension to the PC business, rivaling the performance of standard workstations. This performance will be fueled by a new family of support chip sets. Included in this collection is the first GaAs chip aimed at the PC market. The GaAs-based part is a cache-controller chip set from Vitesse Semiconductor [And93]. There are two chip sets in Vitesse's cache-controller family, the VP945/946 intended for use with fast 486 processors and the VSP951/952 intended for use with the Pentium and OverDrive processor for the 486DX2 used with the Pentium. The VPS951/952 chip set provides a 33-MHz, 64-bit system interface designed to emulate the Pentium's bus. When used with the OverDrive processor, the chip supports pipelined Pentium processor bus cycles rather than the conventional 32-bit, 486 cycles. Both cache controllers (the VSP945 and 951) integrate all the cache-management functions for bus arbitration between the processor and main memory.

Off-the-shelf GaAs ICs for microprocessor applications include devices ranging from clock ICs to memories. For example, Vitesse (and Thomson-CSF) offers the VS12-G422T 256×4-bit SRAM. The memory IC includes TTL-compatible inputs and outputs, and is pin compatible with industry-standard silicon SRAMs that use -422 and -122 designators [Wri92]. One can specify the Vitesse IC with speeds of 4, 5, or 6 ns. The fast-access speed makes the RAM useful in cache memory, signal processing, and video applications.

TriQuint Semiconductor has concentrated its microprocessor-related products around clock circuits and programming logic. The Ga1210E clock doubler and two-phase clock generator and the GA1486 clock generator and buffer are used in an application with an Intel 486 microprocessor. TriQuint's GA22V10 programming logic device offers pin compatibility with industry-standard 22V10s and comes in 5.5-, 6-, and 7.5-ns operating speed [Wri92].

Texas Instruments, planning to be a player in the GaAs market, wants to offer a GaAs microprocessor. U.S. DARPA has funded Texas Instruments and McDonnell Douglas to produce a space-based signal processor system. Signal-processing microprocessor systems are required in military applications where they must operate at very high data rates, over a wide range of temperatures, consume little power, and be radiation hard [Roc88].

The goal for a GaAs microprocessor is a chip of no more than 10,000 FETs, a 200-MHz clock rate, and a 100 to 200 MIPS. One of the major differences between CMOS or NMOS silicon technology and E-MESFET GaAs technology is the ratio of off-chip memory access time to on-chip access time. With the GaAs approach it is necessary to minimize the number of off-chip access. As a result, when designing a GaAs microprocessor, it is crucial to increase the size of on-chip storage to reduce the need for off-chip communication. One way to minimize the impact of off-chip memory access time is to utilize pipelining techniques for both memory and the microprocessor [Gil84]. The fixed instruction format in RISC approach, leads to a highly simplified instruction decoder, reducing the number of transistors for its implementation and thus increasing the speed of the decoder. Therefore it is the only significant control that can represent 95% of the FET count for implementing the data path.

Different design approaches are available with GaAs technologies. McDonnell Douglas has used the JEET technology approach to fabricate a 4-bit GaAs microprocessor circuit that contains 2K transistors and intends to develop a 32-bit RISC based microprocessor [Roc88]. Texas Instrument is implementing a 32-bit microprocessor. This 32-bit microprocessor is fabricated using GaAs heterojunction integrated injection logic (HI^2L) technology. The main advantage of HI^2L is a good control of logic gate threshold voltage over temperature and the fact that a NAND gate requires only one transistor, which permits a high gate density. The overall system would be built with a maximum gate count of 10,000 gates.

Texas Instruments has developed a 200-MIPS GaAs microcessor [Whi88], fabricated with a GaAs/AlGaAs heterojunction bipolar process designed to operate at 200 MHz [Yua86; Eva87]. The microprocessor is part of a chip set. A basic microcomputer system using this set includes a CPU, two identical cache memories for instructions and operands, two identical memory management units (MMU) to operate the cache memories, and a processor board interface (PBI), as shown in Fig. 14.21. The system can be expanded to include up to six floating-point coprocessors. Two types of coprocessors are being developed to increase double-precision computations for addition/subtraction and multiplication/division. The CPU contains more than 12,000 heterojunction integrated injection logic gates and 256 I/O pins. This design was being shrunk to less than 300-mils sq. by using 1.5 μm design rules.

The design was based on a DARPA microprocessor without the interlocked pipe state core instruction set architecture (ISA) and modified to include some hardware/software trade-off for performance improvement. The MIPS core ISA can be defined as an intermediate level between hardware-oriented machine instruction set and a high-level language. Once a program is compiled into the core ISA code, a hardware-dependent translator transforms the code into the machine instruction set and performs any necessary optimizations. Translation programs to convert other machine codes into core ISA are available.

Figure 14.22 shows the CPU data path as it is related to the six-stage pipeline. During I1, an instruction address is sent from the program counter and is latched into the instruction cache memories. Upon completion of I2, the returning instruction data is latched back into the CPU Instruction Register. During Ex, the instruction is decoded, the operands are fetched from the Register File, and ALU computes the result,

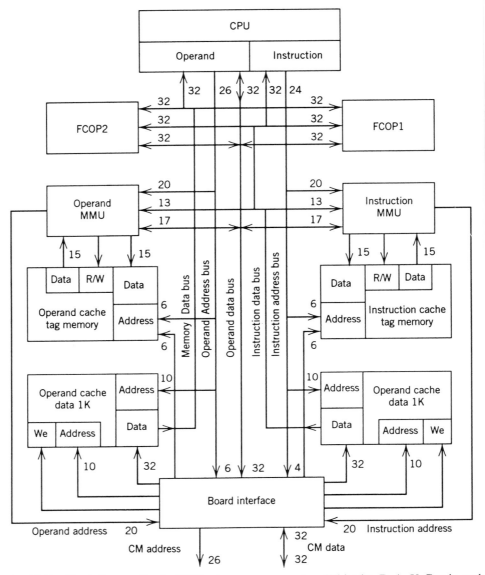

Fig. 14.21 Block diagram for the RISC microprocessor system. Whitmire, D. A., V. Gercia, and S. Evans, *ISSCC 88*, pp. 34–35, IEEE (1988)]

and stores it in Result 1 Register. During M1, operand addresses are sent to the operand cache for load and store instructions. For store instructions, the operand data is also dispatched at this time. For load instructions, the operand is latched into the CPU Result 3 Register at the end of M2. The WR pipestage loads the data from Result 3 Register to the Register File.

The MMU provides the virtual addressing and cache memory control functions for the system. The cache memory is configured as 1K × 32b. As data are being fetched

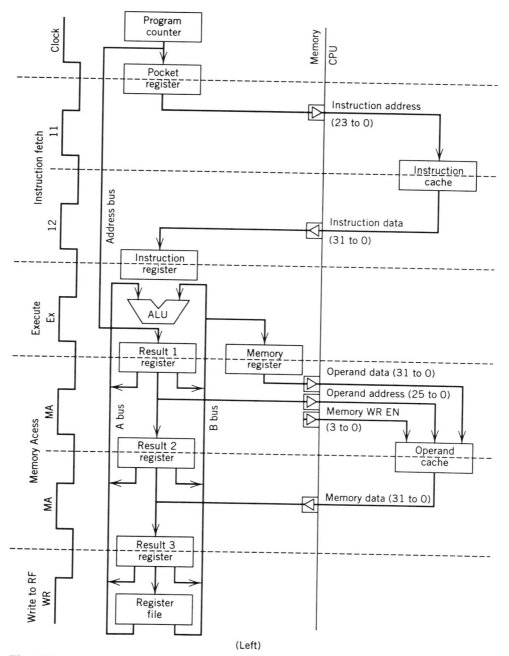

Fig. 14.22 Operation in the six-stage pipeline. [Whitmire, D. A., V. Gercia, and S. Evans, *ISSCC 88*, pp. 34–35, IEEE (1988)]

from the cache, the corresponding MMU calculates whether the CPU address is a hit or a miss in relation to the cache address. If the address is a hit, data returning from the cache to CPU is valid, and the MMU takes no action. If the address is a miss, the MMU issues a memory wait that halts the CPU pipeline for sufficient time to allow the desired block from the main memory to be loaded into the cache. the PBI controls the refresh of the cache memory from the main memory for both the instruction and operand caches.

The CPU data path is 32b wide with a critical path length of 30 gate delays. High-speed gates with 1-mA gate current are selectively used throughout the CPU to reduce critical path delay. Elsewhere 0.25-mA gates were used to conserve total power. Gate speed selection is accompanied by a programmable contact mask where the injector resistor contact taps are selected on a gate-by-gate basis. Gate delays of 160 ps are required for the critical path to achieve the 5-ns cycle time. Recent data have indicated that 250-ps gate delay can be achieved at 2 mw/gate.

14.6 DIGITAL PACKAGING

Ceramic packages are typical for GaAs ICs. Electrical and thermal problems need to be considered in packaging designs. The applications of GaAs ICs are toward the high-pin count area (256 pins or higher). The leaded chip carrier (LCC) and leadless chip carrier (LDCC) have both been developed. An example is given in Fig. 14.23 [Vit91]. LCCs were developed for hybrid and surface-mount applications, with the objective of reducing the board size. Multilayer ceramic packaging, tape-automated bonding (TAB), and multichip packaging (MCP) have been developed.

14.6.1 Multilayer Ceramic Packaging

Multilayer ceramic technology (MLC) has proved to be well suited for application to high-speed packages for GaAs ICs [Smi90]. These packages were initially described in 1985 [Smi85] and have proved to be effective in digital and mixed analog/digital applications involving clock rates into the 1 to 2-GHz frequency range. Multiple packages enable chips to be mounted close together, resulting in fast interconnections. (Some typical MLC packages are illustrated in Fig. 14.24.) The MLC44 package is a high-speed multilayer ceramic package developed at TriQuint to support the special requirements of high-performance ICs. The package is designed to handle clock rates up to 4 GHz and fast edge speeds less than 100 ps. Signals are carried on 50-ohm controlled impedance transmission lines from the package leads to bond pads. Excellent signal isolation is provided by the use of multiple ground lines. Capacitive power planes and decoupling capacitors minimize switching noise on power supplies. The package is shown in Fig. 14.24. [Tri89].

14.6.2 Multichip Packaging

A GaAs 16 × 16-bit parallel multiplier utilizing multichip packaging technology has been developed by Sumitomo [Sek90]. This multichip approach was taken in an effort to realize GaAs ULSIs with high yield and reliability, using multiple smaller-scale integrated circuits. The device is composed of four GaAs 8 × 8-bit expandable parallel

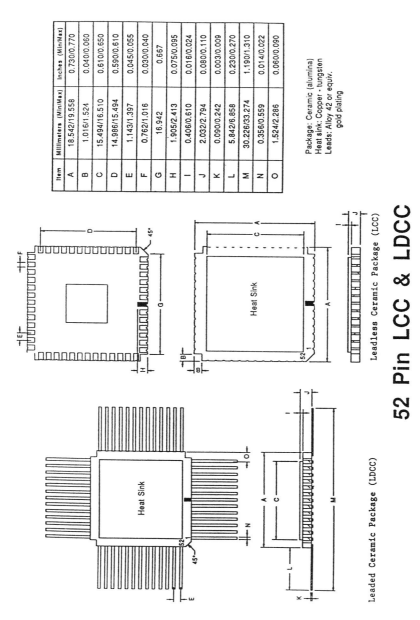

Fig. 14.23 52 Pin LCC 7 LDCC. (Ref: Vitesse 1991 Product Data Book. Vitesse Corp.)

Fig. 14.24 Typical high-speed MLC packages. The packages are normally mounted cavity down, with leads bent in a gull wing, as shown. [Smith, D. H., and R. M. Savara, in *GaAs IC Symp. Tech. Digest*, p. 203, IEEE (1990)]

multipliers and a multichip package (MCP). The developed MCP is composed of five layers of alumina ceramic which include 50-Ω strip lines.

Knowing the strip-line measurement results, the MCP was designed to maintain the waveform of the ECL level and to minimize the propagation delay. Figure 14.25 provides a diagram of the interconnects in the MCP, where XL and YL are the lower eight bits, and XH and YH are the highest eight bits of X and Y.

Figure 14.26(*a*) shows the branches that exist in the input signal lines. Computer transient simulations were done to determine the optimally matched circuit, yielding the results in Fig. 14.26(*b*).

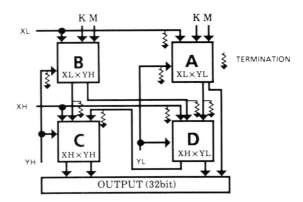

Fig. 14.25 Interconnect diagram (8-bit signals in each line). [Sekiguchi, T., S. Sawada, T. Hirose, M. Nishiguchi, N. Shiga, and H. Hayashi, *GaAs IC Symp. Tech. Digest*, p. 199, IEEE (1990)]

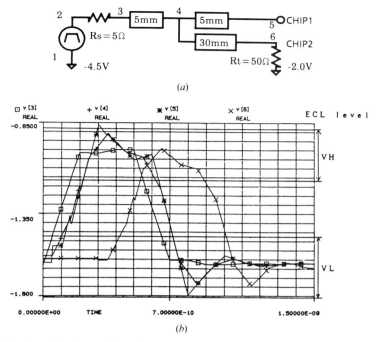

Fig. 14.26 Design of the input signal line branch. (*a*) Simulated circuit; (*b*) wave form at points (3–6). [Sekiguchi, T., S. Sawada, T. Hirose, M. Nishiguchi, N. Shiga, and H. Hayashi, *GaAs IC Symp. Tech. Digest*, p. 199, IEEE (1990)]

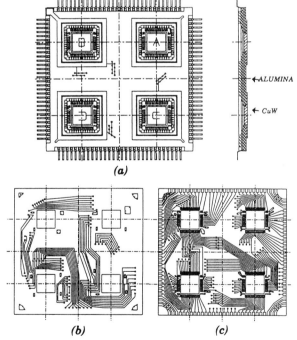

Fig. 14.27 Multichip package (MCP). [Sekiguchi, T., S. Sawada, T. Hirose, M. Nishiguchi, N. Shiga, and H. Hayashi, *GaAs IC Symp. Tech. Digest*, p. 199, IEEE (1990)]

558 GaAs DIGITAL INTEGRATED CIRCUITS

Figure 14.27 gives the detailed design of the MCP, which is coomposed of five layers of alumina ceramic, 120 outer leads, and CuW die attachment bases providing heat sinking. It includes two layers for signal lines presented in (b) a, d c), situated between two ground planes. NiCr thin film 50-Ω internal termination resistors are formed on the layer just above the upper-signal layer. The standard line pitch and its minimum value are 0.45 and 0.27 mm, and the total number of 0.2 mm via holes is 464.

14.6.3 Special Packaging

To minimize signal path lengths, Cray Computer designers eliminated the ceramic package. Instead 3-mil gold posts are welded to the GaAs chips ($0.151^2 \times 0.008$ in. with 52 contact points). The posts are about the same thickness as the PC board. The 5-mil gold-plated holes on the PC boards are then mated with posts, as shown in Fig. 14.28 [Wat92].

Because of differences in the TCE, packaging GaAs chips often calls for new solutions. Paul R. Jay of Bell Northern Research notes that the eutectic die attach is often avoided because of a TCE mismatch. BNR has devised special ceramic packages to accommodate high-speed chips requiring electrical terminations as close as possible to the die. This packaging is found in BNR's 2.4 Gb/s "FiberWorld" transmission products.

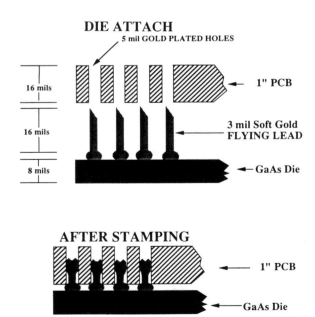

Fig. 14.28 Cross section of chip packaging method showing chip attachment to PCB. [Watts, H. R., Cray Computer Corporation (1992)]

REFERENCES

[Lev82] H. M. Levy, R. E. Lee, and R. A. Sandler, *IEEE Trans. Electron Devices* **29** (1982): 1687.

[Lev83] Levy, H. M., and R. E. Lee, *Electron. Let.* **19** (1983): 155–157.

[Tuy74] Tuyl, R. V., and C. A. Liechti, *IEEE J. Solid State Circuits*, **9** (Oct. 1974): 269–276.

[Tuy77] R. V. Tuyl, and C. A. Liechti, *IEEE J. Solid State Circuits* **12** (Nov. 1977): 485–496.

[Mun88] Mun, J., in *Gallium Arsenide for Devices and Integrated Circuits*, Macmillan (1988).

[Zuc80] R. Zucca, B. M. Welch, R. C. Eden, and S. I. Long, *IEEE Trans. Electron Devices* **27** (June 1980): 1109–1115.

[Lee80] Lee, F. S., E. Shea, G. R. Kaelin, and B. M. Welch, *GaAs IC Symp.* Las Vegas (1980).

[Miz80] T. Mizutani, N. Kato, M. Ida, and M. Ohmori, *IEEE Trans. Microwave Theory and Techniques* **28** (May 1980): 479–483.

[Vit89] Vitesse Semiconductor Corporation (1989).

[Dey92] Deyhimy, I., Vitesse Semiconductor Corporation (1992).

[Rou94] Rousseau, K. V., *ASIC & EDA*. (May 1994): 28.

[Gil86] Gilbert, B. K., in *VLSI Electronics Microstructure Sci.* **11** (1986): 289–331.

[Gil81] Gilbert, B. K., S. K. Kenue, R. A. Robb, A. Chu, A. H. Lent, and E. E. Swartzlander, Jr., *IEEE Trans. Biomed. Eng.* **28** (1981): 98.

[Goe93] Goering, R., *Electronic Engineering Times*. (August 9, 1993): 68.

[Gre86] Greiling, P. T., and C. F. Krumm, in *VlSI Electronics*, vol. 11, ed. by N. G. Einspruch and W. R. Wisseman Academic Press (1986).

[Tro79] Troeger, G., A. Behle, P. Friebertshauser, K. Hu, and S. Watanabe, *IEDM Digest of Technical Papers*, Sec. 21.7, pp. 497–500 (1979).

[Zul80] Zuleeg, R., and Lehovec, K., *IEEE Trans. Nucl. Sci.* **27** (1980): 1343–1354.

[Zul84] Zuleeg, R., J. K. Notthoff, and G. L. Troeger, *IEEE Electron Dev. Lett.* **5** (1984): 21.

[Shu87] Shur, M., *GaAs Devices and Circuits*, Plenum Press (1987).

[Mim82] Mimura, T., K. Nishiuchi, M. Abe, A. Shibatomi, and M. Kobayashi, *Int. Electron. Devices Meet. Tech. Digest*, pp. 578–581 (1982); *VLSI* vol. 11, p. 349, Fig. 14, ed. by N. G. Einsprusch and W. R. Wissreman, Academic Press.

[Abe83] Abe, M., T. Mimura, K. Nishiuchi, A. Shibatomi, and M. Kobayashi, *IEEE GaAs IC Symp. Tech. Digest* (1983): 158–161.

[Abe86] Abe, M., T. Mimura, K. Nichiuchi, and N. Yokoyama, in *VLSI Electronics*, vol. 11, p. 333, ed. by N. G. Einsprusch (1985): Academic Press.

[Hug88] Hughes, W. A., A. A. Rezazadeh, and C. E. C. Wood, p. 376 (1988). ed. by J. Mun, in *GaAs Integrated Circuits*, p. 376, Macmillan (1988).

[Chan86] Chang, M. F., P. Asbeck, K. C. Wang, G. J. Sullivan, and D. L. Miller, *Elect. Lett.* **22** (1986): 1173.

[Mun88] Mun, J., ed., *Gallium Arsenide Integrated Circuits*, Macmillan (1988).

[Tuy74] Van Tuyl, R. L. and C. A. Liechti, *IEEE J. Solid State Circuits*, **9**, 5 (1974): 269–276.

[Tuy77] VanTuyl, R. L., C. Liechti, R. E. Lee, and E. Gowen, *IEEE J. Solid State Circuits* **12** (1977): 485–496.

[How85] Howes, M. J., and D. V. Morgan, in *Gallium Arsenide, Materials, Devices, and Circuits*, Wiley (1985).

[Lie82] Liechti, C., G. Baldwin, E. Gowen, R. Joly, M. Nanjoo, and A. Podell, *IEEE Trans. Microwave Theory and Techniques* **30** (1982): 998–1006.

[Yam83] Yamamoto, R., A. Higashisaka, S. Arai, T. Tsuji, Y. Takayama, and S. Yano, *IEEE Int. Solid State Circuits Conf. Digest* (1983): 403.

[Ede78] Eden, R. C., B. M. Welch, and R. Zucca, *IEEE J. Solid State Circuits* **13** (1978): 419–426.

[Miz80] Mizutani, T., N. Kato, M. Ida, and M. Ohmori, *IEEE Trans. Microwave Theory and Techniques* **28** (May 1980): 479–483.

[Kat82] Katsu, S., S. Nambu, A. Shimano, and G. Kano, *IEEE Electron Device Lett.* **3**, 8 (1982): 197–199.

[Shi83] Shimano, A., S. Katsu, S. Nambu, and G. Kano, in *Proc. ISSCC 83*, p. 42 (1983).

[Shi84] Shimano, A., S. Katsu, S. Nambu, and G. Kano, ISSCC (1984).

[Lee84] Lee, K., Ph.D. dissertation, University of Minnesota, October (1984).

[Liv80a] Livingstone, A. W., and P. T. J. Mellor, *1980 GaAs IC Symp. Digest*, no. 10 (1980).

[Liv80b] Livingstone, A. W., and Mellor, P. T. J., *IEE Proc. I, Solid-State and Electron Devices* **127**, 5 (1980): 297–300.

[Hai89] Haigh, D., and Everard, J., eds., *GaAs Technology and Its Impact on Circuits and Systems*, IEE Press (1989).

[Lar89] Larson, L. E., in *GaAs Technology and Its Impact on Circuits and Systems*, ch. 7, ed. by D. Haigh, and J. Everard, IEE Press, p. 193 (1989).

[Ede84] Eden, R. C., *1984 IEEE GaAs IC Symp. Tech. Digest*, ch. 7, 11–14 (1984).

[Wel85] Welch, B. M., R. C. Eden, and F. S. Lee, *Gallium Arsenide*, ed. by M. J. Howes and D. V. Morgan, Wiley (1985).

[Nuz80] Nuzillat, G., G. Bert, T. P. Ngu, and M. Gloanec, *IEEE Trans.* **ED-27** (1980): 1102–1109.

[Nuz81] Nuzillat, G., G. Bert, F. Damay-Kavala, and C. Arnodo, *IEEE J. Solid State Circuits*, no. 3 (1981): 226–232.

[Asb84] Asbeck, P., D. Miller, R. Anderson, R. Deming, R. Chen, C. Liechti, and F. Eisen, *1984 IEEE GaAs IC Symp. Tech. Digest*, pp. 133–136 (1984).

[Doz92] Dozier, H., *IEEE Spectrum* (Sept. 1992): 66–68.

[Kim86] Kim, M. E., J. B. Camou, A. K. Oki, K. S. Stolt, and V. M. Mulvey, *1986 IEEE GaAs IC Symp. Tech. Digest* (1986): 163–166.

[Cha88] Chang, M. F., Rockwell Science Center (1988).

[Sat87] Sato, Tai, *ISSCC 87*, p. 143, IEEE Press (1987).

[Dey91] Deyhimy, I., View on Vitesse, vol. 2, no. 3, July (1991).

[Vit92] Vitesse Semiconductor Corp. (1992).

[Kan89] Kanopoulos, N., in *Gallium Arsenide Digital Integrated Circuits*, ed. by N. Kanopoulos, Prentice Hall (1989).

[Tan87] Tanaka, H., H. Yamashita, N. Masuda, N. Matsunaga, M. Miyazaki, H. Yanazawa, A. Masaki, and A. Hashimoto, *ISSCC 87*, pp. 138–139, IEEE Press (1987).

[Ume86] Umemoto, Y., *IEEE Electron Device Lett.* **7**, 6 (1986): 396–397.

[Ino82] Ino, M., et al., *Proc. GaAs IC Symp.* p. 2 (1982).

[Hir86] Hirayama, M., M. Togashi, N. Kato, M. Suzuki, Y. Matsuoka, and Y. Kawasaki, *IEEE Trans. Electron Dev.* **33**, 1 (1986): 104–109.

[Tak87] Takano, S., H. Makino, N. Tanino, M. Noda, K. Nishitani, and S. Kayano, *ISSCC Digest Tech. Papers*, pp. 140–141 (1987).

[Tse87] Tsen, C. T., S. Kuwahara, K. Elliot, L. Salmon, A. Cappon, E. V. Korpinen, E. R. Walton, S. J. Ross, W. Kleinhans, and R. Kezer, *GaAs IC Symp. Tech. Digest*, pp. 181–184 (1987).

[She86] Sheng, N. H., H. T. Wang, S. J. Lee, C. P. Lee, G. J. Sullivan, ad D. L. Miller, *GaAs IC Symp.* p. 97 (1986).

[Ken92] Kenney, K. *EDN*, p. 35, November 5 (1992).

[Hen87] Hendrickson, N., W. Larkins, R. Deming, R. Bartolotti, and I. Deyhimy, *GaAs IC Symp.* (1987).

[Pat80a] Patterson, D. A., and D. R. Ditzal, *Computer Architecture News* **8**, 6 (1980): 25–32.

[Pat80b] Patterson, D. and C. Sequin, *IEEE J. Solid State Circuits* **15** (Feb. 1980): 44–51.

[Wri92] Wright, M., *EDN*, p. 97, March 2 (1992).

[Roc88] Rocchi, M., B. Gabillard, E. Delhaye, and T. Ducourant, in *GaAs Integrated Circuits, Design and Technology*, ed. by J. Mun, Macmillan (1988).

[Gil84] Gilbert, B. K., *Proc. IEEE ICCD 84*, pp. 260–266 (1984).

[Whi88] Whitmire, D. A., V. Gercia, and S. Evans, *ISSCC 88*, pp. 34–35 (1988).

[And93] Andrews, Warren, *Computer Design*, pp. 33–36, Pennwell Pub. (July 1993).

[Yua86] Yuan, C., *ISSCC Digest of Technical Papers*, pp. 74–75 (Feb. 1986).

[Eva87] Evans, S., *IEEE GaAs IC Symp. Tech. Digest*, pp. 109–112 (Oct. 1987).

[Vit91] Vitesse, *Product Data Book* (1991).

[Smi90] Smith, D. H., and R. M. Savara, in *GaAs IC Symp. Tech. Digest*, p. 203 (1990).

[Smi85] Smith, D. H., T. G. Bowman, R. Lind and T. S. Riley, in *IEEE GaAs IC Symp. Tech. Digest*, pp. 151–154 (1985).

[Tri89] TriQuint Semiconductor, Inc. (1989).

[Sek90] Sekiguchi, T., S. Sawada, T. Hirose, M. Nishiguchi, N. Shiga, and H. Hayashi, *GaAs IC Symp. Tech. Digest*, p. 199 (1990).

[Isc92] Iscoff, R., *Semiconductional Int.* p. 60 (Mar. 1992).

15

HIGH-SPEED PHOTONIC DEVICES

15.1 INTRODUCTION TO PHOTONIC DEVICES

Semiconductor light-emitting diodes (LEDs), lasers, and photodetectors are critically important devices in optical fiber communications system [Tsa90] because of their small size and high reliability. Figure 15.1 shows a chart of some possible important applications of these devices. As sources, LEDs and lasers are easy to modulate to encode information. As optical receivers, semiconductor photodetectors have high quantum efficiency in converting the input optical signal back into the original electrical format. The LEDs and laser can be integrated monolithically with other optical and electronic devices to form optoelectronic integrated circuits (OEICs).

This chapter will focus on the high-speed aspects of semiconductor photonic devices. Since the silica fibers have the least loss and chromatic dispersion in the wavelength range of 1.0 to 1.6 µm, the devices considered here will be built out of quaternary III–V compounds. InGaAsP alloy systems that span this wavelength range with InP (0.9 µm) and $In_{0.53}Ga_{0.47}As$ (1.67 µm) at their end compositions. InP is a binary compound semiconductor and is used as the substrate. Thin crystalline layers of InGaAsP, having different compositions, are grown epitaxially over the substrate. The lattice constants of these layers are kept the same as the InP substrate underneath. The different compositions result in different energy band gaps, and hence the layers emit photons with different energy.

The demand for more computer-processing power drives the market of optoelectronic technology for interprocessor applications. An optical interconnect technology based on GaAs OEIC and IC technologies would meet the cost and performance objectives. If a successful OEIC-based technology were to be developed for the computer-system marketplace, other optical components such as modulators and switches could be added in the longer term. The last two sections give a brief introduction on the optical interconnects and switching elements used in computer- and signal-processing systems.

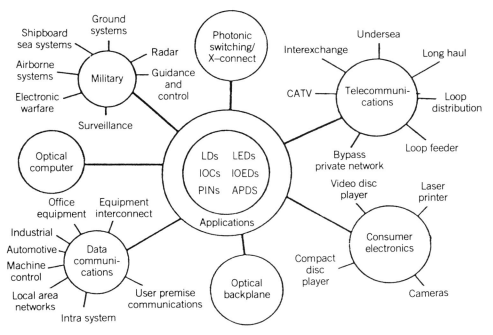

Fig. 15.1 Possible applications of LEDs, semiconductor lasers, and photodetectors in optoelectronic systems. [Sze, S. M., *High-Speed Semiconductor Devices*, p. 587, ed. by S. M. Sze, Wiley-Interscience (1990)]

15.2 LIGHT-EMITTING DIODES

Light-emitting diode (LED) sources can provide a reliable, inexpensive alternative to optical sources besides semiconductor lasers. LEDs utilizes relatively simple driving circuits that do not feed back to control power output, operate over a wide range of temperatures, and have projected lifetimes to two orders of magnitude longer than those of the laser diodes. LEDs are less temperature sensitive than lasers. Long-wavelength InGaAsP LEDs emit near the 1.3-μm wavelength, where silica fibers have low attenuation and minimum dispersion.

15.2.1 Device Structures

There are principally three types of LEDs for surface-emitting diodes, edge-emitting diodes, and superluminescent diodes, as shown in Fig. 15.2 [Tsa90]. In surface-emitting diodes [Uji85] the output beam emits perpendicular to the plane of the layer structure. The emitting diode is typically 25 μm in diameter. The InGaAsP active-layer composition can be varied to provide output wavelengths from about 1 μm to nearly 1.6 μm. Since the chromatic dispersion of the silica fibers is minimum at ~1.3 μm and the emissions from LEDs are broad band, an integral lens is usually formed at the exit surface of InP substrate to improve coupling efficiency to the fibers.

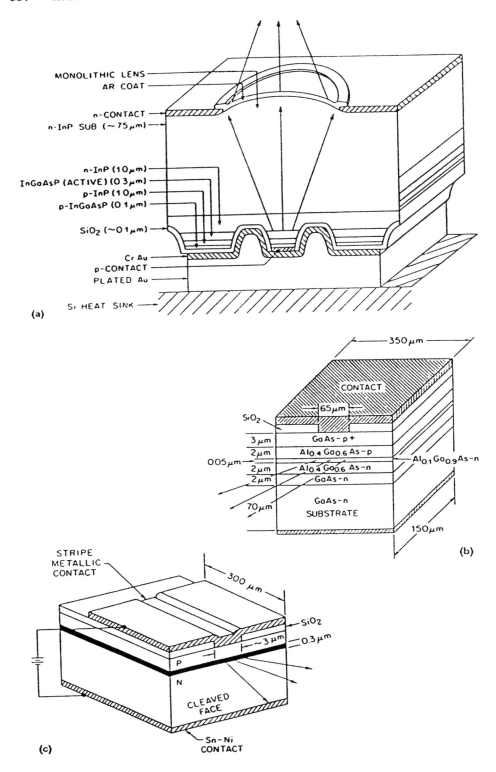

In Fig. 15.2(b) an efficient edge-emitter LED [Kre75; Hor76] emits its beam in a relatively directed beam parallel to the active layer as in a semiconductor laser diode. This improves the coupling efficiency into a fiber, especially into a single-mode fiber (a single-mode fiber guides only the lowest-order optical fiber in its core). To avoid lasing, one end is reflective and the other end has an antireflection coating.

A superluminescent diode [Lee73], Fig. 15.2(c), has one end of the active stripe made optically lossy to prevent reflection and thus supressing lasing. The output beam is emitted from the opposite end. In operation, the current passed through the diode is increased until stimulated emission and amplification occur. There is no feedback nor oscillation because of the high loss at one end. Therefore there is gain in the current-injection region, and output increases rapidly with current due to single-pass amplification. Devices have been made to operate in a pulsed mode to provide a peak output of 60 mW in an optical bandwidth of 6 to 8 nm at 0.87 μm.

15.2.2 Modulation Characteristics

The modulation of the LED output is accomplished by direct modulation of the injected current through the diode according to the following relationship [Liu75; Nam74]:

$$|I(\omega)| = \frac{I(O)}{\sqrt{1 + (\omega\tau)^2}}, \qquad (15.1)$$

where $I(\omega)$ is the intensity of the light output being modulated at an angular frequency ω, $I(O)$ is the intensity of the optical power at zero modulation frequency, and τ is the effective carrier lifetime inside the active medium where photons are generated. The modulation bandwidth, usually defined as the frequency where the response has fallen from 3 dB below its dc value (f_{3dB}), is given by

$$f_{3dB} = \frac{\Delta\omega}{2\pi} = \frac{1}{2\pi\tau}. \qquad (15.2)$$

In the lightly doped active region, and when injection of both electrons and holes occurs, the injection-carrier density Δn is proportional to the current density J and inversely proportional to the thickness of the active layer W_r:

$$\Delta n = \frac{J\tau}{qW}, \qquad (15.3)$$

where $\tau_r = 1/B\Delta n$ is the rediative recombination lifetime, B is the rediative recombination probability, and q is a unit of electrical charge. Using the fact that $\tau \sim \tau_r$, the

Fig. 15.2 (a) Small-area mesa-etched InGaAs/InP surface-emitting LED structure; (b) AlGaAs stripe geometry edge emitter; (c) superluminescent diode with absorbing region. [Sze, S. M., *High-Speed Semiconductor Devices*, p. 587, ed. by S. M. Sze, Wiley-Interscience (1990)]

modulation bandwidth becomes

$$f_{3dB} = \frac{1}{2\pi}\left\{\frac{BJ}{qW}\right\}^{1/2}.$$

Thus the bandwidth of a double-heterostructure (DH) LED with a lightly doped ($\sim 5 \times 10^{17}$ cm^{-3}) active layer—and operated in the bi-molecular recombination region where the electron and hole concentrations are roughly equal—increases as $(J/W)^{1/2}$. Figure 15.3 confirms this relationship [Wad81] in InGaAsP LEDs.

In practical devices parasitic capacitance can impose a limitation on the device modulation bandwidth. The capacitance C is comprised of a space-charge capacitance associated with the p-n junction area and a diffusion capacitance that is related to the carrier lifetime in the small light-emitting area. Figure 15.4 shows the results on AlGaAs and InGaAsP LEDs, where the reciprocal relationship between modulation bandwidth and output power is apparent. The output power P can be related to the bandwidth f_{3dB} by $P \sim (f_{3dB})^{-r}$. For $f_{3dB} < 100$ MHz, $r = 2/3$ and for $f_{3dB} > 100$ MHz, $r = 4/3$. The output power of the AlGaAs LEDs is about a factor of two higher than that of the InGaAsP LEDs at all bandwidths.

The LED-based lightwave systems at the wavelength of minimal dispersion (near 1.3 µm) of silica graded-index fibers can have many applications, including data links that offer advantages of simplicity, reliability, and economy over the laser-based systems [Lee82; Sau83]. Data rates up to a few hundred megabits per second and repeater spacings over tens of kilometers are achievable with LEDs and photodetectors for both single-mode and multiple-mode fiber systems.

15.3 SEMICONDUCTOR LASERS

In 1917 the concept of stimulated optical emission was described by Albert Einstein. Maiman demonstrated the first operation laser—the ruby laser [Mai61] in 1959. In

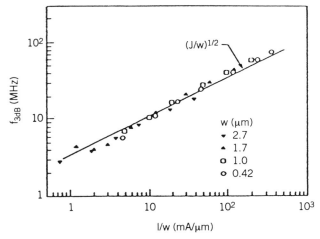

Fig. 15.3 The modulation bandwidth as a function of (J/w) for InGaAsP LEDs. [Sze, S. M., *High-Speed Semiconductor Devices*, p. 587, ed. by S. M. Sze, Wiley-Interscience (1990)]

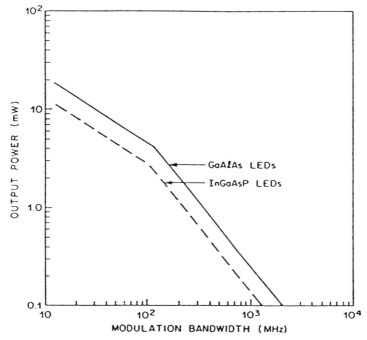

Fig. 15.4 Output power and bandwidth of LEDs at 0.8- and 1.3-μm wavelengths. [Sze, S. M., *High-Speed Semiconductor Devices*, p. 587, ed. by S. M. Sze, Wiley-Interscience (1990)]

1961 Basov et al. [Bas61] suggested that stimulated emission of photons could be produced in semiconductors by the recombination of carriers injected across a *p-n* junction. The first working *p-n* junction semiconductor lasers came into existence in 1962 [Hal62; Nat62; Qui62]. The simple *p-n* junction has been replaced by heterostructure containing several semiconductor layers of different compositions in 1969 [Alf69; Kre69]. Semiconductor lasers are similar to other lasers in that the emitted radiation has spatial and temporal coherence [Sze81]. In this section we consider the basic laser physics and the development of III–V compound semiconductor lasers.

The compound semiconductors can be formed in crystalline solid solutions. The notation used is $A_xB_{1-x}C$ for ternary and $A_xB_{1-x}C_yD_{1-y}$ for quaternary compounds throughout this book, where A and B are the group III elements and C and D are the group V elements. With the growing importance of the light-wave communication and integrated optoelectronic technologies, a great deal of interest has been generated in the AlGaAs (0.83 to 0.9 μm) and InGaAsP (1.1 to 1.6 μm) double-heterostructure (DH) current-injection lasers. The epitaxial growth technique has been changed from LPE to MBE or MOCVD.

In the AlGaAs/GaAs DH system the heterostructure laser system is well developed. With the advancement of MBE systems, the quantum well, quantum wire, and quantum box lasers have been developed. The range of wavelengths available from lattice matched AlGaAs/GaAs conventional DH lasers and quantum well (QW) heterostructure lasers is $\lambda = 0.88-0.65$ μm, defined by the band edges and the intrusion of large-mass indirect conduction band minima [Col90].

After the AlGaAs/GaAs system, the best-studied semiconductor heterostructure laser materials system is the InGaAsP–InP. The range of wavelengths available from lattice-matched InGaAsP/InP conventional DH lasers and QW heterostructure lasers is $\lambda = 1.1$–$1.6\,\mu\text{m}$, defind by the band edges and the heterostructure discontinuity between InGaAsP and InP. These wavelength ranges are sufficient to cover many important applications such as $\lambda = 1.55\,\mu\text{m}$ for optical fiber links, and high-power laser arrays at $\lambda = 0.82\,\mu\text{m}$ for diode pumped Nd:YAG lasers. In Fig. 15.5 [Co190] the wavelength ranges supported by the lattice-matched heterostructure laser systems described above are shown as solid lines, with an obvious gap in the wavelength range of $\lambda = 0.88$–$1.1\,\mu\text{m}$. There are a number of important applications that require laser emission in this range, especially for rare earth Er-doped optical fiber amplifiers. We will describe the strained-layer QW heterojunction lasers in Section 15.3.3.

15.3.1 Basic Semiconductor Laser Physics

Stimulated Emission

Absorption, spontaneous emission, and stimulated emission are the three basic transition processes that relate to laser operation. Consider two energy levels E_1 and E_2 in an atom, where E_1 is the ground state and E_2 is an excited state, as shown in Fig. 15.6 [Lev63]. From selection rules, any transition between these states involves the emission or absorption of a photon with frequency v_{12} given by $hv_{12} = E_2 - E_1$, where h is the planck's constant. At ordinary temperatures most of the atoms are in the ground state. When a photon of energy exactly equal to hv_{12} impinges on the system, an atom in state E_1 absorbs the photon and goes to the excited state E_2. This is the *absorption process*, Fig. 15.6(*a*). However, the atom in the excited state is unstable. After a short time, without any external stimulus, it makes a transition to the ground state, giving off a photon of energy hv_{12}. This process is called *spontaneous emission*, Fig. 15.6(*b*). The lifetime for spontaneous emission varies considerably ranging from 10^{-9} to 10^{-3} s,

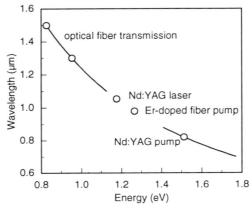

Fig. 15.5. The wavelength ranges supported by conventional lattice-matched heterostructure laser systems. The data points correspond to the wavelength of various important applications of heterostructure lasers. [Coleman, J. J., *IEEE IEDM* **90** (1990): 125–128; permission from IEEE]

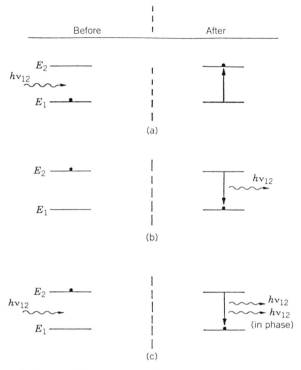

Fig. 15.6 The three basic transition processes between two energy levels E_1 and E_2. The black dots indicate the state of the atom. The initial state is at the left; the final state, after the process has occurred, is at the right. (a) Absorption, (b) spontaneous emission, and (c) stimulated emission. [Sze, S. M., *Physics of Semiconductor Devices*, Wiley-Interscience (1981); Levine, A. K., *Am. Sci.* **51** (1963): 14]

depending on various semiconductor parameters [Sze81]. However, when a photon of energy $h\nu_{12}$ impinges on an atom while it is still in the excited state, the atom is immediately stimulated to make its transition to the ground state and releases a photon of energy $h\nu_{12}$ which is in phase with the incident radiation. This process is called *stimulated emission*, which is shown in Fig. 15.6(c).

Waveguiding

In a DH laser the light is confined and guided by the dielectric waveguide. Figure 15.7(a) [Sze81] shows a three-layer dielectric waveguide with refractive indices n_1, n_2, and n_3, where an active (core) layer is sandwiched between two inactive (cladding) layers. Under the condition

$$n_2 > n_1 \geqslant n_3. \tag{15.4}$$

The ray angle θ_{12} at the layer 1/layer 2 interface at Fig. 15.7(b) exceeds the critical given by Snell's law. A similar situation for θ_{23} occurs at the layer 2/layer 3 interface. Therefore, when the refractive index in the active region is larger than the index of its

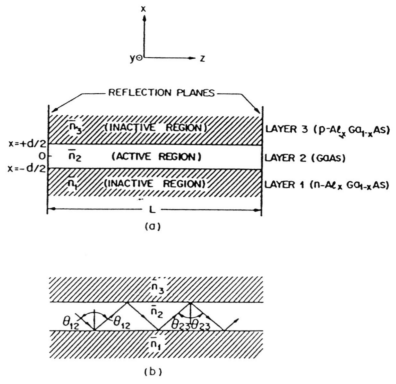

Fig. 15.7 (a) Representation of three-layer dielectric wave guide; (b) ray trajectories of the guided wave. [Sze, S. M., *High-Speed Semiconductor Devices*, p. 587, ed. by S. M. Sze, Wiley-Interscience (1990)]

surrounding layers, as in Eq. (15.4), the propagation of electromagnetic radiation is guided in a direction parallel to the layer interface.

Threshold Current Density

Under thermal equilibrium more atoms occupy the ground states than the excited states. If the photons of energy $h\nu_{12}$ are incident on a simple system described above, where the population of level E_2 is inverted with respect to E_1, stimulated emission exceeds absorption and more photons of energy $h\nu_{12}$ leave the system than enter it. Such a phenomena is called *quantum amplification*.

Figure 15.8 shows the energy versus density of states in a direct band-gap semiconductor. Figure 15.8(a) shows the equilibrium condition at $T = 0\,\text{K}$ for an intrinsic semiconductor in which the shaded area represents the filled states. Figure 15.8(b) shows a situation for an inverted population at 0 K. This population inversion can be achieved with the photoexcitation of photon energy greater than the band gap E_g. The valence band is empty of electrons down to an energy E_{FV}, and the conduction band filled up to E_{FC}. Photons with energy $h\nu$ such that $E_g < h\nu < (E_{FC} - E_{FV})$ will cause downward transition and hence stimulated emission.

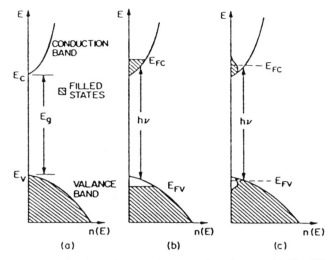

Fig. 15.8 Energy versus density of states in a semiconductor. (a) Equilibrium, $T = 0\,\text{K}$; (b) inverted, $T = 0\,\text{K}$, (c) inverted, $T > 0\,\text{K}$. [Sze, S. M., High-Speed Semiconductor Devices, p. 587, ed. by S. M. Sze, Wiley-Interscience (1990); Nathan, M. I., Proc. IEEE **54** (1966): 1276]

At finite temperatures the carrier distributions will be smeared out in energy, as shown in Fig. 15.8(c). The occupation probability of a state in the conduction band is given by the Fermi-Dirac distribution

$$F_C(E) = \frac{1}{1 + \exp(E - E_{FC}/kT)}, \qquad (15.5)$$

where E_{FC} is the quasi-Fermi level for electrons in the conduction band. A similar expression holds for the valence band.

Consider the rate of photon emission at hv due to a transition from a group of upper states near E in the conduction band to lower states at $(E - hv)$ in the valence band. The rate for this emission is proportional to the product of the density of occupied upper states $n_C(E)F_C(E)$ and the density of unoccupied lower states $n_V(E - hv)$ $[1 - F_V(E - hv)]$. The total emission rate is obtained by integrating over all energies,

$$W_{\text{spont}}(hv) = B \int n_C(E) n_V(E - hv) F_C(E) [1 - F_V(E - hv)] |\langle M \rangle|^2 dE. \qquad (15.6)$$

In a similar manner we can write

$$W_{\text{absorp}}(hv) = B \int n_V(E - hv) n_C(E) F_V(E - hv) [1 - F_C(E)] |\langle M \rangle|^2 dE \qquad (15.7)$$

for the absorption rate. The coefficient B is given by

$$B = \left(\frac{4\pi n q^2 hv}{m^2 \varepsilon_0 h^2 c^3} \right)_{\text{Vol}}, \qquad (15.8)$$

where $\langle M \rangle$ is the matrix element and "Vol" is the volume of the crystal. For a net amplification we require that $W_{spont} > W_{absorp}$. From the above equations we obtain [Sze81]

$$(E_{FC} - E_{FV}) > h\nu. \tag{15.9}$$

Equation (15.9) is a necessary condition for stimulated emission to be dominant over absorption.

In a semiconductor laser the gain g depends on the energy-band structure and is a complicated function of doping levels, current density, temperature, and frequency. The gain can be calculated for a special distribution of energy of states, where both the conduction band and valence band have band tails (Fig. 15.9). It can be given as a function of a nominal current density J_{nom}, which is defined for unity quantum efficiency ($\eta = 1$) as the current density required to uniformly excite a 1-μm-thick active layer. The actual current density is given by

$$J = \frac{J_{nom}d}{\eta}. \tag{15.10}$$

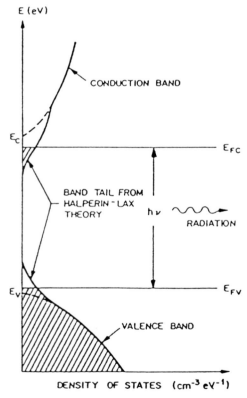

Fig. 15.9 Energy versus density of states where both conduction and valence bands have band tails. [Sze, S. M., *Physics of Semiconductor Devices*, Wiley-Interscience (1981); Halperin, B. I., and M. Lax, *Phys. Rev.* **148** (1966): 722]

As the current is increased, the gain increases until the threshold for lasing is reached; that is, the gain satisfies the condition that a light wave must make a complete transversal of the cavity without attenuation:

$$R \exp[(\Gamma g - \alpha)L] = 1, \tag{15.11}$$

or

$$\Gamma g \text{ (threshold gain)} = \alpha + \left(\frac{1}{L}\right) \ln\left(\frac{1}{R}\right), \tag{15.12}$$

where Γ is the confinement factor for the mth transverse mode perpendicular to the junction plane, α is the loss per unit length from free-carrier absorption and defect-center scattering, L is the length of the cavity, d is the active layer thickness, and R is the reflectance of the ends of the cavity. Tsang [Tsa78] extended this analysis to include the effects of interface recombination at the two GaAs/Al$_x$Ga$_{1-x}$As interfaces and bulk nonradiative recombination due to traps inside the GaAs active layer, to obtain

$$J_{\text{th}} = \left(\frac{d}{\eta}\right)\left(4.5 \times 10^7\right) + \frac{20 \times 10^5}{\Gamma_m}\left[\alpha + \left(\frac{1}{L}\right)\ln\left(\frac{1}{R}\right)\right].$$

To reduce the threshold current density, one can increase η, Γ, L, and R and reduce d and α.

15.3.2 Laser Structures

For high-speed (beyond 1 Gb/s) lightwave applications, semiconductor lasers are the candidate. The modulation speed depends on the intrinsic and extrinsic properties of the lasers. The extrinsic dependence is related to the particular laser structure and geometry employed. Figure 15.10 shows several of the important stripe-geometry laser structures [Bow88] in their end-view section.

Figure 15.10(a) gives the inverted-rib laser structure. The wide-gap waveguide layer underneath the active layer has a riblike structure inside the laser cavity. The difference in thickness produces a larger effective index of refraction in the rib region than on both sides of the rib. This results in waveguiding along the rib structure. The active layer in this structure is planar. Figure 15.10(b) employs the same principle for waveguiding in the lateral dimension.

The etched-mesa buried heterostructure laser shown in Fig. 15.10(c) is quite different from the last two laser structures. In this structure the active layer is first etched into a narrow strip; InP is then regrown to bury the active stripe. Since the surrounding InP has a lower refractive index than the GaInAsP active stripe, strong waveguiding occurs. The wider energy gap of InP than the GaAsInP also confines the injected carriers to the active stripe. This results in more efficient use of injected carriers and leads to a lower lasing threshold.

In Figure 15.10(d) a V-groove is first etched into the substrate; then, by liquid-phase epitaxial growth, the active layer and cladding layers are grown. A crescent stripe is formed directly above the V-groove. This results in an active stripe that is completely buried in the wide band gap InP.

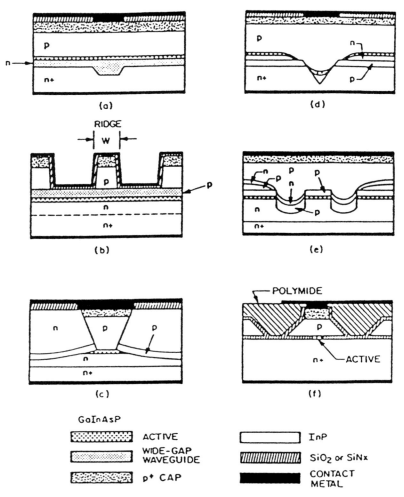

Fig. 15.10 InGaAsP/InP laser structures: (a) inverted-rib; (b) ridge-waveguide, (c) etched-mesa buried heterostructure; (d) channeled-substrate buried heterostructure; (e) double-channel planar buried heterostructure; (f) constricted mesa. [Sze, S. M., *High-Speed Semiconductor Devices*, p. 587, ed. by S. M. Sze, Wiley-Interscience (1990)]

Figure 15.10(e) gives a cross-sectional view of the double-channel planar buried heterostructure laser. The etched channels on both sides of the mesa containing the active stripe define the width of the active stripe. Current confinement is achieved by a reverse-biased *p-n* junction that similar to the structure described in Fig. 15.10(c). The constricted mesa structure shown in Fig. 15.10(f) has its narrow active stripe buried by SiO_2, and the structure is planarized by using polyimide.

Another type of laser structure is the quantum well laser. When the active region is less than 20 nm, quantum size effects are observed. In the quantum well structure, Fig. 15.11(a), the confinement of electrons and holes in one dimension causes a quantization in the allowed energy levels [Ara86]. The density of states changes from

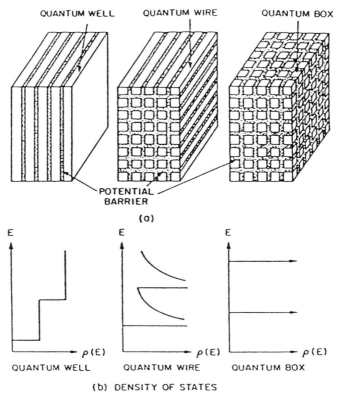

Fig. 15.11 Schematic diagram and density of states for quantum well, quantum wire, and quantum box laser. [Sze, S. M., *High-Speed Semiconductor Devices*, p. 587, ed. by S. M. Sze, Wiley-Interscience (1990)]

parabolic dependence in DH to a steplike structure. If the carriers are confined in two dimensions [quantum wire, Fig. 15.11(b)], or even in three dimensions [quantum box, Fig. 15.11(c)], the peak density of states becomes larger. Such modifications in the density of states of the QW lasers result in reduced threshold-temperature dependence, lower threshold currents, higher modulation frequency, and narrower laser linewidth.

The graded-index waveguide, separate-confinement efficiency heterostructure (GRIN-SCH) improves the optical-confinement efficiency and the carrier-collecting efficiency, as shown in Fig. 15.12. The optical-confinement factor is improved due to the parabolic refractive index profile. This concentrates more optical energy in the active quantum well. The carrier-collecting efficiency is improved due to the change in density of states in the graded layers are reduced. The structure provides the lowest current thresholds ever achieved in any semiconductor laser structures [Tsa90].

The structures just described provide control of the laser's transverse modes. Generally the several longitudinal modes resulting from the Fabry-Perot resonances of the cavity formed by the cleaved laser mirrors have sufficient gain to reach threshold and oscillate simultaneously. The multi-mode operation causes more noise. The single-frequency laser will be stable under high-speed modulation. The key to achieving single

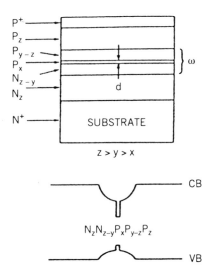

Fig. 15.12 A schematic diagram of a graded-index wave guide separate confinement heterostructure (GRIN-SCH) laser. Energy-band diagram of a GRIN-SCH laser: $N_i P_i$ stand for N, P-$Al_i Ga_{1-i} As$; n, p, stand for n, p-GaAs; CB, conduction band; VB, valence band; P_{y-z} represent the AlAs concentration in P—AlGaAs is varied from $Al_y Ga_{1-y} As$ to $Al_z Ga_{1-z} As$, and likewise for N_{z-y}. [Sze, S. M., *High-Speed Semiconductor Devices*, p. 587, ed. by S. M. Sze, Wiley-Interscience (1990)]

longitudinal-mode operation is to provide adequate gain or loss discrimination between the single desired mode and all the unwanted modes of the laser resonator. There are several ways to achieve this, such as using a short cavity [Lee83], adding an additional in-line, passive-coupled optical cavity [Lio84], using two independently adjustable active-coupled cavities [Tsa85], using the distributed feedback (DFB) and the distributed Bragg reflector (BR) structures [Kog71]. The cleaved-coupled-cavity (C^3) laser also offers the capability of electronically tuning the wavelength of the single longitudinal mode, as shown in Fig. 15.13.

15.3.3 Strained-Layer Quantum Well Heterojunction Lasers

The heterostructure laser system is well developed in the AlGaAs/GaAs system. The range of wavelengths available from lattice-matched AlGaAs/GaAs conventional double heterostructure lasers and quantum well heterostructure lasers is $\lambda = 0.88$–$0.65\,\mu m$, defined by the band edges and the intrusion of the large mass indirect conduction band minima. After the AlGaAs/GaAs system, the best-studied semiconductor heterostructure laser materials system is InGaAsP–InP. The range of wavelengths available from lattice-matched InGaAsP/InP conventional double heterostructure lasers and quantum well heterostructure lasers is $\lambda = 1.1$–$1.6\,\mu m$, defined by the band edges and the heterostructure discontinuity between InGaAsP and InP. These wavelength ranges are sufficient to cover many important applications such as $\lambda = 1.55\,\mu m$ for optical fiber links, and high-power laser arrays at $\lambda = 0.82\,\mu m$ for diode-pumped Nd:YAG lasers. In Fig. 15.5 [Col90], the wavelength ranges supported by the lattice-matched

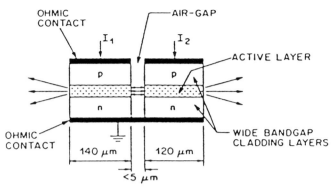

Fig. 15.13 Multiple-element resonators, a 4-mirror configuration in which the gap is formed by etching or by cleaving resulting in a cleaved-coupled-cavity (C^3) laser. [Sze, S. M., *High-Speed Semiconductor Devices*, p. 587, ed. by S. M. Sze, Wiley-Interscience (1990)]

heterostructure laser systems described above are shown as solid lines, with an obvious gap in the wavelength range of $\lambda = 0.88$–1.1 µm. There are a number of important applications that require laser emission in this range, especially for rare earth Er-doped optical fiber amplifiers.

The photo-pumped InGaAs–GaAs strained-layer superlattices have been demonstrated [Osb87] under cw 300-K laser operation, but the indications are that the superlattices are unstable under high excitation, with failure occurring in under an hour. An important key in the continued development of strained-layer InGaAs–GaAs heterostructure lasers was the work by Laidig and coworkers [Lai84] on injection laser diodes. Coleman [Col90] showed the suitability of InGaAs-GaAs strained layer heterostructures for diode lasers at $\mu \sim 1$ µm, and this was the first report of reliable laser operation from strained-layer lasers. The results are exciting, including reduced laser threshold currents, high differential gain, high efficiency, high power, access to wavelengths in the 0.9 to 1.1-µm range, high modulation speeds, reduced linewidth enhancement factors, and excellent reliability.

15.3.4 Surface-Emitting Laser

For the single-frequency application, an alternate approach is to replace the cleaved surfaces used as mirrors with frequency-selective reflectors such as gratings parallel to the junction plane. The grating is a periodic structure that reflects light only when the grating period $\Lambda = q\lambda/2$, where q is an integer. These reflectors are called *distributed Bragg reflectors* (DBR) and the device is known as the DBR laser [Sal91].

Vertical cavity surface-emitting lasers (VCSELs) have recently attracted a lot of interest, with applications including optical recording, optical communication, and optical computing with the VCSEL fabrication of integrated optical structures such as high-density laser arrays or chip-to-chip communication elements with high-speed and high-data rate transmission could be realized.

A VCSEL consists of a semiconductor laser diode sandwiched between two highly reflective mirrors so that the laser oscillation and output occur normal to the wafer. The VCSEL has some distinct advantages over the conventional edge-emitting laser such as a circular output beam, high two-dimensional packing density for arrays, and wafer-testing capability. Three types of top-emitting SELs are reported, as shown in Fig. 15.14 [Hon91]. The design concept has been evolved from a double heterostructure (DH) with a hybrid mirror (thin metal plus a few pairs of semiconductor DBRs) and to a QW structure with a hybrid mirror. The laser performance has low threshold currents under cw operation.

15.4 Pin PHOTODETECTORS

15.4.1 Basic Principles

High-speed and high-sensitivity photodetectors are required for high-bit-rate optical communication systems. Two commonly used photodetectors used in these systems are the pin and avalanche photodiodes (APDs). InP is the substrate material for photodetectors.

The basic principle of operation of a photodiode is the conversion of the incident optical signals into electrical currents by the absorption process in the material employed. In a reverse-biased semiconductor pin photodiode, as illustrated in Fig. 15.15

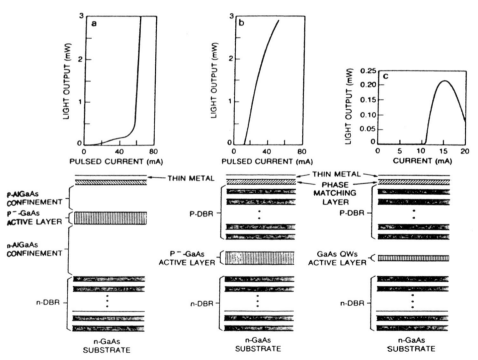

Fig. 15.14 Schematic of the vertial cavity surface emitting laser diodes and their output versus current $(L-I)$ curves: (a) a DH structure with n-DBR as the bottom mirror and a semitransparent Ag as the top mirror; (b) a DH structure using a hybrid reflector consisting of a thin metal and a few pairs of p-DBR as the top mirror; (c) a 3-QW structure with a hybrid reflector as the top mirror. [Hong, M., L. W. Tu, J. Gamelin, Y. H. Wang, R. J. Fischer, K. Tai, and A. Y. Cho, *J. Crystal Growth* **111** (1991): 1052–1056; permission from Dr. Tai]

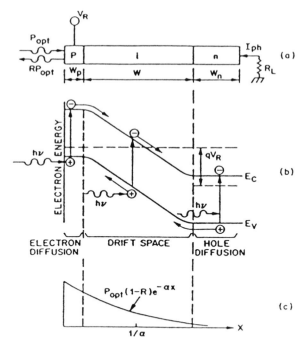

Fig. 15.15 Operation of a photodiode: (a) cross-sectional view of the p–i–n diode; (b) the energy-band diagram under reverse bias; (c) characteristics of carrier generation. [Sze, S. M., *High-Speed Semiconductor Devices*, p. 587 ed. by S. M. Sze, Wiley-Interscience (1990); Melchior, H., in *Laser Handbook* vol. 1, pp. 725–835, ed. by F. T. Arecchi, and E. O. Schulz-Dubois, North-Holland (1972)]

[Mel72], incoming photons are absorbed, and electron-hole pairs are generated primarily in the depleted region. The photogenerated carrier pairs are separated by the high electric field in the depletion region and are collected across the reverse-biased junction. As the carriers traverse the depletion region, a displacement current is induced at the load as the signal current. The structure of pin photodetector consists of n-InP, n^--InGaAs, and p-InGaAs layers. The speed of response of a pin is determined by the drift time across the depletion layer in the n^--InGaAs, the diffusion time outside the depletion regions, and the time it takes to charge or discharge the inherent capacitance of the diode plus the parasitic capacitance, and the trapping at the heterojunctions.

For high-speed operation, the intrinsic layer should be completely depleted, and the field in this layer should be above 50 kV/cm for InGaAs so that the carriers travel at their saturation velocity. Figure 15.16 shows the dependence of velocity on electric field for GaAs and InGaAs lattice-matched to InP substrate [Win82; Pea82; Hil87; Hel82; Bow85; Tsa90].

Figure 15.17 shows the calculated impulse response [Bow87] of a 0.5-μm-thick back-illuminated InGaAs/InP pin for several different wavelengths, that is, different absorption coefficients α. For $\alpha L \gg 1$, excitation is at one edge and results in a rectangular-shaped impulse response. For $\alpha L \ll 1$, the excitation is uniform and results in a triangular-shaped impulse response.

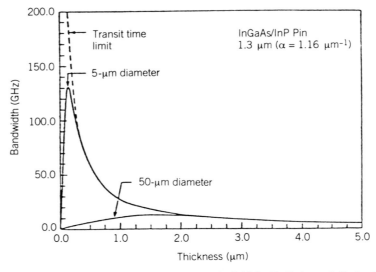

Fig. 15.16 Dependencce of carrier velocity on electric field for InGaAs and GaAs. [Sze, S. M., *High-Speed Semiconductor Devices* p. 587, ed. by S. M. Sze, Wiley-Interscience (1990)]

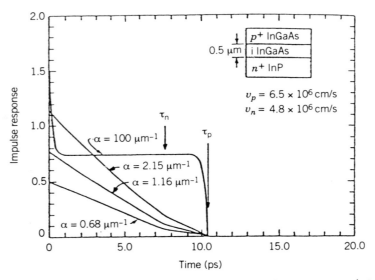

Fig. 15.17 Impulse response of a pin detector for different value of α: $\alpha = 0.68\ \mu m^{-1}$ ($\lambda = 1.55\ \mu m$), $\alpha = 1.16\ \mu m^{-1}$ ($\lambda = 1.36\ \mu m$), $\alpha = 2.15\ \mu m^{-1}$ ($\lambda = 1.06\ \mu m$), ($\theta = 4.8 \times 10^6$ m/s, $\theta_n = 6.5 \times 10^6$ m/s, corresponding to InGaAs. [Sze, S. M., *High-Speed Semiconductor Devices* p. 587, ed. by S. M. Sze, Wiley-Interscience (1990)]

The frequency response has a $\sin(x)/x$ response for $\alpha L \gg 1$. For arbitrary αL with optical excitation from the n side, the frequency response is given by [Bow85]

$$\frac{i(\omega)}{i(0)} = \frac{1}{(1-e^{-\alpha L})} \left(\frac{1-e^{-j\omega\tau_n - \alpha L}}{j\omega\tau_n + \alpha L} + e^{-\alpha L}\frac{e^{-j\omega\tau_n}-1}{j\omega\tau_n} \right) + \frac{1-e^{j\omega\tau_p}}{j\omega\tau_p} + e^{-\alpha L}\left[\frac{1-e^{\alpha L - j\omega\tau_p}}{\alpha L - j\omega\tau_p}\right],$$

(15.13)

where i is the detected current, ω is the angular modulation frequency, $\tau_n = (L/v_n)$ and $\tau_p = (L/v_p)$ are the electron and hole transit times, respectively, and v_n and v_p are their velocities. For high-speed response, very thin intrinsic layers are needed to reduce the transit time. However, this sacrifices the quantum efficiency. In principle, the bandwidths > 200 GHz are possible, but the quantum efficiency will be 11%. The trade-off is plotted in Fig. 15.18, where contours of constant bandwidth are plotted in the area/quantum-efficiency plane.

For the very high speed detector where $\alpha L \ll 1$, the quantum efficiency is given by

$$\eta = (1-R)\alpha L,$$

(15.14)

where R is the Fresnel reflectivity. The bandwidth of a thin detector is given by

$$f_{3dB} = \frac{0.45v}{L},$$

(15.15)

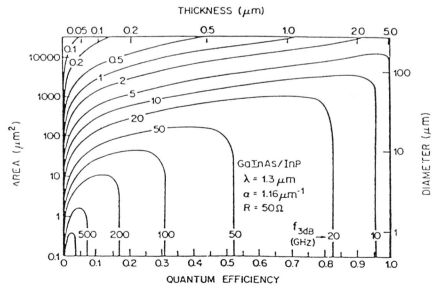

Fig. 15.18 Contours of constant 3-dB bandwidth in the detector area, depletion layer thickness plane: $\alpha = 1.16\,\mu m^{-1}$ for 1.3-μm wavelength, $v_n = 6.5 \times 10^6$ cm/s, $v_p = 4.8 \times 10^6$ cm/s, $\varepsilon = 14.1$. [Sze, S. M., *High-Speed Semiconductor Devices*, p. 587, ed. by S. M. Sze, Wiley-Interscience (1990)]

where we assume that $v_n = v_p = v$. The quantum efficiency of very high speed pins can be increased by collecting the light parallel to the junction plane in a wave guide.

The trapping at the heterojunctions can be serious. This can be improved by incorporating a graded layer at the heterointerfaces. A heterointerface is the interface between two epilayers having different energy band gaps. The effect of leakage current can be reduced by an appropriate design at the high bit rate.

15.4.2 Stability and Output Power of Pin-Avalanche Diodes

As has first been shown by Misawa [Mis66] in a small-signal analysis, pin-avalanche diodes are capable of operating at avalanche resonance because of a nearly frequency independent negative RF conductance, in contrast to Read-type devices. This operation should be advantageous especially for pulsed millimeter-wave oscillators due to high impedance and low losses, which are a result of the disappearance of capacitive currents at resonance. Claassen et al. [Cla82] has showed that this is also true under large-signal conditions. High-voltage modulation can be expected from the flat electric field profile, which should yield high efficiency and output power. The broad-band negative conductance promises devices that will be able to oscillate over a wide frequency range tuned only by the dc current. But little is known about an optimization of the diode dimensions under large-signal conditions with respect to stability, output power, and frequency. A systematic investigation has been carried out by Gaul et al. [Gau91].

Static dc Instability

Early studies of Impatt diodes (Hoe66, Mul68, Bow68] suggested that diodes with short drift regions, compared to the avalanche region (the extreme case is the pin diode), exhibit a negative static dc resistance due to the nonlinear dependence of the ionization coefficient on the electric field. This resistance leads to the formation of filamentary currents and then to the destruction of the device by local overheating.

For calculating the dc characteristics of GaAs pin-avalanche-diodes, the static continuity equations for holes and electrons, together with Poisson's equation, were solved by assuming equal ionization rates for holes and electrons [Cla87] as well as a constant saturated velocity (6×10^6 cm/s) for both charge carrier types. The static differential dc resistance of the pin-diodes consists of a negative contribution of the intrinsic region r_i (solid lines in Fig. 15.19), which is almost independent of the current for short diodes ($w = 100$ nm, the width of the i region), but a strongly increasing function of dc current density for long diodes ($w = 1000$ nm), since it is generally a strongly increasing function of w. Both the field distortions arising from space-charge effects at high current densities, and the nonlinear field-dependence of the ionization coefficient lead to a reduced breakdown voltage. To ensure static stability, the space-charge resistance of the depleted n and p regions r_n and r_p, the ohmic resistances of the n and p layers, and the contact resistances have to compensate the negative differential dc resistance of the i region. A stabilization by an external resistance is not possible as for the formation of filamentary currents by the negative differential dc resistance only. The space-charge resistance of the p- and n-cladding layers $r_n + r_p$ is shown in Fig. 15.19 for a typical doping concentration of 2×10^{18} cm^{-3} on both sides.

Suppose that there is a total value of $r_c = 5 \times 10^{-6}$ Ω-cm^2 for the contact and the ohmic resistances of the highly doped regions. Then the 1000-nm diode will exhibit

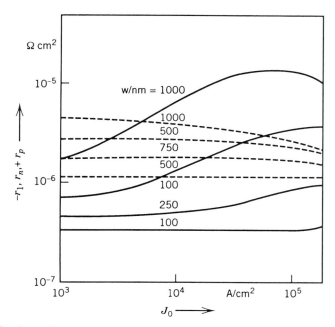

Fig. 15.19 Absolute value of the negative static differential dc-resistance of the intrinsic region r_i (*solid lines*) and positive space-charge resistance $r_n + r_p$ (*dashed lines*) of the highly doped zones versus dc-current density. The doping-concentration of the p and n layers is 2×10^{18} cm^3. [Gaul, L., and M. Classen, *AEü* **45**, 2 (1991): 126–130; permission from the author and S. Hirzel Verlag GmbH and Co.]

a negative dc resistance for current densities higher than $\sim 15\,\text{kA/cm}^2$, whereas the 500-nm diode will be statically stable for all dc current densities. Thus for the sake of static dc stability the width of the i region must not be larger than approximately 500 nm.

Dynamically Induced dc Instability

Static dc stability prevents the diode from destruction. If a statically stable diode develops at large-signal RF amplitudes (a dynamic dc instability), the ensuing filamentary currents will not destroy the diode but distort the oscillation and by this limit the possible RF-amplitude.

For the investigation of pin-avalanche diodes under large-signal conditions, the continuity equations for electrons and holes, together with Poisson's equation, have been solved in the space-charge region by a one-dimensional partial differential computer program [Gib73]. The material parameters were the same as in the static case, whereby additionally a low-field mobility of 450 cm^2/Vs was introduced (for the sake of simplicity mobility was assumed to be equal for electrons and holes). In the carrier-generation term tunneling was also introduced. The formula used for the dependence of the tunnel generation rate on the electric field was adapted from Kane [Kan59] despite its restricted applicability to GaAs. Diffusion was neglected in this study. In the simulation the device was voltage driven, $U(t) = U_0 + U_1 \sin(2\pi f t)$, where $U(t)$ is the applied voltage, U_0 the dc voltage, U_1 the RF amplitude, f the frequency, and t the time.

Figure 15.20 shows the dc characteristics of the space-charge region of a 250-nm pin diode for four different amplitudes U_1. The doping-concentration of the p and n layers is $2 \times 10^{18}\,\text{cm}^{-3}$, and the operating frequency is 60 GHz. At low amplitudes ($U_1 \sim 0$ V) the current-voltage characteristic exhibits a positive slope as in the dc case. With increasing RF amplitude, which in the figure can be seen clearly, an additional negative differential resistance is induced. This behavior was common to all investigated diodes at all frequencies (30 to 135 GHz) as long as tunneling did not become important.

The reason for the above-mentioned behavior is that at low and medium RF amplitudes ($U_1 = 5$ V and 8 V in Fig. 15.20), an increase in the dc current density shifts the carrier generation process toward the edges of the i region because of the space-charge effect. This changed distribution of the generated carriers delays the depletion of the diode during the negative half-cycle. The remaining carriers at the end of the period enhance the generation process in the next positive voltage swing such that a lower dc voltage is sufficient to maintain periodicity.

If very high RF amplitudes are applied, the electric field in the mid i region falls below the value necessary for saturated velocities during the negative half-cycle. The carriers are trapped in the diode, and the depletion is delayed. This effect is enhanced by the rising dc current density because the increasing space-charge lowers the field in the mid i region and may even change its sign. The trapping of carriers results in a strong current increasing the back-bias effect, as can be seen in Fig. 15.20 for the curve $U_1 = 11$ V.

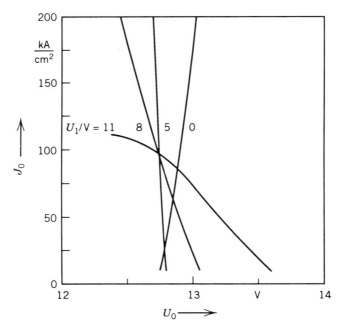

Fig. 15.20 Characteristics of dc-current voltage for different RF amplitudes U_1 (width of the i region is 250 nm, and the frequency is 60 GHz). The dots (●) indicate the current necessary for large-signal avalanche resonance. [Gaul, L., and M. Classen, *AEü* **45**, 2 (1991): 126–130; permission from the author and S. Hirzel Verlag GmbH and Co.]

In Fig. 15.21 the calculated differential dc resistance r_d of the space-charge region is depicted against the modulation depth $m = U_1/U_{0ss}$ (where U_{0ss} is the small-signal dc voltage) for several widths w of the i region. The current is always chosen such that avalanche resonance occurs at the operating frequency of 60 GHz. The dynamically induced negative dc resistance is a strongly increasing function of the modulation depth and of the width w of the i region. The positive increase of r_d for $w = 100$ nm and $m > 0.6$ indicates the onset of tunneling. This effect occurs only at low frequency in short devices [Mis72] that are stable anyway.

In Fig. 15.22 the maximum modulation for dynamically stable operation at avalanche resonance with $r_c = 5 \times 10^{-6}$ Ω-cm^2 is depicted against the width w of the i layer for the frequencies 60, 90, and 135 GHz. As could be expected from Fig. 15.21, the possible modulation decreases strongly with w. The curve for 135 GHz ceases at $w \sim 300$ nm because no avalanche resonance was found with dc current densities up to 500 kA/cm^2 at the RF amplitudes necessary for a dynamically unstable operation.

Optimization of Output Performance

In the previous section we mentioned that short i regions of pin diodes should be advantageous to large modulation depth. However, the device width must not be too small, since short diodes exhibit a lower breakdown voltage that limits the maximum RF voltage. Short device tunneling is likely to occur, and this will distort the RF power as it reduces the phase delay of the current generation. Finally, the conduction current

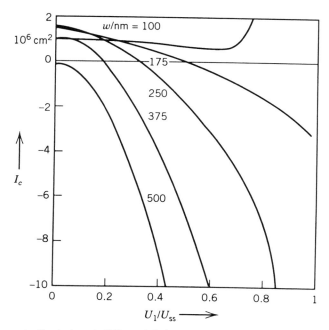

Fig. 15.21 Dynamically induced differential dc-resistance r_d of the space-charge region at avalanche resonance versus modulation depth U_1/U_{0ss} at 60 GHz for different widths w of the i region. [Gaul, L., and M. Classen, *AEü* **45**, 2 (1991): 126–130; permission from the author and S. Hirzel Verlag GmbH and Co.]

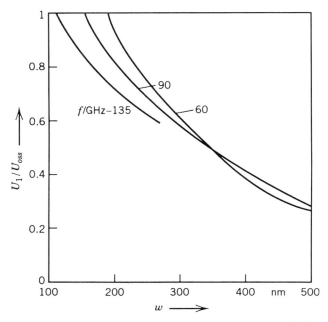

Fig. 15.22 Maximum modulation depth versus i layer width for dynamically stable operation at avalanche resonance with $r_c = 5 \times 10^{-6}$ ohm-cm^2 (r_c is the diode internal series resistance). [Gaul, L., and M. Classen, *AEü* **45**, 2 (1991): 126–130; permission from the auther and S. Hirzel Verlag GmbH and Co.]

is generated in short diodes at high amplitudes during a very short time. This high current pull leads to a substantial voltage drop at the contact resistance of the device such that the avalanche process is quenched before the end of the positive half-swing. The delay of the conduction current with respect to the voltage is again reduced, leading to a smaller negative conductance. Taken together, these effects result in a steep falloff of the maximum possible output power toward smaller device lengths, especially at lower frequency, as can be seen in Fig. 15.23 for 60 and 90 GHz where the maximum output power density is plotted against the width w of the i region. For 135 GHz the fall off will set in below the 100-nm i region width. This effect has not been simulated, since the energy relaxation [Mai83; Dal89] and deadspace effects [Cla87], which is not accounted for in the device model.

The output power for the larger i layer widths in Fig. 15.23 is partly reduced by the falling permissible voltage amplitude, which ensures a dynamically stable operation. Even the small-signal negative RF conductance of the pin diodes decreases with the increasing width w of the i layer, since field distortions are enhanced by the high current densities necessary for avalanche resonance in the investigated frequeny range. With the increasing RF amplitude, the negative RF conductance becomes saturated faster in diodes with long i layers as compared to those with shorter diodes. At RF voltages where carriers are trapped during the negative voltage swing, the negative RF conductance decreases rapidly as a result of both the strongly enhanced back-bias effect (with the consequences explained above) and a breakdown in the conduction current as the RF voltage passes its minimum. Then, to generate RF power, a high current must be

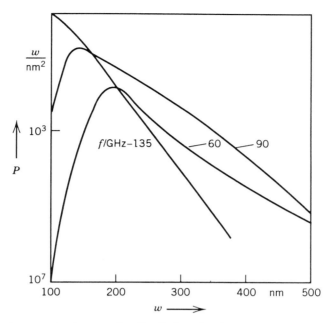

Fig. 15.23 Maximum output power density P at avalanche resonance versus the width w of the i region at 60, 90, and 135 GHz. (1991): [Gaul, L., and M. Classen, *AEü* **45**, 2 (1991): 126–130; permission from the auther and S. Hirzel Verlag GmbH and Co.]

achieved. Due to the space-charge induced field distortions the trapping occurs with increasing i layer widths at the lower modulation depth.

The output power degradation for long i layers in Fig. 15.23 is for 60 GHz devices, mainly caused by the amplitude restriction due to the onset of the dynamic dc instability; at 135 GHz the trapping-induced steep decrease of the negative RF conductance is the limiting mechanism. At the 90-GHz stability criteria, RF conductance degradation and trapping are equally important.

As can be seen from Fig. 15.23, a clear optimum exists for the width of the i layer at each frequency. While the falloff in output power toward smaller devices seems to be unavoidable, the falloff toward longer diodes, which is mainly due to space-charge effects, may be overcome to achieve still higher output power. This is possible when the space-charge of the movable charge carriers is compensated by a πv-doping. One effect of the πv-doping is that the negative conductance is enlarged and the decrease toward longer devices is diminished. Since the πv-doping provides an additional space-charge resistance, the static and dynamic dc stability is improved as well.

15.5 AVALANCHE PHOTODIODES

Long-wavelength avalanche photodiodes (APDs) are utilized in the long-span optical fiber transmission system [Tsa90]. The best receiver performance has been obtained by combining these APDs with low-noise GaAs MESFET preamplifiers. Avalanche photodiodes are operated at high reverse-bias voltages where the avalanche multiplication takes place [Sze81].

Many of the early long-wavelength lightwave transmission experiments utilized Ge APDs, which has a high dark current, unfavorable ratio of ionization coefficients, and a relatively low absorption coefficient at 1.55 μm. The best performance of the structure fabricated by III–V compounds is the separation, absorption, and multiplication (SAM) region APD [Tsa81] (Fig. 15.24), which was first developed to eliminate the tunneling component of the dark current in InGaAs APDs. Since lightwave systems have progressed to higher and higher bit rates, the emphasis of research on these APDs has shifted toward improving the frequency response.

The frequency response of the SAM-APDs is usually poor due to carrier trapping at the interface. Figure 15.25 shows the bandwidth of a back-illuminated InP/InGaAsP/InGaAs mesa-structure SAGM-APD versus the dc avalanche gain M_0. At high gain the frequency response approaches a constant gain-bandwidth limit of 70 GHz. The roll-off at lower gains is due to a combination of three physical effects: the RC time constant, the transit time of carriers through the space-charge region, and charge accumulation at the heterojunction interfaces. An approximate expression for the

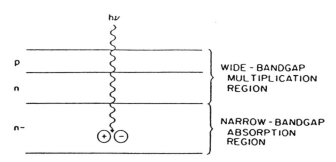

Fig. 15.24 Avalanche photodiode structure with separate absorption and multiplication regions (SAM-APD). [Sze, S. M., *High-Speed Semiconductor Devices*, p. 587, ed. by S. M. Sze, Wiley-Interscience (1990); Tsang, W. T., in *High-Speed Semiconductor Devices*, p. 587, ed. by S. M. Sze, Wiley-Interscience (1990)]

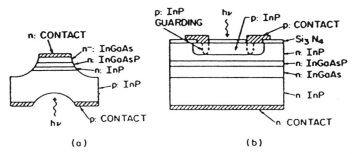

Fig. 15.25 (*a*) Mesa-structure; (*b*) planar-structure SAGM-APDs. [Sze, S. M. *High-Speed Semiconductor Devices*, p. 587, ed. by S. M. Sze, Wiley-Interscience (1990); Tsang, W. T., in *High-Speed Semiconductor Devices*, p. 587, ed. by S. M. Sze, Wiley-Interscience (1990)]

frequency dependence of the gain that includes these effects is given by

$$\frac{M(\omega)}{M_0} = \frac{2\sin(\omega W/v)}{\omega W/v} \frac{1}{\sqrt{1+(\omega RC)^2}} \frac{1}{\sqrt{1+(\omega/e_h)^2}} \frac{1}{\sqrt{1+(\omega \tau_m M_0)^2}}, \quad (15.16)$$

where $M(\omega)$ and M_0 is the gain at ω and dc, respectively. $M_0 \tau_m$ is the avalanche buildup time, and e_h is the hole emission rate over the heterojunction barrier. The first term is an approximate expression for the frequency response of a transit time-limited pin having a depletion width W. The second term is the RC contribution, and the third term is due to the residual interface trapping. The regenerative nature of the avalanche process is covered in the last term. As the gain increases, this term eventually dominates giving rise to a constant gain-bandwidth product shown in Fig. 15.25.

There are other APD structures that, as least in theory, improve performance by artificially increasing the ratio of the ionization coefficients for holes and electrons. Figure 15.26 shows a structure incorporating a superlattice into multiplication region [Cap82]. The difference between the conduction and valence band discontinuities enhances the electron ionization rate relative to that of holes.

15.6 HYBRID INTEGRATION

15.6.1 Grafted-film Process

In optoelectronics there is a need to integrate many devices on a single chip to obtain the cost and performance benefits over hybrids. However, the integration of optoelectronic components into optoelectronic integrated circuits require interconnecting electrically as well as optically several different types of devices with dissimilar structures. These differences have hindered progress in the field.

A reasonable compromise between full monolithic integration and hybrids is hybrid integration [Yi-Y90]. Yi-Yan chose the material system for a particular device based

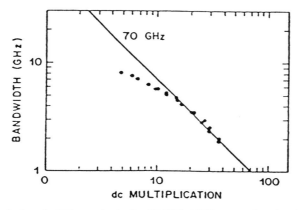

Fig. 15.26 The gain-bandwidth product of a SAGM-APD grown by chemical beam epitaxy, measured at $\lambda = 1.3\,\mu$m. [Sze, S. M., *High-Speed Semiconductor Devices*, p. 587, ed. by S. M. Sze, Wiley-Interscience (1990)]

on performance alone and used thin film technology to make the device and to interconnect it with other parts of the circuit on the final substrate. A new approach in thin film technology is epitaxial lift-off (ELO). An epitaxially grown film of a semiconductor is selectively removed from its grown substrate and reattached to a new substrate by the van der Waals force [Yab87], where it is used to fabricate devices.

ELO allows us to place a high-quality single crystalline semiconductor film on the new substrate without regard to lattice matching or chemical compatibility of the two materials. Materials can be mixed for OEICs, such as GaAs and $LiNbO_3$ or InP and glass.

15.6.2 Device Fabrication

The ELO process consists of two steps: detaching the epitaxial film from the growth substrate and reattaching it to a new substrate. To detach the film, an AlAs sacrificial layer with the GaAs–AlGaAs material system is used (Fig. 15.27). For the InP–InGaAs material system, substrate etching is used with a lattice-matched InGaAs etch stop. For both methods of detachment, the reattachment is through van der Waals forces.

Before detaching the film, the sample is covered with a wax such as Apiezon W. The wax gives mechanical support of thin film during the time it is free of any substrate; its stress lightly bows the wafer during the etching of the sacrificial layer, allowing reaction products to escape more easily. All but a ~ 0.5 to 1-mm periphery of the sample is masked with standard photoresist. The sample is selectively etched except at the bottom-most portion of the desired epitaxial layer, as shown in Fig. 15.28. This

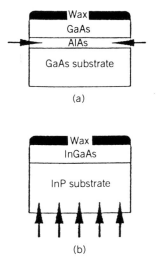

Fig. 15.27 Schematic of the two processes for ELO. The surface is waxed in both cases for mechanical support of the freed film. For the GaAs/AlGaAs material system (*a*), AlAs is used as the sacrificial layer, typically 50-nm thick and selectively etched with HF. For the InP-$In_{0.53}Ga_{0.47}As$ material system (*b*), it is more convenient to etch the InP substrate with HCl. The layer of $In_{0.53}Ga_{0.47}As$ also serves as an etch stop. [Chan, W. K., A. Yi-Yan, and T. J. Gmitter, Grafted semiconductor optoelectronics, *IEEE J. Quantum Electronics*, **27**, 3 (1991): 717–725; permissions from Bellcore and IEEE]

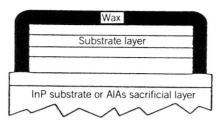

Fig. 15.28 Masking layers that are sensitive to the lift-off etchant. If the film to be lifted-off includes one or more layers that can be etched in the lift-off etchant (a high Al content AlGaAs layer in the GaAs case or an InP or quaternary layer in the InP case), all but the bottom-most layer is etched from the periphery. If the bottom-most layer can be attacked by the etchant, a dummy layer that can later be removed must be included to protect that layer. Wax is applied to cover this ledge, and it protects the sensitive layers from the sides. [Chan, W. K., A. Yi-Yan, and T. J. Gmitter, Grafted semiconductor optoelectronics, *IEEE J. Quantum Electronics*, **27**, 3 (1991): 717–725; permissions from Bellcore and IEEE]

layer must be resistant to the lift-off or etch-off etchant. The photoresist mask is then removed, and the Apiezon W is applied, covering the sidewalls of all the upper layers and the top of the exposed bottom layer. Once waxed, the sample is immersed in HF overnight to etch the sacrificial AlAs layer, or in HCl for 1 to 2 h to etch the InP substrate. When the epitaxial film is freed from the growth substrate, it is removed from the acid, rinsed with deionized water, and placed on the new substrate. The wax provides enough strength and rigidity to the film that it can be handled with tweezers. The transferred film can be moved to the desired location and oriented appropriately by sliding it on the water film.

Once the film is properly placed, the excess water is gently squeezed out and blotted up with filter paper. The assembly is subsequently left to dry overnight with a small weight on the film. In the drying process, attractive forces between the substrate and the semiconductor squeeze the water to the edges by a Poiseuille flow. As the water escapes, the separation between the lifted-off film and the new substrate decreases until the short range, attractive van der Waals forces can hold the two together.

The ELO process does not degrade the minority carriers's lifetime in GaAs–AlGaAs double heterostructures [Yab87]. This indicates that few electrically active defects are introduced by the lift-off process and suggests that the ELO film can be used for making high-quality devices.

15.6.3 Bonding by Atomic Rearrangement

The bonding by atomic rearrangement (BAR) technique has been invented to realize high-quality heteroepitaxy for lasers and optoelectronics [Lo91]. Motivated by the long-wavelength integrated optoelectronics, Lo et al. have attempted to bond InP/InGaAsP 1.5-μm wavelength lasers on GaAs substrate. As shown in Fig. 15.29, the layers for lasers were first grown in an InP(100) substrate by low-pressure (76 torr) MOCVD. After InP and GaAs wafer cleaning in buffered HF, the two wafers were put face to face and loaded into the reactor. The cleavage planes of the wafers were carefully aligned with an error of less than 0.2°. The sample was heated to 650°C in MOCVD

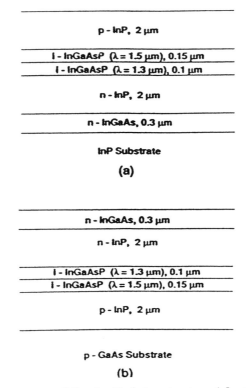

Fig. 15.29 Epitaxial structures of the double-heterostructure 1.5-μm wavelength lasers (a) before and (b) after wafer bonding. [Lo, Y. H., R. Bhat, D. M. Hwang, M. A. Koza, and T. P. Lee, *Appl. Phys. Lett.* **58**, 18 (1991): 1961; permissions from Dr. Lo and Americal Institute of Physics]

chamber and held at that temperature for 30 min with a hydrogen flow to remove native oxides on both substrates and to bond the wafers.

The mechanism of bonding is explained as follows: After the desorption of native oxide, the surfaces of GaAs and InP become strongly reactive. On a free GaAs or InP surface, surface reconstruction takes place to reduce the surface energy. However, when the two surfaces are close enough, new chemical bonds are formed, since the energy is lower than that for separately reconstructed surfaces. The bonding occurs uniformly across the entire surface due to the high surface mobility of the In atoms that fill up the microscopic gaps and holes. On the other hand, the Ga atoms in GaAs stay at their original positions so that no material intermixing takes place. Consequently the interface dislocation is very well confined.

15.7 OPTICAL INTERCONNECTS

The demand for more computer-processing power continues to increase. This demand is being satisfied at a single location through increasingly powerful uniprocessors and through multiprocessor complexes. Improvements both in the hardware technology and in the system organization (architecture) contribute to increases in computing power. On the hardware side, increases in transistor speed, IC complexity, and packaging density have minimized the time needed to process instructions and data. However, these advances have generated a secondary problem. It is becoming increasingly difficult to provide enough electrical-wiring speed and density to take full advantage of the high speed of logic circuits, and still keep noise and crosstalk on the lines to acceptable levels. On the architectural side, parallel processing and the sharing of memory and storage between processors are gaining popularity as ways to increase the aggregate computing power of a system.

An interprocessor network is a means of connecting the elements in such a multiprocessor environment. Unlike a data communications network, an interprocessor network must make connections quickly between the processors so that it does not slow down the computing operations shared by the processors. The interprocessor networks may extend to hundreds of meters in length. The performance and cost objectives of interprocessor network are difficult to meet with electrical wiring.

15.7.1 Network Requirements

Optimizing the cost and performance of optoelectronic technology for interprocessor applications leads to a different chip and package technology than those being developed by the telephone and data communications industry [Leh89]. In these interprocessor networks the processors at the nodes may be mainframes or workstations with processing rates ranging from tens to hundreds of MIPS. These processors must transfer data blocks fast enough, say, in microseconds.

An interprocessor network may reside between boards in one equipment frame, between frames in a computer complex, or even between clusters of frames in a building. The desire for microsecond response times over the network will keep distances short (generally less than 1 km). The electrooptic technology used in such a network should be compatible with the high packaging density, the power supply, and the cooling requirements of processor and memory frames. The network reliability is very critical. The functions that adapt the processor bus to the optical network contribute to the robustness, besides conditioning the data for transmission.

In designing the required network interface functions, there is a trade-off between the detrimental impact of the time taken to execute a function and the enhancement of network features. These features include reducing I/O count, improving reliability, and gaining flexibility. Typical requirements for an optimized optical link technology can be derived from these network interface requirements [Cro90; Lan88].

15.7.2 Optoelectronic Transducers

Quantum well lasers are attractive optoelectronic (OE) components because they meet the requirements of 1 gigabit/s modulation speed and high electrical-to-optical conversion efficiency. GaAs/AlGaAs lasers have been fabricated with drive-current requirements of only a few milliamperes, energy conversion efficiencies of over 40%, and modulation frequencies in the multigigahertz range. These optical sources are less power-consuming than the electrical counterparts.

Another possible optical source is the InGaAsP LED array. Although the LED is less efficient and slower than the laser, it has the advantages of a simpler structure, no need for device bias control, and longer demonstrated lifetime. The strained-layer, quantum well, surface-emitting laser, currently at the research stage, has demonstrated many attractive features for optical interconnection.

Photodiodes are integrated on the same chip with the receiver electronics to reduce the optical receiver's susceptibility to electrical noise pickup from closely packaged digital circuits without bulky and expensive shielding. With integration, the high-gain amplifier can be placed only 10 to 20 µm away from the photodiode, minimizing an electrical-noise–input-coupling loop.

15.7.3 Optoelectronic Interfaces

Optical interfaces are key to the success of optical interconnects. In the European Community's ESPRIT optical interconnections for VLSI and electronic systems (OLIVES) program [Par91], the focus is on the construction of four demonstrators of optical interconnections at the module, back plane, multichip module, and chip levels. One of the efforts is to develop the technology for monolithic integration of multiquantum well (MQW) modulators on CMOS circuitry.

One example of the MQW modulator design is described as follows [Goo91]: The two principal requirements for the modulator design are (1) that the modulator must operate correctly with the drive voltage levels available from standard high-performance silicon circuits, typically a 5 V swing, and (2) that the flip-chip hybridization approach adopted must have a modulator geometry in which the modulators lie face down on the silicon for electrical connection. Two potentially attractive material systems can be identified that satisfy these requirements: the InGaAs/InP MQW system (operating at 1.55 µm) and the strained AlGaAs/InGaAs MQW system (operating at ~1-µm wavelength).

Figure 15.30 shows the structure of the inverted asymmetric Fabry-Perot (ASFP) electroabsorption modulator, which comprises an InGaAs/InP MQW absorbing region in conjunction with an InGaAlAs/InP multilayer mirror. This approach uses a low reflective semiconductor, with a small number of periods and a broader optical bandwidth than the 95% mirrors required for the conventional ASFP device, with correspondingly relaxed fabrication tolerances.

15.7.4 Monolithic Integration of Functions

GaAs MESFET IC technology is a strong candidate for integrating all the functions of a link adapter onto a single chip. The IBM NEXUS project demonstrates a chip design, package, and link module assembly technique based on manufacturable technologies in GaAs MESFETs and lasers [Ewe91]. The link operates at 1 GBd and consists of a hybrid optoelectronic transmitter module and a monolithic integrated optoelectronic receiver module. The transmitter module contains a GaAs MESFET IC, a four-element AlGaAs laser array, and a four-fiber array mounted on a multilayer thin film wiring carrier. The receiver module contains a GaAs MESFET OEIC and a four-fiber array and is also mounted on a multilayer thin film wiring carrier.

The integrated circuits in the transmitter and receiver modules are fabricated using a 1-µm self-aligned refractory-gate GaAs MESFET enhancement/depletion process with 2.5 levels of global wiring [Mag87]. The chips contain a mix of analog and

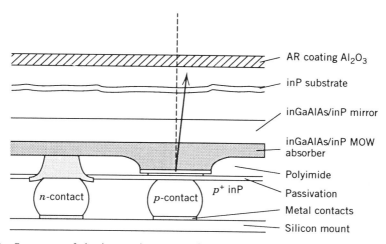

Fig. 15.30 Structure of the inverted asymmetric Fabry-Perot electroabsorption modulator. [Goodwin, M. J., A. J. Moseley, M. Q. Kearley, R. C. Morris, C. J. G. Kirkby, J. Thompson, R. C. Goodfellow, and I. Bennion, *IEEE J. Lightwave Tech*, **9**, 12 (1991): 1764–1773; permission from IEEE]

digital circuits. The analog circuits are designed using a minimum number of off-chip components to reduce package complexity. The digital circuits are implemented using both fully differential source-coupled logic and super-buffer logic (SBL) [Ewe88]. The differential logic provides the minimum delay and delay skew.

The receiver IC (Fig. 15.31) consists of an array of four receivers, a phase-lock loop (PLL) retiming circuit, and a 1:10 deserializer circuit. The four receivers can be operated independently to evaluate crosstalk in the high-speed front-end portion of the chip. The receiver chip (Fig. 15.32) measures 3×4 mm and has 54 I/O for signal and power. The four receivers and photodetectors are at the top of the chip. The retiming circuit occupies the lower right quadrant of the chip, with the deserializer in the remaining area. A ground ring surrounds the active area of the chip to facilitate high-speed testing and packaging.

The transmitter IC (Fig. 15.33) contains four laser modulator and bias control circuits, a phase-lock loop clock synthesizer, and a 10:1 serializer circuit. The serializer circuit contains a ten-bit ring counter driving a 10:1 data multiplexer. The transmitter chip (Fig. 15.34) measures 3×4 mm and has 54 I/O for signal and power. The laser modulator and bias circuit are at the top center of the chip. The PLL is in the lower right quadrant with the serializer in the lower left quadrant. The logic circuits use a 1.5-V power supply, while a +3-V supply is used in the analog laser control circuits. The chip contains approximately 800 NOR-3 equivalent gates, operates to 1 Gb/s, and nominally dissipates 400 mW.

15.7.5 High-Density Packaging

Modules for the OEIC chip and its complement of passive optical and electrical components can take advantage of the base technologies developed by the IC industry. Multicomponent IC carriers demonstrate 1 gigabit/s speed and processed electrical

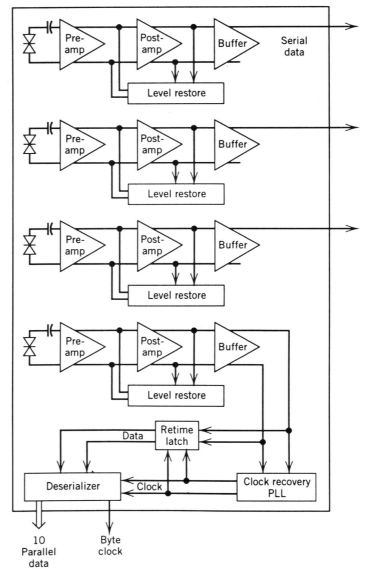

Fig. 15.31 Receiver OEIC block diagram with a four-channel photodiode receiver array, clock recovery PLL, and deserializer circuits. [Ewen, J. F., K. P. Jackson, R. J. S. Bates, and E. B. Flint, *IEEE J. Lightwave Tech.* **9**, 12 (1991); permission from IEEE]

wiring (on the carriers), but their optical wiring and packaging trails fall far behind their electrical counterparts. Two important needs for OEIC packaging are (1) planar-processed optical coupling between chips, chip carriers, and off-carrier optical transmission.

A number of candidate lightguide technologies are used for integrated optics. Multichip carriers can be as large as 10 cm, so lightguide loss should be less than a

OPTICAL INTERCONNECTS   **597**

Fig. 15.32 Receiver chip photograph showing the four detectors and receivers (*top center*), the PLL retiming circuit (*lower right*), and the deserializer. [Ewen, J. F., K. P. Jackson, R. J. S. Bates, and E. B. Flint, *IEEE J. Lightwave Tech.* **9**, 12 (1991); permission from IEEE]

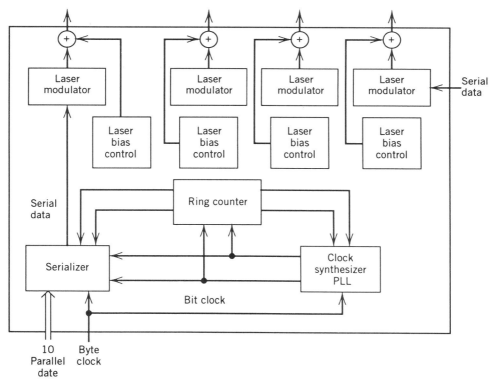

Fig. 15.33 Transmitter IC block diagram with four laser modulators and bias control circuits, clock synthesizer PLL, and serializer circuit. [Ewen, J. F., K. P. Jackson, R. J. S. Bates, and E. B. Flint, *IEEE J. Lightwave Tech.* **9**, 12 (1991): permission from IEEE]

Fig. 15.34 Transmitter chip photograph showing the four laser modulator and bias circuits (*top center*), PLL clock synthesizer (*lower right*), and serializer circuit (*lower left*). [Ewen, J. F., K. P. Jackson, R. J. S. Bates, and E. B. Flint, *IEEE J. Lightwave Tech.* **9**, 12 (1991); permission from IEEE]

few tenths of 1 dB/cm; branching and bends in the guide should have losses less than 1 dB. Once fabricated, the lightguide must maintain its low loss through all subsequent carrier processing, including the fabrication of electrical wiring and the chip-attachment soldering procedures. The materials used must be compatible with the carrier's temperature coefficient of expansion over the IC's operational temperature range.

Both dielectric and polymer materials are candidates for lightguides. They can be directly deposited onto a chip-carrier substrate and then planar processed, using photolithography and etching techniques to make lightguides, passive optical components, and features for mechanical alignment. An interesting class of polymer for these applications are the polyimides because of their popularity as a low-dielectric insulator for multilayer electronic wiring and their stability at high temperatures ($>350°C$). Honeywell has demonstrated polyimide lightguides etched with reactive ion beams that have 0.3-dB/cm loss and stability to 200°C. The polyimide and silica light-guide technologies have not yet been integrated into a multichip module with OE components.

15.7.6 Future Directions

The computer industry is increasingly making use of optical link technology, and it is likely be a high-volume user of link adapter components. However, the industry does not seem to be taking a leadership role in developing cost/performance-optimized link technology. A more optimistic indication for the future comes from recent OE industry meetings between technology users and suppliers. An optical interconnect technology based on GaAs OEIC and IC technologies and on planar-processed optical and electrical wiring could potentially meet cost and performance objectives of the interprocessor network application. Manufacturing tools and processes must still be developed.

The emergence of standards activities for high-speed computer networks (e.g., the ANSI X3T9.3 Fiber Channel Standard) is under development. Such standards could facilitate the development of an optimized technology for optical network adapters because standards would increase the volume of interface components used and amortize the cost of OEIC development. Recently Vitesse Semiconductor and AMD have jointly developed the 1.25 GHz G-TAXI chip set of support the fiber distributed data interface (FDDI) and the high-performance parallel interface (HIPPI) standards [Leo91].

The next few years will be critical for OEIC development. For data communications the evolution from discrete to integrated components will likely be driven by the need for the higher-performance levels of the B-ISDN networks. For interprocessor network applications the network technologists might take a lead in the development of a highly integrated optical technology. If a successful OEIC-based technology were to be developed for the computer-system marketplace, other optical components such as modulators and switches would probably be added in the longer term. We could then anticipate that the range of applications would broaden to encompass optical storage, printing, and display.

15.8 QUANTUM WELL OPTICAL MODULATORS

Nonlinear optical properties of multiple quantum wells (MQWs) have potential applications in electrooptical devices. There is considerable interest in optical modulators for fiber transmission systems, as well as for interconnect and signal-processing systems. The quantum well structure has attracted much interest in optical modulators because of its large field-induced change in the index associated with the excitonic quantum-confined Stark effect (QCSE), which is larger than in bulk material [Woo88].

15.8.1 Principles of Quantum Well Optical Modulators

Figure 15.35 shows the band diagram of a quantum well. Electrons and holes in the materials tend to become localized in the region of the low band-gap material where their potential energies are lowest. The motion of electrons and holes in the quantum well is affected by their confinement. According to the laws of quantum mechanics [Din75], the energies of the particles cannot be equal to the minimum energies of their respective wells. Since their position is localized in the region of the well, the Heisenberg uncertainty principle requires that the particles have a nonzero momentum uncertainty [Mid92]. This translates into a "zero-point energy," which displaces the ground state energy of a particle from the bottom of the well. These zero-point energies are very important because it is their modulation with electric field that makes MQW modulators possible.

Light is absorbed by the quantum well when a photon of sufficient energy excites an electron from the valence band into the conduction band, making an electron-hole pair. The lowest energy absorption occurs when the photon creates an electron-hole pair bound together in a state called an *exciton*. By looking at Fig. 15.35(a), one sees that the photon energy that create this exciton is just the sum of the bulk band gap of

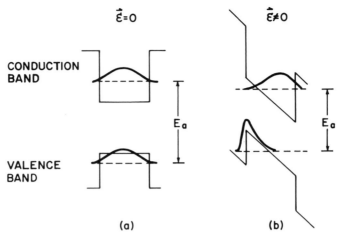

Fig. 15.35 Schematic view of the electroabsorption effect in MQWs. (*a*) The energy of the valence-band maximum and conduction-band minimum as a function of position for a quantum well (the energy levels of the ground states for electrons and holes are shown, along with wavefunction envelops); (*b*) the effect of an electric field applied perpendicular to the quantum well layers. [Wood, T. H., *IEEE J. Lightwave Tech.* **6**, 6 (1990): 743–756; permission from IEEE and AT&T]

the semiconductor making up the well and the zero-point energies of the electron and hole, with a correction for the exciton-binding energy. The energy needed to create the exciton E_a is given by

$$E_a = E_g + E_{e1} + E_{h1} - B, \tag{15.17}$$

where E_g is the bulk band gap of the well material, E_{e1} and E_{h1} are the zero-point energies of the electron and hole, respectively, and B is the binding energy of the exciton.

The electrical field dependence of the optical absorption near the band edge has been extensively studied [Mil84]. There is a distinct difference of the field dependence of exciton absorption for fields parallel and perpendicular to the quantum well layers. In the parallel-field case the excitons broaden and disappear at fields of about 10^4 V/cm; in the perpendicular configuration, however, the excitons shift to lower energies by as much as 2.5 times their zero-point energy, remaining resolved up to 10^5 V/cm. This unique effect, which modifies the zero-point energies of the particles, is due to the confinement of the carriers which inhibits the exciton field ionization. This effect, dubbed the quantum-confined Stark-effect (QCSE), arises because the potential wells seen by the particles in the two panels of Fig. 15.35 are different. When the field is increased, the electron and hole wavefunctions relax into the displaced wavefunctions shown in Fig. 15.35(*b*). These modified wavefunctions are lower in energy than the original wavefunctions, so the zero-point energies for the electron and hole are reduced.

This change in zero-point energies results in a decrease in the effective band gap of the quantum well. As the electric field applied to the wells increases, the absorption peaks shift toward lower energy. Figure 15.36 shows the schematic structure of a

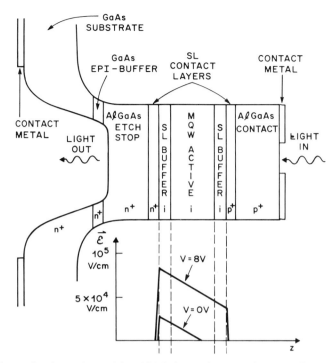

Fig. 15.36 Schematic of sample used in MQW absorption experiments. The quantum wells, in the "MQW active" layer, are in the center of a *p-i-n* diode, which is back biased. The electric-field profile, calculated in the depletion approximation at two different applied voltages, is shown in the lower half of the figure. Light propagates perpendicular to the MQW layers. [Wood, T. H., *IEEE J. Lightwave Tech.* **6**, 6 (1990): 743–756; permissions from IEEE and AT&T]

"transverse" modulator. In the center of the sample there is a region labeled "MQW active" which contains a set of 50 GaAs quantum wells, each 95 Å thick, separated from each other by $Al_{0.32}Ga_{0.68}As$ barriers, 98 Å thick. In order to apply the electric field to the wells, they are fabricated in the undoped region of a *p-i-n* diode. The back biased *p-i-n* diode makes it possible to apply large fields to the wells without significant leakage current. The electric field profile at two applied voltages is shown at bottom of Fig. 15.36. At 0 V applied, the field is close to zero for the MQWs, but at an applied back-bias voltage of 8 V, the field increases to approximately 7×10^4 V/cm.

15.8.2 Symmetric Self-electrooptic Effect Devices

Recent developments in optical computer components and systems are providing motivation for the development of computer architecture for optical implementation [Web92]. The hardware consists of a number of GaAs chips containing large arrays of self-electrooptic effect devices (SEEDs) interconnected with regular arrays of light beams [Dic90]. SEED is a nonlinear device consisting of hundreds of alternating GaAs and AlGaAs layers sandwiched in the intrinsic region of the *p-i-n* diode. These layers act as quantum wells, trapping electrons and holes close by. The devices operate essentially as reverse-biased diodes. When the applied electric field is increased across the device, there is large shift of the absorption edge to the lower photon energy. This "quantum confined Stark effect" can be employed to make a high-speed electro-absorption modulator. The S-SEED behaves as an optical transistor, performing in a NOR function by allowing one light beam to switch another. SEEDs can be employed as gates, memories, or optical flip-flops.

REFERENCES

[Tsa90] Tsang, W. T., in *High-Speed Semiconductor Devices*, p. 587, ed. by S. M. Sze, Wiley (1990).
[Uji85] Uji, T., and J. Hayashi, *Electron. Lett.* **21** (1985): 418.
[Kre75] Kressel, H., and M. Ettenberg, *Proc. IEEE* **63** (1975): 1360.
[Hor76] Horikoshi, Y., Y. Takanashi, and G. Iwane, *Jap. J. Appl. Phys.* **15** (1976): 485.
[Lee73] Lee, T. P., C. A. Burrus, Jr., and B. I. Miller, *IEEE J. Quant. Electron.* **9** (1973): 820.
[Liu75] Liu, Y. S., and D. A. Smith, *Proc. IEEE* **63** (1975): 542.
[Nam74] Namizaki, H., M. Nagano, and S. Nakahara, *IEEE Trans. Electron. Dev.* **21** (1974): 688.
[Wad81] Wada, O., S. Yamakoshii, M. Abe, Y. Yishitoni, and T. Sakwai, *IEEE J. Quant. Electron.* **17** (1981): 174.
[Lee82] Lee, T. P., *Opt. Laser Tech.* **14** (1982): 15.
[Sau83] Saul, R. H., *IEEE Trans. Electron Dev.* **30** (1983): 285.
[Bas61] Basov, N. G., O. N. Kroklin, and Y. M. Popov, *Sov. Phys. JETP* **13** (1961): 1320.
[Hal62] Hall, R. N., G. E. Fenner, J. D. Kingsley, T. J. Soltys, and R. O. Carlson, *Phys. Rev. Lett.* **9** (1962): 366.
[Nat62] Nathan, M. I., W. P. Dumke, G. Burns, F. H. Dill, and G. J. Lasher, *Appl. Phys. Lett.* **1** (1962): 62.
[Qui62] Quist, T. M., R. H. Rediker, R. J. Keyes, W. E. Krag, B. Lax, A. L. McWhorter, and H. J. Zeiger, *Appl. Phys. Lett.* **1** (1962): 91.
[Mai61] Maiman, T. H., *Phys. Rev.* **123** (1961): 1145.
[Alf68] Alferov, Zh. I., V. M. Andreev, V. I. Korol'kov, E. L. Portnoi, and D. N. Tret'yakov, *Fiz. Tekh. Poluprov.* **2** (1968): 1016.
[Kre69] Kressel, H., and H. Nelson, *RCA Rev.* **30** (1969): 106.
[Sze81] Sze, S. M., *Physics of Semiconductor Devices*, Wiley (1981).
[Col90] Coleman, J. J., *IEEE IEDM* **90** pp. 125–128 (1990).
[Tsa78] Tsang, W. T., *Appl. Phys. Lett.* **33** (1978): 245.
[Bow88] Bowers, J. E., and Pollack, M. A., in *Optical Fiber Telecommunications II*, ed. by S. E. Miller and I. P. Kaminow, Academic Press (1988).
[Ara86] Arakawa, Y., and A. Yariv, *J. Quant. Electron.* **22** (1986): 1887.
[Lee83] Lee, T. P., C. A. Burrus, R. A. Linke, and R. J. Nelson, *Electron Lett.* **19** (1983): 82.
[Lio84] Liou, K. Y., C. A. Burrus, R. A. Linke, I. P. Kaminow, S. W. Granlund, C. B. Swan, and P. Besomi, *Appl. Phys. Lett.* **45** (1984): 729.
[Tsa85] Tsang, W. T., *Semiconductors and Semimetals*, vol. 22B, pp. 257–373, ed. by R. K. Willardson and A. C. Beer, Academic Press (1985).
[Kog71] Kogelink, H., and C. V. Shank, *Appl. Lett. Phys.* **18** (1971): 152.
[Nat66] Nathan, M. I., *Proc. IEEE* **54** (1966): 1276.

[Hal66] Halperin, B. I., and M. Lax, *Phys. Rev.* **148** (1966): 722.
[Lev63] Levine, A. K., *Am. Sci.* **51** (1963): 14.
[Col90] Coleman, J. J., *IEEE IEDM*, p. 126 (1990).
[Mat74] Matthews, J. W., and A. E. Blakeslee, *J. Cryst. Growth* **27** (1974): 118.
[Osb87] Osbourn, G. C., in *Semiconductor and Semimetals*, vol. 24, p. 459, ed. by R. K. Willardson and A. C. Beer, Academic Press (1987).
[Lai84] Laidig, W. D., P. J. Caldwell, Y. F. Lin, and C. K. Peng, *Appl. Phys. Lett.* **44** (1984): 653.
[Shi89] Shieh, C., J. Mantz, H. Lee, D. Ackley, and R. Englemann, *Appl. Phys. Lett.* **57** (1989): 843.
[Yor89] York, P. K., K. J. Beernick, G. E. Fernandez, and J. J. Coleman, *Appl. Phys. Lett.* **54** (1989): 499.
[Sal91] Saleh, B. E. A., and M. C. Tiech, *Fundamentals of Photonics*, Wiley (1991).
[Hon91] Hong, M., and L. W. Tu, *J. Cryst. Growth* **111** (1991): 1052–1056.
[Bow87] Bowers, J. E., and C. A. Burrus, *IEEE J. Lightwave Tech.* **5** (1987): 1339.
[Mel72] Melchior, H., in *Laser Handbook*, vol. 1, pp. 725–835, ed. by F. T. Arecchi and E. O. Schulz-Dubois, North-Holland (1972).
[Win82] Windhorn, T. H., L. W. Cook, and G. E. Stillman, *J. Electron. Mat.* **11** (1982): 1065.
[Pea82] Pearsall, T. P., ed., *GaInAsP Alloy Semiconductors*, Wiley (1982).
[Hil87] Hill, P., J. Schlafer, W. Powazinik, M. Urban, W. Eichen, and R. Olshansky, *Appl. Phys. Lett.* **50** (1987): 1260.
[Bow85] Bowers, J. E., C. A. Currus, and R. J. McCoy, *Electron. Lett.* **21** (1985): 812.
[Hel82] Hellwege, K. H., ed., *Landolt-Bornstein Numerical Data and Functional Relationship in Science and Technology*, vol. 17a, ed. by O. Madelung, Physics of Group IV Elements and III–V Compounds, pp. 532–533, Springer-Verlag, Berlin (1982).
[Sze81] Sze, S. M., *Physics of Semiconductor Devices*, Wiley-Interscience (1981).
[Mis66] Misawa, T., *IEEE Trans. Electron Dev.* **13** (1966): 137–143.
[Hoe66] Hoefflinger, B., *IEEE Trans. Electron Dev.* **13** (1966): 151–158.
[Gau91] Gaul, L., and M. Claassen, *AFU* **45**, 2 (1991): 126–130.
[Mul68] Muller, M. W., and H. Guckel, *IEEE Trans. Electron Dev.* **15** (1968): 560–568.
[Bow68] Bowers, H. C., *IEEE Trans. Electron Dev.* **15** (1968): 343–354.
[Cla87] Claassen, M., H. Grothe, R. Pierzina, and W. Harth, *AEU* **41** (1987): 380–381.
[Gib73] Giblin, R. A., E. F. Scherer, and R. L. Wierich, *IEEE Electron Dev.* **20** (1973): 404–418.
[Kan59] Kane, E. O., *J. Phys. Chem. Solids* **12** (1959): 181–188.
[Mis72] Misawa, T., *Solid State Electron.* **15** (1972): 457–465.
[Mai83] Mains, R. K., G. I. Haddad, and P. A. Blakey, *IEEE Electron Dev.* **30** (1983): 1327–1338.
[Dal89] Dalle, C., and P. A. Rolland, *Int. J. Numerical Modelling: Electronic Networks, Devices and Fields* **2** (1989): 61–73.
[Tsa81] Tsang, W. T., *Appl. Phys. Lett.* **39** (1981): 786
[Tsa90] Tsang, W. T., in *High-Speed Semiconductor Devices*, Wiley (1990).
[Nis79] Nishida, K., K. Taguchi, and Y. Matsumoto, *Appl. Phys. Lett.* **35** (1979): 251–253.
[Cap82] Capasso, F., W. T. Tsang, A. L. Hutchinson, and G. F. Williams, *Appl. Phys. Lett.* **40** (1982). 38.
[Cha91] Chan, W. K., A. Yi-Yan, and T. J. Gmitter, Grafted semiconductor optoelectronics, *IEEE J. Quantum Electronics* **27**, 3 (1991).
[Yi-Y90] Yi-Yan, A., W. K. Chan, T. J. Gmitter, and M. Seto, Semiconductor-grafted integrated optics, in *Integrated Photon. Res. Tech. Digest*, Paper MI1 (1990)

[Yab87] Yablonovitch, E., T. Gmitter, J. P. Harbison, and R. Bhat, *Appl. Phys. Lett.* **51** (1987): 2222–2224.

[Lo91] Lo, Y. H., R. Bhat, D. M. Hwang, M. A. Koza, and T. P. Lee, *Appl. Phys. Lett.* **58**, 18 (1991): 1961.

[Leh89] Leheney, R. F., *IEEE Circuits and Devices* **5,** 3 (1987): 38.

[Cro90] Crow, J. D., *Critical Review of Optical Science and Technology*, vol. 35: *Optical Computing*, SPIE Optical Engineering Press (1990).

[Lan88] Lane, T. A., Digital system applications of optical interconnections, *Proc. SPIE O-E/Fiber Lase* (Sept. 1988).

[Cro91] Crow, J. D., *IEEE Circuits and Devices* (Mar. 1991).

[Leo91] *Electronic Design*, p. 149, September 26 (1991).

[Woo88] Wood, T.H., *IEEE J. Lightwave Technol.* **6** (1988): 743–757.

[Mil84] Miller, D. A. B., D. S. Chemla, and T. C. Damen, *Phys. Rev. Lett.* **53**, 22 (1984).

[Mil85] Miller, D. A. B., *Phys. Rev. B* **32**, 2 (1985).

[Frö85] Fröhlich, D., A. Nöthe, and K. Reimann, *Phys. Rev. Lett.* **55** (1985): 1335.

[Smi82] Smith, P. W., *Bell Sys. Tech. J.* **61** (1982): 1975.

[Kam88] Kamiya, T., *O Plus* **E-98** (1988): 63.

[Shi89] Shimizu, A., and K. Fujii, Ext. Abst. *21st Conf. on Solid State Devices and Materials*, pp. 313–316, Tokyo (1989).

[Dic90] Dickinson, A. and A. Huang, Optics in Complex Systems, *Proc. SPIE*, vol. 1319, pp. 177–178 (1990).

[Web92] Weber, J., *Byte*, p. 169, September (1992).

[Mid92] Midwinter, J. E. and Y. L. Guo, *Optoelectronics and Lightwave Tehnology*, Wiley Interscience (1992).

INDEX

Aberration, 198
   chromatic aberration, 198
   spherical aberration, 198
Absorption coefficient, 221
Absorption edge, 17, 205
Acetone, 170
Activation energy, 24
Adduct, 84
Adjoint system, 516
Advanced air-to-air missile (AARAAM), 522
Aharonov–Bohm effect, 460, 461
Alkyl, 75, 83
AlGaAs, 1
Alpha, 3
Aluminum (Al), 57
Aluminum nitride, 147, 153
Amine, 173
   trithanol amine, 173
   gaseous amine, 173
Amplifier, 483
Anadigics, 3
Anderson's electron affinity rule, 16
Anderson localization, 227
Angle-resolved ultraviolet photoelectron spectroscopy (ARUPS), 64
Anisotropic, 21
Antimony (Sb), 57
Antireflection coating (ARC), 172
Arrhenius-type temperature dependence, 134
Arsenic-to-gallium ratio, 84
Arsine (AsH_3), 96, 97
Asynchronous transfer mode (ATM), 549
AT&T Microelectronics, 5
Atomic layer epitaxy (ALE), 56, 94

Atomic plane doping, 98
Atomic radii, 11
Auger electron spectrometer (AES), 58, 63, 65, 308
Autocorrelation function, 249
Automatic wafer register technique, 331
Avalanche photodiode (APD), 578, 582
   separation, absorption, and multiplication (SAM-APD), 588
   SAGM-APD, 588
Avantek, 3

Backgating effects, 352
Backscattering, 183, 188
Ballistic transport, 239
Ball-up effect, 313
Band curvature, 21
Band structure, 7
Bardeen limit, 282
Bardeen model, 281
Basis, 7
Bass cell, 83
Beam-blanking plates, 181
Beam-deflection system, 181
Beam-forming system, 181
β-elimination, 86
B-ISDN network, 599
Bleaching characteristic, 178
Bloch function, 206, 209
BNR (Bell Northern Research), 558
Boiling point, 135
Boltzmann constant, 353
Boltzmann distribution, 25
Boltzmann transport equation, 27, 233

Bonding by atomic rearrangement (BAR), 591
Bond-orbital model, 206, 215
Bonds, 10–12
Boric oxide encapsulant, 43
Bragg's law, 61
Bravais lattice, 7
Bridgman technique, 36
　horizontal Bridgman, 36
　vertical Bridgman, 36
Brillouin zone, 11, 12, 206, 209, 224
Brimide, 127
Bromine, 127
Buffered FET logic (BFL), 528, 532
Buffer oxide etch (BOE), 150
Built-in potential, 337, 343
Burgers vector, 388
Butterfly adder chip, 395

Cadence, 530
Camel diode, 445
Capacitance-voltage measurement, 294
Capacitive coupled logic (CCL), 538
Capacitor diode-coupled FET logic (CDFL), 538
Carbon, 46, 47, 79, 80, 85, 413
　carbon doped, 413
Cascade Design Automation, 530
CATV, see Common-antenna television
Cellular phones, 524
Charge control model, 377
Charge-injection logic, 456
Charge-injection transistor (CHINT), 454
Chemical beam epitaxy (CBE), 56, 90, 92
　photo-CBE, 93
Chemical-reaction limited etching, 119, 120
Chemical solution, 84
Chemisorption, 68
Child's law, 138
Chloride transport, 94
Chlorine, 125, 135
Chlorobenzene, 189
Chlorocarbons, 125, 135
Chromium, 33
Cleaning, 116
Cleanliness, 116
Clean room, 116
Cleaved-coupled-cavity (C^3 laser), 576
Clock-and-data-recovery (CDR), 549
Collision avoidance system, 524
Common-antenna television (CATV), 524
Compaq, 550
Complex multiplier chip, 395
Compounds, 65
　ternary, 65
　quaternary, 65
Copper, 47
Conduction band, 13, 18, 23
Conductivity, 33, 204
　ambipolar conductivity, 33

Contact resistance, 279
Convex, 550
Coulombic binding, 220
Coupled bands, 213
Covalent bonds, σ and π, 94
Cox–Strack method, 315
CPU, 551
Cracker cells, 60
CRAY-3, 524
Critical-point energies, 15
Crosslink, 188
Cryoshrouds, 58
Crystal structure, 7
Crystal defects, 7
Current-mode logic (CML), 532
Current-voltage measurement, 294
Curtice model, 335, 350, 351

Dangling bonds, 282
DARPA, see Defense Advanced Research Projects Agency
Data conversion circuit:
　analog-to-digital (ADC), 547
　digital-to-analog (DAC), 547
de Broglie wavelength, 2, 225, 254
Deep level transient spectroscopy (DLTS), 43, 352
Defect(s), 42
　morphology defects, 71
　oval defects, 71–73, 104, 408
　point defects, line defects, volume defects, 43
Defense Advanced Research Projects Agency (DARPA), 5, 549
Deformation potential, 235, 272
　optical, 235
Deionized water (DI), 40
Delta-doped, 99, 101, 391
Density matrix, 252
Density of states, 14, 23, 26
Department of Defense (DOD), 5
Depletion-mode field-effect transistor, 530
Desorption rate, 66, 67
Diamond lattice, 9, 18
Diazonaphthoquinone, 178
Dichloropropyl acrylate (DCOPA), 193
Dielectric assisted lift-off (DAL), 394
Diethylarsine (DEAs), 80
Diethylgalliumchloride (DEGaCl), 94
Diethyltelluride (DETe), 87
Diethylzinc (DEZn), 87
Differential linearity, 548
Diffraction effects, 169
Diffraction limited, 167
Diffusion-limited etching, 119, 120
Dimer, 60
Dimethylgallium (DMG), 102
Dimethylzinc (DMZn), 87
Direct broadcast satellite (DBS) receiver, 524
Direct-coupled FET logic (DCFL), 529, 534

Direct gap, 13, 16
Direct wafer writing (DSW), 180
Dislocation, 42
Distributed Bragg refractors (DBR), 576
Distributed feedback (DFB), 577
Dope, 27
Doped-base transistor, 447
Double barrier structure, 2
Double heterojunction (DH) lasers, 57, 567
Drain I-V collapse, 383
Drift-diffusion equation, 239
Dry etching, 123
Dual-PAC imaging, 178, 179
Dynamic load line, 503
DX center, 224

E-beam lithography, 5
Ebers–Moll model, 415
Effective mass, 18, 20
    band edge mass, 22, 226
    effective-mass equation, 209
    optical effective mass, 22
Effective mass filter, 226, 227
Effusion cells, 58
Einstein's relation, 240, 241
Electric field, 30
Electron, 29
    drift velocity, 29, 30
Electron affinity, 15, 366
Electron cyclotron resonance (ECR), 104, 115, 138, 139
Electron cyclotron resonance plasma CVD, 145
Electronegativity, 288
Electronic warfare, 521
Electron wave interference, 478
Electrooptic effects, 3
Electroreflectance, 14
Ellipsoids, 13
EL2 centers, 33, 42, 44–46, 48, 50, 352
Emitter-coupled logic (ECL), 523
Empirical pseudopotential model (EPM), 14
Energy band, 11, 12
Energy gap, 11
Enhancement-mode field-effect transistor, 531
Envelop function, 206
    envelop function approximation, 206, 207
Epitaxial lift-off (ELO), 590, 591
Epoch/GaAs, 530
Error checking and correction chip (ECC), 542
Etching, 119
    dry, 123
    focused ion beam etching, 140
    ion beam milling, 136, 139
    laser-assisted dry etching, 140
    laser-assisted wet etching, 121
    plasma, 123
    preferential, 119, 120
    reactive ion beam etching (RIBE), 123, 136
    reactive ion etching (RIE), 123, 128
    wet, 119
Etch masking process, 165
Etch-pit densities, 7
Ethoxyethanol, 189
Ethyl-cellosolve acetate (ECA), 330
Excimer laser, 93
Exposure, 173
    flood exposure, 173, 178

Fabry-Perot, 575
Face-center cube (FCC), 9
Faraday rotation, 21
Fermi-Dirac integral, 23
Fermi-Dirac statistics, 203
Fermi energy, 23, 24
Fermi golden rule, 229, 244
Fermi wavelength, 459
Feynman path integral, 252, 256, 258
Fiber channel (FC), 549
Fiber distributed data interface (FDDI), 599
Fick's law, 248
Field-effect transistors, 1
Field-emission, 287
Field-emission source, 182
Figure of merit, 399, 494
Finite-temperature effect, 252
Flicker noise, 361
Flip-flop, 522
    D-type, 522, 523
    T-type, 522
Fluctuation-dissipation theorem, 241
Foundry, 3
Fourier series, 209
Fourier transform, 210, 211
Fraunhofer diffraction approximation, 473
FURY family, 541

Gain, 490
    maximum available gain (MAG), 361, 491
    maximum stable gain (MSG), 491
    transducer power gain, 490
Gallium, 72
Gallium arsenide (GaAs), 1
Gallium chloride (GaCl), 96, 97
Gallium oxide, 72
Gate array, 529, 540
Gate priority, 325
Gaussian proximity equation, 183
Gaussian round-beam approach, 181, 182
Gaussian shape-beam approach, 181, 182
General Electric, 3
Germanium, 9
Glass, 154
    borophosphosilicate, 154
    phophosilicate, 154
Glow discharge, 141
Glycidyl methacrylate-co-ethyl acrylate, 193
GPS receiver, 524

Graded-index waveguide, separate-confinement efficiency structure (GRIN-SCH), 575
Gradient method, 346
Grafted-film process, 589
Graphite heater, 159
G-TAXI chip, 599
Gummel–Poon model, 415
Gunn effect, 238

Hall effect, 27, 31, 32
Hall factor, 31, 32, 36
Hamiltonian, 207
Harmonic-balance method, 513
Harris, 3
Heaviside function, 204
Helium temperatures, 17
Hermitian, 211
Heteroepitaxial process, 1
Heterojunction bipolar transistor (HBT), 2, 37, 57, 399
  collector-up, 401, 402
  emitter-up, 401, 402
Heterojunction integrated injection logic (HI^{2}L), 551
Heteropolar bonds, 12
H-GaAs-III processing, 530
High-electron-mobility transistor (HEMT), 2, 37, 57, 365
  depletion-mode, 365
  enhancement-mode, 365
  inverted HEMT, 383
  pseudomorphic HEMT, 365, 389
  pulsed-doped HEMT, 391
High-temperature processing, 154
High-performance parallel interface (HIPPI) standards, 599
Hisenberg uncertainty principle, 599
Hole, 14
  heavy hole, 14, 18, 26, 225
  light hole, 14, 26
Homopolar bonds, 12
Hopping, 225, 227
Horizontal gradient freeze, 37
Hot electron, 240
Hot-electron spectroscopy, 449
Hot-electron transistor, 445
HP-FET model, 349, 350, 351
Hughes, 3
Hybride, 83
Hybrid integrated circuit (HIC), 483
Hydrogen (H_2), 83
Hydrogen chloride (HCl), 96
Hydrogen fluoride (HF), 151
Hydrogen selenide (H_2Se), 86

IC-CAP, 349
Ideality factor, 286, 343
Image charge, 284

Image force lowering, 284
Image reversal, 164, 173
Imidazole, 173
Impurities, 7
Indene carboxylic acid group, 173, 178
Indirect gap, 13, 16, 17
Indium chloride (InCl), 96
Indium phosphorous (InP), 2
Induced-based transistor, 451
InGaAs, 1
Injection lasers, 3
Instruction set architecture (ISA), 551
Intermolecular interactions, 205
Intracollisional field effect (ICFE), 273
Intrinsic absorption, 13
Intrinsic carrier density, 35
Intrinsic reflectance, 13
Iodide (I), 127
Ion implantation, 11
Iron (Fe), 47
Iso-propyl alcohol, 117
ITT Defense, 3, 4

Jastrzebski model, 335, 340
Junction field-effect transistor (JFET), 1, 323, 530, 531

k·p perturbation, 211
k-space, 203
Kaufman source, 137
Kinetic energy, 29
Knudsen cells, 56, 59
Kohn–Luttinger, 217
Kronig–Penney potential, 222

Laminar flow, 116
Lanthanum boride (LaB_6), 181, 183
Laser interferometers, 198
Lateral surface superlattice (LSSL), 469, 471
Lattice vibrations, 12
Lethal concentration (LC_{50}), 78, 80, 82
Level shifting diodes, 529
Levenberg–Marquardt method, 519, 520
Lewis acid, 84
Lewis base, 84
Lift-off, 165, 327
Light, 164
  ultraviolet, 164
Light-emitting diodes (LED), 563
  edge-emitting diodes, 563
  superluminescent diodes, 563, 565
  surface-emitting diodes, 563
Light path length, 93
Liquid-encapsulated Czochralski (LEC), 36, 44
  magnetic liquid-encapsulated Czochralski (MLEC), 41, 48
Liquid-encapsulated Kyropoulous (LEK), 36
Liquid phase epitaxy (LPE), 55

Lithography, 116
   electron-beam lithography, 166, 180
   ion-beam lithography, 180, 193, 195–197
   optical lithography, 180
   x-ray lithography, 180
Litton, 3, 4
Load-lock, 58
Load-pull method, 518
Localization, 225
Lookup table model, 349
Low pinch-off voltage FET logic (LPFL), 539
Low-pressure injection compounding, 39
Low temperature processing, 154

M/A-COM, 4
Mach–Zehnder interferometer, 468
Magnetic susceptibility, 21
Magnetization, 204
Magnetoabsorption, 472
Magnetocapacitance, 472
Magnetoconductivity, 205
Magnetoresistance, 204, 205
Manganese (Mn), 47
Many-body interactions, 252
Master-slave flip-flop, 531
Materka and Kacprzak model, 340
Matching network, 127
Matthiessen's rule, 29
Maximum available gain (MAG), 361
Maxwell–Boltzmann momentum distribution, 250
MBE phase diagram, 69
McCamant model, 335, 341
McDonnell Douglas, 551
Medium-scale integration (MSI), 528
Melting points, 12
Memory management unit (MMU), 551
MESFET, 1, 323, 531
Mesoscopic structure, 469
Metal-based transistor, 445
Metal-insulator-semiconductor (MIS), 151, 153
Metallization, 1, 279
Metalorganic chemical vapor deposition (MOCVD), 55, 75, 82
Metalorganic molecular beam epitaxy (MOMBE), 75, 79
Metal-oxide-metal-oxide-metal transistor (MOMOM), 446
Metal-semiconductor contacts, 279
Methacrylic acid, 189
Methane (CH_4), 79
Methanol, 117, 171
Methylmethacrylate, 189
Methylisobutylketone (MIBK), 189, 330
Microwave Journal, 3, 5
Microwave landing system (MLS), 524
Microwave monolithic integrated circuits (MMIC), 5, 365, 483

Miller indices, 9, 10
Millimeter wave, 7
MIMIC, 5
Miniband(s), 204, 219, 222
Miniband conduction, 225
Minimaxima, 18, 19
Mobility, 28
   electron mobility, 81
   Hall mobility, 28, 29
Modulation-doped, 228
Modulation-doped field-effect transistor (MODFET), 2
Modulation doping, 366
Molar concentration of reactant gas, 93
Molar extinction coefficient, 93
Molecular beam epitaxy (MBE), 2, 55, 57
Molecular layer eitaxy (MLE), 2, 101
Moll–Ross–Kroemer relation, 412
Molybdenum (Mo), 87, 135
Moment equation, 252, 261
Momentum matrix element, 19
Monazoline, 173
Monoethylarsine (MEA), 106, 108
Monomethylgallium (MMG), 102
Motorola, 530
Monte Carlo method, 27, 240, 263
Multilayer ceramic packaging, 554
Multichip packaging (MCP), 554

NASA Search and Rescue program, 524
Negative differential conductivity, 238
Negative differential resistance (NDR), 423
Newton's law, 242, 244
Noise, 323
   $1/f$ noise, 352
Noise equivalent circuit, 353
Noise figure, 354
Noise margin, 531
Noise match, 495
Noise theory, 353
Nonparabolicity, 18, 20, 21
Nonthreshold logic (NTL), 532
NORAND, 456
Nyquist's relation, 240

Occupation factor, 205
Octahedron, 11
OEIC, 37, 590
Ohmic contact, 279
   alloyed, 279, 309
   nonalloyed, 279, 310
Ohmic priority, 325
1-dB-gain compression point, 501
Optical density, 93
Optical interconnect, 592
Optical stepping, 165, 167
Optical transient current spectroscopy (OTCS), 43

Optically detected electron nuclear double resonance (ODENDOR), 49
Orthogonalized plane wave (OPW), 13
Oxygen, 86

Palladium, 89
Parallelepiped unit cell, 8
Parameter extraction, 323, 343
Parasitic effects, 323
Pattern-control system, 181
Pattern-generation system, 181
Pauli exclusion principle, 252, 266
Peak-to-valley current ratio (PVCR), 430
Penumbral effect, 192
Perpendicular transport, 225
Persistent photoconductivity (PPC), 365, 383, 452
Phase-array antenna, 524
Phase-array radar, 353
Phase-lock loop (PLL), 595
Phase-shifting mask, 175
Phonon-assisted tunneling, 227
Phonons, 30, 31
  acoustic phonons, 30
  optical phonons, 31
Phosphine (PH_3), 77, 83
Phosphorus (P), 57
Photoassisted single molecular layer epitaxy, 94
Photodecomposition, 93
Photoluminescent (PL) intensity, 57, 81, 366
Photonic devices, 3, 562
Photoresist, 133, 167
  dual-tone photoresist, 173, 175
  DUV resist, 167
  electron-beam resist, 167
  mid-UV resist, 167
  multilevel resist (MLR) technique, 171
  negative resist, 167, 168
  positive resist, 167, 168
  quasi-multilevel resist system, 172
  single layer resist (SLR) technique, 171
  x-ray resist, 167
Photoresponse measurement, 295
Pinch-off, 333, 344
Pinhole, 152
Pin photodetectors, 578
Planar-doped barriers, 445
Plasma-enhanced chemical vapor deposition (PECVD), 115, 140
Plasma etching, 115
Podell's noise model, 360
Poisson's ratio, 388
Polybutene-1 sulfone (PBS), 188
Poly(3-butenyltrimethylsilane sulfone) (PBTMSS), 189
Polyimide, 574
Polymethyl methacrylate (PMMA), 167, 172, 188, 189, 330, 461

Polymethyl methacrylate and methacrylate acid copolymers, P(MMA-MAA), 330
Post bake, 169
Poential-effect devices, 3
Power-added efficiency, 418, 503
Prescaler ICs, 523
Primitive basis, 8
Primitive vectors, 8
Principle of least action, 256
Printing, 165
  contact printing, 165, 166
  projection printing, 165
  proximity printing, 165, 167
Processor board interface (PBI), 551
Proximity effect, 171
Pseudopotential, 205
Pull-down (PD) transistor, 533
Pull-up (PU) transistor, 533
Pumps, 88
  cryo-, 88
  diffusion, 88
  turbomolecular, 88
Pyrolytic boron nitride (PBN), 58

Quadrupole mass spectrometer (QMS), 64, 97
Quantum confinement (QC), 202
Quantum confined Stark effect (QCSE), 599, 600
Quantum dissipation, 256
Quantum dots, 3
Quantum-effect devices, 2
Quantum Hall effect, 205
Quantum interference device, 468
Quantum Liouville equation, 255, 269
Quantum Monte Carlo transport, 256, 269
Quantum point contact, 473
Quantum-well-base transistor, 451
Quantum well box (QWB), 202, 468, 469
Quantum well optical modulator, 599
Quantum well wires (QWW), 3, 202, 468, 469

Raman scattering, 366
Rapid thermal annealing, 154, 310
Rapid thermal processing, 154
Raytheon, 3, 4
Raytheon model, 335
Reactive ion beam etching (RIBE), 123
Reactive ion etching (RIE), 123
Real-space transfer devices, 454
Recessed channel technology, 327
Reciprocal lattice, 209
Recombination center, 410
Reduced instruction set computer, *see* RISC microprocessor
Reflectance, 13
Reflective high-energy electron diffraction (RHEED), 60, 63
Registration, 186
Relative humidity (RH), 117

Relaxation time, 233
Residual gas analyzer (RGA), 104
Resonant-tunneling transistors, 2, 423–425
Resonant-tunneling hot-electron transistor (RHET), 2, 3
RISC microprocessor, 2, 524, 550
Room temperature absorption, 17
Room temperature reflectance, 16, 17
Runout, 166, 192
Rutherford back scattering (RBS), 289

Sampled-analog IC, 352
Scanning electron microscope (SEM), 60, 308
Scattering, 27, 228
  acoustic phonon, 28
  Brooks–Herring, 29, 30
  deformation potential, 30
  intervalley scattering, 237
  ionized impurity, 28, 30, 31
  nonpolar optical mode, 28, 31
  piezoelectric, 31
  polar optical mode, 28, 31, 236
  remote scattering mechanisms, 228
Scattering parameter (S-parameter), 341, 342, 488
Scattering potential, 230
Schottky barrier, 279
Schottky contact, 2
Schottky diode FET logic (SDFL), 528, 533
Schottky limit, 283
Schottky model, 280
Schrödinger equation, 203, 231
Scumming, 170
Secondary ion mass spectrometer (SIMS), 58, 63
Selenium (Se), 86
Self-aligned implantation for n^+-layer technology (SAINT), 325, 326
Self-bootstrapping technique, 352
Self-consistent time-dependent potential variation, 252
Semiconductor grade, 116
Semiconductor laser, 566, 567
Semi-insulating, 1, 33
Sense amplifier circuits, 547
Sense and Destroy Armor (SARARM), 522
Sensitivity analysis, 514
Sidegate, 352
Silicide, 279
  refractory metal, 279, 289
  transition metal, 291
Silicon (Si), 1, 8
Silicon nitride (Si_3N_4), 147
Silicon oxide (SiO_2), 147
Silicon oxynitride (SiO_xN_y), 147
Simplex method, 347
Simulated annealing, 346, 347
Size quantization, 204
Slip line, 155
Smith chart, 491

Soft bake, 169
Source-coupled FET logic (SCFL), 523, 535
Space group symmetry, 206
SPARC microprocessor, 524
Specific heat, 204
Spectrum analyzer, 523
Sphalerite, 9, 13, 18
Spin-orbit coupled orbitals (SOBO), 216
Spin-orbit split energy, 13, 213
Split-level e-beam exposure technique, 330
Spread spectrum communication, 529
Standing waves effect, 173
Static random access memory (SRAM), 2, 528, 531
Statistical weight, 21, 25
Statz model, 336, 339, 350, 351
Sticking coefficient, 60
Stimulated emission, 568
Stranski–Krastanov mechanism, 388
Stripping, 170
Super-buffer logic (SBL), 595
Superlattice, 205, 219
Superlattice unit cell (SUC), 217
Surface photoabsorption (SPA), 94
Symmetric self-electrooptic effect devices (SEED), 601
Symmetry, 7
Synopsis, 530

Tantalum (Ta), 87, 88
Tape-automated bonding (TAB), 173
TCE, 558
Tellurium (Te), 87
Temperature-programmed desorption (TPD), 94
Tertiarbutylarsine (TBAs), 80
Tetramer, 60, 89
Texas Instruments, 3, 4, 551
Thermal deblocking chemistry, 175
Thermal field-emission (TFE), 183
Thermal speed, 27
Thermally stimulated current (TSC), 43
Thermionic emission, 284
Thermionic field emission, 286
Thermoelectric power, 204
Threading dislocation, 388
Three-level Kane model, 19
Tight-binding approximation, 206
Transconductance, 345
Transfer-matrix method, 217
Transition frequency, 418
Translational symmetry, 7
Transmission line model (TLM), 313, 316, 379
Transmitter/receiver (T/R) module, 483
Transport, 27
  electron transport, 27
  hole transport, 27
Traps, 43
  deep traps, 352

Traps (*Continued*)
  electron traps, 43
  hole traps, 43
Trichloroethane, 117
Triethylaluminum (TEA), 93
Triethylantimony (TESb), 77
Triethylgallium (TEGa), 79, 87, 93, 106, 108
Triethylindium (TEIn), 76, 79, 87
Triethylphosphine (TEP), 77
Triisobutylaluminum (TIBA), 93, 106
Trimethylaluminum (TMAl), 76
Trimethylantimony (TMSb), 77
Trimethylarsenic (TMAs), 77
Trimethylgallium (TMGa), 75, 79, 87
Trimethylindium (TMIn), 87
TriQuint, 3, 4
TRW, 3, 4
T-shaped cross-sectional gate, 329
Tungsten filament, 137
Tungsten-halogen lamp, 158
Tunneling, 227
Tunneling hot-electron transfer amplifier (THETA), 448, 449
TVRO module, 520
Two-dimensional electron gas (2DEG), 365, 374

UHF/VHF, 524
Ultra-high-vacuum (UHV) evaporators, 56, 57
Unit cell, 7, 206
UTMOST III, 349

Valence band, 13, 17
Valence band model (VBM), 219
Van der Waals force, 590
Van der Ziel model, 338
Vapor-liquid-solid (VLS) mechanisms, 73
Vapor phase epitaxy (VPE), 44, 55, 96

Vapor transport epitaxy (VTE), 56, 104
Varian GEN-II system, 330
Velocity-field characteristics, 30
Vertical cavity surface-emitting laser (VCSEL), 577
V-groove, 573
Via, 318
Vibrational spectrum, 11
Viewlogic, 530
Vitesse, 529, 530
VLSI, 530
Voltage-controlled current source (VCCS), 517
Voltage-controlled devices, 1
Voltage standing wave ratio (VSWR), 499

Wafer, 3
Wafer distortion, 155
Waveguiding, 569
Wave vector, 13, 18
Westinghouse, 4
White noise, 241
WKB approximation, 424
Wigner distribution function, 252, 259
Work function, 281

X-ray, 190, 191
X-ray photoelectron spectroscopy (XPS), 14, 60, 308
  soft x-ray photoelectron spectroscopy measurement (SXPS), 297

YIG-tuned oscillator buffer amplifier, 523

Zinc (Zn), 47
Zincblend, 9, 206
Zone-center-band extrema, 18, 19